Receptor-Mediated
Targeting of Drugs

NATO ASI Series

Advanced Science Institutes Series

A series presenting the results of activities sponsored by the NATO Science Committee, which aims at the dissemination of advanced scientific and technological knowledge, with a view to strengthening links between scientific communities.

The series is published by an international board of publishers in conjunction with the NATO Scientific Affairs Division

A	Life Sciences	Plenum Publishing Corporation
B	Physics	New York and London
C	Mathematical and Physical Sciences	D. Reidel Publishing Company Dordrecht, Boston, and Lancaster
D	Behavioral and Social Sciences	Martinus Nijhoff Publishers
E	Engineering and Materials Sciences	The Hague, Boston, and Lancaster
F	Computer and Systems Sciences	Springer-Verlag
G	Ecological Sciences	Berlin, Heidelberg, New York, and Tokyo

Recent Volumes in this Series

Volume 75—Photoreceptors
edited by A. Borsellino and L. Cervetto

Volume 76—Biomembranes: Dynamics and Biology
edited by Robert M. Burton and Francisco Carvalho Guerra

Volume 77—The Role of Cell Interactions in Early Neurogenesis: Cargèse 1983
edited by A.-M. Duprat, A. C. Kato, and M. Weber

Volume 78—Organizing Principles of Neural Development
edited by S. C. Sharma

Volume 79—Regression of Atherosclerotic Lesions
edited by M. Rene Malinow and Victor H. Blaton

Volume 80—Mechanisms of Gastrointestinal Motility and Secretion
edited by Alan Benett and Giampaolo Velo

Volume 81—Dynamics of Biochemical Systems
edited by Jacques Ricard and Athel Cornish-Bowden

Volume 82—Receptor-Mediated Targeting of Drugs
Edited by G. Gregoriadis, G. Poste, J. Senior, and A. Trouet

Series A: Life Sciences

Receptor-Mediated Targeting of Drugs

Edited by

G. Gregoriadis

Royal Free Hospital School of Medicine
London, England

G. Poste

Smith Kline & French Laboratories
Philadelphia, Pennsylvania

J. Senior

Royal Free Hospital School of Medicine
London, England

and

A. Trouet

International Institute of Cellular and Molecular Pathology
and Université Catholique de Louvain
Brussels, Belgium

Plenum Press
New York and London
Published in cooperation with NATO Scientific Affairs Division

Proceedings of a NATO Advanced Study Institute on
Receptor-Mediated Targeting of Drugs, held June 20–July 1, 1983, in
Cape Sounion, Greece

Library of Congress Cataloging in Publication Data

NATO Advanced Study Institute on Receptor-mediated Targeting of Drugs (1983:
Ákra Soúnion, Greece)
    Receptor-mediated targeting of drugs.

    (NATO ASI series, Series A, Life sciences; v. 82)
    "Proceedings of a NATO Advanced Study Institute on Receptor-mediated
Targeting of Drugs, held June 20–July 1, 1983, in Cape Sounion, Greece."—T.p.
verso.
    Bibliography: p.
    Includes index.
    1. Drug receptors—Congresses. 2. Drugs—Vehicles—Congresses. 3.
Liposomes—Congresses. I. Gregoriadis, Gregory. II. Title. III. Series. [DNLM: 1.
Receptors, Drug—congresses. 2. Drugs—administration & dosage—congresses.
QV 38 N2785r 1983]
RM301.N378   1983   615.7   84-18183
ISBN-13: 978-1-4684-4864-1   e-ISBN-13: 978-1-4684-4862-7
DOI: 10.1007/978-1-4684-4862-7

PREFACE

Conventional attempts to control cell behaviour and function
are often marred by the toxicity of the drugs used, their premature
waste or inactivation or by their inability to interact with or
reach target sites efficiently. New trends in pharmacology emphasize the development of methods for the optimization of drug action,
for instance by the delivery of drugs, enzymes, hormones, antigens,
genetic material, ets. through carrier systems selectively to relevant cellular and subcellular sites. A wide assortment of carriers
was discussed in the first NATO Advanced Studies Institute (ASI)
"Targeting of Drugs," the proceedings of which were published by
Plenum Press in 1982 (eds., G. Gregoriadis, J. Senior and A. Trouet).

This book, containing the proceedings of the 2nd NATO ASI
"Receptor-Mediated Targeting of Drugs" held again at Cape Sounion,
Greece during 20 June-1 July 1983, deals with drug delivery through
systems possessing ligands which can recognize and interact with
receptors on the target's surface. Receptor-recognizing carriers
that have recently given promise of realistic expectations in
targeting include monoclonal antibodies, certain proteins that home
to specific cells in the body and liposomes either as such (when
they can trigger specific reactions in the reticuloendothelial
system) or in association with targeting macromolecules. Each of
these are discussed extensively by leading authorities, in terms of
applications in biology and medicine and related methodologies.

We wish to express our appreciation to Drs. G. Deliconstantinos,
A. Delitheos, A. Evangelopoulos, M. Maragoudakis and P. Thorpe who,
as members of the international or local committees provided valuable
advice and help throughout the planning stages of this Institute.
The ASI meeting was held under the sponsorship of NATO.

May 1984   Gregory Gregoriadis
                                        George Poste
                                        Judith Senior
                                        André Trouet

v

CONTENTS

DRUG TARGETING IN HUMAN CANCER CHEMOTHERAPY 1

Y.-J. Schneider, J. Abarca, E. Aboud-Pirak, R. Baurain,
F. Ceulemans, D. Deprez-De Campeneere, B. Lesur,
M. Masquelier, C. Otte-Slachmuylder, D. Rolin-van Swieten
and A. Trouet

TOXIC CONJUGATES OF EPIDERMAL GROWTH FACTOR AND
ASIALOFETUIN 27

H.R. Herschman, D. Cawley and D.L. Simpson

SACCHARIDE RECEPTOR-MEDIATED DRUG DELIVERY 53

M.M. Ponpipom, R.L. Bugianesi, J.C. Robbins, T.W. Doebber
and T.Y. Shen

STUDY OF THE TRANSFERRIN RECEPTOR USING A CYTOTOXIC
HUMAN TRANSFERRIN-RICIN A CHAIN CONJUGATE 73

V. Raso and M. Basala

HOW DO PROTEIN TOXINS KILL CELLS? 87

S. Olsnes, K. Sandvig, A. Sundan, K. Eiklid and A. Pihl

ROLE OF THE B-CHAIN IN THE CYTOTOXIC ACTION OF ANTIBODY-
RICIN AND ANTIBODY ABRIN CONJUGATES 105

D. McIntosh and P. Thorpe

MONOCLONAL ANTIBODIES AS CELL TARGETED CARRIERS OF
COVALENTLY AND NON-COVALENTLY ATTACHED TOXINS 119

V. Raso and M. Basala

LONG-TERM KINETICS OF ANTIBODY-TOXIN CONJUGATES
SHOW RESISTANT CELLS 139

R.J. Youle and D.M. Neville, Jr.

SIGNIFICANCE OF THE KINETICS OF IMMUNOTOXIN CYTOTOXICITY 147

F.K. Jansen, H.E. Blythman, B. Bourrie, D. Carriere,
P. Casellas, D. Dussossoy, O. Gros, J.C. Laurent,
M.C. Liance, P. Poncelet, G. Richer and H. Vidal

THE USE OF RICIN A CHAIN-CONTAINING IMMUNOTOXINS TO
KILL NEOPLASTIC B CELLS 179

E.S. Vitetta and J.W. Uhr

SELECTIVE CYTOTOXICITY OF RICIN A CHAIN-ANTI CARCINO-
EMBRYONIC ANTIBODY CONJUGATES TO HUMAN ADENOCARCINOMA
CELLS 187

T.W. Griffin, L.R. Haynes and L.V. Levin

DEVELOPMENT OF MONOCLONAL ANTIBODIES WITH SPECIFICITY
FOR HUMAN EPITHELIAL CELLS 201

J. Taylor-Papadimitriou and A.B. Griffiths

ATTEMPTED THERAPY OF A HUMAN XENOGRAFT COLONIC CANCER
HT29 USING ^{131}I-LABELLED MONOCLONAL ANTIBODIES 235

A.A. Epenetos

FATE OF LIPOSOMES IN VIVO: CONTROL LEADING TO TARGETING 243

G. Gregoriadis, J. Senior, B. Wolff and C. Kirby

CONTENTS

INTERACTIONS OF LIPOSOMES WITH LIPOPROTEINS AND
LIVER CELLS IN VIVO AND IN VITRO STUDIES 267

G. Scherphof, J. Damen, J. Dijkstra, F. Roerdink
and H. Spanjer

ENDOCYTOSIS OF LIPOSOMES AND INTRACELLULAR FATE
OF ENCAPSULATED MOLECULES: STRATEGIES FOR ENHANCED
CYTOPLASMIC DELIVERY 297

R.M. Straubinger, K. Hong, D.S. Friend, N. Duzgunes
and D. Papahadjopoulos

LIPOSOMES IN LEISHMANIASIS: THE LYSOSOME CONNECTION 317

C.R. Alving, J.S. Weldon, J.F. Munnell and W.L. Hanson

THROMBOGENIC POTENTIAL OF PHOSPHATIDYLSERINE-CONTAINING
LIPOSOMES 333

G.B. Humphrey, L.V. Allen, R. Blackstock, P.C. Comp,
L.E. De Bault, C.T. Esmon, A. Harriman, F.A. Holloway,
H.F. Krous, M. Mojarad, H.L. Stone and R. Wierimaa

THE USE OF LIPOSOMES IN DIAGNOSTIC IMAGING 347

V.J. Caride

UNEXPECTED TISSUE DISTRIBUTION OF LIPOSOMES COATED WITH
AMYLOPECTIN DERIVATIVES AND SUCCESSFUL USE IN THE TREATMENT
OF EXPERIMENTAL LEGIONNAIRES' DISEASES 359

J. Sunamoto, M. Goto, T. Iida, K. Hara, A. Saito and
A. Tomonaga

ALTERATION OF LIPOSOMAL-SURFACE PROPERTIES WITH SYNTHETIC
GLYCOLIPIDS 373

M.M. Ponpipom and T.Y. Shen

ANTIBODY-BEARING LIPOSOMES AS PROBES OF RECEPTOR-
MEDIATED ENDOCYTOSIS 393

L.D. Leserman, D. Aragnol, J. Barbet, P. Machy and
A. Truneh

ANTIBODY-TARGETED LIPOSOMES: DEVELOPMENT OF A CELL-
SPECIFIC DRUG DELIVERY SYSTEM 407

T. Heath, K. Bragman, K. Matthay, N.G. Lopez and
D. Papahadjopoulos

TARGETING AND LYSOSOMAL HANDLING OF POLYMETHACRYLAMIDE-
OLIGOPEPTIDE CONJUGATES 417

J.B. Lloyd, R. Duncan, J. Kopecek and P. Rejmanova

DRUG TARGETING IN CANCER THERAPY 427

G. Poste

PHOTO 475

CONTRIBUTORS 477

INDEX 485

DRUG TARGETING IN HUMAN CANCER CHEMOTHERAPY

Y.-J. Schneider, J. Abarca, E. Aboud-Pirak,
R. Baurain, F. Ceulemans, D. Deprez-De Campeneere,
B. Lesur, M. Masquelier, C. Otte-Slachmuylder,
D. Rolin-van Swieten and A. Trouet

International Institute of Cellular and Molecular
Pathology and Université Catholique de Louvain
(Laboratoire de Chimie Physiologique), 75, Avenue
Hippocrate, B 1200, Brussels

INTRODUCTION

Drug targeting aims to restrict the access of pharmacological
agents to selected cells. Theoretically, such a method should on
one hand, decrease unsuitable side effects resulting from an interaction of the drugs with non target cells and, on the other hand,
enhance the pharmacological activity by increasing the proportion
of the administered drug found within the target cells. Our conceptual approach to this problem consists in linking drugs through
a covalent bond to macromolecular carriers which are recognized by
receptors or antigens present at the cell surface of the target
cells and thereafter endocytosed to allow the release of the drug
after hydrolysis of the covalent linkage by lysosomal enzymes.
However, to be efficient, the drug carrier conjugate should fulfill
several criteria (Trouet et al., 1982a): (i) the drug carrier conjugate must be inactive as such but the drug must be released in a
pharmacologically active form after endocytosis of the conjugate by
the target cells; (ii) the linkage between the drug and the carrier
must be stable in the plasma and the extracellular spaces but be
hydrolysed within lysosomes after endocytosis of the conjugate;
(iii) the drug carrier conjugate should permeate through the
anatomic barrier which could separate the site of administration
from the target; it must also be recognized as specifically as
possible by receptors or antigens present on the plasma membrane
of target cells; finally it must be endocytosed and gain access to
lysosomes; (iv) the drug carrier conjugate should be non toxic,

1

non immunogenic and be biodegradable to avoid cellular overload during
long term repetitive treatment; (v) production of drug carrier conjugates in the amounts and conditions (sterility, apyrogenicity,
stability) required for clinical use should be rather easy to achieve.

We actually developed this approach in an attempt to link antitumour drugs such as anthracyclines or Vinca alcaloids to macromolecules. These drugs have the ability to block tumour cell multiplication but unfortunately they also kill normal cells with high
division rate such as hematopoietic cells or intestinal epithelial
cells. They further affect physiological functions of other normal
cells, being respectively cardio- and neurotoxic. For instance,
daunorubicin (DNR) has been covalently bound to serum albumin
through a succinylated tetrapeptide arm consisting of ala-leu-ala-
leu, and the activity of the conjugate completely overshadowed that
of free DNR for the treatment of the murine L1210 leukemia (Trouet
et al., 1982b). The validity of this conjugation method and of the
drug carrier concept has also been demonstrated by linking an anti-
malarial drug, primaquine, to asialofetuin, a glycoprotein selectively
recognized and endocytosed by hepatocytes. Compared to free primaquine, the conjugate had significantly increased chemotherapeutic
activity against the hepatocytic stage of murine malaria (Trouet et
al., 1983).

At the present time, our main efforts are devoted to the search
of carriers which could be used for the treatment of human diseases
and in particular for cancer and infectious diseases. We currently
envisage different types of carriers which could fulfill the criteria
listed above.

The first group consists of neoglycoproteins and glycopeptides.
Since the pioneering work of Ashwell and Morell (1974) it has become widely accepted that exposed sugar residues on glycoproteins
serve as signals for recognition by plasma membrane receptors.
Among others, it is well established that hepatocytes display at
their sinusoidal membrane a receptor which promotes the uptake and
endocytosis of glycoproteins with complex glycanic moieties exposing
galactose residues after removal of terminal sialic acid (Ashwell
and Harford, 1982). On the other hand, it has been reported that
macrophages have at their cell surface a receptor specific for
glycoproteins with terminal mannose or N-acetyl-β-glucosamine
(Stahl and Schlessinger, 1980). Accordingly, molecules with terminal
galactose or mannose residues in an appropriate spatial configuration
could be used to target selectively drugs to hepatocytes and macrophages respectively. They could find application in the case of
viral or protozoal infection of hepatocytes or of diseases implicating macrophages such as for instance parasitic, inflammatory and
intracellular infectious diseases. In addition, delivery of immuno-
stimulating substances to macrophages could be achieved after
linkage to such carriers.

As already mentioned we have used a primaquine-asialo fetuin conjugate for the treatment of the exoerythrocytic stage of malaria (Trouet et al., 1983). On the other hand, Fiume et al. (1982) reported on the targeting of antiviral drugs to hepatocytes after linkage to a galactose terminated glycoprotein. We believe however that isolation and preparation of human asialoglycoproteins in the amounts required for clinical application would be very difficult. Therefore, we have envisaged, as it will be discussed below, to synthesize neoglycoproteins. Another possibility could consist in the use of glycopeptides which will have enough selectivity for the target receptor. These glycopeptides could for instance be obtained by pronase digestion of natural glycoproteins (Baenziger and Fiete, 1979) or by chemical synthesis (Robbins et al., 1981).

In the case of tumour cells, another type of carrier could be used to target drugs. It is based on the selectivity of the antigen-antibody reaction and on the hypothesis that tumour cells express more or less specific antigens on their plasma membrane. The first type of antigen to which antibodies could be raised are organotypic ones, which could restrict the uptake of drug-antibody conjugate to one cell type but would not restrict the access to the tumour cells only. A second type of target consists in tumour associated antigens. In this category, one might include antigens that are present at the cell surface of embryonic cells and reappear in the course of the malignant transformation or alternatively antigens that are present at the plasma membrane of normal cells in a limited number but which considerably increases on tumour cells. In some other cases, viral antigens appear at the cell surface of infected and transformed cells. The third type of target would be tumour-specific antigens. Although largely controversial, it is believed that specific antigens could appear during the malignant transformation and be expressed at the cell surface of the tumour cells. Antibodies to these antigens would represent the ideal drug delivery vehicle system since the uptake of the drug could be perfectly restricted to the target cells. One example of a functional tumour-specific antigen is the idiotypic part of surface immunoglobulin present on some B leukemic or lymphoma cells.

In this paper we describe our search of possible carriers which could be used for targeting drugs to human hepatoma and breast carcinoma cells. We also briefly report on the preparation and pharmacological characterization of a conjugate between dauno-rubicin and one of these carriers. Finally we emphasize the potential problems that could be encountered during the clinical utilization of these conjugates.

DRUG CARRIERS FOR HUMAN HEPATOMA

Although hepatocarcinoma or primary liver cancer is relatively rare in Western Europe and America it is the most frequent type of cancer throughout Africa and Asia. From a clinical point of view this type of tumour has a very poor prognosis. Chemotherapy which is in many cases the only therapeutic approach, is however considerably restricted by the toxicity of the drugs currently used.

We have envisaged different potential macromolecular carriers in order to restrict as much as possible the access of antitumour drugs only to hepatoma cells. Monoclonal antibodies to human alpha-foetoprotein (AFP) and to the surface antigen of hepatitis B virus (HBsAg) could achieve this goal. However, as we have recently described (Rolin-van Swieten et al., 1984), anti-AFP IgG is not recognized selectively by three different human hepatoma cell lines, two of which are secreting the antigen in the culture medium and could not therefore be used as drug carrier. In contrast, our results have shown that anti-HBsAg IgG is selectively bound and taken up by PLC/PRF human hepatocarcinoma cells. Furthermore, cell fractionation data have indicated that after incubation at $37^{o}C$, the antibody is accumulated within lysosomes wherein it is progressively digested. Therefore, experiments are now in progress to link drugs to this antibody and to evaluate the chemotherapeutic activity of the conjugate.

As already discussed, it is well known from the work of Ashwell and Hartford (1982) that hepatocytes expose at their sinusoidal membrane a receptor which recognizes selectively sugar moieties with terminal galactose and which promotes their endocytosis and lysosomal degradation. Accordingly and as long as hepatoma cells would keep expression of this receptor at their plasma membrane, the use of drug-carrier conjugates selectively recognized by this receptor could allow a selective delivery of the drug to the hepatocytes and hepatoma cells and restrict its access to the liver only.

However, before envisaging the use of such carriers, we have developed methods to detect the presence of the receptor on sections obtained from human hepatoma biopsies. Tissue sections were incubated in the presence of an asialofetuin-horseradish peroxidase conjugate prepared according to Wall et al. (1980). In the case of receptor-positive cells, the specific binding of the conjugate was completely inhibited after incubation of a subjacent section of the same tissue in the presence of a 150 fold excess of galactosylated serum albumin, another ligand of the receptor (see below). In both cases, the binding of the conjugate was revealed by cytochemical reaction of peroxidase. Typical results of positive tumour cells are illustrated at Fig. 1. Whereas hepatocytes and hepatocarcinoma cells are strongly stained by reaction products of peroxidase activity, they remain unstained after competition with unconjugated ligands. In another case (Fig. 2) we had the opportunity to obtain liver biopsies from a patient with a hepatocarcinoma, six months before his death and immediately post mortem as well as a necropsy

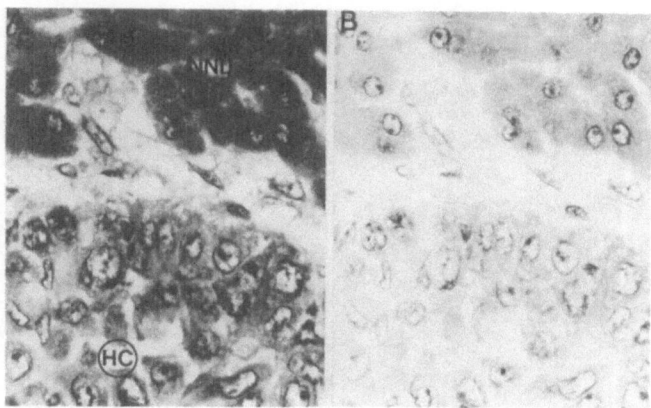

Fig. 1. Immunoperoxidase reaction of liver sections marked by an
         asialofetuin-horseradish peroxidase conjugate. Ethanol
         fixed, paraffin embedded sections from human liver with
         hepatocarcinoma were incubated for 1 h at 37°C with
         50 μg/ml of an asialofetuin-horseradish peroxidase con-
         jugate in the presence (B) or absence (A) of a 150-fold
         excess of galactosylated serum albumin. After washings,
         the sections were reincubated for 2 min in the presence
         of 0.5 mg/ml of diaminobenzidine and 0.01% H_2O_2. Mayer's
         hematoxylin counterstain; X950 real magnification; NNL:
         non neoplastic liver; HC: hepatocarcinoma.

from a pulmonary metastasis. Staining by reaction products
indicated the persistence of receptors on both normal and malignant
liver cells. Metastatic cells had also kept the receptor although
less stained. At the present time, preliminary investigations on
tissues from different patients indicate that out of 10 biopsies,
7 were positive for the presence of the receptor for glycoprotein-
bearing terminal galactose.

    In order to further investigate the potential usefulness of
drug carrier conjugates specific for this receptor, we took advantage of the observation of Schwartz et al. (1982) who reported that
the Hep G2 cells, a human hepatoblastoma cell line, capture
selectively ^{125}I-labelled asialoorosomucoid. Considering however
possible clinical application, we decided to use a carrier selectively recognized by the receptor that can further be prepared in
large amounts. On the basis of previous data (Quintart et al.,
1981; Limet et al., 1982) we have selected galactosylated serum
albumin (gal-SA).

    Terminal galactose residues are linked to human serum albumin

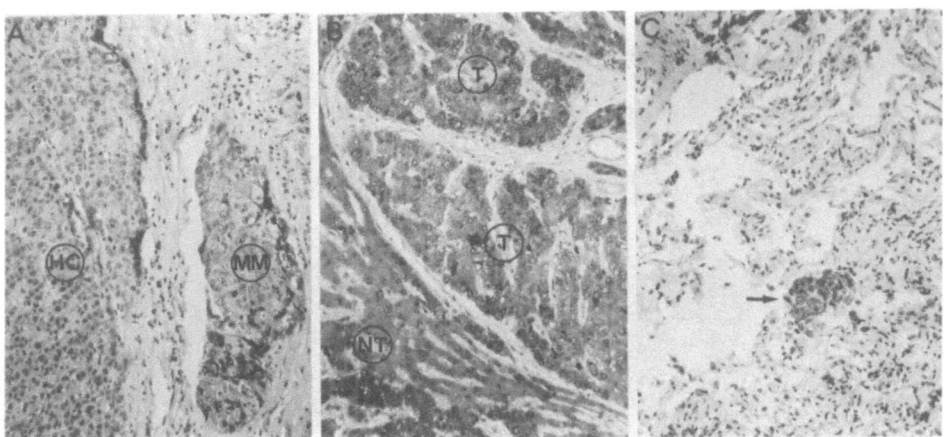

Fig. 2.   Immunoperoxidase reaction of different tissue sections
          from the same patient, marked with asialofetuin-horse-
          radish peroxidase conjugate. A. Section from initial
          liver biopsy showing hepatocarcinoma (HC) and a vascular
          vessel micrometastasis (MM); B. Liver necropsy section
          including tumour (T) and non-tumour (NT) cells; C. Lung
          necropsy section with a pulmonary metastasis (arrow).
          Same experimental procedure as in Fig. 1. Real magnif-
          ication X203.

by reductive lactosamination of lysines following the procedure
described by Wilson (1978). In brief, albumin in a sodium phosphate buffer pH 9.0 is successively treated with lactose and
NaBH$_3$CN. After incubation at 37°C the neoglycoprotein is freed
from any non-covalently bound sugar residue and from the remaining
reagents by extensive dialysis first against 0.1 M acetate buffer
pH 4.0 and then against PBS. Tritium-labelled glycoproteins have
been prepared using the same lactosamination procedure but in this
case, the intermediate imino derivative is treated for 1 h at 37°C
with NaB^{3}H$_4$ before adding NaBH$_3$CN. The number of galactose residues
covalently linked to the protein is determined by the phenol-H$_2$SO$_4$
method (Dubois et al., 1956) and a complementary determination of
the number of free amino group left in the protein is also run using
either TNBS (Habeeb, 1966) or fluorescamine (Udenfriend et al., 1972).
We have prepared by this method a series of gal-SA bearing a mean
number of galactose ranging from 14 to 36. The uptake of these neoglycoproteins by cultured rat liver cells was determined. Cells
were incubated in culture medium containing 1 μg/ml of ^{3}H-gal-SA for
24 h at 37°C. The uptake of labelled material by the hepocytes
(i.e. accumulated by the cells or released in the culture medium
under the form of labelled degradation products soluble in trichloroacetic acid) were assayed and the results are shown in Fig. 3.

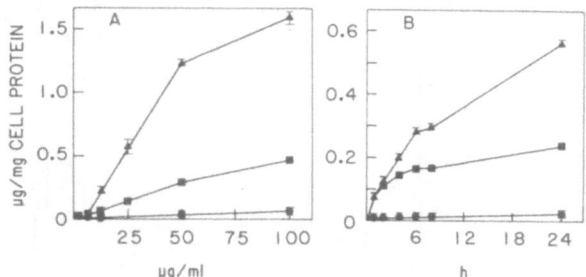

Fig. 3. Cultured human hepatoma Hep G2 cells were incubated in
25 cm³ flasks (ca. 1.5 mg cell protein) in 1.2 ml of
culture medium either for 24 h at 37°C with different
concentrations of ³H-labelled gal-SA or native serum
albumin or for different durations at 37°C with 10 μg/ml
of ³H-labelled gal-SA or native serum albumin. At the
end of the experiment the culture medium was removed and
analyzed for the presence of labelled degradation products
soluble in trichloracetic acid. The cells were washed and
analyzed for the presence of labelled material and for
protein content. Mean of 3 independent experiments are
given. ■ : Cell associated labelled material after incubation with gal-SA; ▲, ● : total uptake of gal-SA (▲)
or native serum albumin (●) i.e. cell associated labelled
material plus digestion products in culture medium.

It appears that a mean number of galactose residues equal to or
higher than 24 is required in order to achieve almost complete uptake of the neoglycoprotein.

As indicated in Fig. 3, cultured Hep G2 cells take up ³H
labelled gal-SA specifically as compared to ³H labelled native
serum albumin, in a process saturable with extracellular concentration (Fig. 3, A) and continuous with the incubation time (Fig.
3, B). At half the saturation of the specific uptake (0.35 μM) the
amounts of labelled material accumulated by the cells correspond to
0.14 μg/mg cell protein after 24 h incubation at 37°C; in addition,
labelled material amounting to 0.30 μg/mg is found in the culture
medium in the form of degradation products soluble in trichloroacetic acid. After isopycnic centrifugation of the particulate
fraction from Hep G2 cells incubated for 16 h with ³H-gal-SA, a
large proportion of the label equilibrates at the same densities as
N-acetyl-β-glycosaminidase, a marker enzyme of lysosomes, whereas
after incubation with native serum albumin, the label is found
throughout the gradient (Fig. 4).

Daunorubicin was linked to gal-SA via a succinylated tetrapeptide spacer arm. The coupling reaction occurs between the amino

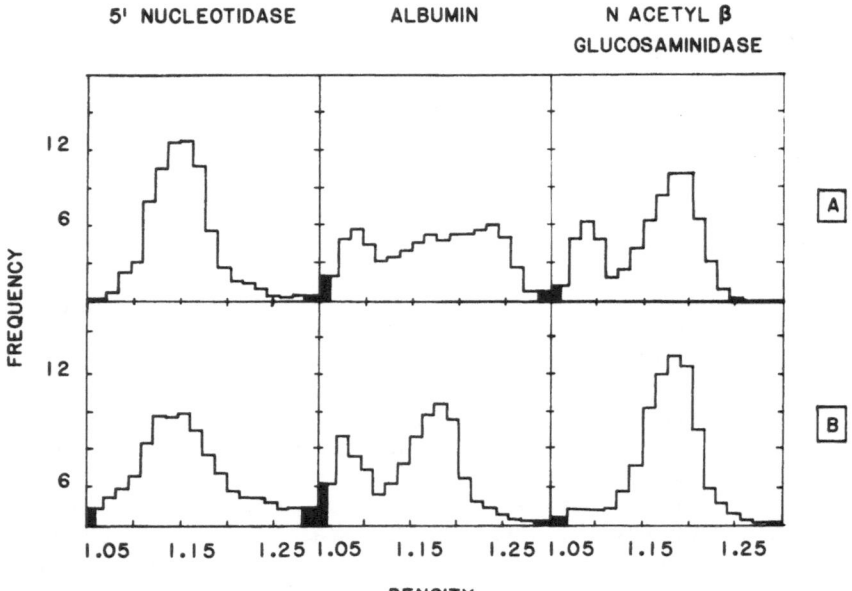

Fig. 4.   Isopycnic centrifugation of MLP fractions prepared as in
          Limet et al. (1982) from cultured Hep G2 hepatoma cells
          incubated for 16 h at 37°C with ³H-labelled native serum
          albumin (A) or gal-SA (B). MLP fractions contained from
          40 to 80% of marker enzymes and respectively 14.6 (A) and
          18.5 (B) % of ³H-labelled material. Results are presented
          under the form of normalized histograms.

group of lysine residues of the protein and the carboxylic group of
the succinylated tetrapeptide-DNR derivatives as summarized at
Fig. 5. The peptide ala-leu-ala-leu-DNR derivative is prepared
stepwise starting with DNR and the trityl protected N-hydroxysuccinimide ester of leucine (Anderson et al., 1967). The peptide
derivative is further succinylated by reacting with succinic anhydride. In order to perform the coupling reaction the carboxylic
group is transformed into the N-hydroxysuccinimide ester using a
mixed anhydride procedure. An excess of 0.6 to 0.8 of the ester
(calculated on the basis of the amount of free amino group in the
protein) is then added to gal-SA (2 mg protein/ml 0.1 M borate
buffer pH 9.0). A filtration on Ultrogel ACA 44 is carried out in
order to separate the drug-protein conjugate from higher molecular
weight aggregates as well as from any free peptide-DNR or remaining
reactants. Subsequent treatment of the conjugate with active
charcoal allows the removal of non-covalently bound drug. The
protein content of the conjugate is assayed by the Lowry method
whereas the drug is quantified by fluorometry after separation by

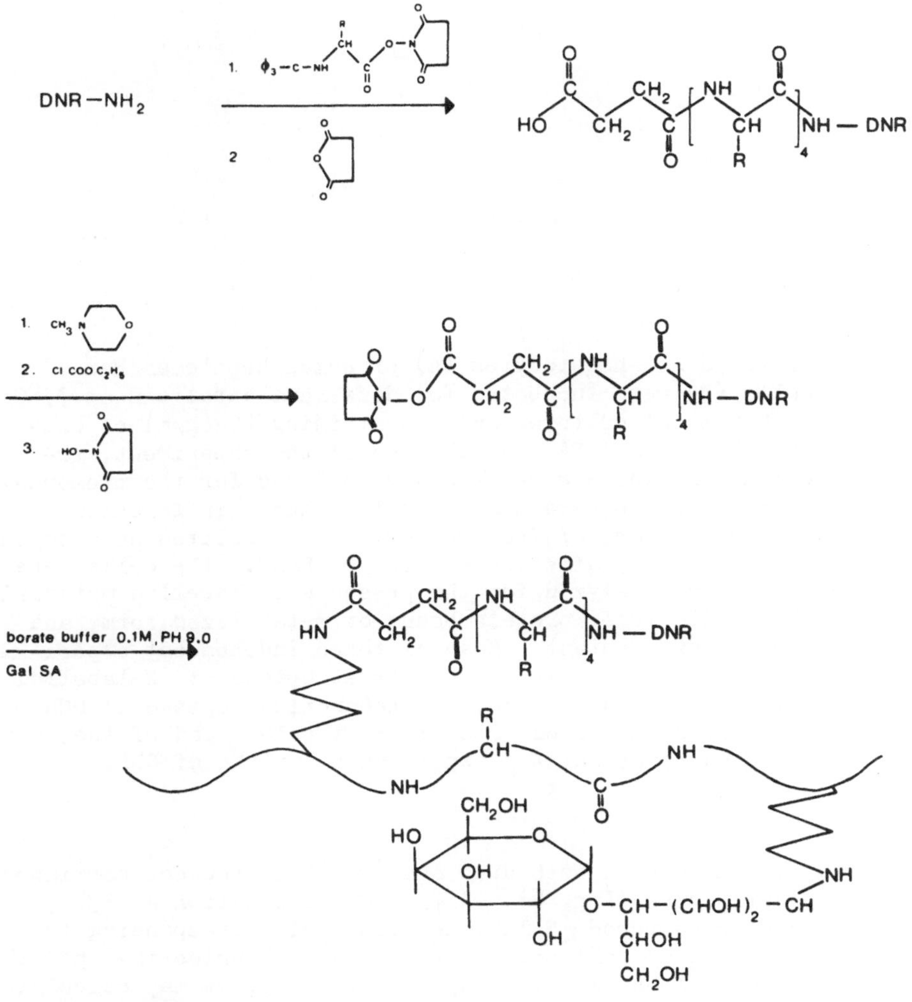

Fig. 5. Chemical reaction involved in the covalent linkage of DNR to gal-SA.

HPLC. The DNR to protein molar ratio ranges from 5 to 8 molecules of DNR/molecule of gal-SA and less than 1% of DNR was found unbound to the protein.

The uptake of DNR-^{3}H-gal-SA by both cultured rat hepatocytes and Hep G2 hepatoma cells has been studied. The amounts of fluorescent material as analyzed by HPLC (drugs) and of ^{3}H label (protein) associated with the cells or released in culture medium in the form of degradation products (protein) or free or metabolized drug were

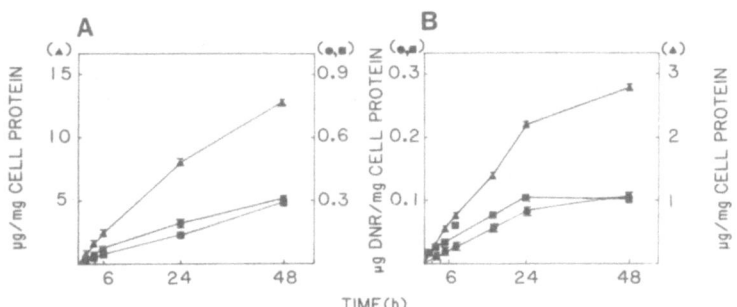

Fig. 6.   Cultured rat hepatocytes (A) or human hepatomas Hep G2
cells (B) were incubated for different durations at 37°C
in 4.5 ml of culture medium containing 40 μg/ml of DNR-
³H-labelled gal-SA. At the end of the experiment, the
culture medium was removed and analyzed for the presence
of labelled degradation products soluble in trichloroacetic acid and of free DNR or its metabolites as assayed
by fluorometry after separation by HPLC. The cells were
washed and analyzed for the presence of labelled material,
DNR in its free, protein bound or metabolized form, and
for protein content. Mean of three independent experiments   S.D. are given. ▲ : total uptake of ³H-labelled
gal-SA as defined above; ● : theoretical uptake of DNR
calculated on the basis of that of gal-SA and of the molar
ratio DNR/protein; ■ : experimental uptake of DNR.

recorded. The uptake of both drug and ³H label proceeds continuously
in the two cell types (Fig. 6). After 48 h incubation at 37°C,
hepatocytes have captured ³H labelled material corresponding to
11 μg of gal-SA/mg of cell protein (Fig. 6A). Considering that the
molar ratio DNR/gal-SA of this conjugate was 3.5, we may calculate
that theoretically, the cells should have captured 0.28 μg DNR/mg
cell protein, which is very close to the experimental value (0.30
μg/mg). Hep G2 cells also take up the conjugate but in smaller
amounts. This is related to the fact that these cells expose about
1/3 of the number of receptors present on hepatocytes (Schwartz
et al., 1981). After 48 h incubation, hepatoma cells have captured
2.8 μg of ³H gal SA/mg cell protein (Fig. 6B). On the basis of the
molar ratio 3.5 this corresponds to a theoretical level of 80 ng
DNR/mg as compared to an experimental value of 104 ng/mg. These
results indicate that the DNR-gal-SA conjugate is taken up by both
hepatocytes and hepatoma cells in amounts related to the number of
receptors exposed at the plasma membrane. By analogy to experiments
carried out with gal-SA, these results strongly suggest that the
conjugate is then endocytosed and that it gains access to lysosomes.
Therein the tetrapeptide arm is cleaved by hydrolytic enzymes and

the drug liberated. DNR either remains accumulated within the cells
or is released in the culture medium. Cell fractionation experiments are now in progress to further confirm the processing of the
conjugate.

The tissue distribution in mice and biliary excretion in rats
of a DNR-^{3}H labelled gal-SA have been compared with those of free
DNR and of DNR linked to ^{3}H-labelled native human serum albumin.
Both conjugates and DNR were given i.v. (0.5·mg DNR/kg). The concentrations of free and protein-bound DNR were determined by
fluorometry after separation by HPLC; the amounts of radioactive
material and the presence of labelled degradation products soluble
in trichloroacetic acid were also assayed. Free DNR disappeared
from the plasma with a half-life of 1.5 min and up to 60% of the
injected dose was found in the liver after 10 min. Thereafter DNR
escaped biphasically from this organ with half-life values of 20
min and 18 h. After i.v. injection of DNR into rats, one third of
the injected dose was excreted in the bile within 24 h. After i.v.
administration of DNR-gal-SA into mice, both the drug and the ^{3}H-
label disappeared from the plasma with a half-life of 12 min. There
was no significant difference in the plasma clearance curves of the
drug and of the radioactive label. In the plasma, no free DNR was
detected and no more than 2% of the radioactive label was soluble
in trichloroacetic acid. These results suggest that the conjugate
remains stable in the bloodstream as previously described (Trouet
et al., 1982b) and that DNR is cleared from the plasma bound to its
carrier. The conjugate was taken up essentially by the liver, as
illustrated in Fig. 7. However, the fate of DNR and ^{3}H label were
different. Within 30 min, 70% of the injected drug was taken up by
the liver and released from its carrier. Thereafter the free DNR
was eliminated from this tissue with a half-life of 2 h. The ^{3}H
label was taken up more slowly and it disappeared from the liver as
fragments soluble in trichloroacetic acid with a half-life of 33 h.
In the spleen, kidney and heart, less than 1% of both the drug and
of ^{3}H label were found. When compared to free DNR administered at
the same dose, after injection of DNR-gal-HSA, the concentration of
the drug in the heart and the kidney was 5 to 10 times lower. After
i.v. injection into rats of DNR-gal-SA the drug released in the liver
was rapidly excreted in the bile. After 24 h, the total biliary
excretion of the drug was 52% of the injected dose, in contrast with
6% of the ^{3}H label. When DNR conjugated to native serum albumin
labelled with ^{3}H by reductive methylation of lysine residues was
given i.v. to mice (0.5 mg DNR/kg), the protein-bound DNR and ^{3}H-
labelled material disappeared more slowly and to a lesser extent
from the plasma. Lower proportions of protein bound DNR and ^{3}H
label were taken up by the liver. Free DNR and ^{3}H-labelled material
soluble in trichloroacetic acid appeared in lower amounts in this
tissue. These experiments indicate that after hydrolysis in the
liver, the ^{3}H-labelled lysines did not remain accumulated within
this tissue but were rapidly excreted.

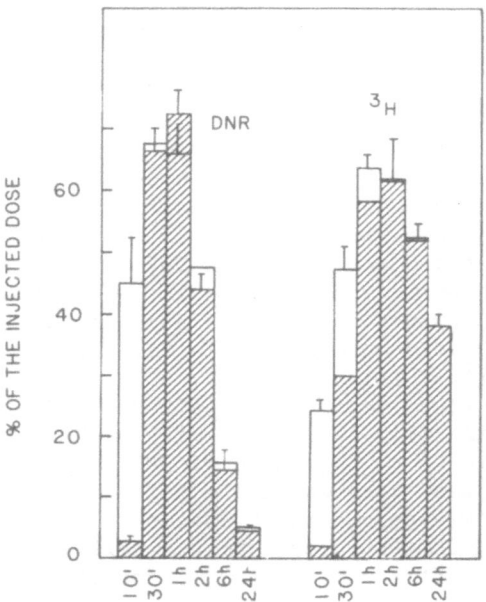

Fig. 7.   Liver accumulation of total DNR (free and protein bound)
and free DNR and of ^{3}H label, soluble or not soluble in
trichloroacetic acid after i.v. administration to mice of
DNR-^{3}H-labelled gal-SA (0.5 mg DNR/kg body weight). □ :
total drug or ^{3}H label insoluble in trichloroacetic acid;
▣ : free drug or ^{3}H label soluble in trichloroacetic acid.
The results are expressed as % of the injected dose.
Means of two independent experiments ± S.D. are given.

These experiments demonstrate that, in-vivo, DNR-gal-SA is
selectively taken up by the liver wherein free active DNR is rapidly
released after digestion of the conjugate or of the peptide arm
linking it to the neoglycoprotein. The toxicity towards normal
tissues, mainly the heart, should be considerably reduced, 5 to 10
times less DNR being found in the heart, kidney and spleen after
administration of DNR linked to gal-SA. In addition, the toxicity
towards normal hepatocytes should also be very low as a result of
the rapid biliary excretion of DNR liberated from the conjugate in
the hepatocytes and therefore the selectivity and activity towards
the hepatoma cells unable to excrete the liberated DNR should be
increased.

In parallel to a comparison of the pharmacokinetic papameters
of DNR in its free and conjugated form we have started to evaluate
the toxicity and pharmacological properties of these drugs. We have
first determined the LD_{50} and MTD values of DNR free or linked to

Table 1. Comparative toxicity of daunorubicin, free (DNR) or
          conjugated to galactosylated human serum albumin
          (DNR-gal-SA), after i.v. injection into mice.

| Drug | LD_{50} (mg/kg) | | MTD (mg/kg) |
	After 30 days	After 60 days	
DNR	28.0	28.0	13.6
DNR-gal-SA	61.2	56.9	22.8

Female DBA_2 mice received a single i.v. injection of DNR
free or linked to gal-SA. LD_{50} (lethal dose for 50% of
mice) was evaluated from 9 different doses using 10 mice
per dose (probit scale-linear regression). MTD (maximal
tolerated dose) was calculated by linear regression from
the percentage of weight loss on day 5 as a function of
the given dose, taking into account 5% weight loss.

gal-SA after a single i.v. injection in mice. As reported in
Table 1, in terms of LD_{50} the conjugate is at least twice less
toxic than free DNR, after 30 as well as after 60 days of observation. A complete histopathologic study of the mice treated with
the conjugate as compared to free DNR at different time intervals
after i.v. administration, is presently under progress. When comparing the MTD values, defined as the doses inducing 5% weight loss
on day 5 after injection, DNR-gal-SA is tolerated 1.7 times better
than DNR.

Preliminary experiments were also performed in order to
evaluate the chemotherapeutic effectiveness of the DNR-gal-SA conjugate on human hepatomas. Therefore, tumour fragments obtained
from post mortem biopsies of patients with hepatomas were implanted
under the renal capsule of mice according to the method developed
by Bogden and coworkers (1979). Table 2 gives the data obtained in
3 different cases as well as the histopathologic diagnoses. To
ascertain the persistence of tumour cells at the end of the experiments, histopathologic examination of the implanted graft fragments
was carried out directly on the mouse kidney. In view of our finding that after a 6 day assay infiltration of inflammatory cells
caused by host versus graft reaction could affect the final measurements of the graft size, we decided to end the assay on day 5 (cf
Hep 3) and modified therefore also the drug treatment schedule.
The mice bearing human hepatoma fragments were treated i.v. with

Table 2. Chemotherapeutic activity of daunorubicin (DNR),
          free or conjugated to galactosylated human serum
          albumin (DNR-gal-SA), after i.v. injection into
          mice implanted with human hepatoma fragments under
          their renal capsule.

Drug[b]	Dose (mg DNR/kg) per injection	Change in tumour size (in mm)[a]		
		HEP 1[c]	HEP 2[d]	HEP 3[e]
Control	–	+ 0.26	+ 0.12	+ 0.04
DNR	12	+ 0.18	– 0.10	+ 0.03
DNR-gal-SA	19	– 0.06	– 0.18	– 0.12

[a] Human hepatoma fragments (1 mm^3) were implanted on day 0
under the renal capsule of female BDF1 mice; initial and
final graft measurements and evaluation of the results
were done according to Bogden et al. (1979). Final day
of experiment was day 6 for Hep 1 and 2, and day 5 for
Hep 3.

[b] Drugs were given i.v. on days 2 and 4 for Hep 1 and 2,
and on days 1 and 2 for Hep 3.

[c] Hep 1: patient with hepatocarcinoma.

[d] Hep 2: patient with hepatocarcinoma and cholangiocarcinoma.

[e] Hep 3: patient with hepatocarcinoma.

DNR either free or conjugated to gal-SA or native serum albumin,
given at equitoxic doses. In the first case (Hep 1) a mixture of
gal-SA and the free tetrapeptide derivative of DNR was also administered as a control (not shown). Our results indicate that for
Hep 1, only the DNR-gal-SA conjugate induced tumour regression
whereas neither free DNR nor the mixture of protein and drug were
active. The experiment on Hep 3 clearly shows that only the conjugate prepared with gal-SA was effective whereas the conjugate prepared with native serum albumin was completely inactive (not shown).
The case of Hep 2 is less clear, perhaps due to the different histopathologic diagnosis but here also DNR-gal-SA appeared to induce
the greatest tumour regression. The experimental model we have used
for the chemotherapeutic evaluation of our drug-carrier conjugate,
potentially selective for human hepatoma cells, could however be
different from the clinical situation of hepatomas. The implantation

of very small tumour fragments (1 mm^3) under the renal capsule as
compared to the whole intact liver of the mouse is a very difficult
model since hepatocytes also recognize glycoproteins exposing
galactose residues. After i.v. administration of the conjugate
prepared with gal-SA, the mouse liver will take up significant
amounts of the drug-carrier conjugate and the preliminary therapeutic effects we have observed at the level of hepatoma fragments
implanted on the kidney seem therefore promising.

All these results strongly suggest that galactosylated serum
albumin to which daunorubicin has been linked by a tetrapeptide arm
is a promising conjugate for chemotherapy of human hepatoma cells
that have kept expression of the receptor for galactose terminated
glycoproteins. In particular, our data have shown that this conjugate could restrict the access of the drugs to normal hepatocytes
and tumour cells eliminating almost totally the uptake of drug by
other cell types. Once the drug-carrier conjugate has been bound
by the receptor, it is endocytosed and the drug is liberated after
digestion of the carrier and the tetrapeptide arm. In normal hepatocytes the drug is rapidly excreted into the bile. Preliminary
experiments indicate that, in-vivo, the conjugate is more active on
human hepatoma fragments implanted in mice than free DNR. In-vitro
experiments are now in progress to evaluate the cytotoxic action of
DNR conjugated to gal-SA on hepatoma cells.

DRUG CARRIERS FOR HUMAN BREAST CARCINOMA

In the course of our studies on carriers able to direct
selectively antitumour drugs to human breast cancer cells, we have
prepared antibodies which bind specifically to the plasma membrane
of cells from primary tumours and of metastasis derived from them.

To start with, we have limited our efforts to antibodies
directed against antigens which are organ-specific. An interesting
model for studying the nature of human breast cell surface membranes
had been proposed by Ceriani et al. (1980, 1983). During lactation
the lipid products of milk produced by the inner organelles of the
acinar cell approach the apical membrane where they are packed with
fragments of the plasma membrane of this domain. Once in the lumen
the membrane-surrounded lipids are referred to as milk fat globules
(MFG). Biochemical and morphological evidence indicates that the
membranes associated with MFG (referred to as MFGM) are derived
indeed from the apical plasma membrane of mammary epithelial cells.

We have prepared MFGM from fresh human milk samples. The procedure takes advantage of the fact that MFG, due to their low
density, can be recovered following a low speed centrifugation on
top of all other milk constituents. Thermic shock accompanied by
churning liberates the fats and the membrane fraction can be

collected after high speed centrifugation. SDS-polyacrylamide
electrophoresis of the proteins associated with those membranes
reveals a multiband polypeptide pattern. Experiences involving
interaction of ^{125}I-labelled concanavalin A with blotted electrophoresis patterns of MFGM proteins indicate that most of them are
glycosylated. Antiserum to human MFGM was prepared in goat.
Indirect immunohistochemical studies using a peroxidase conjugated
second antibody have shown that this polyclonal antibody was able
to react with sections from human normal adult resting breast.
Staining was located on the apical membrane of the epithelial cells

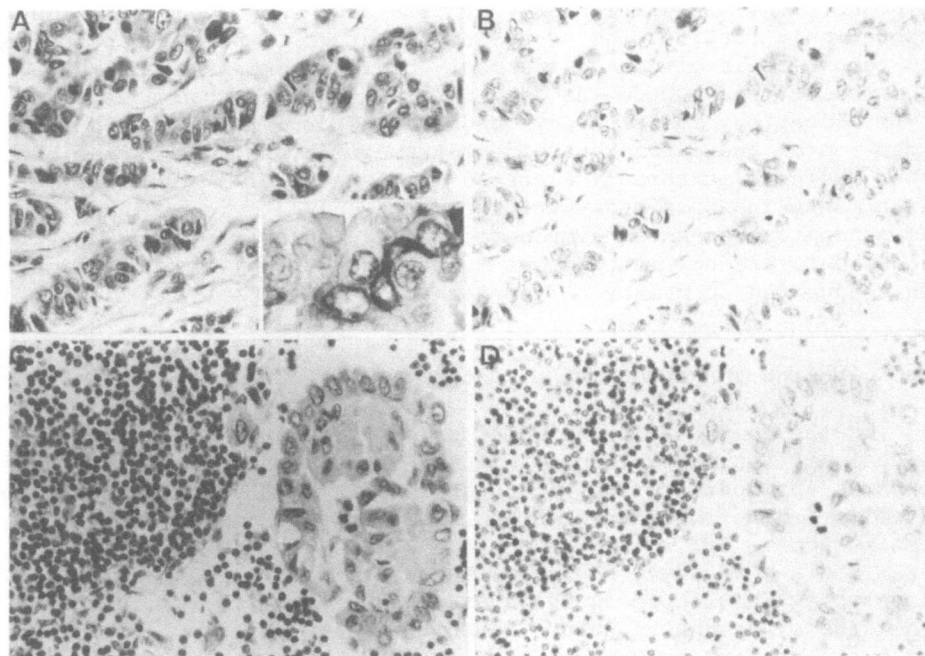

Fig. 8. Indirect immunoperoxidase reaction of human duct breast
         primary carcinoma (A, B) or lymphatic axillary node sus-
         pected of metastasis (C, D) marked with goat polyclonal
         anti-human MFGM IgG. Formalin-PBS fixed, paraffin em-
         bedded tissue sections were incubated for 30 min at 37°C
         with goat anti-MFGM antiserum diluted 40 times (A, C) or
         goat control antiserum (same dilution) (B, D) washed and
         reincubated for 30 min at 37°C with peroxidase labelled
         rabbit anti-goat antiserum diluted 30 times. After
         washings, the sections were incubated for 3 min in the
         presence of 0.5 mg/ml diaminobenzidine and 0.01% H_2O_2.
         Mayer's hematoxylin counterstain; X515 real magnification.
         Insert in A: X1500 real magnification.

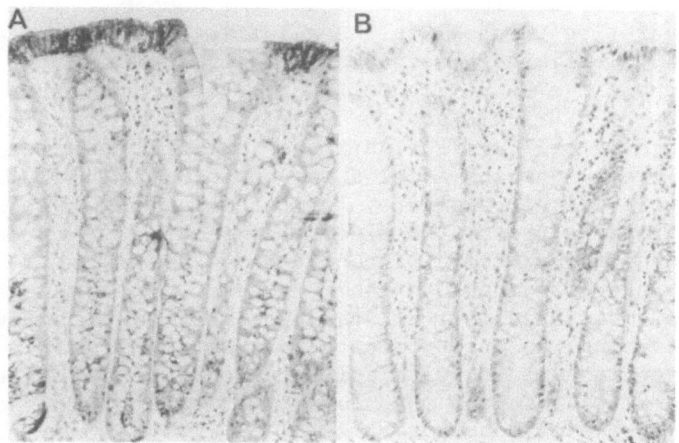

Fig. 9. Indirect immunoperoxidase reaction of human intestinal
mucosa marked with polyclonal goat anti-human MFGM IgG
before (A) and after (B) immunoadsorption of the antiserum
on human colon immobilized on sepharose 2B. Same experimental protocol as in Fig. 8. X203 real magnification.

lining the ducts and the terminal ductules. Secretion within the
ducts and the terminal ductules was often stained as well. At the
lobular level, reaction was located at both membrane and cytoplasm.
Myoepithelial cells were completely unstained. Breast carcinoma
sections were positively stained as well as metastatic cells in the
lymph nodes (Fig. 8). The extent of reactivity varied within the
different histopathologic cases examined. The specificity of this
antibody against different human tissues has been screened. Negative
results were obtained with connective, muscle and nervous tissues
as well as with blood and lymphatic vessels and cells. However,
reactivities were observed with epithelial cells from the proximal
tubules of the kidney, biliary canaliculi of the liver, pyloric
glands, intestinal glands, folliculi cells of the thyroid, epithelia
of bronchial and pancreatic acini. Almost all cross reactions were
eliminated by an adequate adsorption on membranes prepared from
human colon, stomach and trachea mucosa, liver and kidney immobilized by reaction with cyanogen bromide activated sepharose without
affecting the specific reaction with breast carcinoma or derived
metastatic cells. An example is illustrated in Fig. 9. Part A
indicates the interaction of the polyclonal antiserum with human
colon sections before adsorption on immobilized membranes of human
colon mucosa whereas part B illustrates the disappearance of staining afterwards on a serial section from the same tissue.

The fusion of SP 207 non-immunoglobulin secreting murine
myeloma cell line with splenic lymphocytes of mice immunized with

MFGM resulted in the production of a large number of hybridoma
cultures secreting immunoglobulins reactive with MFGM on a solid
phase ELISA. Preliminary selection of hybridoma was done by
eliminating all IgM secreting cultures. To further define the
reactivities of hybridoma and to determine whether the antibodies
produced bind to cell surface antigens, an ELISA test using cultured
MCF7 human breast carcinoma cells was performed. The 27 selected
hybridoma were further cloned by limited dilution and then injected
i.p. into pristane-pretreated mice. The resulting ascites were
screened on sections from different human tissues. As it was observed with the polyclonal antibody, none of the monoclonal antibodies reacted with cells from non-epithelial origin. However,
positive staining was observed with primary breast carcinoma and
metastatic cells but also with epithelial cells from different
tissues and with different monoclonal antibodies. For example,
reaction patterns of monoclonal IgG_3 14B2 with a primary breast
carcinoma or a lymph node metastatis are illustrated in Fig. 10.

    The binding and endocytosis of the polyclonal antibody purified
by immuno-adsorption on MFGM have been studied on human breast
carcinoma MCF7 cells. Incubation at $4°C$ in the presence of 100 μg
of ^{3}H-labelled IgG/ml of culture medium has shown that the amount
of IgG fixed by the cells was 70-fold higher than that of control
IgG (Fig. 11A). Kinetic studies at $37°C$ in the presence of 10 μg
of specific antibody/ml of extracellular medium have shown that
accumulation of labelled material reaches a plateau after 3-4 h
incubation, which amounts to the equivalent of 0.78 μg per mg of
cell protein (Fig. 11B). After 24 h incubation, digestion products
corresponding to the equivalent of 0.18 μg/mg cell protein were
released in the extracellular medium, which suggests that the antibody gained access to lysosomes. Similar experiments are now in
progress to study uptake of different monoclonal antibodies by the
MCF7 cells. In addition, cell fractionation techniques are used to
establish their cellular localisation.

    These preliminary experiments are very promising and support
the idea that antibodies and, in particular, monoclonal antibodies
could be used to target drugs. Therefore, experiments have been
performed to link DNR to IgG. Using the same method as that
described in Fig. 5, we reacted IgG from a different origin (poly
or monoclonal origin and from different animals) at 1.5 mg/ml in a
0.1 M borate buffer pH 9.0 containing 0.3 M NaCl with a 0.2 to 0.3
excess relative to the lysine content of the antibody of the
activated ester of DNR. The drug-antibody conjugate was further
purified by gel filtration on Ultrogel ACA 3-4 and treated with
active charcoal. The protein and drug content were determined as
previously described. Preliminary experiments indicate that more
aggregated protein is formed with IgG than with gal-SA. On the
other hand, DNR/antibody molar ratios higher than 3-4 were not
observed for the monomeric form of IgG. Attempts to increase this

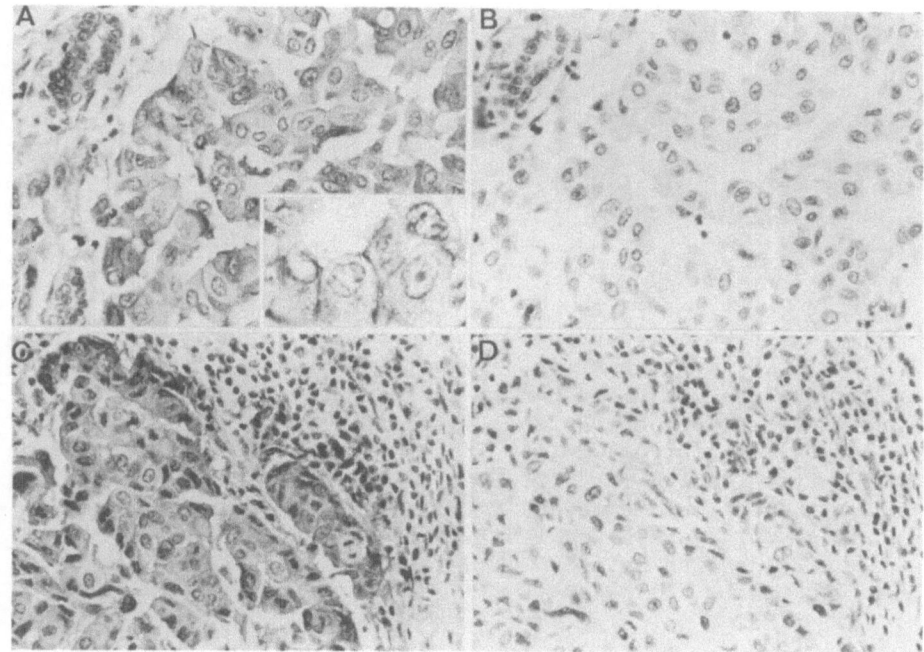

Fig. 10. Indirect immunoperoxidase reaction of an infiltrating ductal human breast carcinoma (A, B) or of a lymph node metastasis of the same tumour (C, D) marked with a mouse monoclonal antibody (14B2) to human MFGM (A, C) or with control mouse IgG (B, D). Same experimental protocol as in Fig. 8 except that the monoclonal antibody concentration was 3 μg/ml and that the second antibody was a peroxidase conjugated rabbit anti-mouse IgG. X515 real magnification.

molar ratio by raising the excess of DNR over the protein lysines (from 0.15 to 0.4) resulted in lower recovery of the IgG in its monomeric form.

GENERAL CONCLUSIONS

The experiments reported in this paper strongly suggest that the concept of drug targeting could soon find application in the case of human cancer. In particular, carriers have been identified which could limit the access of antitumour drug to one cell type. Methods have also been developed to link to macromolecules a drug such as daunorubicin by a tetrapeptide arm that is stable in blood and extravascular space but that can be hydrolysed by lysosomal enzymes after endocytosis of the drug carrier conjugate. In-vitro

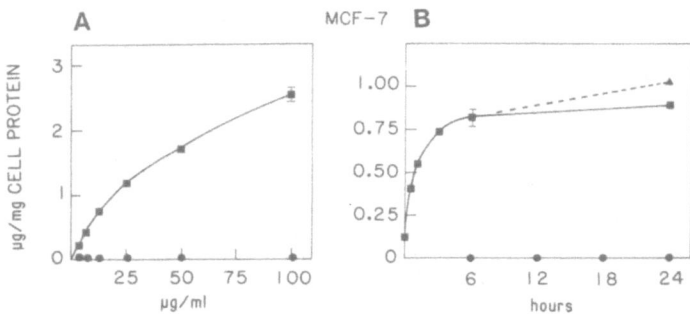

Fig. 11.   Cultured human breast carcinoma MCF7 cells were incubated
in 25 cm³ flasks (ca. 1 mg of protein) in 1.5 ml of
culture medium either at 4°C for 5 h with different concentrations of ³H-labelled goat polyclonal anti-MFGM IgG
(A) or control IgG (A) or at 37°C with 10 μg/ml of both
types of IgG for different durations (B). At the end of
incubation, media and cells were process as in Fig. 3.
■: cell associated labelled material; ▲,● : uptake of
³H-labelled anti-MFGM IgG as defined in Fig. 3.

and in-vivo pharmacological experiments are in progress to evaluate
the chemotherapeutic activity of these conjugates.

Although all these results appear very promising, there are
several problems that could be encountered during treatment of
patients and that should be carefully considered. There are first
purely technical problems that result from the difficulties to
obtain conjugate preparations in the amounts and conditions required
for clinical use. In the case of ligands recognized by receptors
specific for sugar moieties, we believe that in addition to synthetic
neoglycoproteins, one should also consider the use of glycopeptides
which could be obtained by enzymatic digestion from natural glycoproteins, synthesized by chemical methods or prepared by genetic
engineering. In the case of antibodies, recent progress in the
monoclonal biotechnology should allow production of the required
amounts of antibody carriers.

Another possible limitation of the use of drug carrier conjugates is the extent of their passage through anatomic barriers.
This problem should be carefully investigated since it could limit
the access of the drug to the tumour cells.

The conjugates could also induce side effects. For example,
liver toxicity could result from the uptake of DNR-gal-SA by normal
hepatocytes. However our pharmacokinetic data indicate that a
large proportion of injected free DNR is also captured by the liver
but that the drug is rapidly excreted into the bile. In addition

to cytotoxicity for tumour cells, drug-antibody conjugates could be toxic for other cells by which they would be taken up. Nevertheless, so far as the antibody had been carefully screened for cross re- actions with normal cells and more particularly with those for which the drug could be toxic, this complication should be circumvented. On the other hand, in the case of drug-neoglycoprotein conjugates, immunogenicity could be encountered despite the fact that human serum albumin is used to prepare gal-SA. This problem could probably be solved by the use of small glycopeptides. In the case of monoclonal antibodies, immunogenicity could be a problem since they are produced in mice or rats. Since however DNR and many other anticancer drugs are immunosupressive, this complication may be avoided. Furthermore, preliminary phase I clinical trials indicate that the DNR-gal-SA conjugate is well tolerated by patients with hepatoma even during repetitive treatments over more than 6 months (4 days perfusion separated by two weeks). The conjugate does not induce appearance of antibodies in the plasma nor affect significantly hematological, kidney or liver functions.

Probably the most crucial problem concerns the number of drug molecules required to gain cytotoxicity in the tumour cells. Therefore, a key question is to determine whether drug carrier con- jugates will be able to deliver enough drug molecules to achieve the killing of all tumour cells. One has indeed to realize that the number of receptors or antigens that will recognize the drug carrier conjugate and promote its endocytosis is limited resulting therefore in a saturable uptake of the conjugates. Accordingly, experiments should be done to investigate the influence of the pharmacokinetics of conjugate administration on tumour cell killing. However, there are some possibilities to increase the pharmacological activity of drug carrier conjugates. The first one consists in increasing the duration of the conjugate infusion to maintain a stable plasma concentration over extended periods of time, as we have actually done in preliminary clinical trials with the DNR-gal- SA conjugate on patients bearing hepatomas. Another method could consist of increasing the number of drug molecules attached to the carrier. On the basis of our present experience, we know however that molar ratios higher than 15 to 20 DNR molecules/gal-SA or 3 to 5 DNR/antibody promote aggregation of the carrier. Loading of albumin or of an inert non toxic polymer (Lloyd et al., this book) with the drug before linkage to the carrier could be a possible way to solve that problem. Finally, a third and probably the most attractive method could consist in preparing conjugates with drugs which gain cytotoxicity at lower intracellular concentrations.

In view of the amounts of drug required to achieve tumour cell killing, the problem of the selection of monoclonal antibodies should perhaps be reconsidered. Theoretically antigens of interest in drug targeting should be present only at the cell surface of tumour cells, to gain specificity, and must be endocytosed to permit

access of the conjugate to lysosomes. Since in addition the antigens must allow the uptake of enough drug molecules to kill the
tumour cells, the selection of an antibody raised to an antigen
totally specific to the target cell, but present in very limited
extent, could not be the best approach. It could indeed be more
efficient to select an antibody to an antigen that is present at a
higher concentration on the plasma membrane of different cell types
with the exception of cells for which the drug or the antibody can
be cytotoxic. On the other hand on the basis of our previous observations (Schneider et al., 1979a, b; 1981) as well as current
concepts of receptor mediated endocytosis, one should remember that
antigens and receptors, after endocytosis and access to lysosomes,
are very often recycled back to the cell surface, acting therefore
as some sort of endless moving belt able to pick up large amounts
of drug carrier conjugates.

Another problem of cancer chemotherapy that could also be
encountered with this conceptual approach concerns resistance of
tumour cells to antitumour drugs. A method which could circumvent
it, consists in the use of a mixture of drug carrier conjugate with
different classes of drug for which there is no cross resistance.

Tumour cell heterogeneity will most probably also be an important problem to consider. To achieve complete chemotherapeutic
activity, it is indeed crucial that all tumour cells, both in
primary and metastatic tumours, maintain expression of the receptor
or antigens promoting recognition and endocytosis of the drug carrier
conjugates. Experiments are now in progress to test tumour cell
heterogeneity in the case of hepatoma and breast carcinoma cells.
We believe however that to gain cytotoxicity towards all tumour
cells, a mixture of different types of carriers or cocktail of
monoclonal antibodies will be required.

A last possible restriction of the use of antibodies as drug
carriers results from the presence of antigens in the serum of the
patients. These antigens can be secretion products of the tumour
cells, such as alpha-foeto-protein or can have been shed from the
tumour cell surface such as for example milk fat globule membrane
antigens in the case of human breast carcinoma (Ceriani et al.,
1982). Such antigens could cross react with the drug carrier conjugate before it reaches the target cells and be cleared as immune
complexes, suppressing therefore chemotherapeutic activity. This
problem could however be overcome by evaluating the concentration
of the circulating antigen and injection of unconjugated antibody
before administration of the conjugated drug.

ACKNOWLEDGEMENTS

The excellent technical assistance of Mrs. B. Agneessens-

Hennau, M. Debroux-Dechambre, C. Decamps-Michel, P. Delvaux-de Ville de Goyet, Mrs. T. Aerts, Mr. H. Leduc, E. Verstraeten and M. Wauters is gratefully appreciated. This work was supported by Rhône-Poulenc Santé (Paris, France), the Belgian Caisse Générale d'Epargne et de Retraite (CGER), the Ministère de la Région Bruxelloise and the IRSIA.

REFERENCES

Anderson, G.W., Zimmerman, J. and Callahan, F.M., 1964, A reinvestigation of the mixed carbonic anhydride method of peptide synthesis, J. Amer. Chem., 86:1839.

Ashwell, G. and Morell, A.G., 1974, The role of surface carbohydrates in the hepatic recognition and transport of circulating glycoproteins, Adv. Enzymol., 41:99.

Ashwell, G. and Harford, J., 1982, Carbohydrate-specific receptors of the liver, Ann. Rev. Biochem., 51:531.

Baenziger, J.V. and Fiete, D., 1979, Structure of the complex oligosaccharides of fetuin, J. Biol. Chem., 234:789.

Bogden, A.E., Haskell, P.M., Lepage, D.J., Kelton, D.E., Cobb, W.R. and Esber, H.J., 1979, Growth of human tumour xenografts implanted under the renal capsules of normal immunocompetent mice, Exptl. Cell Biol., 47:281.

Ceriani, R.L., Sasakri, M., Peterson, J.A. and Blanck, E.W., 1980, Mammary epithelial cell identification by means of cell surface antigens, in: "Cell Biology of Breast Cancer", McGrath, C.M., Breuman, M.J. and Prick, M.A., eds., Academic Press, New York.

Ceriani, R.L., Sasakri, M., Sussman, M., Waren, W.M. and Blanck, E.W., 1982, Circulating human mammary epithelial antigens in breast cancer, Proc. Natl. Acad. Sci. USA, 79:5420.

Ceriani, R.L., Peterson, J.A., Lee, Y.Y., Moncada, R. and Blanck, E.W., 1983, Characterization of cell surface antigens of human mammary epithelial cells with monoclonal antibodies prepared against human milk fat globules, Som. Cell Gen., 9:415.

Dubois, M., Gilles, K.A., Hamilton, J.K., Rebers, P.A. and Smith, F., 1956, Colorimetric method for determination of sugars and related substances, Anal. Chem., 28:350.

Fiume, L., Busi, C. and Mattioli, A., 1982, Lactosaminated human serum albumin as hepatotropic drug carrier: rate of uptake by mouse liver, FEBS Lett., 146:42.

Habeeb, A.F.S.A., 1966, Determination of free amino groups in proteins by trinitrobenzene sulfonic acid, Anal. Biochem., 14:328.

Limet, J.N., Quintart, J., Otte-Slachmuylder, C. and Schneider, Y.-J., 1982, Receptor-mediated endocytosis of hemoglobin-haptoglobin, galactosylated serum albumin and polymeric IgA by the liver, Acta Biol. Med. Germ., 41:113.

Quintart, J., Limet, J. and Baudhuin, P., 1981, Receptor-mediated

endocytosis of glycosylated derivatives of bovine serum
albumin: targeting based on sugar recognition, Prot. Biol.
Fluids, 29:389.

Robbins, J.C., Lam, M.H., Tripp, C.S., Bugianesi, R.L., Pompipom,
M.M. and Shen, T.Y., 1981, Synthetic glycopeptide substrates
for receptor-mediated endocytosis by macrophages, Proc. Natl.
Acad. Sci. USA, 78:7294.

Rolin-van Swieten, D., Schneider, Y.-J. and Trouet, A., 1984,
Drug targeting with monoclonal antibody for human hepatomas,
Prot. Biol. Fluids, 31:791.

Schneider, Y.-J., Tulkens, P., de Duve, C. and Trouet, A., 1979a,
The fate of plasma membrane during endocytosis. I. Uptake
and processing of non specific and anti-plasma membrane and
control immunoglobulins by cultured fibroblasts, J. Cell
Biol., 81:449.

Schneider, Y.-J., Tulkens, P., de Duve, C. and Trouet, A., 1979b,
The fate of plasma membrane during endocytosis. II. Evidence
for recycling (shuttle) of plasma membrane constituents, J.
J. Cell Biol., 81:466.

Schneider, Y.-J., de Duve, C. and Trouet, A., 1981, The fate of
plasma membrane during endocytosis. III. Evidence for incomplete breakdown of immunoglobulins in lysosomes of c
cultured fibroblasts, J. Cell Biol., 88:380.

Schwartz, A.L., Fridovich, S.E., Knowles, B.B. and Lodish, H.F.,
1981, Characterization of the asialoglycoprotein receptor
in a continuous hepatoma line, J. Biol. Chem., 256:8878.

Schwartz, A.L., Fridovich, S.E. and Lodish, H.F., 1982, Kinetics
of internalization and recycling of the asialoglycoprotein
receptor in a hepatoma cell line, J. Biol. Chem., 257:4230.

Stahl, P.D. and Schlessinger, P.H., 1980, Receptor-mediated pinocytosis of mannose N-acetylglucosamine terminated glycoproteins and lysosomal enzymes by macrophages, Trends Biochem.
Sci., 5:194.

Trouet, A., Baurain, R., Deprez-De Campeneere, D., Masquelier, M.
and Pirson, P., 1982a, Targeting of antitumour and anti-
protozoal drugs by covalent linkage to protein carriers,
in: "Targeting of Drugs", G. Gregoriadis, J. Senior and
A. Trouet, eds., Plenum Press, New York.

Trouet, A., Masquelier, M., Baurain, R. and Deprez-De Campeneere, D.,
1982b, A covalent linkage between daunorubicin and proteins
that is stable in serum and reversible by lysosomal hydro-
lases, as required for a lysosomotropic drug-carrier conjugate: in-vitro and in-vivo studies, Proc. Natl. Acad. Sci.
USA, 79:626.

Trouet, A., Pirson, P., Baurain, R. and Masquelier, M., 1983, Cell
targeting of primaquine by association with liposomes and by
covalent linkage to asialofetuin, in: "Handbook of Experimental Pharmacology", Peters, W. and Richards, W.H.G., eds.,
Springer Verlag, Heidelberg.

Udenfriend, S., Stein, S., Bohlen, P., Dairman, W., Leimburzer, W.

and Weigele, M., 1972, Fluorescamin: a reagent for assay of amino acids, peptides, proteins and primary amines in the picomole range, Science, 178:871.

Wall, D.A., Wilson, G. and Hubbard, A.L., 1980, The galactose-specific recognition system of mammalian liver: the route of ligand internalization in rat hepatocytes, Cell, 21:79.

Wilson, G., 1978, Effect of reductive lactosamination on the hepatic uptake of bovine pancreatic ribonuclease A dimer, J. Biol. Chem., 253:2070.

TOXIC CONJUGATES OF EPIDERMAL GROWTH FACTOR AND ASIALOFETUIN

Harvey R. Herschman, Daniel Cawley,* and
David L. Simpson +

Department of Biological Chemistry, and
Laboratory of Biomedical and Environmental Sciences
UCLA Center for the Health Sciences
Los Angeles, California 90024 USA

*Present address: Department of Microbiology and
Immunology
Box 8093
Washington University Medical School
St. Louis, Missouri 63110

+Present address: Department of Biochemistry
Meharry Medical College
1005 D.B. Todd Boulevard
Nashville, Tennessee 37208

INTRODUCTION

Our laboratory has, for a number of years, been interested in
the role and mode of action of polypeptide growth factors. These
agents bind to specific cell surface receptors and initiate a
cascade of biological responses leading either to cell division or
expression of a new phenotype as a consequence of ligand-receptor
interactions. In many cases one of the most notable initial responses is ligand-induced endocytosis of the receptor-ligand complex,
with eventual degradation of the ligand and, in some cases, degradation of the receptor as well. As one of several approaches
(Herschman et al., 1982) to the study of polypeptide hormone/growth
factor induced responses, we have constructed heteroconjugates of
binding ligands with protein toxins or their enzymic portions.

CHOICE OF BINDING LIGANDS

Epidermal growth factor (EGF) is a small polypeptide (53 amino

27

acids, 6045 daltons) whose sequence has been completely determined
(for review see Carpenter and Cohen, 1979). It is easily purified
in large quantity from murine salivary glands (Savage and Cohen, 1972)
and is extremely stable. EGF has no free lysine groups; the only
free amino group is at the N terminus. This reactive amino group
has been derivatized with a variety of agents without impairing
the biological activity of the molecule. Consequently we felt it
likely the EGF could be derivatized, coupled to toxins and still
retain substantial binding capability. EGF induces rapid and
extensive endocytosis of the ligand-receptor complex, with extensive
degradation of bound ligand and down-regulation (i.e., loss of
cell surface binding activity) of the EGF receptor (Carpenter
and Cohen, 1979). It is a potent mitogen for murine 3T3 cells
Rose et al., 1975), human fibroblasts (Carpenter and Cohen, 1979),
and a variety of other types of cells. Our laboratory has been interested in the mechanisms of the EGF-mediated mitogenic response for
some time (Herschman et al., 1982). Construction of EGF-toxin heteroconjugates would, we felt, be useful in elucidating both the biology
and the biochemistry of EGF and the mode of entry and cytoplasmic
activation of ligand-toxin conjugates.

Desialylated, or asialo, serum glycoproteins are rapidly
removed from the circulation by a liver-specific receptor for
desialylated (galactose-terminated) glycoproteins. The receptor is
present, in large numbers, uniquely on mammalian hepatocytes. The
asialoglycoprotein receptor (ASGP) has been characterized and
purified to homogeneity by affinity chromatography on Sepharose-
bound asialoglycoproteins (Kawasaki and Ashwell, 1976). The bound
ligand is rapidly endocytosed by hepatocytes and degraded in the
lysosomes, with kinetics very similar to those observed for EGF.
In contrast to the EGF system, however, the ASGP receptor is recycled
to the plasma membrane, where it can function in another round of
of ligand-receptor internalization (Steer and Ashwell, 1980);
ligand-induced internalization does not, in this case, lead to
receptor degradation. Because (1) we could prepare the binding
ligand, asialofetuin (ASF), in large quantity from commercially
available starting materials, (2) the ASGP receptor is restricted
to a single cell type and is present on hepatocytes in substantial
numbers, (3) the ligand receptor complex is rapidly taken into the
cells by endocytosis, and (4) the receptor is returned to the membrane - in contrast to the EGF receptor - we felt this would be
an excellent system with which to compare and contrast the biochemistry of toxin conjugates of the EGF system.

Toxins:

Ricin is a toxic protein isolated from the castor bean. The
active protein is composed of two polypeptides, joined together
by a disulfide bond. One of these polypeptides, the ricin B chain,
binds to galactose-containing glycoproteins and/or glycolipids on the

the cell surface (Sandvig et al., 1976). The other polypeptide, the
ricin A chain (RTA) is an enzyme that catalytically inactivates the
larger of the two ribosomal subunits (Olsnes et al., 1975). The
precise mechanism of ribosomal inactivation is not known, and is
currently the subject of active investigation (reviewed by Olsnes and
Pihl, 1982a). The two chains of ricin must be joined through a
disulfide bond in order to elicit substantial toxicity; the B chain
alone can bind to cell surfaces but is unable to kill because it is
devoid of any enzyme activity. The A chain alone is only weakly
toxic, since it cannot bind to cells and cannot enter the cytoplasm
to attack susceptible ribosomes. It is, however, a potent inhibitor
of protein synthesis in cell-free systems. Thus A chain alone
possesses intrinsic enzymatic activity; the substantially reduced
toxicity of RTA relative to inact ricin (four to five orders of magnitude) is due to the inability of RTA to bind, be internalized and
reach the cytoplasm to gain access to the ribosomes. Toxic concentrations of ricin in vivo are in the range of nanograms to micrograms
per kilogram. In culture, the ED50 (the dose required for fifty
percent inhibition of protein synthesis and/or viability) is in the
range of 10^{-13}-10^{-11}M.

Diphtheria toxin (DT) is produced by Corynebacterium diphtheriae as a single polypeptide, but is susceptible to limited proteolytic cleavage which generates two fragments (DTA and DTB) held
together by a single disulfide bond. DTB is a binding component;
DTA catalytically transfers ADP-ribose from NAD to elongation factor
2, irreversibly inactivating the enzyme and thus shutting down protein synthesis (Honjo et al., 1968). Like ricin, the two fragments
of DT must be joined via the disulfide group in order for toxicity
to be expressed; DTB can bind but not intoxicate cells, DTA cannot
gain access to its cytoplasmic target. Rodent cells are resistant
to the action of DT, although their EF2 can be inactivated in-vitro
by DTA-mediated ADP-ribosylation. The resistance apparently lies
in the inability of DT to bind and/or be productively internalized
in rodent cells.

CONSTRUCTION AND ISOLATION OF EGF AND ASF HETEROCONJUGATES

EGF was purified by the method of Savage and Cohen (1972).
Fetuin was purchased from Sigma, and desialylated with agarose bound
Clostridium perfringens neuraminidase (Cawley et al., 1981). Ricin
was purified as described by Cawley et al., (1978); A chain was
isolated by mercaptoethanol elution of Sepharose-bound ricin and
subsequent re-chromatography on Sepharose (Cawley et al., 1980).
DTA, prepared as described by Chung and Collier (1977), was the gift
of Dr. John Collier.

In order to synthesize disulfide-linked conjugates of EGF and
ASF we prepared modified proteins containing a pyridine-dithiopro-
pionate group. The thiopyridyl moiety is an excellent leaving

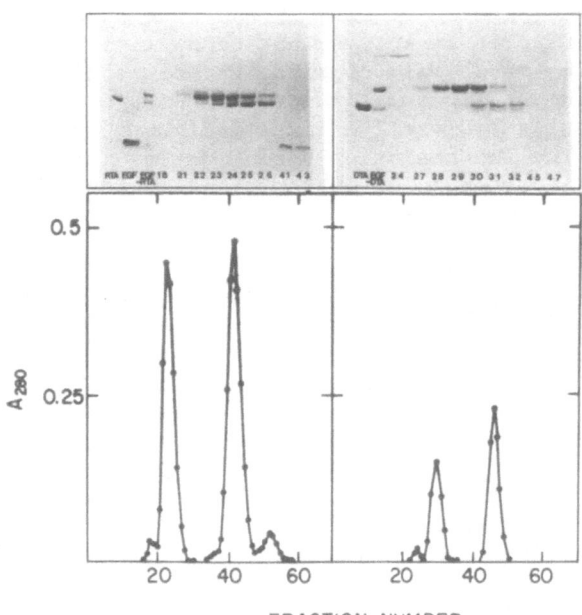

Fig. 1.   Sephadex G-75 chromatography of EGF-DTA and EGF-
RTA reaction mixtures. The gels were 13.5% polyacrylamide and 15% polyacrylamide for EGF-RTA and
EGF-DTA respectively. The numbers indicate the
sample fractions from the chromatography columns.
EGF-RTA is on the left, EGF-DTA on the right. Left:
Slots contain (1) 5μg RTA, (2) 18μg EGF-PDP, (3)
10 μg of EGF-RTA reaction mixture prior to chromatography. Right: Slots contain (1) 10μg fragment
A, (2) 9μg of the EGF-DTA reaction mixture prior
to chromatography (Cawley et al., 1980).

group for thiol-disulfide interchange reactions. To prepare the
derivatized ligands, EGF and ASF were incubated with excess N-
succinimidyl, 3-2' (pyridyldithio) propionate (SPDP). The PDP
derivatives were separated from the excess SPDP by gel filtration
chromatography. We also prepared the PDP derivative of intact
fetuin by a similar procedure.

To prepare the RTA and DTA derivatives of EGF-PDP, fetuin-PDP
and ASF-PDP the toxin A chain were reduced with dithiothereitol
desalted, and mixed with the various ligands to undergo thiol-disulfide exchange reactions. In general, the PDP-derivatized binding

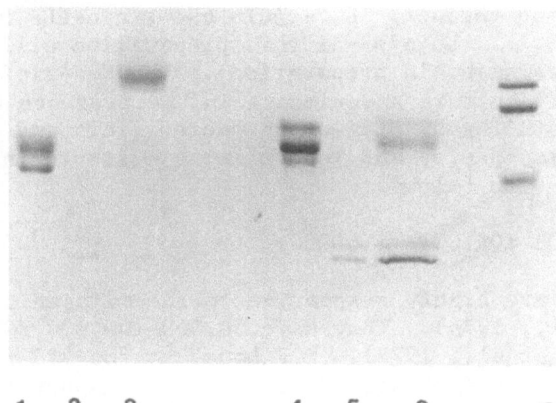

Fig. 2.   SDS gel electrophoresis of ASF-RTA conjugate.
          Samples in lanes 1-3 were not reduced, samples
          in lanes 4-7 were reduced in 150 mM 2- mercap-
          toethanol. (1,4),15μg ASF; (2,5), 5μg RTA;
          (3,6), 10μg ASF-RTA purified by two cycles of
          Sephadex G-100 chromatography; (7) molecular
          weight standards (BSA, Mr = 67,000; catalase,
          Mr = 60,000; ovalbumin, Mr = 44,000; cytochrome
          C, Mr = 13,500). The two forms of RTA (Mr =
          30,000 and 33,000) are apparent in lanes 2, 5,
          and 6. Following reduction, ASF-RTA (lane 6)
          is cleaved to RTA (lane 5) and ASF (Lane 4)
          (Simpson et al., 1982).

ligands were in slight molar excess. The products were then separated from the reactants by chromatography on Sephadex G75 (EGF) or
Sephadex G100 (ASF, fetuin). Details of the preparation of EGF-RTA,
EGF-DTA, ASF-RTA and ASF-DTA have been described (Cawley et al.,
1980; Cawley et al., 1981; Simpson et al., 1982). The purification
of EGF-RTA and EGF-DTA are shown in Fig. 1, characterization of
purified ASF-RTA is shown in Fig. 2.

    To construct a conjugate between EGF and intact ricin we der-
ivatized ricin with SPDP to prepare a ricin-PDP derivative with an
average of 1.2 PDP molecules per ricin molecule. EGF-PDP was
reduced in dithiothreitol to prepare EGF-SH, recovered by chromatography on Sephadex G-15, and mixed in equimolar amounts with ricin-
PDP. After the thiol-disulfide exchange reaction had proceeded, the

reactants were passed over a Sepharose 4B column and the unreacted
EGF-SH was washed through. EGF-ricin and ricin-PDP were then eluted
with 10 mM lactose. Details of this preparation will be described
(Cawley and Herschman, in preparation). The EGF-ricin preparation
was used in cell culture experiments in the presence of lactose (50
mM) to block both the binding of unreacted ricin and the conjugate
through ricin receptors, and thus eliminate its toxin action via the
ricin receptor (see below).

TARGET CELLS FOR TOXICITY STUDIES

3T3 cells are highly responsive to the mitogenic activity of
EGF (Rose et al., 1975). They have 50,000-100,000 EGF receptors per
cell (Aharonov et al., 1978). EGF bound to the 3T3 receptor is
rapidly endocytosed into coated vesicles, transported to the lysosome,
and degraded (Fine et al., 1981). The time course of internalization
and degradation of EGF by 3T3 cells is illustrated in Fig. 3. We
have also isolated several 3T3 variant cell lines that do not possess
the ability to bind EGF (Pruss and Herschman, 1977; Butler-Gralla and
Herschman, 1981). These cells serve as excellent controls to eval-

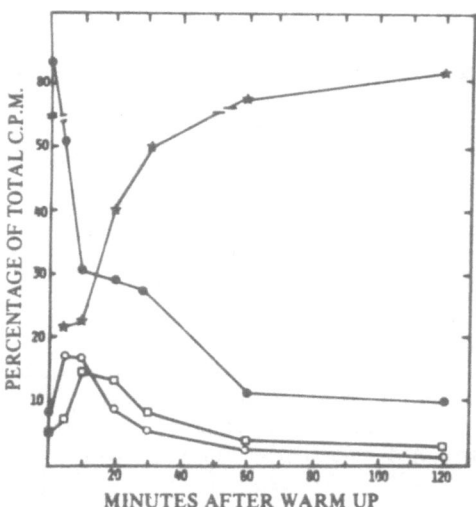

Fig. 3.   Distribution of ^{125}I-EGF in sucrose gradients.
^{125}I-EGF was bound to 3T3 cells at 4° for 60
min. Cells were washed, placed at 37° for the
times shown, then harvested and analysed by
centrifugation for distribution of label.
Plasma membrane, (●); coated vesicles, (○);
Lysosomes, (□); TCA soluble counts in supernatant, (✱) (Fine et al., 1981).

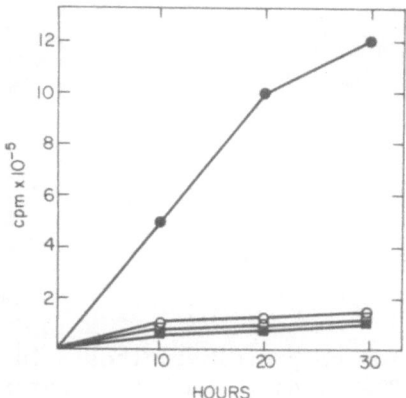

Fig. 4. Degradation of ASF and fetuin by 24 h rat
hepatocyte cultures. Approximately 1.5×10^6
hepatocytes or 6×10^5 3T3 cells, grown in 35mm
multiwell dishes, were exposed to 2ml of medium
containing either ^{125}I-fetuin or ^{125}I-ASF (0.2
μg/ml; 4.4 nM). Undegraded protein was precipitated from the medium by 2% phosphotungstic
acid: 10% trichloroacetic acid and the radioactivity was measured. Hepatocytes plus [125]I-
ASF, (●); hepatocytes plus ^{125}I-fetuin, (○);
3T3 cells + ^{125}I-ASF, (■); 3T3 cells plus [125]I-
ASF, (□). Hepatocytes degraded ^{125}I-ASF at a
rate of 1-2 fmol/min/10^6 cells. 3T3 cells degraded this reagent at a rate less than 0.1 fmol/
min/10^6 cells. Iodinated fetuin was not degraded (less than 0.1 fmol/mm/10^6 cells) by
either cell type (Herschman et al., 1982).

uate the EGF receptor-mediated portion of the toxicity of EGF conjugates. A431 cells are a line of human cervical carcinoma cells
with an extraordinarily high number of EGF receptors ($1-2 \times 10^6$ EGF
receptors per cell; Fabricant et al., 1977).

To evaluate the activity of the fetuin and ASF derivatives it
was necessary to prepare primary hepatocyte cultures. We utilized
the collegenase perfusion method (Attie et al., 1979) to prepare rat
hepatocyte cultures, and demonstrated (Cawley et al., 1981) that
these preparations could internalize and degrade ^{125}I-asialofetuin
(Fig. 4).

TOXICITY OF THE RTA DERIVATIVES OF EGF AND ASF

We first tested the toxicity of EGF-RTA on 3T3 cells. EGF-RTA

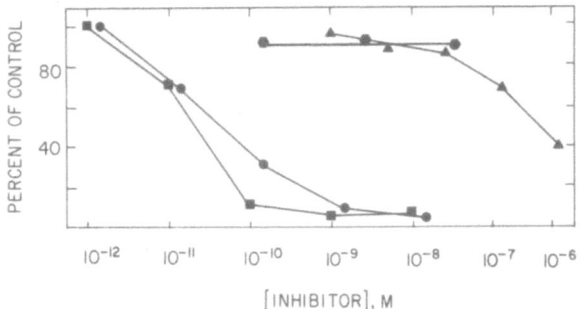

Fig. 5.   Inhibition of protein synthesis in 3T3 cells
by EGF-RTA, ricin, and RTA. Approximately
3×10^4 cells were seeded into small scintillation vials and incubated 24 h prior to
toxin addition. Fresh medium containing the
indicated toxin concentration was added.
After 24 h the medium was removed and the
cells were incubated for 2 h in fresh medium
containing labeled amino acids. Samples were
processed for scintillation spectrometry as
described by Mohring and Mohring (1977).
Data are plotted as percent of untreated controls. Ricin, (■); EGF-RTA, (●); RTA, (▲);
equimolar mixtures of EGF and RTA, (◕)
(Cawley et al., 1980).

was a potent toxin on these cells (Fig. 5), approaching the activity
of ricin in some experiments. When RTA is joined via a disulfide
bond to either RTB or EGF it is a far more potent toxin than when
present alone. RTA must be joined via a disulfide bond, however;
EGF present with RTA, but not covalently bound, did not enhance the
toxicity of RTA.

If the toxicity of the EGF-RTA conjugate is mediated by the EGF
receptor it should be blocked by the simultaneous addition of underivatized EGF. This was, indeed, the case; EGF could competitively
block the toxicity of EGF-RTA (Fig. 6). We would also predict that
cells unable to bind EGF would have a reduced sensitivity to the
EGF-RTA conjugate, while cells with a large number of EGF receptors
might be unusually sensitive to this conjugate. This was, indeed,
the case (Table 1). A431 and 3T3-TNR-2 cells were nearly equal to
3T3 in their sensitivity to ricin. A431 cells, with a large number
of receptors, were quite sensitive to EGF-RTA. In contrast, 3T3-
TNR-2 cells, with no EGF-binding activity (Butler-Gralla and
Herschman, 1981), were relatively insensitive to EGF-RTA, and were

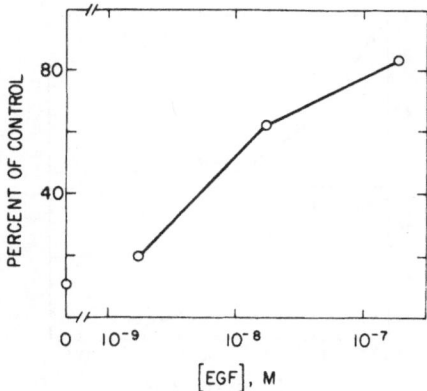

Fig. 6. EGF blocks the inhibition of protein synthesis
in 3T3 cells caused by EGF-RTA: EGF (2.9×10^{-9}M)
and EGF were added simultaneously to the cells.
Protein synthesis was measured, as described in
the legend to Fig. 5, after 24 h (Cawley et al.,
1980).

Table 1. Inhibition of Protein Synthesis in Various Cell Lines by
Ricin and EGF-RTA.

Cell Line	Number of EGF Receptors per Cell ($\times 10^{-7}$)	ED_{50} (M) of Protein Synthesis Inhibition	
		EGF-RTA	Ricin
3T3-NR-2	not detectable	$> 3 \times 10^{-8}$	2.5×10^{-12}
CHO	21	3.6×10^{-8}	2.5×10^{-12}
3T3	66	1.0×10^{-10}	5.0×10^{-12}
Vero	110	6×10^{-11}	4×10^{-13}
A-431	1300	2×10^{-11}	5×10^{-12}

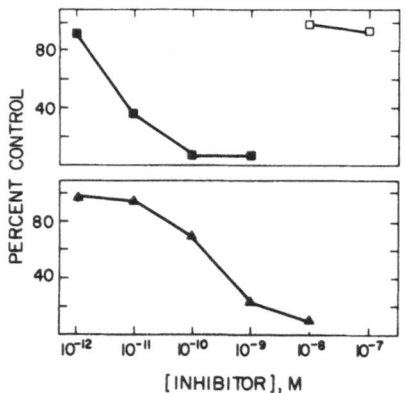

Fig. 7.   Inhibition of protein synthesis by ASF-RTA in
          cultured hepatocytes. Top: Ricin, (■), RTA,
          (□); Bottom: ASF-RTA, (▲). Protein syn-
          thesis assayed as described in the legend to
          Fig. 5, after 24 h (Simpson et al., 1982).

killed by this conjugate only at high concentrations. This inhibition was not antagonized by EGF (Cawley and Herschman, unpublished).
EGF-RTA is a potent toxin only on cells possessing EGF receptors.

     ASF-RTA is a potent toxin on cultured rat hepatocytes (Fig. 7).
RTA, as expected, was not appreciably toxic on these cells. When
ASF was mixed with equimolar amounts of RTA no toxicity was observed
(data not shown); ASF must be joined to RTA by a disulfide bond to
affect cytotoxicity. If the toxicity of ASF-RTA is mediated by ASGP
receptors then only asialoglycoproteins should block this toxicity.
This was, indeed, the case; neither orosomucoid or fetuin could
block the toxicity of ASF-RTA. In contrast, the desialylated derivatives of either serum glycoprotein were effective competitive inhibitors (Fig. 8).

TOXICITY OF THE DTA DERIVATIVES OF ASF, FETUIN, AND EGF

     The DTA derivative of ASF was also toxic to cultured rat hepato-
cytes (Fig. 9). In contrast, the parent diphtheria toxin molecule
was not able to kill these cells, since rodent cells are highly
resistant to this toxin. If the toxicity of the ASF-DTA conjugate
is mediated by the ASGP receptor, then the potency of glycoprotein
antagonists should reflect their ability to bind to the receptor.
This was the case; orosomucoid and fetuin could not block the toxicity of ASF-DTA, while their asialoglycoprotein derivatives could
do so (Fig. 10). Moreover, removal of the terminal galactosyl

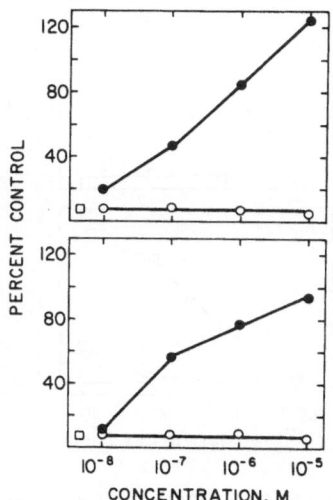

Fig. 8.   Inhibition of ASF-RTA toxicity on cultured
          hepatocytes by asialoglycoproteins. Upper
          panel: ASF-RTA, (□); ASF-RTA + ASF, (●);
          ASF-RTA + fetuin, (○). Lower panel: ASF-
          RTA, (□); ASF + asialoorosomucoid, (●);
          ASF + orosomucoid (○).

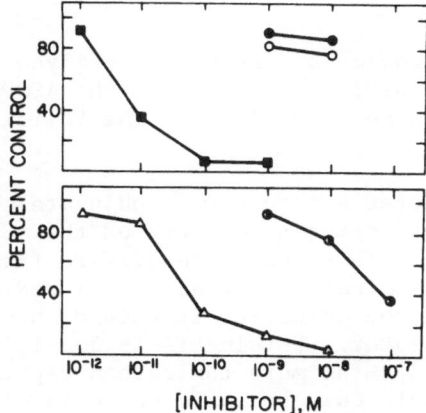

Fig. 9.   Inhibition of protein synthesis in cultured
          rat hepatocytes by ASF-DTA and diphtheria
          toxin. Upper panel: ricin, (■); diphtheria
          toxin, (●); DTA, (○). Lower panel: ASF-DTA,
          (△); fetuin-DTA, (◑). (Simpson et al., 1982)

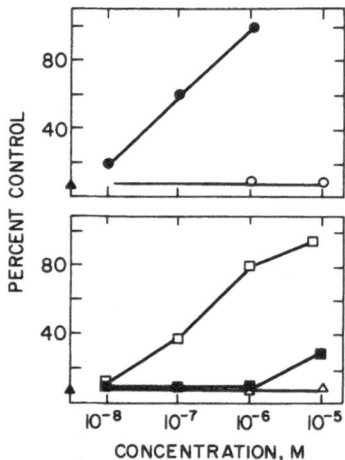

Fig. 10. Inhibition of ASF-DTA toxicity on cultured hepa-
        tocytes by serum glycoproteins and their asialo
        and asialoagalacto derivatives. Top panel: 10^{-9}M
        ASF-DTA, (▲); ASF-DTA + fetuin, (○) ASF-DTA +
        ASF. (●); Lower panel: ASF-DTA, (▲); ASF-DTA +
        orosomucoid, (△); ASF-DTA + asialoorosomucoid,
        (□); ASF-DTA asialoagalactoorosomucoid, (■)
        (Cawley et al., 1981).

residue from asialoorosomucoid generated a glycoprotein, asialoagal-
actoorosomucoid, incapable of binding to the ASGP and thus no longer
capable of serving as an inhibitor of the toxicity of the ASF-DTA
conjugate.

    We also constructed a fetuin-DTA conjugate, to provide another
tool to examine the receptor specificty of the ASF conjugates on
cultured hepatocytes. This conjugate differs from the ASF-DTA conjugate only in the terminal sialic acid residues of the oligosaccharide chains of the molecule; it should, however, be unable to
bind to the ASGP receptor. Fetuin-DTA is 300-fold less toxic than
ASF-DTA (Fig. 9), and is no more toxic than diphtheria toxin or DTA.
Finally, ASF-DTA should only be toxic on cells with ASGP receptors.
Two other primary culture preparations and three permanent cell lines
were insensitive to the action of ASF-DTA (Table 2). The ASF-DTA
conjugate exerts well over 99% of its toxic effect on cultured
hepatocytes via the ASGP receptor.

    When the EGF-DTA conjugate was tested on 3T3 cells we were sur-
prised to find that this conjugate was completely ineffective as a

Table 2. Inhibition of Protein Synthesis in Various Cell Types by ASF-DTA

Cell Type	ED_{50} (pM)
Vero	>10,000 (NI)[1]
3T3	>10,000 (NI)
CHO	>10,000 (NI)
Primary Rat Hepatocytes	50
Primary Rat Heart Cells	>10,000 (NI)
Primary Bovine Aortic Epithelium Cells	>10,000 (NI)

[1]NI, no inhibition

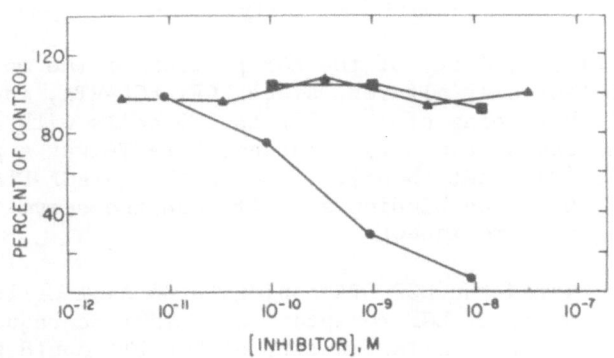

Fig. 11. Inhibition of protein synthesis in 3T3 cells by EGF-DTA. EGF-RTA, (●); diphtheria toxin, (■); EGF-DTA, (▲) (Cawley et al., 1980).

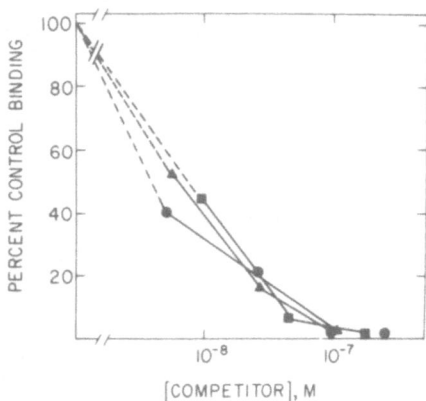

Fig. 12. EGF, EGF-RTA, and EGF-DTA competition for
binding of ^{125}I-EGF to 3T3 cells. ^{125}I-EGF
concentration was 30 ng/ml. Incubation was
at 4°C for 1.5 h. The competitors and ^{125}I-
EGF were added simultaneously. Non-specific
binding (binding in the presence of a 50-fold
excess of unlabeled EGF) was 20%, and has
been subtracted from the total binding to
give specific binding. EGF, (●); EGF-DTA,
(■); EGF-RTA, (▲) (Cawley et al., 1980).

toxin (Fig. 11), in contrast to the EGF-RTA conjugate (Fig. 7). The
EGF-DTA conjugate was able to catalyze the ADP ribosylation of EF2;
thus the lack of toxicity of this conjugate was not the result of
inactivation of its enzymatic activity.

The binding activity of the EGF portion of the molecule was
also not substantially altered, since EGF, EGF-RTA, and EGF-DTA
could all block binding of ^{125}I-EGF to 3T3 cells with equivalent
molar effectiveness (Fig. 12). Indeed, EGF-DTA was a potent mitogen
for 3T3 cells (data not shown). Thus, although EGF-DTA was not toxic
on 3T3 cells, both the binding activity and the enzymatic activity
of the conjugate were intact.

We then tested the EGF-DTA conjugate on A431 cells, since they
have a large number of EGF receptors and might consequently show
greater sensitivity. While toxicity of EGF-DTA could be observed on
A431 cells, there was still a two-order of magnitude difference in
sensitivity of 3T3 and A431 cells to EGF-RTA vs. EGF-DTA (Table 3).
The EGF-DTA conjugate was a much less potent toxin than the EGF-RTA
conjugate on both 3T3 and A431 cells.

The differential toxicities of the RTA and DTA conjugates of

Table 3. Toxicity of EGF-RTA and EGF-DTA on A431 Cells

Cell Type	ED$_{50}$ (M)	
	EGF-RTA	EGF-DTA
3T3	5×10^{-11}	$>3 \times 10^{-8}$
A431	7×10^{-12}	5×10^{-10}

EGF are in striking contrast to the essentially equivalent toxicities of the RTA and DTA conjugates of ASF on hepatocytes. Perhaps the difference in potency of RTA and DTA conjugates is a function of the target cell. To determine if this was the case we tested EGF-RTA and EGF-DTA on hepatocyte cultures. We first demonstrated that primary hepatocyte cultures had substantial numbers of EGF receptors. There were, in fact, about three times as many EGF receptors per hepatocyte as there were on 3T3 cells (data not shown). In contrast to what we observed for 3T3 cells and A431 cells, EGF-DTA was as toxic on primary rat hepatocyte cultures as EGF-RTA (Fig. 13). The EGF conjugates were, in fact, somewhat more toxic than the ASF conjugates on hepatocytes. In order to facilitate this comparison the data are replotted in a common Figure (Fig. 14). Note that the two EGF conjugates are almost equivalent to ricin in their toxicity on primary hepatocytes.

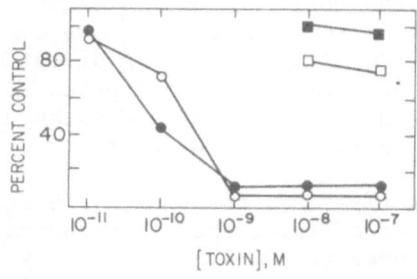

Fig. 13. Inhibition of protein synthesis in cultured hepatocytes by EGF- RTA and EGF-DTA. RTA, (■); DTA, (□); EGF-RTA, (○); EGF-DTA,(●) (Simpson et al., 1982).

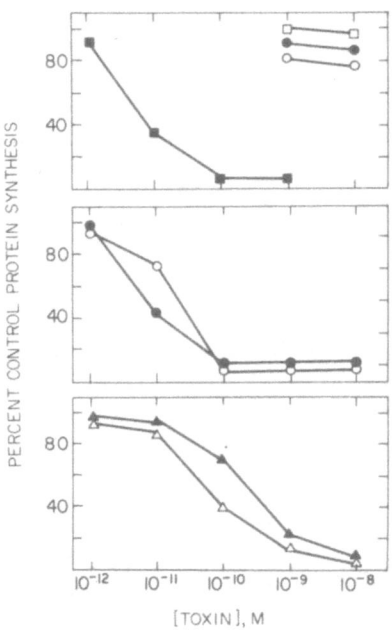

Fig. 14. Inhibition of protein synthesis in cultured
rat hepatocytes by toxins, A chains, and
toxic conjugates. Top panel: Ricin, (■);
RTA, (□); DTA, (○); diphtheria toxin, (●).
Middle panel: EGF-RTA, (○); EGF-DTA (●).
Lower panel: ASF-RTA, (△); ASF-DTA, (▲)
(Herschman et al., 1982).

TOXICITY OF RICIN, EGF-RTA AND ASF-RTA AS A FUNCTION OF AGE OF HEPATOCYTE CULTURES

During the course of our studies it appeared that the sensitivity of hepatocytes to the toxic conjugates decreased with time in
culture. To characterize this phenomenon more completely we examined
the toxicity of ricin, EGF-RTA and ASF-RTA at three different times
after plating (Fig. 15). Cultures were exposed to various doses of
toxins at 4, 28 and 52 h after plating and assayed for inhibition of
protein synthesis 24 h after toxin administration.

RTA was only slightly toxic throughout the experiment. Ricin
sensitivity increased as cultures aged. In contrast, the sensitivity of hepatocyte cultures to EGF-RTA declined by a factor of 25
during the course of the experiment. The sensitivity to EGF-RTA was
even more extensively shifted; by 28 h the ED_{50} had increased 50-

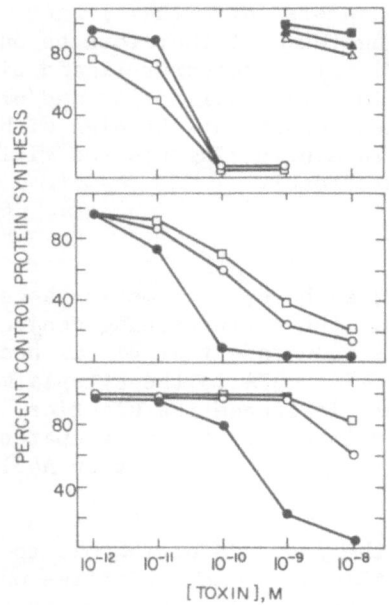

Fig. 15. Inhibition of protein synthesis by ricin, RTA, ASF-RTA and EGF- RTA in hepatocyte cultures of increasing age. Toxins were added to hepatocytes in culture for 4, 28 or 52 h. After a 24 h exposure to toxin, protein synthesis was assayed. (Top) Ricin added to 4 (●), 28 (○) or 52 (□) h cultures. RTA added to 4 (▲), 28 (△) or 52 (■) h cultures. (Middle) EGF-RTA added to 4 (●), 28 (○) or 52 (□) h cultures. (Bottom) ASF-RTA added to 4 (●), 28 (○) or 52 (□) h cultures (Simpson et al., 1982).

fold. At the last point in the experiment the cells were essentially resistant to ASF-RTA, despite their increased sensitivity to ricin. Binding studies (data not shown) demonstrated the EGF receptors did not appreciably diminish during culture, while ASGP receptors increased 2.4-fold at 28 h, then decreased to 50% of the initial value at 52 h. Loss of receptors for both ligands thus occurred much more slowly than loss of sensitivity to the RTA conjugates.

The increase in sensitivity to ricin of cultured hepatocytes with time in culture demonstrates that the ribosomes remain sensitive to RTA throughout this period. In contrast, sensitivity to ASF-RTA and EGF-RTA are reduced dramatically over the first 24 h, without any

loss in cell-surface receptors for either ligand. The decrease in
sensitivity to the conjugates must therefore be due to alterations
in events which occur distal to receptor-ligand binding; i.e., to
steps in the sequestration, internalization and/or cytoplasmic penetration of the A chains. Additional studies with cultured hepatocytes should be useful in elucidating how the ricin A chain in these
conjugates gains to the cytosol.

TOXICITY OF EGF-RICIN

    Whole ricin conjugates have, in some instances, been reported
to be more effective than the corresponding conjugates prepared with
RTA. The suggestion has been made that RTB is necessary, at some
postreceptor step, to deliver RTA to the cytoplasm. To determine if
the presence of RTB would influence the kinetics of internalization
or the potency of EGF-toxin conjugates we prepared an EGF-ricin conjugate and compared its biological effects on A431 and 3T3 cells
with that of EGF-RTA.

    EGF-RTA-ricin (in the presence of lactose to block ricin B chain
binding) showed essentially the same toxicities on A431 cells in the
end-point assay employed (Fig. 16). Lactose was able to completely
block the toxicity of ricin. EGF-RTA and EGF-ricin (in the presence
of lactose) also had essentially identical toxicities on 3T3 cells
(data not shown). EGF was able to block the toxicity of EGF-ricin

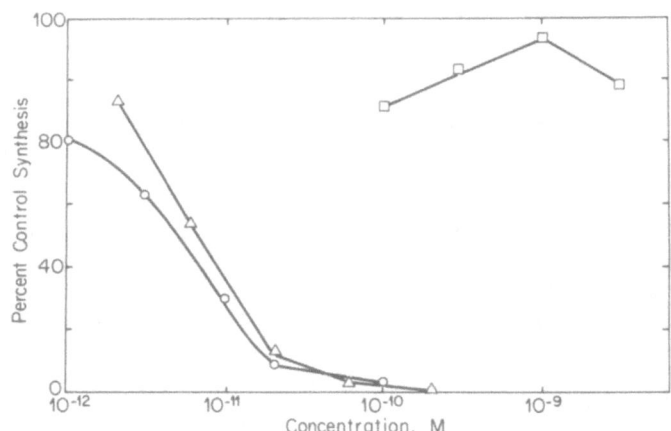

Fig. 16. Inhibition of protein synthesis in A431 cells
         by EGF-RTA and EGF-ricin. Data are expressed
         for equivalent concentrations of EGF in the
         two conjugates. EGF-RTA, (O); EGF-ricin +
         50 mM lactose, (△); EGF, (□) (Cawley and
         Herschman, in preparation).

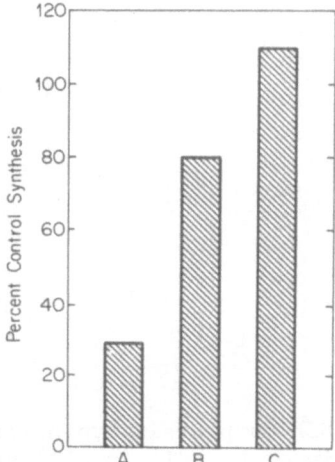

Fig. 17. EGF blocks inhibition of protein synthesis by
EGF-ricin in 3T3 cells. Cells were exposed to
10^{-9}M EGF-ricin in the presence of 50 mM lactose
for 20 h with (A) no further addition, (B) 10^{-8}M
EGF, (C) 10^{-7}M EGF (Cawley and Herschman, in
preparation).

on 3T3 cells (Fig. 17). In contrast, EGF and lactose were unable
to block the toxicity of EGF-ricin on A431 cells (Fig. 18). This
surprising result will be discussed more fully below.

We also measured the rate at which EGF-ricin and EGF-RTA are
internalized in A431 cells (defined as antibody inaccessibility)
and the rate at which the two conjugates gain access to the cytoplasm (by measuring the onset of protein synthesis inhibition).
There were no significant differences in the rate of sequestration
or cytoplasmic appearance of EGF-ricin and EGF-RTA. In the case of
EGF conjugates the B chain of ricin does not appear to facilitate
either of these processes.

CONCLUSIONS

The pathways which convey toxic conjugates from the cell surface to the cytoplasm might be expected to reflect the hybrid nature
of these molecules. Certainly studies with antagonists have made it
clear that toxin-hybrid proteins can bind to the relevant ligand
receptors. However, there is very little information about the
nature of the steps distal to receptor binding which permit the
introduction of the toxin A chains to the cytoplasm. Indeed, we

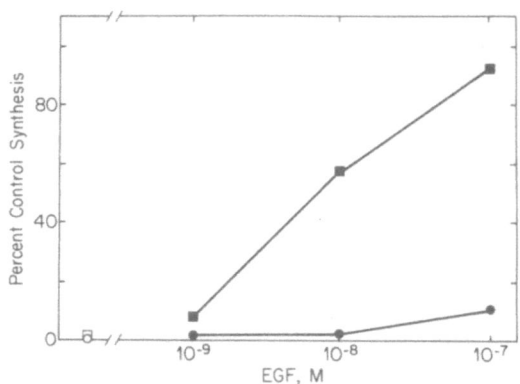

Fig. 18. Effect of EGF on protein synthesis inhibition
         by EGF-ricin or EGF-RTA in A431 cells. EGF-RTA
         and EGF-ricin were present at 2×10^{-10}M; the
         latter in the presence of 50 mM lactose. The
         experiments were performed as described in the
         legend to Fig. 16. Data have been corrected
         for a diminution of protein synthesis of about
         50% caused by EGF alone on A431 cells at 10^{-7}
         and 10^{-8}M EGF. EGF-RTA, ($\square$); EGF-RTA + EGF,
         ($\blacksquare$); EGF-Ricin, ($\circ$); EGF-ricin + EGF, ($\bullet$)
         (Cawley and Herschman, in preparation).

know only slightly more about the internalization and cytoplasmic
penetration of the parent toxins. Recent evidence suggests that the
acidification of endocytotic vesicles results in a conformational
change in diphtheria toxin fragment B. The low pH conformation is
thought to form membrane channels through which an extended fragment
A might pass. Agents which increase lysosomal and vesicular pH,
such as NH_4Cl and chloroquine, block the action of diphtheria toxin.
Interestingly, chloroquine enhances ricin's cytotoxic action. Receptor-mediated endocytosis is thought to be a "productive" pathway for
ricin (one which can ultimately convey the ricin A chain to the
cytoplasm). However, the mechanism by which the A chain (or toxin)
crosses the membrane to reach the cytoplasm is not known. Given that
there is so little information on the post-receptor steps in the
action of ricin, it is not surprising that we have a poor understanding of the analogous mechanisms which operate in the case of ricin
and ricin A chain conjugates.

    Epidermal growth factor and desialylated glycoproteins both
enter the cell via a clustering of receptors, invagination of coated
pits and vesicular endocytosis of the ligand and receptor. The bulk

of the ligand is, in both cases, taken to the lysosome, where it is degraded. We presume that the bulk of the EGF and ASF conjugates are also taken into cells via the EGF and ASGP endocytotic pathways. However, if this mode of entry is a part of the "productive" internalization, we must explain how RTA or DTA can subsequently escape from the endocytotic vesicle to reach the cytoplasm. Chloroquine enhances the activity of the EGF-RTA conjugate (Cawley et al., 1980) and EGF-DTA conjugate (Cawley and Herschman, unpublished), suggesting that inhibition of lysosome function or the alkalinization of endocytotic vesicles can increase the chance that RTA or DTA will gain access to the cytosol. Whether this is due to prevention of degradation of the A chains in the lysosome or facilitation of processes active in the cytoplasmic transport of the A chains is not known. EGF-RTA, EGF-ricin and ricin are sequestered by cells very rapidly, as measured by the rate at which they become inaccessible to antibody. An initial EGF-conjugate or ricin exposure of 30 min at 37° at a concentration of 3×10^{-9}M is sufficient to cause greater than 80% inhibition of protein synthesis in A431 cells assayed at a 20 h "end point". Since inhibition of protein synthesis begins only 1.5 to 2 h after exposure of the cells to toxins, those processes which subsequently convey the A chain to the cytoplasm appear to be rate-limiting on the pathway from cell surface receptor to cytosol. We suggest that the function of the ricin B chain or the conjugate binding protein (e.g., EGF, ASF) is to promote efficient endocytosis of the toxin or conjugate and that the RTA plays a predominant role in the processes which convey the toxins or RTA to the cytoplasm.

The hydrophobic amino and carboxyl terminal regions of RTA (Funatsu et al., 1978) might participate in the membrane traversal steps along the cytotoxic pathways of ricin, EGF-RTA and EGF-ricin. DTA has no such hydrophobic arms (De Lange et al., 1979). Conjugates of DTA are often much less active than the corresponding RTA conjugates (Uchida et al., 1980; Olsnes and Pihl, 1982b). It is perhaps not surprising that DTA does not associate with membranes, given the role now perceived for DTB in the cytoplasmic transport of DTA. EGF-DTA may be a poor toxin on 3T3 and A431 cells because the DTA cannot escape endocytotic vesicles. Why, then, is EGF-DTA a potent toxin on primary rat hepatocytes? DTA may inhibit the normal EGF endocytotic process by disrupting receptor aggregation or down-regulation in 3T3 or A431, but not in cultured hepatocytes. Alternatively, the difference in toxicity of EGF-DTA on these cells may be the result of very different pathways of internalization and/or intracellular processing of the EGF-receptor complex in hepatocytes vs. the other two cell types. A pathway of EGF internalization may exist in hepatocytes that permits DTA to enter the cytoplasm; this pathway may be absent in 3T3 and A431 cells. We think it a very interesting, and unappreciated, possibility that different cell types may internalize the same ligand (e.g., EGF) in different ways.

What role, if any, does the ricin B chain have in the "prod-

uctive" internalization of ligand-ricin conjugates? Youle et al.,
(1981) suggested that an intracellular B chain receptor interaction
might be necessary for toxic conjugates whose ligands are "en route
to intracellular compartments exclusive to the cytosol" and that
conjugates of such ligands would require the B chain to be effective
toxins. Both ASGP and EGF are predominantly taken up be endocytosis
and degraded in lysosomes. Therefore one might predict, and indeed
Youle et al., (1981) did so for ASGP conjugates, that RTA conjugates
of ASGP or EGF would not be very toxic. However, we have shown that
ASF-RTA and EGF-RTA are potent toxins. If the ricin B chain were
required, we agree that these conjugates should not be toxic. To
pursue this point further we compared the toxicity of EGF-ricin and
EGF-RTA, and did not observe any significant differences in toxic
potency, kinetics of sequestration, or time of penetration to the
cytoplasm for EGF-ricin vs. EGF-RTA.

Houston (1982) has shown that cell-surface bound RTB promote the
internalization of RTA. We suggest that the B chain of ricin may
facilitate endocytosis of the molecule, but not penetration through
the membrane to the cytoplasm. In those cases where the binding
ligand does not induce rapid endocytosis the intact ricin conjugate,
as a consequence of B chain-facilitated endocytosis, may be a more
potent conjugate than the RTA derivative. In cases where the binding
ligand initiates rapid endocytosis, such as EGF or ASF, this aspect
of the intact ricin molecule may be redundant and RTA conjugates will
be as potent as intact ricin conjugates. We would predict three
cases in which A chain conjugates would be poor toxins compared with
their B chain analogs: (1) the ligand in the conjugate does not
undergo efficient endocytosis; (2) the presence of the A chain inter-
feres with the endocytosis of a ligand which is normally taken up
efficiently; (3) the conjugate undergoes efficient endocytosis but
is delivered rapidly and singularly to lysosomes (note that ricin is
very resistant to proteolysis but the A chain is sensitive). It is
our view that there is no need, based upon existing evidence, to
postulate an intracellular role for the ricin B chain.

The inability of a combination of lactose and EGF to block the
activity of EGF-ricin on A431 cells is puzzling, since lactose can
block the action of ricin and EGF can block the toxicity of EGF-RTA
on these cells. Moreover, EGF and lactose can block the toxicity of
EGF-ricin on 3T3 cells. We postulate that the EGF-ricin conjugate
is capable of binding simultaneously to EGF and ricin B chain receptors in a way which precludes blocking by antagonists. We suggest
that a brief association of EGF (in conjugate) with EGF receptors,
an association which normally would be displaced by competing EGF,
is stabilized by subsequent binding of ricin (in conjugate) to B
chain receptors. In this scheme lactose is a poor inhibitor of
ricin B chain binding because the latter has been brought to the
cell surface, where the effective concentration of competing glycoconjugates is very high, compared with the bulk solvent. The diva-

lent conjugate/receptor(s) complex would then be internalized. Why does this not happen on 3T3 cells? One potential explanation rests upon the observation that the dissociation of EGF from 3T3 cells is much more rapid than from A431 cells. It is possible that the bivalent complex cannot form in the shorter lifetime of the transient EGF-receptor complex on 3T3 cells. The possibility that whole toxin conjugates can bind to toxin receptors complicates the interpretation of the mechanism of action of conjugates using intact toxins. It may not always be possible to prevent the binding of the toxin B chain in the conjugates, even with agents which are capable of blocking the binding of free toxin under the same conditions.

A rational basis for designing conjugates for use in the re-search lab and perhaps in chemotherapy requires that we understand how they make their way from cell-surface receptor to the cytoplasm. In the course of elucidating those pathways we undoubtedly will add considerably to both our knowledge of receptor-mediated endocytosis of various ligands and our understanding of how protein toxins cross membranes.

ACKNOWLEDGEMENTS

This work was supported by NIH grant GMS24797 and contract number DE AM03 76 SF00012 from the Department of Energy. DBC was the recipient of a National Research Service Award (F32 GM07489).

REFERENCES

Aharonov, A., Pruss, R.M., and Herschman, H.R., 1978,Epidermal growth factor: Relationship between receptor regulation and mitogenesis in 3T3 cells, J. Biol. Chem., 253:3970.
Attie, A.D., Weinstein, D.B.,Freeze, H.H., Pittman, R.C. and Steinberg, D., 1979, Altered catabolism of desialylated low-density lipoprotein in the pig and in cultured rat hepatocytes, J. Biochem., 180:643.
Butler-Gralla, E., and Herschman, H.R., 1981, Variants of 3T3 cells lacking mitogenic response to the tumor promoter tetradecanoyl-phorbol-acetate, J. Cell. Physiol., 107:59.
Carpenter, G., and Cohen, S., 1979, Epidermal growth factor, Ann. Rev. Biochem., 48:193.
Cawley, D.B., Hedblom, M.L., and Houston, L.L., 1978, Homology between ricin and Ricinus communis agglutinin: Amino terminal sequence analysis and protein synthesis inhibition studies, Arch. Biochem. Biophys., 190:744.
Cawley, D.B., Simpson, D.L., and Herschman, H.R., 1981, Asialoglycoprotein receptor mediates the toxic effects of an asialofetuin-diphtheria toxin fragment A conjugate on cultured rat hepatocytes, Proc. Natl. Acad. Sci USA., 78:3383.
Cawley, D.B., Herschman, H.R., Gilliland, D.G., and Collier, R.J., 1980, Epidermal growth factor-toxin A chain conjugates: EGF-

ricin A is a potent toxin while EGF-diphtheria fragment A is non-toxic, Cell., 22:563.

Chung, D.N., and Collier, R.J., 1977, The mechanism of ADP-ribosylation of elongation factor 2, Biochim. Biophys. Acta., 483:248.

De Lange, R.J., Williams, L.S., Drazin, R.E., and Collier, R.J., 1979. The amino acid sequence of fragment A, an enzymically active fragment of diphtheria toxin, J.Biol. Chem., 254:5838.

Fabricant, R.N., DeLarco, J.E., and Todaro, G.J., 1977, Nerve growth factor receptors on human melanoma cells in culture., Proc. Natl. Acad. Sci. USA., 74:565.

Fine, R.E., Goldenberg, R., Sorrentino, J., and Herschman, H.R., 1981, Subcellular structures involved in internalization and degradation of epidermal growth factor, J. Supramol. Struct., 15:235.

Funatsu, G., Yoshitake, S., and Funatsu, M., 1978, Primary structure of the chain of ricin D, Agric. Biol. Chem., 42:501.

Gilliland, D.G., Steplewski, Z., Collier, R.J., Mitchell, K.F., Chang, T.H., and Korprowski, H., 1980, Antibody-directed cytotoxic agents: Use of monoclonal antibody to direct the action of toxin A chains to colorectal carcinoma cells, Proc. Natl. Acad. Sci. USA., 77:4539.

Herschman, H.R., Sorrentino, J., Butler-Gralla, E., and Cawley, D., 1982, Isolation and characterization of variants of 3T3 cells deficient in a proliferative response to specific mitogens in: "Maturation Factors and Cancer," M. Moore, ed., Raven Press, New York.

Honjo, T., Nishizuka, Y., Hayaishi, D., and Kato, I., 1968, Diphtheria toxin-dependent adenosine diphosphate ribosylation of aminoacyl-transferase II and inhibition of protein synthesis, J. Biol. Chem., 243:3553.

Houston, L.L., 1982, Transport of ricin A chain after prior treatment of mouse leukemia cells with ricin B chain, J. Biol. Chem., 257:1532.

Hubbard, A.L., Wilson, G., Ashwell, G., and Stuckenbrok, H., 1979, An electron microscope autoradiographic study of the carbohydrate recognition systems in rat liver. I. Distribution of iodine-125 labeled ligands among the liver cell types, J. Cell Biol., 83:47.

Kawasaki, T., and Ashwell, G., 1976, Chemical and physical properties of a hepatic membrane protein that specifically binds asialoglycoproteins, J. Biol. Chem., 251:1296.

Moehring, T.J. and Moehring, J.M., 1977, Selection and characterization of cells resistant to diphtheria toxin and Pseudomonas exotoxin A: Presumptive translational mutants, Cell., 11:447.

Olsnes, S., and Pihl, A., 1982a, Toxic lectins and related proteins in "The Molecular Actions on Toxins and Viruses," Van Heyningen, S., and Cohen, P., eds., Elsevier/North Holland, Amsterdam.

Olsnes, S., and Pihl, A., 1982b, Chimeric Toxins, Pharmac. Ther., 15:335.

Pruss, R.M., and Herschman, H.R., 1977, Variants of 3T3 cells lacking mitogenic response to epidermal growth factor, Proc. Natl. Acad. Sci. USA 74: 3918.

Rose, S.P., Pruss, R.M., and Herschman, H.R., 1975, Initiation of 3T3 fibroblast cell division by epidermal growth factor, J. Cell. Physiol. 86:593.

Sandvig, K., Olsnes, S., and Pihl, A., 1976, Kinetics of the binding of the toxic lectins abrin and ricin to surface receptors of human cells, J. Biol. Chem. 251:3977.

Simpson, D.L., Cawley, D.B., and Herschman, H.R., 1982, Killing of cultured hepatocytes by conjugates of asialofetuin and EGF linked to the A-chain of ricin or diptheria toxin, Cell 29:469.

Steer, C.J., and Ashwell, G., 1980, Studies on a mammalian hepatic binding protein specific for asialoglycoproteins, J. Biol. Chem. 255:3008.

Uchida, T., Mekada, E., and Okada, Y., 1980, Hybrid toxin of the A chain of ricin toxin and a subunit of Wisteria Floribunda lectin, J. Biol. Chem. 255:6687.

Youle, R.J., Murray, G.J., and Neville, D.M., Jr., 1979, Ricin linked to monophosphopentamannose binds to fibroblast lysosomal hydrolase receptors, resulting in a cell-type specific toxin, Proc. Natl. Acad. Sci. USA 76:5559.

Youle, R.J., Murray, G.J., and Neville, D.M., Jr., 1981, Studies on the galactose-binding site of ricin and the hybrid toxin man-6P-ricin, Cell 23:551.

SACCHARIDE RECEPTOR-MEDIATED DRUG DELIVERY

M.M. Ponpipom, R.L. Bugianesi, J.C. Robbins,
T.W. Doebber and T.Y. Shen

Merck Sharp & Dohme Research Laboratories
Rahway, New Jersey 07065 U.S.A.

INTRODUCTION

Since the pioneering work of Ashwell and Morell on the in-vivo plasma clearance of desialylated serum glycoproteins, [1] the concept has become widely accepted that exposed sugar residues on glycoproteins serve as determinants for in-vivo (i.e., clearance) and in-vitro (i.e., uptake) recognition. Carbohydrate-mediated endocytosis in mammals has since become the subject of intensive studies.[2,3] Mammalian hepatic receptors are known to be specific for terminal D-galactose of desialylated serum glycoproteins;[4] whereas in avian liver, the receptors recognize terminal 2-acetamido-2-deoxy-D-glucose residues.[5] Hepatocytes also contain a receptor that binds glycoproteins specifically through L-fucose in $\alpha(1\rightarrow3)$ linkage to 2-acetamido-2-deoxy-D-glucose.[6] Macrophages and Kupffer cells have been shown to bind glycoproteins and synthetic neoglycoconjugates that have D-mannose and 2-acetamido-2-deoxy-D-glucose in the exposed non-reducing position.[7,8] The uptake of some lysosomal enzymes by fibroblasts involves recognition of D-mannose 6-phosphate.[9]

Maynard and Baenziger recently examined the specificity and kinetics of endocytosis of iodinated glycopeptides and glycoproteins by isolated rat reticuloendothelial cell preparations.[10] Glycoproteins and glycopeptides that were recognized by this system contain within their oligosaccharide moiety the common structural unit shown in Fig. 1, which they hypothesized to be the minimum structure required for binding and endocytosis. Our approach in drug delivery is to use small synthetic glycolipids[11-13] and glycopeptides[14-16] which are recognized by cell-surface receptors, as ligands for delivery of pharmaceuticals to target tissues and organs. This report will deal mainly with small synthetic glycopeptides that can bind specifically

53

to receptors on macrophages and cells or the reticuloendothelial system.

CLUSTER GLYCOSIDES

    In our initial screening studies, an old sample of N-lipoyl-β-D-mannopyranosylamine[17] was unexpectantly found to be a potent inhibitor of ^{125}I-labeled Man-BSA conjugate uptake by rat alveolar microphages. The active ingredient was subsequently shown to be a polymeric material that was formed gradually from the monomer upon storage at room temperature (Figure 2). A freshly prepared sample of N-lipoyl-β-D-mannopyranosylamine was only weakly active, and lipoic acid itself was totally inactive. Photolysis of the monomer in methanol gave a polymeric material (Figure 2), the bulk of which had a K_{av} of 0.63 on Sephadex G-100 that corresponded to a molecular weight of ~10000 (i.e., n~18-20) and gave 90% inhibition at 10μM. The enhanced potency of the polymer suggests the importance of multi-

Fig. 2.   Irradiation of N-lipoyl-β-D-mannopyranosylamine.

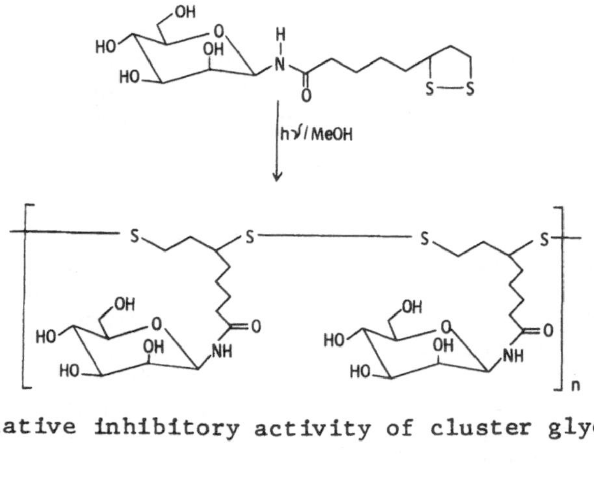

Fig. 3.   Relative inhibitory activity of cluster glycosides.

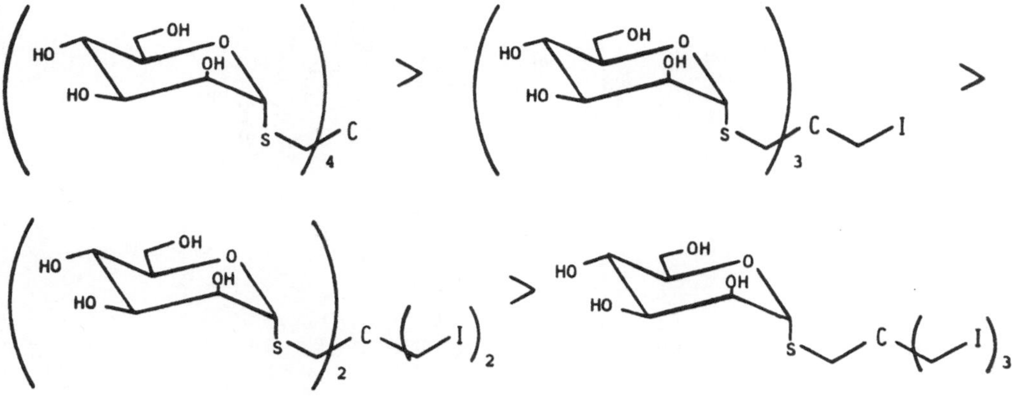

Fig. 3.   Relative inhibitory activity of cluster glycosides.

valency or clustering of D-mannose residues in recognition. This is
supported by the observed relative inhibitory potency of the synthetic saccharides shown in Fig. 3.   At 20 μ M, the tetra-($K_i \simeq$ 17 M),
tri-, and di-mannosyl derivatives gave 72%, 59%, and 25% inhibition,
respectively, whereas the monomer was inactive. The importance of
multivalency or clustering of saccharide residues in recognition by
cells or binding proteins has also been suggested by other workers.
18-20   Thus, a systematic effort to introduce various numbers of
D-mannose residues into convenient peptide backbones, such as lysine,
dilysine, and oligolysine, was carried out.

SYNTHETIC GLYCOPEPTIDES

    The synthesis of mannose-lysine conjugates, as exemplified by
Man3Lys2, is outlined in Fig. 4. Amidination of L-lysyl-L-lysine
with 2-imino-2-methoxyethyl 1-thio-α-D-mannopyranoside[21,22] gave
the amidine-linked Man3Lys2; whereas condensation of the p-nitro-
phenyl ester of carboxyalkyl tetra-O-acetyl-1-thio-α-D-mannopyran-
oside with L-lysyl-L-lysine, followed by deacetylation, gave the
amide-linked Man3Lys2.[14]

    The chirality of the peptide backbone does not seem to play a
major role in inhibition (Fig. 5). The K_i values of both the L and
D isomers of Man2Lys are about the same ($K_i \simeq$ 6.2 and 7.7 μM, respectively). On the other hand, the K_i value of the amidine-linked
Man2Lys ($K_i \simeq$ 100 μ M) is rather high. Since amidines are known to

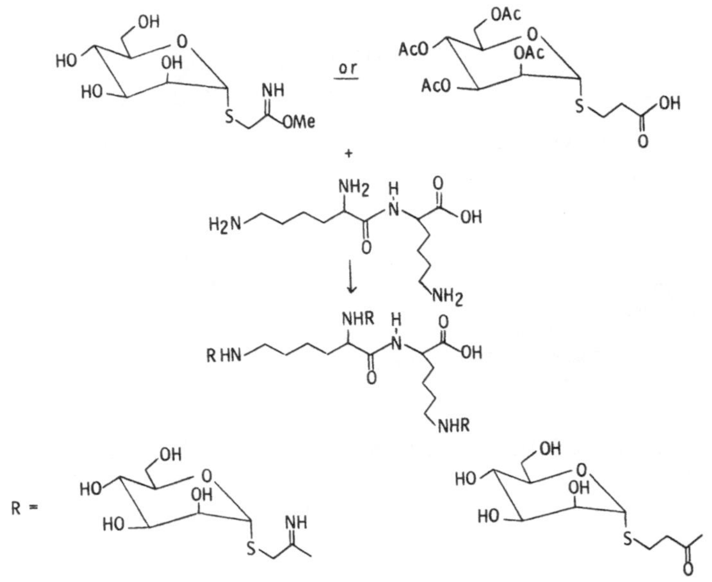

Fig. 4.   Chemical synthesis of mannose-lysine conjugates.

be labile in solution [23] and the fact that the amidine-linked Man_2Lys did not give correct elemental analysis, its K_i value ($100 \mu M$) would have to be interpreted with caution.   To avoid these problems, only the stable amide-linked mannose-lysine conjugates were studied in more detail, and will be discussed here.

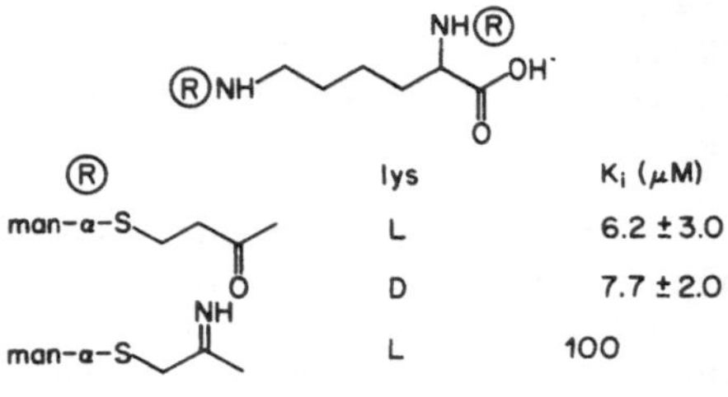

(R)	lys	K_i (μM)
man-α-S	L	6.2 ± 3.0
	D	7.7 ± 2.0
man-α-S	L	100

Fig. 5.   Inhibition of ^{125}I-labeled mannose-BSA binding to macrophages.

Table 1. Inhibition of ^{125}I-labeled mannose-BSA uptake.

Ligand	K_i (μM)
Man_4Lys_3	2.6 ± 0.7 (3)
Man_3Lys_2	3.9 ± 1.1 (3)
Man_2Lys	6.2 ± 3.0 (3)
$Man_3Lys_2NH_2$	8.0
Man_3Lys_2BH	9.3 ± 4.7 (5)
Man_3Lys_2Raf	18.0 ± 2.8 (3)
Man_3Lys_2Chol	4.0
Man_3Lys_2Dex	6.0

Inhibition constants were determined as shown in Fig. 6. Values listed are mean $\pm$ SD (number of experiments). By student's t test, 18 μM is significantly (P$<$0.05) higher than the other values; the other differences are not statistically significant.

Table 2. Inhibitory activity of synthetic glycopeptides.

Ligand	Conc.	Inhibition of Binding,[a] %		
		10 μM	50 μM	100 μM
Man_3Lys_2		43	81	89
Gal_3Lys_2		0	0	2
Fuc_3Lys_2		9	-	45

[a]Inhibition of ^{125}I-labeled mannose-BSA binding to alveolar macrophages. Each value represents the mean of triplicate determinations in one experiment.

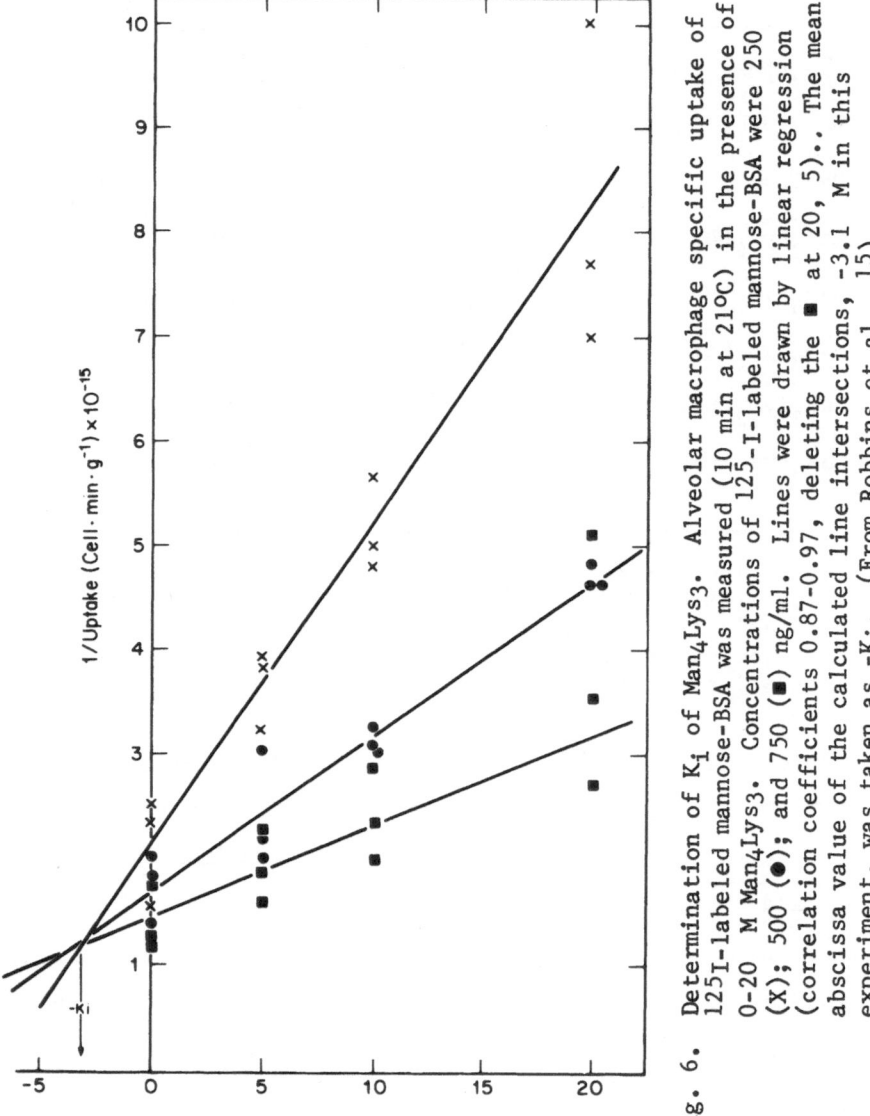

Fig. 6.   Determination of K_i of Man_4Lys_3. Alveolar macrophage specific uptake of ^{125}I-labeled mannose-BSA was measured (10 min at 21°C) in the presence of 0–20 M Man_4Lys_3. Concentrations of $^{125}-I$-labeled mannose-BSA were 250 (X); 500 (●); and 750 (■) ng/ml. Lines were drawn by linear regression (correlation coefficients 0.87–0.97, deleting the ■ at 20, 5).. The mean abscissa value of the calculated line intersections, –3.1 M in this experiment, was taken as $-K_i$. (From Robbins et al., 15).

The inhibitory potency of mannose-lysine conjugates appears to increase with increasing number of D-mannose residues (Table 1) : Man_4Lys_3 > Man_3Lys_2 > Man_2Lys. This finding demonstrates again the importance of multivalency of saccharides for ligand-receptor inter- actions. The carbohydrate specificity of the macrophage receptor is clearly demonstrated by the inactivity of the D-galactose analog, Gal_3Lys_2 (Table 2). This is in accord with Stahl and his co-workers' observation that D-galactose-terminated glycoproteins are not bound by alveolar macrphages.[18] However, in contrast to their recent suggestion[7,24] that the macrophage receptor has broad specificity and also recognizes L-fucose-terminated glycoconjugates, the syn- thetic L-fucose analog, Fuc_3Lys_2, was only moderately active (Table 2).

The synthetic mannose-lysine conjugates were all found to be competitive inhibitors of rat alveolar macrophage uptake of [125]I- labeled Man-BSA, as shown in Fig. 6 for Man_4Lys_3. Results of sev- eral experiments are summarized in Table 1. The carboxy function of Man_3Lys_2 (Fig. 7) can be used for direct attachment to drugs or it can be converted into other functional groups such as the - aminohexyl analog, $Man_3Lys_2NH_2$ (Fig. 7), prior to coupling with therapeutic agents. A number of compounds, e.g., Bolton-Hunter reagent, raffinose, dexamethasone and cholesterol, can be attached to Man_3Lys_2 without substantial loss of affinity for macrophages (Table 1). It is noteworthy that the glycoconjugate (M_r > 10^6, con- taining 2.5% L-lysine and 4.6% D-mannose), prepared from ω-aminohexyl Man_3Lys_2 and meningococcal group C polysaccharide (data not shown), was still inhibitory with an activity of 82% per mg per ml, whereas the polysaccharide itself was totally inactive.

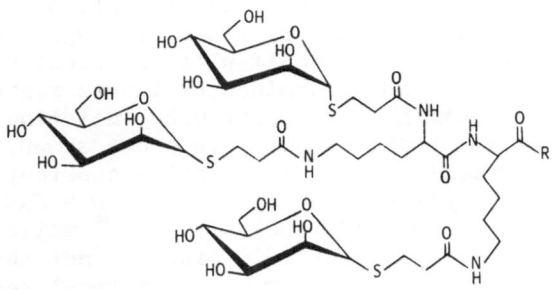

R = OH, Man_3Lys_2
R = $NH(CH_2)_6NH_2$, $Man_3Lys_2NH_2$

Fig. 7.   Structures of Man_3Lys_2 and $Man_3Lys_2NH_2$.

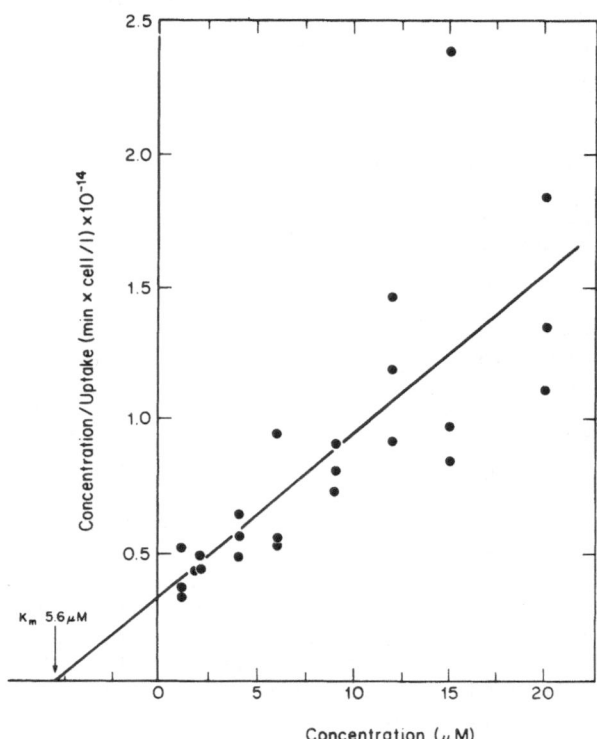

Concentration (μM)

Fig. 8.   Binding of ^{125}I-labeled Man₃Lys₂BH. Alveolar
          macrophages were incubated at 0°C with [125]I-
          labeled Man₃Lys₂BH at 8.7x10⁶ cpm/ml (≤0.3μM)
          and various amounts of unlabeled Man₃Lys₂BH to
          give the desired total concentrations, and
          80 μl (1.1x10⁶ cells) was centrifuged at 30
          min. Shown is a Woolf plot of specific binding,
          with the line determined by linear regression
          (correlation coefficient 0.86, deleting an
          aberrant point at 40, 12.6 - not shown). This
          experiment gave a dissociation constant K_d
          -1x(abscissa intercept) of 2.7μ M and a max-
          imum binding (1/slope) of 5.2x10⁵ molecules
          per cell. Non-specific binding (not shown)
          was a constant 0.57 ± 0.03% of total radio-
          activity at all concentrations.[15]

IN-VITRO STUDIES

The demonstration of the uptake of synthetic glycopeptides into macrophages, as distinct from binding at the cell surface, is necessary for good targeting of therapeutic agents to macrophages. A labeled ligand was thus prepared for binding and uptake studies by reaction of $Man_3Lys_2NH_2$ with ^{125}I-labeled Bolton-Hunter reagent. This ligand,[125]I-labeled Man_3Lys_2BH, binds specifically (i.e., inhibited by mannan) to the alveolar macrophages. The specific binding at 0°C is shown in Fig. 8 as a Woolf plot.[25] There were 6.4x 10^5 binding sites per cell, with an apparent dissociation constant (K_d) of 2.4 μ M. Warming the incubation mixture to 37°C increased the amount of ^{125}I-labeled Man_3Lys_2BH specifically associated with cells (Table 3). All of the labeled ligand bound specifically at 0°C could be removed by EDTA, showing its cell-surface location, whereas the labeled ligand that associated with the cells at 37°C was not affected by EDTA.[15] Thus, incubation at 37°C not only increased the mannan-sensitive cellular uptake of[125]I-labeled Man_3Lys_2 BH but also permitted the cells to sequester most of it in a location, presumably internal, where it could not be released by EDTA. The specific uptake of the labeled ligand by the alveolar macrophages at 21oC was saturable, with a Michaelis-Menten constant (K_m) of 6.4 μ M and a maximum velocity (ν max) of 1.7x10^5 molecules per cell (Fig. 9). With the present experimental precision, the values of K_d and K_m for ^{125}I-labeled Man_3Lys_2BH are not statistically different from the K_i value of the unlabeled ligand (Student's $\underline{t}$ test, P>0.05).

Table 3. Uptake of ^{125}I-labeled Man_3Lys_2BH

Post-Incubation Addition	Specific Cell-Associated Label (cpm) Incubation	
	0°C	37°C
NaCl	3,623*	6,088
EDTA	-255	5,751

Rat alveolar macrophages were incubated for 20 min at 0°C or 37°C in 90 μl of medium with $\sim$2 μM ^{125}I-labeled Man_3Lys_2BH, then chilled on ice. Fifteen microliters of 0.85% saline or 100 mM EDTA was added and 80 μl of the mixture (5.6x10^5 cells) was centrifuged through oil after 10 min. Values shown are cell pellet radioactivity minus non-specific cpm obtained with mannan (2.5 mg/ml) present throughout the incubations. This non-specific binding was 7994 cpm at 0°C and 10,268 cpm at 37°C.

*Approximately 2x10^5 molecules per cell.

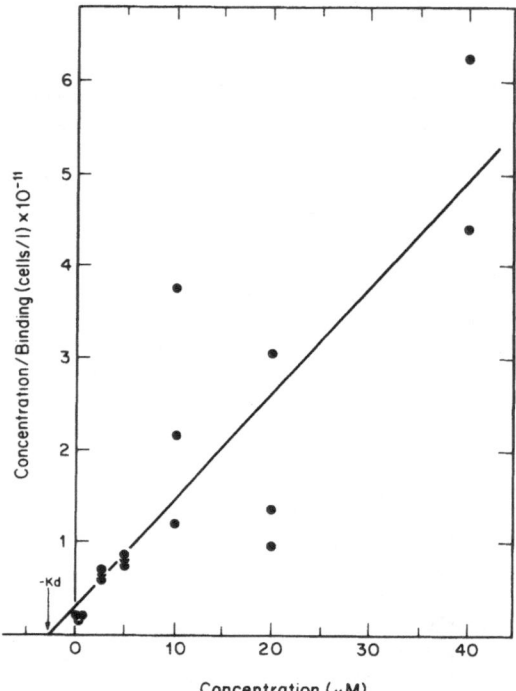

Fig. 9.   Uptake of ^{125}I-labeled Man$_3$Lys$_2$BH. This experiment was similar to the binding measurement shown in Fig. 8 but was run at higher temperature to permit uptake. Cells were incubated 10 min at 21°C with ^{125}I-labeled Man$_3$Lys$_2$BH (9.42x10^4 cpm/mol at all concentrations) and 80 μl (6.9x10^5 cells) was centrifuged. The plot (correlation coefficient 0.78) indicates a K_m of 5.6 μM and a ν max (1/slope) of 0.99x10^5 molecules per min per cell. Mannan-resistant uptake (not shown) was a constant 0.27 $\pm$ 0.01% of total radioactivity at all concentrations.[15]

(^{3}H)Raffinose covalently coupled to plasma proteins has been shown to be a useful, radioactive tracer for detecting the tissue and cellular sites of catabolism of long-lived, circulating proteins.[26] The sucrose portion of raffinose is resistant to lysosomal hydrolysis, and does not readily diffuse from lysosomes. In addition, the D-fructosyl group is not known to serve as a recognition marker for carbohydrate-mediated, clearance process.[27] For these reasons, $Man_3Lys_2NH_2$ was also labeled with (^{3}H)raffinose via reductive amniation of 6"-aldehydo-(^{3}H)raffinose to give Man_3Lys_2(^{3}H)Raf whose label should remain in lysosomes even after enzymic digestion.[28] Thioglycollate-elicited mouse peritoneal macrophages in culture took up $Man_3Lys_2NH_2$(^{3}H)Raf much faster and to a much greater extent than they took up (^{3}H)raffinose (Fig. 10).

B-GLUCOCEREBROSIDASE: IN-VITRO AND IN-VIVO STUDIES

A potential application of the macrophage ligand, Man_3Lys_2, is in the area of enzyme replacement therapy. Gaucher's disease offers a good target, since macrophages, the cells affected with the pathological accumulation of the glycolipid β-glucocerebroside, exhibit

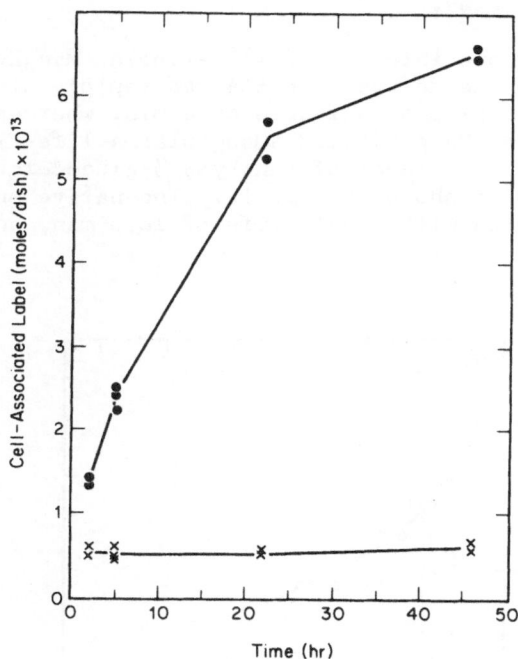

Fig. 10. Uptake of Man_3Lys_2(^{3}H)Raf. Thioglycollate-elicited mouse peritoneal macrophages were cultured in the presence of 14 nM Man_3Lys_2 (^{3}H)Raf(●) or 10 nM(^{3}H)raffinose (X), and cell-associated radioactivity was measured. The calculation of mol from cpm is based on initial specific activities, assuming no metabolism.[15]

a high affinity for glycoproteins having terminal D-mannose and 2-acetamido-2-deoxy-D-glucose residues. Previous studies by Brady's group showed that the uptake of infused β-glucocerebrosidase in-vivo by Kupffer cells could be greatly enhanced by treatment of the native enzyme with glucosidase to expose 2-acetamido-2-deoxy-D-glucose or D-mannose residues.[29] Thus, it was anticipated that the uptake of β-glucocerebrosidase by macrophages would be increased by covalent attachment of Man3Lys2 to the native enzyme, a procedure with certain advantages over the glycosidase treatment.

The ligand Man3Lys2 was coupled to β-glucocerebrosidase with the aid of 1-ethyl-3-(3-dimethylaminopropyl)carbodiimide hydrochloride. After 17 h at 4°C, there was a 60% recovery of total enzyme activity but a full recovery of enzyme specific activity with the synthetic substrate. Man3Lys2-β-glucocerebrosidase contained 8-9 moles of Man3Lys2 per mole of 67,000 molecular weight enzyme subunit as determined by both fluorescamine and phenol-sulfuric acid assays.[16] The conjugate Man3Lys2-BSA was also prepared as a model for in-vivo clearance studies.

Upon intravenous infusion of ^{125}I-labeled Man3Lys2-BSA into an anesthetized rat, the derivatized BSA was rapidly cleared from the circulation with a plasma half-life of 4 min, whereas the underivatized ^{125}I-labeled BSA exhibited a long plasma life-time (Fig. 11). The in-vivo plasma clearance of Man3Lys2-β-glucocerebrosidase and the native enzyme is shown in Fig. 12. The native enzyme was cleared from the circulation with a half-life of 18.5 min, and 29% still

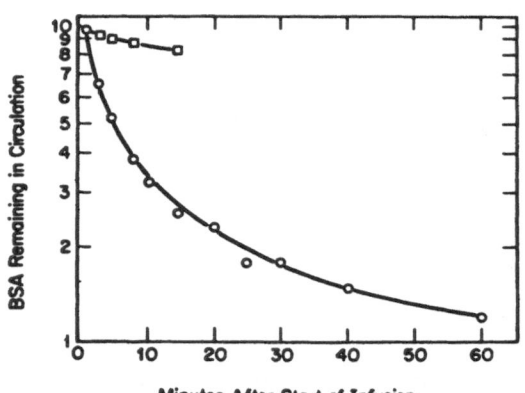

Fig. 11. In-vivo plasma clearance of ^{125}I-labeled Man3Lys2-BSA and ^{125}I-BSA. □-□ , BSA; o-o, Man3Lys2-BSA with a half-life of 4 min.

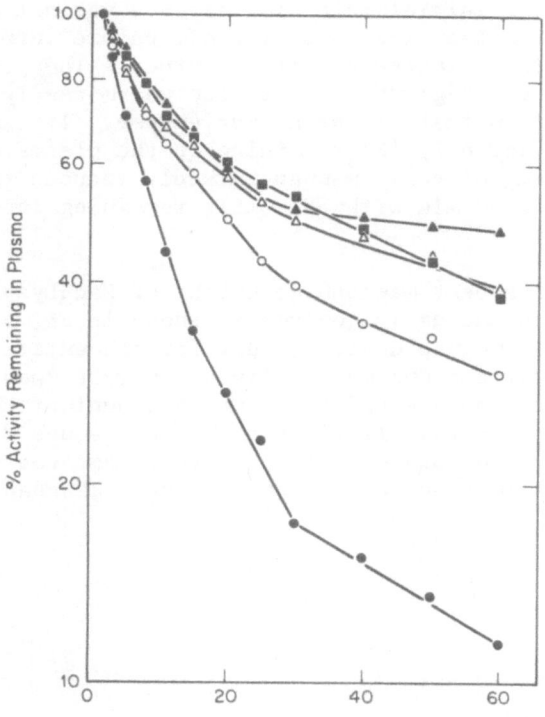

Fig. 12. In-vivo plasma clearance of native and Man$_3$Lys$_2$-B-glucocerebrosidase. One-half ml of β-glucocerebrosidase-containing solution was infused in 70 s, and the first blood sample was taken 20 s later. The circulating levels of infused enzymes 90 s after the start of the infusion was at least 15 times the level of endogenous β-glucosidase in the plasma and 70-90% of the total activity infused (correcting for plasma inhibition). All clearance studies were performed in duplicate. O—O, native enzyme; $\triangle$—$\triangle$, native enzyme + 3 mg of mannan; ▲—▲, native enzyme with circulation to liver tied off; ●—●, Man$_3$Lys$_2$-B-glucocerebrosidase; ■—■, Man$_3$Lys$_2$-B-glucocerebrosidase + 3 mg of mannan.[16]

present in the plasma at 60 min. A coinfusion of 3 mg of yeast
mannan slightly retarded the clearance to give a plasma half-life
of 28 min, but the initial rate of clearance was not affected. If
the circulation to the liver was tied off before infusion, the
clearance rate was retarded a little further. The plasma clearance
of infused Man_3Lys_2-B-glucocerebrosidase was markedly faster and
more extensive than that of the native enzyme. If exhibited a half-
life of 8.5 min and only 12% remaining in the plasma at 60 min. A
coinfusion of 3 mg of yeast mannan markedly reduced the clearance
to a half-life of 34 min with 38% still remaining in the plasma at
60 min.[16]

The time-dependent macrophage uptake of Man_3Lys_2-B- glucocerebrosidase and the native enzyme is shown in Fig. 13. The uptake
of the native enzyme was nearly linear for the entire 60 min while
that of the derivatized enzyme was linear only between 20 and 60 min.
By 60 min, the macrophage had taken up an amount of the derivatized
enzyme equal to more than 11 times their endogenous B-glucocerebrosi-
dase level while the uptake of the native enzyme was slightly less
than 7 times the macrophage endogenous level. Mannan inhibited 80%

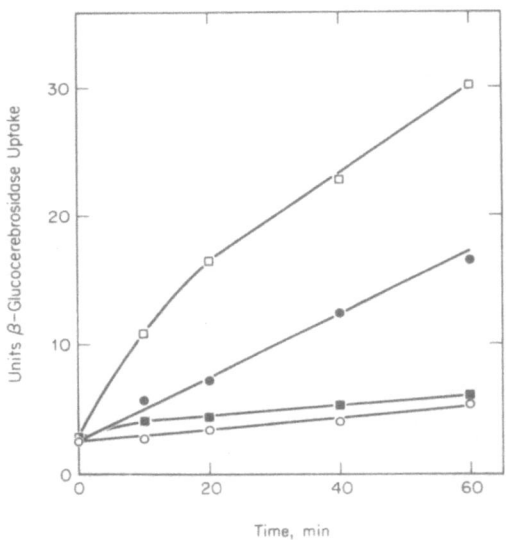

Fig. 13. Time-dependent macrophage uptake of native
         (●—●) and Man_3Lys_2-β-glucocerbrosidase
         (□—□). Macrophages were incubated with
         4300 units of enzyme for indicated times.
         O—O, native enzyme + mannan; ■—■, deriva-
         tized enzyme + mannan.[16]

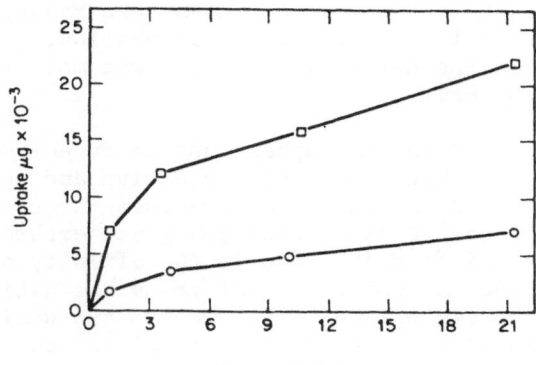

Fig. 14. Enzyme concentration dependence of macrophage
uptake of native (O—O) and Man₃Lys₂-β-glucocerebrosidase (□—□). Time of incubation for
macrophage uptake was 20 min. Only the mannan-
sensitive uptake is shown which is the total
uptake minus the non-specific (mannan-insensi-
tive) uptake.[16]

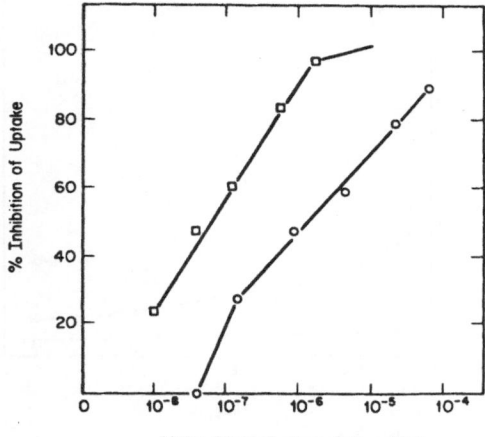

Fig. 15. Inhibition of macrophage uptake of native (□—□)
and Man₃Lys₂-β-glucocerebrosidase (O—O) by
increasing concentrations of mannose-BSA. Macrophages were incubated with 3100 units of enzyme
and the indicated concentration of mannose-BSA
for 20 min. The percent inhibition by mannose-
BSA of the mannan-sensitive uptake is plotted.[16]

and 87% of the uptake of the native enzyme and Man_3Lys_2-β-glucocere-
brosidase, respectively. At 20 min of incubation, the net mannan-
sensitive uptake of the derivatized enzyme was more than 3 times
that of the native enzyme.

The dependence of the macrophage uptake on enzyme concentration
is shown in Fig. 14. The uptake of the native and derivatized enzyme
exhibited saturation kinetics. The K_m values for uptake were estimated at 0.88 μM for native and Man_3Lys_2-β-glucocerebrosidase, respectively, indicating a 4-fold increase in the affinity of the derivatized enzyme for the macrophages relative to the native enzyme. The
Vmax for the macrophage uptake of the native and derivatized enzyme
were determined to be 0.42 and 0.86 $pmol/h/5x10^5$ cells, respectively.[16]

The inhibition of macrophage uptake of the native and derivatized enzyme by D-mannosyl BSA was also compared (Fig. 15). Higher

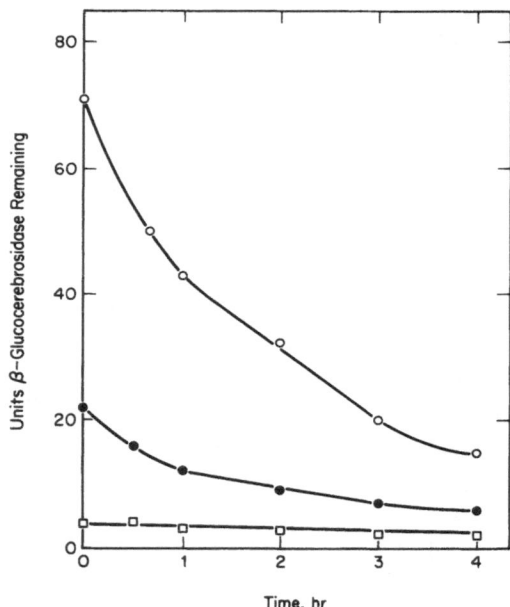

Time, hr

Fig. 16. Stability of native ($\bullet$—$\bullet$) and Man_3Lys_2-β-gluco-
         cerebrosidase (O—O) endocytosed by macrophages.
         Macrophages were incubated with enzyme for 1 h
         at 37°C. For the preincubation, the ratio of
         enzyme to cells was 3800 units and 3000 units/
         $3.4x10^5$ cells for native and Man_3Lys_2-β-gluco-
         cerebrosidase, respectively. □—□ , endogenous
         β-glucosidase in control macrophages.[16]

concentrations of mannose-BSA were required to inhibit the macrophage uptake of Man3Lys2-β-glucocerebrosidase relative to that of
the native enzyme. This again shows the higher affinity of the
derivatized enzyme for macrophages, as compared to the native enzyme.

     The stability of the native and derivatized enzymes in macro-
phages was also examined. Fig. 16 shows the absolute amount of enzyme in control and enzyme-loaded macrophages at increasing times
of incubation. The loss of activity of both enzyme preparations in
macrophages was quite rapid with half-lives of 1 h and 1.5 h for the
native enzyme and Man3Lys2-β-glucocerebrosidase, respectively. However, becasue much more derivatized enzyme than native enzyme was
taken up by the macrophages during the preincubation, the macrophage
content of Man3Lys2-β-glucocerebrosidase was 4 to 5 times that of
the native enzyme through 4 h of incubation.[16]

CONCLUSIONS

     A small molecular weight synthetic glycopeptide, Man3Lys2, has
been shown to be a good substrate for the macrophage D-mannose-spec-
ific glycoprotein uptake system. This and related ligands may be
useful in the selective delivery to macrophages of antigens, adjuvants, anti-inflammatory drugs, antiparasitic agents, and other pharmaceutical compounds. Analogous ligands (Fig. 17) may also be useful
for delivery of such agents to other target cells that may contain
distinctive uptake systems. Using the small synthetic glycopeptide,
Man3Lys2, chemically coupled to human placental β-glucocerebrosidase,
we have demonstrated the increased delivery of the derivatized enzyme to macrophages both in-vivo and in-vitro.

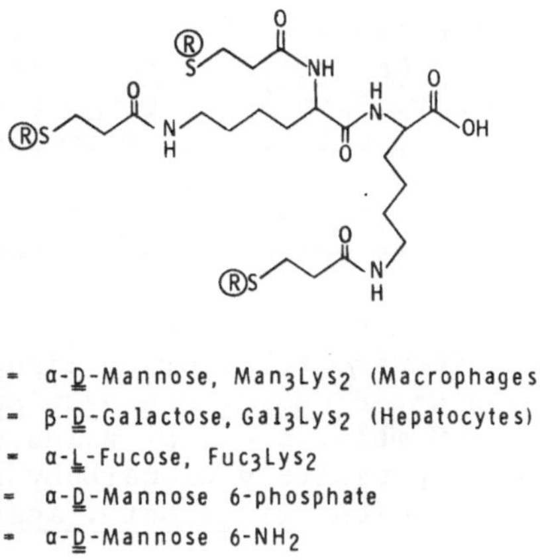

Ⓡ = α-D-Mannose, Man3Lys2 (Macrophages)
Ⓡ = β-D-Galactose, Gal3Lys2 (Hepatocytes)
Ⓡ = α-L-Fucose, Fuc3Lys2
Ⓡ = α-D-Mannose 6-phosphate
Ⓡ = α-D-Mannose 6-NH2

Fig. 17. Structures of synthetic glycopeptides.

ACKNOWLEDGMENTS

     The authors thank M.H. Lam, C.S. Tripp, and M.S. Wu for their
technical assistance.

REFERENCES

1.   G. Ashwell and A.G. Morell, The role of surface carbohydrates
     in the hepatic recognition and transport of circulating glyco-
     proteins, Adv. Enzymol., 41:99 (1974).
2.   E.F. Neufeld and G. Ashwell, Carbohydrate recognition systems
     for receptor-mediated pinocytosis, in "The Biochemistry of
     Glycoproteins and Proteoglycans," Plenum Press, New York (1980).
3.   J.M. Besterman and R.B. Low, Endocytosis: A review of mechan-
     isms and plasma membrane dynamics, Biochem. J., 210:1 (1983).
4.   G. Ashwell and J. Harford, Carbohydrate-specific receptors of
     the liver, Annu. Rev. Biochem., 51:531 (1982).
5.   J. Lunney and G. Ashwell, A hepatic receptor of avian origin
     capable of binding specifically modified glycoproteins, Proc.
     Natl. Acad. Sci. USA, 73:341 (1976).
6.   J.-P. Prieels, S.V. Pizzo, L.R. Glasgow, J.C. Paulson, and R.L.
     Hill, Hepatic receptor that specifically binds oligosaccharides
     containing fucosy α 1→3 N-acetylglucosamine linkages, Proc.
     Natl. Acad. Sci. USA, 75:2215 (1978).
7.   P.D. Stahl and P.H. Schlesinger, Receptor-mediated pinocytosis
     of mannose/N-acetylglucosamine-terminated glycoproteins and
     lysosomal enzymes by macrophages, Trends Biochem. Sci., 144
     (1980).
8.   C. Tietze, P. Schlesinger, and P. Stahl, Mannose-specific endo-
     cytosis receptor of alveolar macrophages: Demonstration of two
     functionally distinct intracellular pools of receptor and their
     roles in receptor recycling, J. Cell Biol., 92:417 (1982).
9.   W.S. Sly and H.D. Fischer, The phosphomannosyl recognition
     system for intracellular and intercellular transport of lyso-
     somal enzymes, J. Cellular Biochem., 18:67 (1982).
10. Y. Maynard and J.V. Baenziger, Oligosaccharide specific endo-
     cytoses by isolated rat hepatic reticuloendothelial cells,
     J. Biol. Chem., 256:8063 (1981).
11. M.M. Ponpipom, R.L. Bugianesi, and T.Y. Shen. Cell surface
     carbohydrates for targeting studies, Can.J. Chem., 58:214
     (1980).
12. M.S. Wu, J.C. Robbins, R.L. Bugianesi, M.M. Ponpipom, and T.Y.
     Shen, Modified in-vivo behavior of liposomes containing syn-
     thetic glycolipids, Biochim. Biophys. Acta., 674:19 (1981).
13. M.R. Mauk, R.C. Gamble, and J.D. Baldeschwieler, Targeting of
     lipid vesicles: Specificity of carbohydrate receptor analogues
     for leukocytes in mice, Proc. Natl. Acad. Sci. USA, 77:4430
     (1980).
14. M.M. Ponpipom, R.L. Bugianesi, J.C. Robbins, T.W. Doebber, and
     T.Y. Shen, Cell-specific ligands for selective drug delivery

to tissues and organs, J. Med. Chem., 24:1388 (1981).

15. J.C. Robbins, M.H. Lam, C.S. Tripp, R.L. Bugianesi, M.M. Ponpipom, and T. Y. Shen, Synthetic glycopeptide substrates for receptor-mediated endocytosis by macrophages, Proc. Natl. Acad. Sci. USA, 78:7294 (1981).

16. T.W. Doebber, M.S. Wu, R.L. Bugianesi, M.M. Ponpipom, F.S. Furbish, J.A. Barranger, R.O. Brady, and T.Y. Shen, Enhanced macrophage uptake of synthetically glycosylated human placental β-glucocerebrosidase, J. Biol. Chem., 257:2193 (1982).

17. M.M. Ponpipom, R.L. Bugianesi, and T.Y. Shen, Novel analogs of glycopeptides, Carbohydrate Res., 82:141 (1980).

18. P.D. Stahl, J.S. Rodman, M.J. Miller, and P.H. Schlesinger, Evidence for receptor-mediated binding of glycoproteins, glyco-conjugates, and lysosomal glycosidase by alveolar macrophages, Proc. Natl. Acad. Sci. USA, 75:1399 (1978).

19. T. Kawasaki, R. Etoh, and I. Yamashina, Isolation and character-ization of a mannan-binding protein from rabbit liver, Biochem. Biophys. Res. Commun., 81:1018 (1978).

20. R. Kornfeld and S. Kornfeld, Comparative aspects of glycoprotein structure, Annu. Rev. Biochem., 45:217 (1976).

21. Y.C. Lee, Synthesis of some cluster glycosides suitable for attachment to proteins or solid matrices, Carbohyd. Res., 67:509 (1978).

22. Y.C. Lee, C.P. Stowell, and M.J. Krantz, 2-Imino-2-methoxyethyl 1-thioglycosides: New reagents for attaching sugars to proteins, Biochemistry, 15:3956 (1976).

23. R.H. De Wolfe, Kinetics and mechanisms of reations of amidines, in "The Chemistry of Amidines and Imidates", S. Patai, Ed., Wiley, New York (1975).

24. V.L. Shepherd, Y.C. Lee, P.H. Schlesinger, and P.D. Stahl, L-Fucose-terminated glycoconjugates are recognized by pinocytosis receptors on macrophages, Proc. Natl. Acad. Sci. USA, 78:1019 (1981).

25. D.D. Keightley and N.A. Cressie, The Woolf plot is more reliable than the Scatchard plot in analysing data from hormone receptor assays, J. Steroid Biochem., 13:1317 (1980).

26. J. Van Zile, L.A. Henderson, J.W. Baynes, and S.R. Thorpe, (^{3}H)Raffinose, a novel radioactive label for determining organ sites of catabolism of proteins in the circulation, J. Biol. Chem., 254:3547 (1979).

27. R. Wattiaux, in G.A. Jamieson and D.M. Robinson, Eds., Mammalian Cell Membranes, Vol.2, Butterworths, London (1977).

28. M.M. Ponpipom, R.L. Bugianesi, and J.C. Robbins, Synthesis and carbon-13 n.m.r. spectroscopy of Man$_3$Lys$_2$-raffinose conjugate, Carbohyd. Res., 107:142 (1982).

29. F.S. Furbish, C.J. Steer, N.L. Krett, and J.A. Barranger, Uptake and distribution of placental glucocerebrosidase in rat hepatic cells and effects of sequential deglycosylation, Biochim. Biophys. Acta, 673:425 (1981).

STUDY OF THE TRANSFERRIN RECEPTOR USING A CYTOTOXIC HUMAN

TRANSFERRIN-RICIN A CHAIN CONJUGATE

Vic Raso and Marylu Basala

Department of Pathology

Harvard Medical School and the
Dana-Farber Cancer Institute

INTRODUCTION

The A chain of ricin which catalytically inactivates eukaryotic ribosomes inhibits cellular protein synthesis only after entering the cell. Although attachment of pure A chain to cell membranes has not been demonstrated (Raso, 1981), it is toxic at high concentrations indicating that cells can inefficiently internalize this 30,000 MW protein. Cytotoxicity is enhanced 10^3-10^5-fold however by recombining A chain either with its natural B chain subunit or with alternative carriers which deliver it to the cell surface and facilitate access to ribosomes within the cytoplasm. Thus, cytotoxins possessing different cell receptor selectivities can be designed by coupling A chain to antibody or ligand carriers with appropriately chosen specificity characteristics. This controlled cytotoxic action has stimulated interest from the standpoint of eliminating undesirable cells for therapeutic reasons but is equally important for cell biology studies.

The iron transport protein transferrin binds to specific cell surface receptors (Jandl and Katz, 1963; Van Bockxmeer and Morgan, 1979) and delivers Fe^{+++} across the plasma membrane. This bilobal 80,000 MW glycoprotein (Gorinsky et al., 1979) can carry two ferric ions in nonidentical binding sites (Aisen and Listowsky, 1980). Its membrane-situated receptor is a 180,000 MW dimeric glycoprotein composed of two apparently identical subunits each with the capacity to bind a transferrin molecule (Schneider et al., 1982). Upon attaching to these cell receptors, transferrin relinquishes its iron either while still residing on the surface (Jandl and Katz, 1963) or after penetration by an endocytotic mechanism (Aisen and Listowsky,

73

1980; Octave et al., 1981). Neither transferrin itself nor its receptor appear to be degraded during this process but are recycled to be used again (Karin and Mintz, 1981; Dautry-Varsat et al., 1983; Klausner et al., 1983).

Ricin A chain has been disulfide linked to 2-pyridyldithio-propionate-substituted human transferrin to produce a conjugate which is capable of specifically binding to cellular transferrin receptor sites. This conjugate is an efficient cytotoxin indicating that its A chain component does cross the membrane to reach the ribosomes. The selective cytotoxic pressure of transferrin-ricin A chain (TF-A chain) has allowed for the isolation of variant cells with altered receptor properties. This specific TF-A chain cytotoxin plus receptor modified cell lines should be useful in the study of transferrin mediated iron delivery pathways and their role in cell function and proliferation.

RESULTS AND DISCUSSION

The scheme for producing disulfide coupled TF-A chain conjugate (Fig. 1) involves substitution of human transferrin with the heterobifunctional crosslinking reagent, SPDP (Carlson et al., 1978).

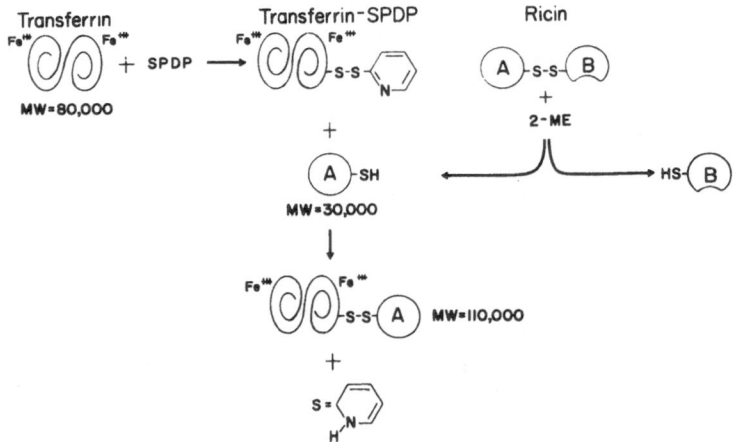

Fig. 1. Synthesis of the TF-A chain conjugate. Reduced
         ricin A chain was reacted with 2-pyridyldithio-
         (propionate)-substituted human transferrin to
         produce a disulfide linked transferrin-ricin A
         chain conjugate (TF-A chain).

This reagent reacts with amino groups on the protein, allowing
addition of three 2-pyridyldithiopropionate substituents per transferrin molecule. Ricin A chain was isolated from the whole toxin by
reduction with 2-mercaptoethanol followed by affinity chromatography
on acid treated-Sepharose which removes the B chain by virtue of its
galactose binding site (Raso et al., 1982). Mixing modified transferrin with a two-fold molar excess of A chain promotes a disulfide-
thiol exchange reaction which yields the desired conjugate (Fig. 1).
Conjugate having a 1:1 molar ratio of A chain to transferrin, MW
110,000, was isolated on a Sephacryl S-300 sizing column. Species
having 2 A-chain moieties were also produced but were not used in
this study.

    An obvious prerequisite for conjugate cytotoxicity is effective
binding to the cell surface. Monitoring this can be complicated by
the rapid internalization of transferrin receptors and their bound
ligands (Van Renswoude et al., 1982). To ensure exclusive measurement of intact surface localized TF-A chain, cells were maintained
at 4°C during exposure to conjugate and a fluorescein labeled anti-
ricin A chain probe was used for detection. Human leukemia CEM cells
which express ample amounts of transferrin receptor (Sutherland et
al., 1981), were treated in this manner and surface bound TF-A
chain was quantified using flow cytofluorimetry (Fig. 2). When
excess native transferrin was included during treatment with conju-

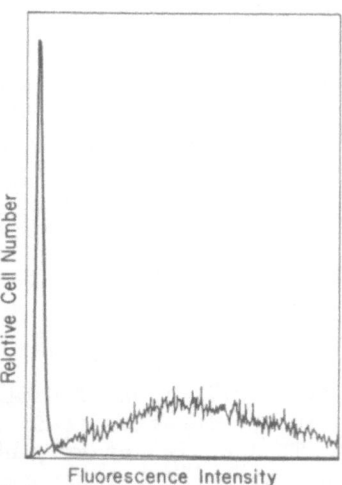

Fig. 2. Surface localization of TF-A chain on CEM cells.
        CEM cells treated with either PBS or TF-A chain,
        were washed, exposed to fluorescein-labeled anti-
        ricin A chain and analyzed for surface fluores-
        cence using a FACS I.

gate, A chain was no longer found on the cells. Since A chain
itself shows no ability to attach to cells, the transferrin portion
of this conjugate fulfilled its delivery function and blocking with
native transferrin confirmed specific interaction with receptors.

   The ability of membrane situated TF-A chain conjugate or at
least its A chain component to internalize and inactivate ribosomes
can be monitored by the resultant inhibition of cellular protein
synthesis. This parameter was measured as a function of time after
addition of TF-A chain (10^{-7}M) to CEM cells and provided a kinetic
analysis of cell kill (Fig. 3) (Youle and Neville, 1982). At this
concentration, all available receptor sites were filled and linear
first order inhibition of leucine incorporation with time was
observed, giving a t $_{1/2}$ max of 6 h. This value is similar to those
obtained for antibody-A chain systems (Youle and Neville, 1982) and
may reflect the efficiency of translocation processes which eventually permit A chain to contact ribosomes.

   In accordance with its ability to inhibit cellular protein
synthesis, TF-A chain dramatically affected the growth of CEM cells
cultured in serum-free RPMI 1640 media with 1% BSA plus insulin and
selenium (Fig. 4). Concentrations of 10^{-9}M - 10^{-7}M gave complete
lysis by 48 h, leaving no identifiable intact cells upon microscopic
examination. The lower level of 10^{-10}M TF-A chain was cytostatic

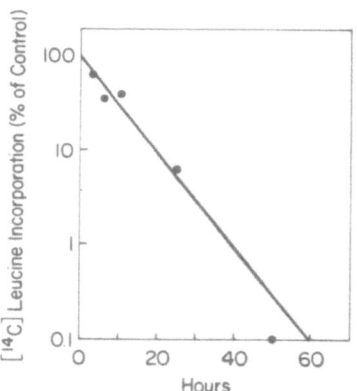

Fig. 3. Kinetics of inactivation of protein synthesis.
         CEM cells in leucine-free media were treated
         with 10^{-7}M TF-A chain for varying lengths of
         time, pulsed with ^{14}C-L-leucine, harvested and
         assayed for incorporation of radioactvity.

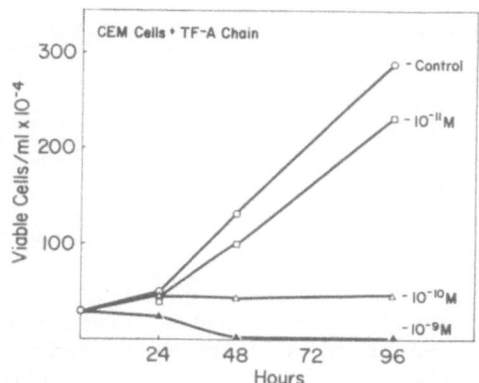

Fig. 4.   Cytotoxicity of the TF-A conjugate.   CEM cells
          plated at 30×10^4 cells/ml were grown in media
          alone, or media containing the designated con-
          centrations of TF-A chain.   The cultures were
          assayed daily for viable cells.

while 10^{-11}M produced only a minor change in the growth rate.   The
concentration of conjugate required to reduce growth to half the
normal level was $ID_{50}=3 \times 10^{-11}$M (Fig. 5) while 10,000-fold higher
levels of A chain alone or A chain plus uncoupled transferrin were
required for comparable inhibition.   This supports the carrier role
served by covalently linked transferrin since no synergistic action
was noted upon simultaneous addition of the two separated molecules.
Blockage of cytotoxicity by anti-transferrin or anti-A chain antibodies confirmed the requirement of both moieties for action on
cells.

No significant difference in ID_{50} value was noted between apo
TF-A chain pre-loaded with Fe^{+++} (Fig. 5) nor did the inclusion of
2×10^{-4} ferric ammonium citrate in the culture medium affect the
cytotoxic potency of TF-A chain.   While this finding is consistant
with previous results demonstrating equivalent receptor affinity
for apo and holo forms of transferrin, (Ward et al., 1982; Karin
and Mintz, 1981) it might reflect the presence of trace amounts of
iron in the media which can convert apo to holo TF-A chain (Dautry-
Varsat et al., 1983).   Further experiments using chelators to
remove iron from the medium may clarify this point.

Fig. 5 also illustrates that a 2-fold increased ID_{50} was
obtained for TF-A chain on human CEM cells in media containing 10%
fetal calf serum in place of 1% BSA.   This partial blockage of

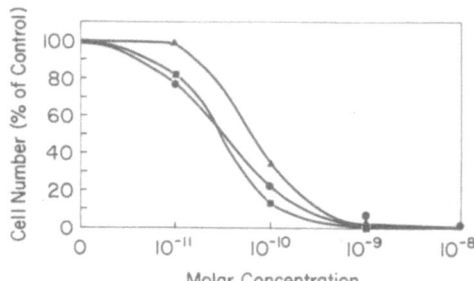

Fig. 5.   Cytotoxicity dose response curves for apo and
          holo TF-A chain, and apo TF-A chain in media
          containing 10% fetal calf serum. CEM cells
          cultured in serum-free media received the des-
          ignated comcentrations of apo (●──●) or holo
          (■──■) TF-A chain. Apo TF-A chain was also
          added to CEM cells cultured in media containing
          10% FCS (▲──▲). Viable cells were counted
          after 4 days and values expressed as percent of
          control (cells in media alone).

toxic action provided a sensitive indication of the interaction of
heterologous bovine transferrin with human receptors, which may not
be detected by other means (Ward et al., 1982). Indeed, bovine
transferrin in fetal calf serum is the presumed source of iron for
cultured cells so that binding to these receptors is expected. CEM
cells do grow well without serum but require 1% BSA (Fig. 4) and
this suggests that a low level of transferrin contaminating the BSA
could be supporting their growth.

The effect of human transferrin on TF-A chain toxicity was
studied to verify their competitive interaction for cellular transferrin receptors. Media without 1% BSA supported cell growth when
supplemented with 10^{-9}M human transferrin but these cells were
killed by addition of an equimolar level of TF-A chain (Table 1).
Under these conditions enough receptor sites were available for both
transferrin and toxin binding making it unlikely that the observed
cytotoxicity was due to antagonism of the normal iron transport
functions of transferrin. Simultaneous exposure of CEM cells to
10^{-9}M TF-A chain plus 10^{-7}M unmodified human transferrin completely

Table 1. Blockage of TF-A Chain Cytotoxicity by Human Transferrin

Treatment	Cells/mlx10^{-4}	
	Start	4 Days
CEM cells + 10^{-9}M TF	30	91
+ 10^{-9}M TF + 10^{-9} TF-A chain	30	1
+ 10^{-7}M TF	30	115
+ 10^{-7}M TF + 10^{-9}M TF-A chain	30	98

abrogated the effects of conjugate showing that receptor-mediated cytotoxicity was precluded when all sites were occupied by the natural ligand.

Transferrin receptor densities vary on different cell lines and many, such as the Burkitts lymphoma lines Daudi and Raji, have only low amounts as determined by reactivity with monoclonal anti-receptor antibody (Sutherland et al., 1981). The ID$_{50}$ level for TF-A chain on these two lines was about an order of magnitude higher than for CEM cells, possibly reflecting their low receptor levels. Interestingly, these cells proliferate well despite this relatively weak expression of transferrin receptors and this raises questions regarding the significance of variations in receptor density.

An often fruitful approach for determining the functional importance of receptor expression involves the examination of variant cell lines which are receptor deficient, possess receptors incapable of binding ligand or which have an abnormal processing pathway for the receptor-ligand complex. Appropriate ligand-toxin conjugates have been used to select insulin receptor (Miskimins and Shimizu, 1981) and mannose 6-phosphate receptor (Robbins et al., 1981) modified cell lines. In order to have a valid comparison of the influence of transferrin receptor density on cell growth, selection of variants within a single cell population was undertaken using the TF-A chain cytotoxin.

Prolonged exposure to high concentrations of TF-A chain results in complete loss of cells (Fig. 2) therefore CEM cells were acclimatised to grow in the presence of gradually increasing doses of TF-A chain. Cells started in 10^{-11}M TF-A chain eventually grew at

10^{-8}M levels of the cytotoxin by elevating concentrations in step-wise 2-fold increments over several months. The derived cell line was 1000-fold resistant to TF-A chain, having an $ID_{50}=3\times10^{-8}$M compared to an $ID_{50}=3\times10^{-11}$M for the parent CEM cells. Equivalent dose response curves for both the resistant and control CEM lines were obtained using whole ricin which binds to surface glycoproteins via its carrier B subunit. This indicated that protection from TF-A chain toxicity was based upon the mode of attachment and entry into the cell not on insensitivity to A chain.

Quantification of transferrin receptor on the CEM lines further implicated involvement of the transferrin carrier in the mechanism underlying resistance. TF-A chain resistant cells displayed a much lower receptor density than control CEM cells as measured by indirect immunofluorescence using monoclonal anti-receptor anti-

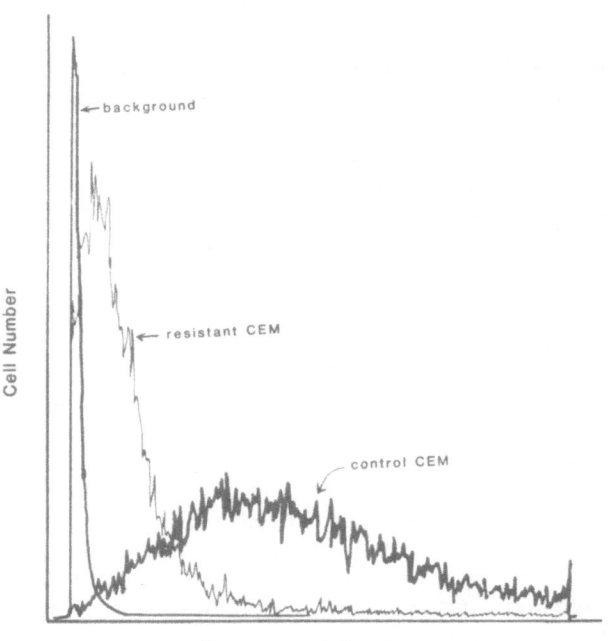

Fig. 6.   Transferrin receptor levels on control and TF-A
          chain resistant CEM cells. Cytofluorometric
          analysis was performed on control and TF-A chain-
          resistant CEM cells which had been treated sequen-
          tially with monoclonal anti-transferrin receptor
          antibody (OKT9) and a fluorescein labeled goat
          anti-mouse reagent.

body and cytofluorometric analysis (Fig. 6) (Sutherland et al., 1981).
Furthermore it appeared that the receptors remaining on the resistant
cell line had little or no capacity to bind human transferrin
($5x10^{-7}M$), goat anti-human transferrin and fluoresceinated rabbit
anti-goat antibodies, control CEM cells displayed bright fluorescence
by cytofluorimetry while levels obtained for the TF-A chain resistant
line were no higher than background. Direct binding of fluoresceinated human transferrin to both parent and resistant CEM lines was also
examined by video intensification microscopy (Willingham and Pastan,
1978) using live cells at 4°C to prevent internalization. Fig. 7

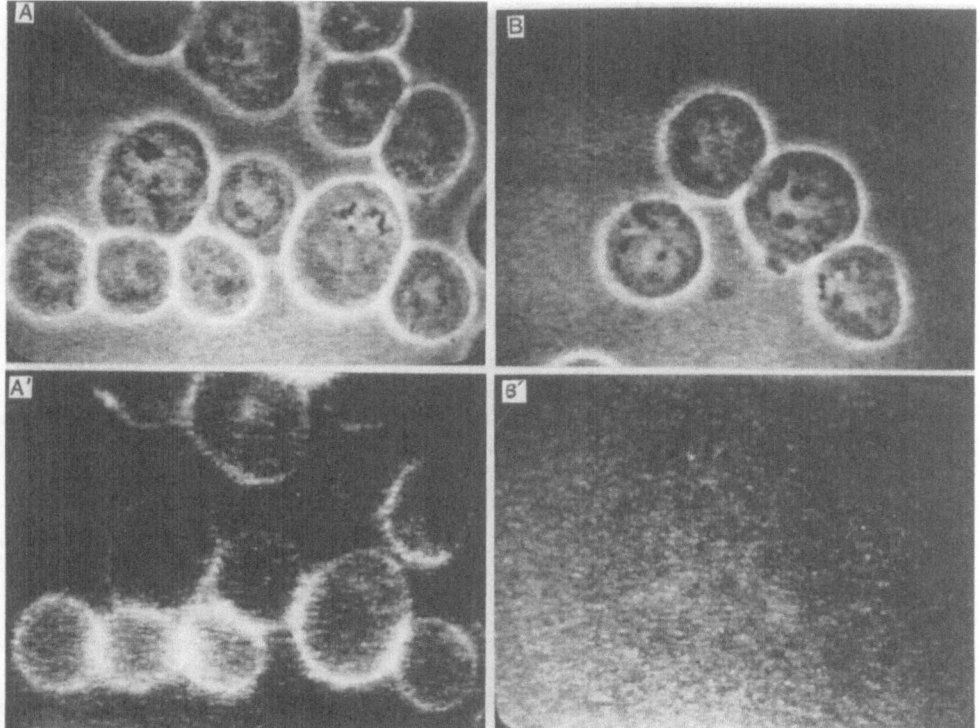

Fig. 7. Photographs of sequential video recordings of phase or
epifluorescent microscopic images of living control and
TF-A chain-resistant CEM cells with fluoresceinated
human transferrin.
A: control cells, phase mode;
A′: control cells, fluorescence mode;
B: resistant cells, phase mode,
B′: resistent cells, fluorescence mode.

shows the diffuse surface and peripheral staining of control CEM cells consistant with membrane localized ligand, and also illustrates the absence of detectable binding to residual receptors on the resistant line.

TF-A chain resistant cells were not totally devoid of specific transferrin binding since they were killed by 10^{-7}M TF-A chain and this cytotoxicity was blocked by 10^{-5}M transferrin. The residual receptors which were detectable with monoclonal antibody may be functionally impaired and could be immature (Trowbridge and Omary, 1981) or low affinity forms (Octave et al., 1981) of transferrin receptor. These possibilities can be tested by comparing the physical properties of internally labeled receptors from the two CEM lines. Information regarding the rates of receptor synthesis, degradation and translocation to the cell surface should define the molecular basis for the paucity of receptor on the TF-A chain resistant line.

The resistant CEM cells retained their receptor-modified phenotype when grown for several months in media without the selective pressure of TF-A chain. They were not replaced by higher receptor density revertants, indicating that the line was stable and at no obvious growth disadvantage under these conditions. Following a 2 day lag interval, the growth pattern of TF-A chain resistant cells paralleled that observed for receptor-rich CEM cells in terms of doubling time and final cell density.

It was important to determine what functional consequences reduced receptor levels had on regulation of cell growth. Control and receptor modified cells were cultured in transferrin-free medium and their proliferative response to addition of iron-loaded transferrin or TF-A chain was evaluated. Control CEM cells remained quiescent in the absence of transferrin but a 10^{-10}M level of the protein allowed for near maximal cell growth. TF-A chain carrying bound Fe^{+++} was lytic at this same concentration. It appeared that any potential stimulatory effects were overcome by its toxic action. A similar pattern was observed for the resistant CEM line but these cells required 100-fold higher doses of each agent for an equivalent response. Growth stimulation by transferrin and TF-A chain cytotoxicity both commenced at 10^{-8}M levels for these cells (Table 2).

The enhancement of cell growth by native transferrin and the cytotoxic effects of TF-A chain began at concentrations which were far below receptor-saturating levels (Table 2). Apparently similar threshold amounts of ligand or cytotoxin must reside on the cell surface before the events which initiate cell growth or toxicity proceed. Furthermore, both of these biological responses correlated with the amount of receptor on the cell since 100-fold higher concentrations of transferrin and TF-A chain were required to affect receptor-deficient cells. This close correspondence between func-

Table 2. Effect of TF and TF-A Chain on Cell Growth in Media Lacking Transferrin

| | Cells/ml x 10^{-4} After 4 Days With: | | | |
| | Control CEM | | Resistant CEM | |
Amount Added	TF	TF-A	TF	TF-A
0	36	36	24	24
10^{-10}M	176	10	32	24
10^{-8}M	206	0	169	8

Cells initially started at 30×10^4/ml.

tional response, receptor level and effective dose of ligand or toxin suggest that both follow a common pathway for part of their journey through the cell.

The presence of high levels of transferrin receptor on many non-hemoglobin synthesizing cells is difficult to reconcile from the standpoint of their iron requirements. The finding that high receptor densities allow cells to respond to lower concentrations of ligand might provide an expanation for these elevated amounts of receptor. This would be a survival benefit for cells in a milieu having reduced transferrin levels. Whether such regions exist in the body (Fairbanks and Beutler, 1977) or this feature is important for neoplastic cell proliferation and metastasis remains to be shown.

It is possible that receptor-bound TF-A chain on the cell surface penetrates the membrane directly to gain access to ribosomes. This would be consistant with the observed blockage of toxicity by anti-transferrin and anti-A chain antibodies. The receptor-mediated endocytotic pathway proposed for transferrin cycling however is a more likely route of toxin entry into the cell. It has been estimated that the round trip for transferrin endocytosis, iron delivery and exocytosis occurs with a halftime of 8 min (Dautry-Varsat et al., 1983). A much longer $t_{1/2}$ of 6 h was required for cell kill by TF-A chain (Fig. 3). This discrepancy could indicate that linkage to A chain slows the internalization process or that conjugate deviates

from the route of native transferrin. Alternatively, a slow leakage
of cytotoxin or cleaved A chain from the endosome or lysosome may
account for damage to ribosomes. If the ligand-receptor complex
remains intact throughout recycling and does not leave closed vesicular compartments (Klausner et al., 1983; Dautry-Varsat et al., 1983),
then a mechanism must exist for passage of iron into the cytosol.
A chain or intact conjugate may exist via a similar route.

It is important to note that ricin A chain linked to a monoclonal antibody directed against transferrin receptor is a potent
cytotoxin (Trowbridge and Domingo, 1981). Furthermore we found that
a hybrid formed between human transferrin and a monoclonal anti-A
chain antibody can non-covalently deliver ricin A chain to mediate
receptor-specific cell killing. Whether these agents as well as
TF-A chain all utilize the same transferrin pathway to affect cells
remains to be elucidated.

The differential dose response of control and receptor modified
cells to transferrin might be explained by several mechanisms
(Fig. 8). According to the law of mass action, the amount of transferrin bound to a cell is dependent upon the concentration of both
ligand and receptor. Insufficient reactant concentrations for
binding may exist when low levels of transferrin are added to receptor deficient cells. Alternatively, if ligand internalization
is facilitated by micro-aggregation of receptors, such cross-linking
should occur more easily on receptor dense cells than on cells with
sparsely distributed receptors. This multiple interaction possibility is intriguing since transferrin receptors are dimers with two
binding sites (Schneider et al., 1982) and transferrin is a bilobal
molecule (Gorinsky et al., 1979) with internal homology, possibly
arising from gene duplication (Aisen and Listowsky, 1980). The high

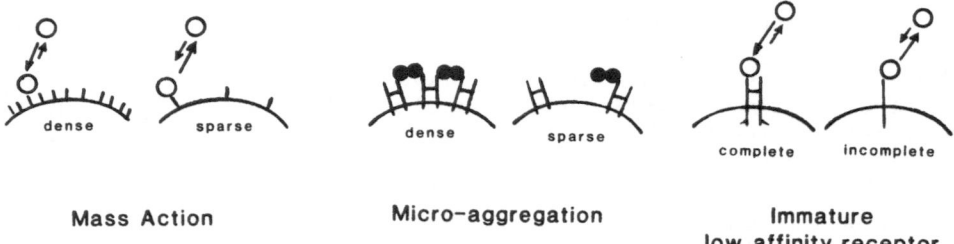

Fig. 8.   Possible ligand receptor interactions.

doses of transferrin required for growth stimulation of the receptor
altered CEM line could also be due to reduced receptor affinity,
perhaps a result of some defect in the post-translational modification of the molecule.

REFERENCES

Aisen, P. and Listowsky, I., 1980, Iron transport and storage
    proteins, Ann. Rev. Biochem., 49:357.
Carlsson, J., Drevin, H. and Axen, R., 1978, Protein thiolation
    and reversible protein-protein conjugation. N-succinimidyl
    3-(2-pyridyldithio) propionate a new heterobifunctional
    reagent, Biochem. J., 173:723.
Dautry-Varsat, A., Ciechanover, A. and Lodish, H.F., 1983, pH and
    the recycling of transferrin during receptor-mediated endo-
    cytosis, Proc. Natl. Acad. Sci. USA, 80:2258.
Fairbanks, V.F. and Beutler, E., 1977, Iron metabolism. in:
    "Hematology," W.J. Williams, ed., McGraw-Hill, New York.
Gorinsky, B., Horsburgh, C., Lindley, P.F., Moss, D.S., Parkar,
    M. and Watson, J.L., 1979, Evidence for the bilobal nature
    of diferric rabbit plasma transferrin, Nature, 281:157.
Jandl, J.H. and Katz, J.H., 1963, The plasma to cell cycle of
    transferrin, J. Clin. Invest., 42:314.
Karin, M. and Mintz, B., 1981, Receptor-mediated endocytosis of
    transferrin in developmentally totipotent mouse teratocar-
    cinoma stem cells, J. Biol. Chem., 256:3245.
Klausner, R.D., Van Renswoude, J., Ashwell, G., Kempf, C.,
    Schechter, A.N., Dean, A. and Bridges, K.R., 1983, Receptor-
    mediated endocytosis of transferrin in K562 cells, J. Biol.
    Chem., 258:4715.
Miskimins, W.K. and Shimizu, N., 1981, Genetics of cell surface
    receptors for bioactive polypeptides: Varients of Swiss/3T3
    fibroblasts to a cytotoxic chimeric insulin, Proc. Natl.
    Acad. Sci USA, 78:445.
Octave, J.N., Schneider, Y.J., Crichton, R.R. and Trouet, A.,
    1981, Transferrin uptake by cultured rat embryo fibroblasts,
    Eur. J. Biochem., 115:611.
Raso, V., 1981, Antibody mediated delivery of toxic molecules to
    antigen bearing target cells, Immunological Rev., 62:93.
Raso, V., Ritz, J., Basala, M. and Schlossman, S.F., 1982, Mono-
    clonal antibody-ricin A chain conjugate selectively cytotoxic
    for cells bearing the common acute lymphoblastic leukemia
    antigen, Cancer Res., 42:457.
Robbins, A.R., Myerowitz, R., Youle, R.J. Murray, G.J. and
    Neville, D.M., 1981, The mannose 6-phosphate receptor of
    chinese hamster ovary cells, J. Biol. Chem., 256:10618.
Schneider, C., Sutherland, R., Newman, R. and Greaves, M., 1982,
    Strucural features of the cell surface receptor for trans-
    ferrin that is recognized by the monoclonal antibody OKT9,
    J. Biol. Chem., 257:8516.

Sutherland, R., Delia, D., Schneider, C., Newman, R., Kemshead, J.
    and Greaves, M., 1981, Ubiquitous cell-surface glycoprotein
    on tumor cells is proliferation-associated receptor for trans-
    ferrin, Proc. Natl. Acad. Sci. USA, 78:4515.
Trowbridge, I.S. and Domingo, D.L., 1981, Anti-transferrin receptor
    monoclonal antibody and toxin-antibody conjugates affect
    growth of human tumor cells, Nature, 294:171.
Trowbridge, I.S. and Omary, M.B., 1981, Biosynthesis of the human
    transferrin receptor in cultured cells, J. Biol. Chem.,
    256:12888.
Van Bockxmeer, F.M. and Morgan, E.H., 1979, Transferrin receptors
    during rabbit reticulocyte maturation, Biochim. Biophys.
    Acta., 584:76.
Van Renswoude, J., Bridges, K.R., Hartford, J.B. and Klausner,
    R.D., 1982, Receptor-mediated endocytosis of transferrin and
    the uptake of Fe in K562 cells: Indentification of a non-
    lysosomal acidic compartment, Proc. Natl. Acad. Sci USA,
    79:6186.
Ward, J.H. Kushner, J.P. and Kaplan, J., 1982, Regulation of HeLa
    cell transferrin receptors, J. Biol. Chem., 257:10317.
Willingham, M.C. and Pastan, I., 1978 The visualization of
    fluorescent proteins in living cells by video intensification
    microscopy (VIM), Cell, 13:501.
Youle, R.J. and Neville, D.M. Jr., 1982, Kinetics of protein
    synthesis inactivation by ricin-anti-thy 1.1 monoclonal anti-
    body hybrids. Role of the ricin B subunit demonstrated by
    reconstitution, J. Biol. Chem., 257:1598.

HOW DO PROTEIN TOXINS KILL CELLS?

Sjur Olsnes, Kirsten Sandvig, Anders Sundan,
Kristin Eiklid and Alexander Pihl

Norsk Hydro's Institute for Cancer Research and
The Norwegian Cancer Society
Montebello, Oslo, Norway

INTRODUCTION

Diphtheria toxin and ricin, the toxins most frequently used in attempts to construct target-specific cytostatic agents, belong to a group of closely related toxic proteins produced by bacteria and plants. The main reason why these toxins are used for this purpose is their extreme toxicity. This is due to the fact that the active moieties of the toxins possess enzymatic activity. A single molecule may inactivate components required for protein synthesis more rapidly than the cell can produce new ones and hence kill the cell in the course of a few hours.

Although these toxins inhibit protein synthesis in all animal cells, their toxicity differs considerably from one cell line to another and from toxin to toxin. The main reason appears to be differences in the efficiency of transfer of the toxins from the cell surface into the cytosol. Another reason is the different extents of toxin binding to the surface of different cells. These two parameters are also very important for the uptake of chimeric toxins. High-efficiency binding can usually be obtained by choosing an appropriate ligand. However, so far little is known about the requirements for the entry of toxin bound to the cell surface. To facilitate the preparation of highly toxic conjugates, more information about the normal entry route of toxins is therefore needed.

STRUCTURE

The structure of the toxins contains the clue to their mechanism of action. All toxins in this group have the same basic structure. They consist of an enzymatically active moiety, A,

87

which is linked through a disulfide bond to a binding moiety, B.
This division of function is most clearly apparent in the plant
toxins, abrin, ricin, modeccin and viscumin (Fig. 1). In these
toxins the binding activity and the enzymatic activity are carried
on two separate polypeptide chains, the length of which varies

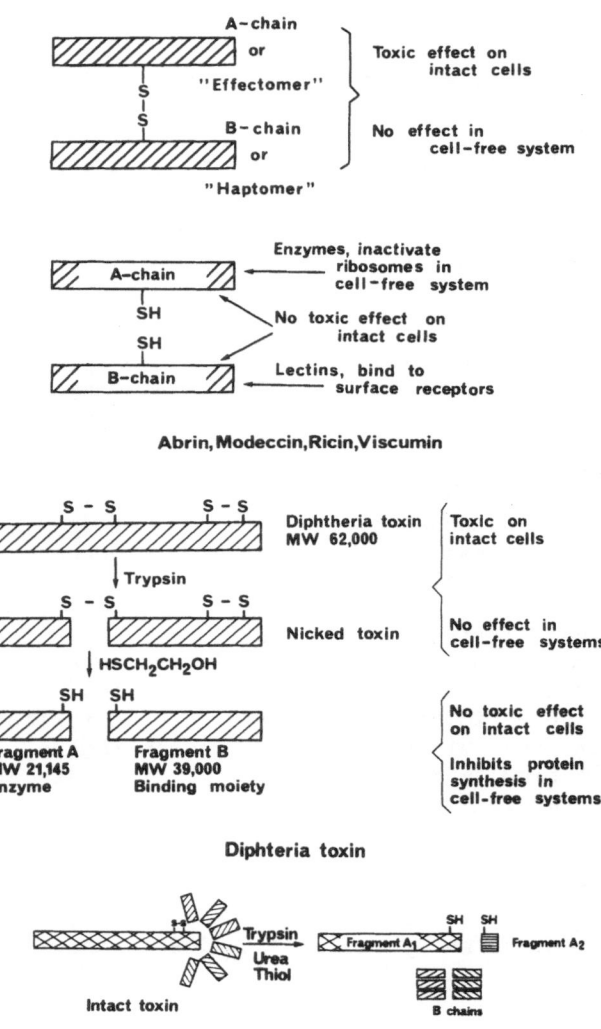

Fig. 1. Schematical structure of toxins.

slightly in the four toxins. In each case the A-chain migrates
slightly more rapidly in SDS-polyacrylamide gels than the B-chains
(for review, see Olsnes and Pihl, 1982).

Among these toxins ricin has been most extensively studied and
its primary structure has been elucidated (Funatsu et al., 1978;
Funatsu et al., 1979). The amino acid sequence of ricin A-chain
contains stretches of hydrophobic amino acids as well as stretches
of hydrophilic amino acids. Low resolution (4Å) X-ray crystallo-
graphic studies of ricin have revealed that the B-chain has a
bilobal structure and that each domain is able to bind galactose.
One of the binding sites was found to be more highly occupied than
the other (Villafranca and Robertus, 1981).

Diphtheria toxin and Pseudomonas aeruginosa exotoxin A are
bacterial toxins with the same intracellular site of action.
Diphtheria toxin which has been studied in greatest detail (Pappenheimer, 1977), is synthesized by Corynebacterium diphtheriae as a
single polypeptide chain. The toxin contains an arginine-ricin
region which is easily split ("nicked") by trypsin-like enzymes, to
yield two polypeptide fragments linked by a disulfide bridge. The
shorter fragment (fragment A) carries the enzymatic activity, whereas
fragment B binds the toxin to cell surface receptors.

Diphtheria toxin fragment A is highly hydrophilic and resistant
to boiling, extreme pH and other denaturing conditions. Fragment B
which is sensitive to such treatments, contains a highly hydrophobic
region which is buried inside the molecule under neutral, non denaturing conditions (Boquet et al., 1976). After denaturation, or
when pH is reduced to 4.5 and lower, the hydrophobic region of the
B-fragment is exposed (Sandvig and Olsnes, 1981).

Lambotte et al., (1980) have made observations which may be a
clue to the mechanism whereby the B fragment facilitates the entry
of the A fragment. Thus they found that diphtheria toxin fragment
B contains two lipid-associating domains. One of these is located
in the highly hydrophilic, 9,000 dalton, N-terminal region. It has
a structure similar to that of the phospholipid head group-binding
domain of human apolipoprotein 1 and can be considered a surface
lipid-associating domain. The second of these domains which is
highly hydrophobic, is located in the middle of fragment B. Its
structure resembles that of the membranous domain of intrinsic
membrane proteins and it can be considered as a transverse lipid-
associating domain. The C-terminal 8,000 dalton region which is
assumed to bind to the receptor, does not show similarities with
lipid-associating domains.

Pseudomonas aeruginosa exotoxin A is also synthesized by the
bacterium as a single polypeptide chain (MW 70,000). Its structure
is less well characterized than that of diphtheria toxin and it is

not clear if the enzymatic function and the binding function are associated with different domains of the protein. The enzymatic activity of Pseudomonas aeruginosa toxin, like that of diphtheria toxin, is strongly increased by treatment with thiols in the presence of denaturing agents (Leppla et al., 1978), suggesting that the enzymatic region is not exposed in the intact native toxin. Enzymatically active fragments of 26,000 MW (Vasil et al., 1977; Chung and Collier, 1977) and 48,000 MW (Sanai et al., 1980) have been identified. A receptor-binding domain has so far not been detected.

The toxin from Shigella dysenteriae has a more complicated structure than those described above. It consists of one A-chain (MW 30,500), linked by non covalent bonds to 6 or 7 B chains (Olsnes et al., 1981). The A-chain is easily split by trypsin into two fragments, A_1 (MW 27,500) and A_2 (MW 3,000), which, in the absence of reducing agents, are linked by a disulfide bridge. The non-covalent linkage of the A chain to the B chains appears to involve primarily the smaller fragment, A_2. The A_1 fragment is an enzyme which inactivates the 60S ribosomal subunits and thus inhibits protein synthesis (Reisbig et al., 1981). Although it has not been directly demonstrated, it is likely that the B chains are responsible for the binding of Shigella dysenteriae cytotoxin to cells.

BINDING

The action of the toxins discussed here can be divided into four separate steps: Binding to the cell surface, endocytic uptake, entry into the cytosol and inactivation of the intracellular target, the ribosomes.

Binding of the toxins to cell surface receptors is the first step in the intoxication. Abrin, ricin, modeccin and viscumin bind to galactose-terminated oligosaccharides which may be attached to a variety of glycoprotein and glycolipid species at the cell surface. The binding as well as the toxic effect on the cells is inhibited by galactose, lactose and glycoproteins with terminal galactose residues. Modeccin receptors are different from the receptors for abrin, ricin and viscumin and they are present on the cell surface in a much smaller number (Olsnes et al., 1978).

The diphtheria toxin receptor appears to be a glycoprotein with molecular weight 153,000 (Proia et al., 1981). Different cell lines vary considerably in their number of diphtheria toxin receptors (4,000-200,000 receptors/cell) (Middlebrook et al., 1978), and in toxin-sensitive cells the number of receptors is roughly related to the sensitivity. However, it is not clear whether resistant cells from mice and rats also have receptors or not (Chang and Neville, 1978; Boquet and Pappenheimer, 1976).

The binding of diphtheria toxin to its receptors through the B fragment is inhibited by adenosine 5'-tetraphosphate and by other polyphosphate compounds (Middlebrook, 1981). It is well established that the toxin has high affinity for NAD (K_d 9 x 10^6M), due to the presence of a binding site on the A fragment (Lory et al., 1980 a, b; Proia et al., 1980). Recent data have shown that the toxin has another site with affinity for phosphate-containing compounds, the P-site (Lory and Collier, 1980), which apparently is located in the C-terminal, cyanogen bromide fragment. This 8,000 daltons cationic part of the B fragment also carries the second SS-bridge.

Although binding of nucleotides and other phosphate-containing compounds to the P-site inhibits binding of diphtheria toxin to its receptors (Proia et al., 1979), this site may not be directly involved in the binding to the receptor. In binding of nucleotides such as ATP, both the NAD site on the A fragment and the P-site on the B fragment appear to be involved. Also polycations interfere with the binding of diphtheria toxin to its receptor, apparently by competing with the cationic P-site for the receptor (Proia et al., 1981). The P-site could also bind to membrane phospholipids and thus stabilize the association of diphtheria toxin with the cells. In fact, diphtheria toxin was found to bind to the phosphate portion of some, but not all kinds of phospholipids in liposomes (Alving et al., 1980). In accordance with this the binding was reduced when the cells were treated with phospholipase C (Moehring and Crispell, 1974), an enzyme which removes phosphate-containing polar groups from phospholipids.

The nature of the receptor for Pseudomonas aeruginosa exotoxin A is not known. Also the nature of the receptor for Shigella dysenteriae cytotoxin is not known in detail. Different cell lines exhibit widely different sensitivities to this toxin. Most cell lines are highly resistant, and tolerate 10^6 times more Shigella toxin than sensitive cells. So far, only primate epithelial cells were found to be sensitive. Unexpectantly, Shigella toxin receptors were found to be present both on sensitive and resistant cells (Eiklid and Olsnes, 1980).

The strength of the binding of the toxins to different cells varies, but it falls within the same range (K_a=10^7-10^{10}M^{-1}) as that of the binding of protein hormones to their receptors. The number of toxin receptors varies greatly (60/cell to 3 x 10^7/cell) with the nature of the toxin and the cell type. Thus, Vero cells have 1.5 x 10^5 diphtheria toxin receptors whereas HeLa cells, which are 100 times less sensitive, have only 5 x 10^3 receptors/cell (Middlebrook et al., 1978; Moynihan and Pappenheimer, 1981).

HeLa cells possess 3 x 10^7 binding sites per cell for abrin and ricin (Sandvig et al., 1976), but much fewer sites for modeccin (2 x 10^5 binding sites per cell) (Olsnes et al., 1978). After

neuraminidase treatment the number of binding sites for abrin,
modeccin and ricin was strongly increased (Nicolson et al., 1975;
Rosen and Hughes, 1977; Sandvig et al., 1978). The K_a for binding
to the new receptors did not differ from that of binding to the
receptors available before the neuraminidase treatment and in most
cases the new receptors were equally efficient in internalizing the
toxins as those normally exposed on the cells.

The binding of Shigella toxin to cell surface receptors is
particularly strong (K_a ~$10^{10}M^{-1}$) and sensitive cells were found
to have a high number of binding sites (~10^6/cell) (Eiklid and
Olsnes, 1980). Also many resistant cells had a high number of binding sites.

ENDOCYTOSIS

After toxin binding to cell surface receptors, the next step
in the intoxication process appears to consist of endocytic uptake
of the bound toxins (Sandvig and Olsnes, 1979; Sandvig and Olsnes,
1982). Gonatas et al., (1975, 1977) studied by electron microscopy
the uptake and distribution of complexes of ricin and horse radish
peroxidase which were visualized as a precipitate of oxidized
diaminobenzidine-osmium black. Upon incubation at 4°C only a continuous peripheral rim, the plasma membrane, was stained. However,
if the cells were washed and subsequently incubated at 37°C, the
staining of the plasma membrane first acquired a patchy pattern and
then diminished in intensity. Clusters of stained vesicles were
found adjacent to the elongated cisternae of the Golgi apparatus.
After 0.5-1 h, vesicles were usually found near the concave (trans)
aspects of the Golgi cisternae and at the edges of the cisternae.
After 3 h at 37°C most of the label was found in the cytoplasm,
particularly in one or two of the parallel cisternae of the Golgi
apparatus. Only after incubation for more than 3 h were a few dense
bodies (lysosomes) labelled. Comparison with the staining pattern
obtained with acid phosphatase indicated that ricin had accumulated
in those vesicles which belong to the GERL apparatus. The pattern
of endocytosis of ricin linked to horse radish peroxidase was definitely different from that obtained with free horse radish peroxidase which was found to be endocytosed into lysosomes and small
vesicles adjacent to the lysosomes. Experiments with ricin-ferritin
complexes and mouse 3T3 cells showed essentially the same pattern as
that described above (Nicolson, 1974; Nicolson et al., 1975b). The
diffuse adsorptive endocytosis of ricin is quantitatively and qualitatively different from that of EGF and LDL which, within minutes,
are endocytosed via coated pits into multivesicular bodies and lysosomes. In contrast to the findings with ricin, ferritin-labelled
Pseudomonas aeruginosa exotoxin A accumulated in coated pits (Fitz-
Gerald et al., 1980). When the incubation temperature was 4°C,
ferritin-labelled Pseudomonas aeruginosa exotoxin A and ^{125}I-
diphtheria toxin remained at the cell surface, but after 2 h at

37°C, most of the surface-bound toxin was internalized (FitzGerald et al., 1980; Dorland et al., 1981).

After the diphtheria toxin has entered endocytic vesicles it appears to retain the ability to penetrate into the cytosol only for a short period of time at 37°C (Draper and Simon, 1980; Sandvig and Olsnes, 1980) and then it is rapidly degraded. Endocytosed abrin, ricin and modeccin on the other hand, appear to be able to enter the cytosol for hours (Sandvig and Olsnes, 1982). This will be discussed in detail below.

Even in the presence of very high concentrations of the toxins described here, a certain lag period (10-60 min) elapses before any effect on the cells can be observed (Olsnes et al., 1976; Youle and Neville, 1979; Moynihan and Pappenheimer, 1981). In contrast, in cell free systems the toxins act almost immediately. The lag time observed in intact cells is at least partly due to the time required for the toxin to be taken up by endocytic vesicles and released into the cytosol. With increasing toxin concentrations the lag time was found to decrease to a certain lower limit (Fig. 2).

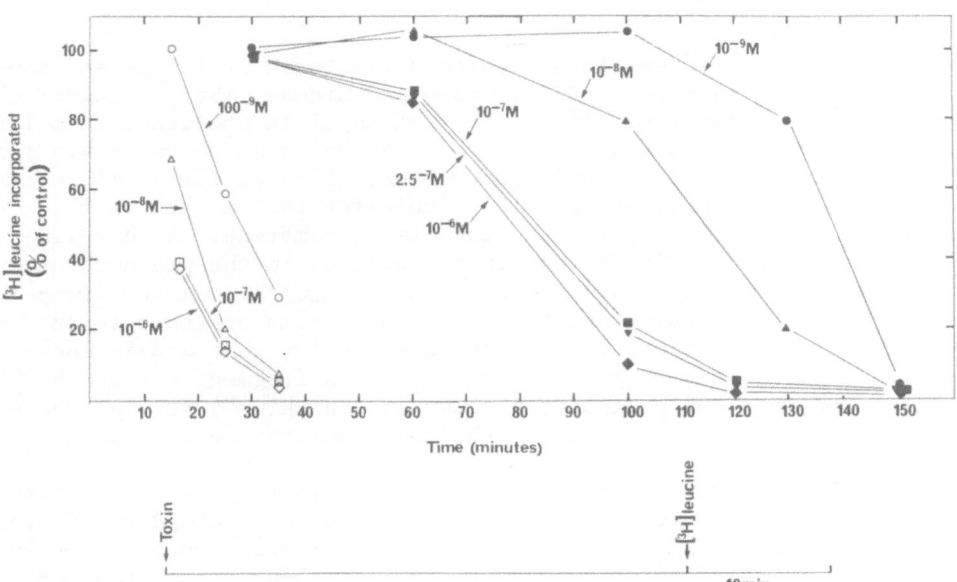

Fig. 2. Rate of protein synthesis inhibition after addition of diphtheria toxin (open symbols) and modeccin (filled symbols) to cells. The indicated concentrations of toxins were added and the cells were incubated for the period of time given on the abscissa. Finally, the rate of protein synthesis during a 10 min interval was measured.

ENTRY

The mechanism whereby the enzymatically active part enters into the cytosol is best understood in the case of diphtheria toxin. In this case low pH is required for toxin entry. It was first observed by Kim and Groman (1965) that NH$_4$Cl protects cells against diphtheria toxin although it does not inhibit toxin binding. Chloroquine and several amines were also found to protect in a similar way (Sandvig et al., 1979; Leppla et al., 1980). Common to NH$_3$, chloroquine and related amines is the fact that they are able to pass membranes in their uncharged form. When they enter vesicles having low pH such as lysosomes and receptosomes, they become protonated. As a result, the pH in the vesicles increases.

Also a variety of other compounds that are able to dissipate pH gradients across membranes, strongly protect against diphtheria toxin (Sandvig and Olsnes, 1982). These include the protonophores FCCP and CCCP as well as the ionophores Br-X537 A, nigericin and monensin (Marnell et al., 1982) which exchange protons for monovalent cations, and compounds like DCCD and tributyltin which inhibit the proton pump of the vesicles (Sandvig and Olsnes, manuscript in preparation).

Cells are most sensitive to diphtheria toxin at low pH and they are almost insensitive at pH 9 (Duncan and Groman, 1969; Middlebrook et al., 1978). The protective effect of NH$_4$Cl is abolished if cells with preadsorbed diphtheria toxin are incubated at pH below 4.6 for a few seconds at 37°C (Sandvig and Olsnes, 1981) or for 30 min at 4°C (Draper and Simon, 1980). This indicates that at low pH the toxin may enter directly through the plasma membrane. Also toxin that has accumulated in intracellular vesicles in the presence of NH$_4$Cl is able to intoxicate the cells once NH$_4$Cl is removed (Draper and Simon, 1980; Sandvig and Olsnes, 1980). With unnicked toxin very little toxic effect is obtained even at low pH (Sandvig and Olsnes, 1981). This suggests that only the A fragment enters the cytosol and that the proteolytic cleavage ("nicking") between the A and B fragment must occur before the toxin enters the cell.

The hydrophobic domain of diphtheria toxin fragment B seems to be important for the entry. This region is exposed at pH below 4.5, but not at neutral pH (Boquet et al., 1976; Sandvig and Olsnes, 1981). When toxin in solution is exposed to pH 4.5, about 90% of the toxic activity is lost. However, if the toxin has first been bound to receptors on the cell surface, the toxic activity is not reduced by low pH. This suggests that diphtheria toxin first binds to cell surface receptors, then it is transferred to a compartment with low pH where the hydrophobic region on the B-chain is exposed. This region may then become inserted into the membrane where it possibly forms a hydrophilic channel through which the A chain can pass.

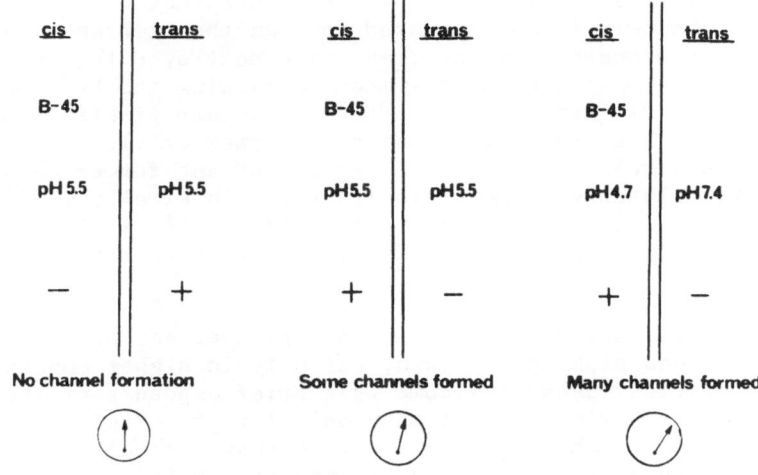

Fig. 3.   Ability of B-45, a fragment of diphtheria
          toxin containing the hydrophobic region, to
          form ion permeable channels in lipid bilayers.
          Channels were formed when a potential was
          applied across the membrane with the positive
          pole at the same side of the membrane as the
          B-45 fragment. A pH-gradient (acidic at the
          cis-side) increased the number of channels
          formed.

Experiments with planar lipid bilayer membrances have shown
that under appropriate conditions diphtheria toxin can indeed insert
itself into the membrane and form channels which are permeable at
least to ions. Such channels were formed when a positive potential
was applied across the membrane and when the pH was low on the cis
side, i.e. the same side as the toxin (Fig. 3). When the incomplete
diphtheria toxin molecule, crm 45, which lacks the C-terminal 17,000
daltons, was used, channels were formed when the pH on the cis side
was 5.5 or lower. Similar results were obtained with B-45, the part
of the B fragment present in crm 45 (Kagan et al., 1981). When the
whole diphtheria toxin was used, a low pH (4.5) was required for
channel formation (Donovan et al., 1981). The channels opened only
when the potential was cis positive and there was evidence of opening and closure of single channels. Channel formation required
that the phospholipids in the membrane were negatively charged
(Kagan et al., 1981). The presence of phosphatidyl inositol phos-

phate in the membrane strongly increased the extent of channel formation. The channels formed appeared to span the membrane since
treatment with pronase from the <u>trans</u> side destroyed the channel
activity. A cyanogen bromide fragment containing the hydrophobic
domain in the B fragment was also found to insert itself into membranes and form ion-conductive channels (Kayser et al., 1981).
B-45 inserted itself into liposomes at low pH and formed channels
permeable to solutes with MW up to 1,500 (Kagan et al., 1981). Such
a permeability requires a diameter of at least 18Å, which is just
enough to allow diphtheria toxin fragment A to penetrate in its
extended form.

   NH$_4$Cl, chloroquine and other amines protect against several
toxins other than diphtheria toxin, but only in higher concentrations
and the protection is not overcome by a brief exposure of the cells
to low pH. Since amines affect many cellular processes, it is conceivable that their ability to protect against some of the toxins
may be unrelated to their ability to increase pH in acidic vesicles.
In the case of <u>Pseudomonas aeruginosa</u> exotoxin A, the effect of
amines differs in different cell types (Mekada et al., 1979;
FitzGerald et al., 1980).

   Interestingly, NH$_4$Cl is more efficient in protecting the very
sensitive L cells and mouse 3T3 cells than the far less sensitive
BHK, HeLa S$_3$ and Vero cells. Treatment of the BHK cells with
trifluoperazine or dansylcadaverine strongly sensitized them to
Pseudomonas toxin. This effect was easily blocked by NH$_4$Cl. This
indicates that the two drugs direct the toxin to a compartment where
the toxin readily enters the cell by a mechanism which can be inhibited by NH$_4$Cl.

   NH$_4$Cl and chloroquine also protect against modeccin (Sandvig et
al., 1979). Also the local anesthetic procaine which likewise is an
amine, did protect (Fig. 4). It is possible that modeccin requires
processing in the lysosomes before entry into the cytosol and that
this processing is inhibited by NH$_4$Cl and chloroquine. A modeccin-
resistant HeLa cell variant tolerating 10^4 times more modeccin than
the parent cells, was not further protected against modeccin by NH$_4$Cl.
This indicates that modeccin can be internalized by two mechanisms,
one efficient and NH$_4$Cl-sensitive, and another NH$_4$Cl-resistant and
inefficient. If so, only the latter one is preserved in the modeccin-
resistant variant. The protonophores FCCP and CCCP, the carboxylic
ionophores monensin, Br-X-537A and nigericin, as well as DCCD and
tributyltin protected strongly against modeccin, supporting the view
that low pH is required for entry (Sandvig and Olsnes, 1982).

   Surprisingly, NH$_4$Cl, chloroquine and methylamine sensitized
cells to abrin and ricin and to a hybrid of <u>Wistaria floribunda</u>
lectin and diphtheria toxin fragment A (Mekada et al., 1981; Sandvig
et al., 1979; Ray and Wu, 1981). This could be related to the fact

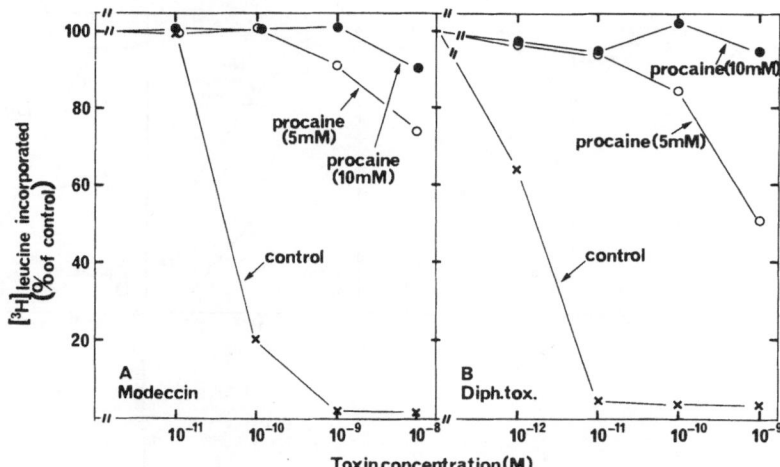

Fig. 4. Ability of procaine to protect cells against
         modeccin and diphtheria toxin. Cells were
         incubated with procaine and toxin as indicated
         for 4 h and then the rate of protein synthesis
         during a 10 min interval was measured.

that these toxins are most active at pH above neutrality. When the
cell culture medium was adjusted to pH 6 or lower, Vero cells were
completely protected against abrin, modeccin and ricin, although
toxin binding and endocytosis were not much reduced. With increasing
pH up to 8.5, the sensitivity to abrin and ricin increased, while
modeccin exerted its maximal effect between pH 7 and pH 8 (Sandvig
and Olsnes, 1982).

The toxins differ in their ion requirement for entry (Fig. 5).
Thus the presence of Ca^{2+} in the medium is required for the entry
of abrin, modeccin, viscumin and Pseudomonas toxin, whereas Cl^- is
required for the entry of diphtheria toxin and modeccin (Sandvig and
Olsnes, 1982). In both cases not only the presence of the ions is
required, but apparently the ions must be able to enter the cells.
So far it is not known which roles the ions play in toxin entry.

It is interesting that Cl^- deprivation of treatment of cells
with compounds that interfere with Cl^- entry, strongly reduced the
ability of the cells to bind diphtheria toxin. Possibly the anion
channel is directly involved in toxin binding. From studies with
turtle bladder, Brodsky et al., (1979) suggested that binding of
Pseudomonas aeruginosa exotoxin A and diphtheria toxin to the cell
membrane may interfere with anion transport in the cells.

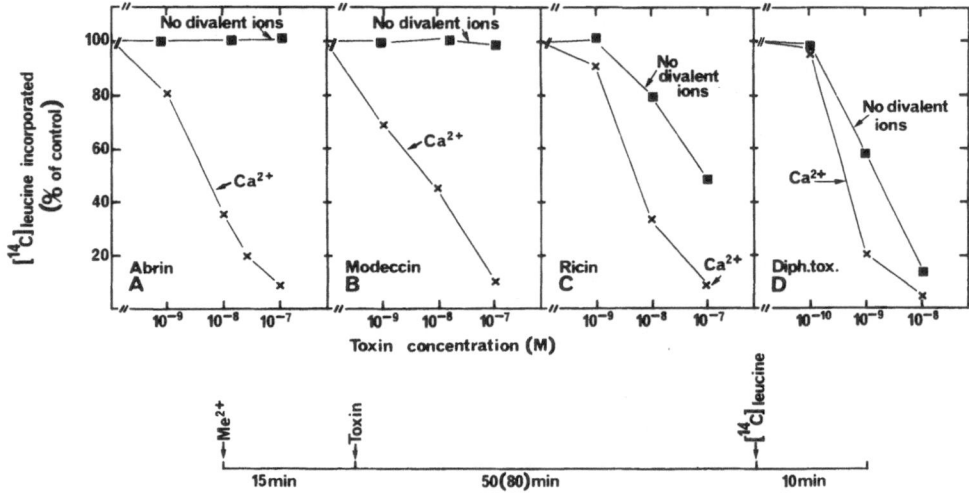

Fig. 5. Ca^{2+} requirement for toxin entry. Vero cells
were incubated in isotonic buffer with and
without 2 mM CaCl$_2$ and with the toxin concentrations indicated. After 50 min (abrin,
ricin, diphtheria toxin) or 80 min (modeccin),
the rate of protein synthesis during a 10 min
interval was measured.

INACTIVATION OF INTRACELLULAR TARGETS

The A chains of the four plant toxins inactivate the 60S ribosomal subunits enzymatically in a still unknown way (Olsnes and Pihl,
1982). As a result, elongation factor 2 (EF2) is unable to bind to
the subunit and protein synthesis stops. Each A chain molecule of
abrin and ricin inactivates pure ribosomes at a rate of 1,500 ribosomes per min. The K$_m$ with respect to ribosomes is about 2 x 10^{-7}M,
which ensures that abrin and ricin A chains act in the cytosol at
close to their maximal rates (V$_{max}$). These results account for the
finding that abrin, ricin and modeccin seem to kill cells even if
only a single molecule enters the cytosol (Eiklid et al., 1980).

The different toxins apparently differ in their precise mechanism of action. Thus, EF2, particularly in the presence of GTP,
protects ribosomes against inactivation by abrin and ricin, whereas
EF2 (in the absence of GTP) sensitizes the ribosomes to modeccin,
possibly by slightly changing the conformation of the ribosomes
(Olsnes and Abraham, 1979). Even when EF2 is present, modeccin acts

more slowly in cell-free systems than abrin and ricin A chains. Since modeccin is as toxic to cells as abrin and ricin, it is possible that modeccin must somehow be activated, e.g. by proteolytic cleavage, to achieve its maximal enzymatic activity. Such cleavage could occur in the cell, before the A chain enters the cytosol. Also Shigella toxin inactivates the 60S ribosomal subunits (Reisbig et al., 1981), but it is not known if the enzymatic modification is the same as in the case of the plant toxins. One A_1 fragment molecule can inactivate at least 40 ribosomes per minute.

The intracellular mechanism of action of diphtheria toxin and Pseudomonas aeruginosa exotoxin A appears to be identical. It consists in enzymatic ADP-ribosylation and consequent inactivation of EF2 (Van Ness et al., 1980). EF2 is required for the translocation of the growing polypeptide chain from the A-site back to the P-site on the ribosome after the peptide bond has been formed. Therefore, when EF2 is inactivated, protein synthesis stops.

The amino acid which is ADP-ribosylated, diphthamide, is an unusual amino acid, which has only been found in EF2. It is probably produced by post-translational modification of a histamine residue. At least three enzymes appear to be involved in the post-translational synthesis of diphthamide (Van Ness et al., 1980).

Prokaryotic and mitochondrial protein synthesis are not inhibited by diphtheria toxin because their elongation factors cannot be ADP-ribosylated (Pappenheimer, 1977). On the other hand, Archebacteria which are considered to be intermediary between prokaryotes and eukaryotes, contain EF2 which is ADP-ribosylated by diphtheria toxin (Kessel and Klink, 1980).

Moynihan and Pappenheimer (1981) have found that diphtheria toxin may ADP-ribosylate 2,000 EF2 molecules per minute. Since in rapidly growing cells about 3,000 new EF2 molecules are synthesized per minute, one A fragment may prevent proliferation. In fact, it has been shown that one molecule of diphtheria toxin can kill a cell (Yamaizumi et al., 1978). The K_m of fragment A for NAD is 1.4 x 10^{-6}M and for EF2 1.5 x 10^{-7}M. This indicates that at the concentrations of EF2 and NAD found in the cytosol, the A fragment acts at close to its maximal rate (V_{max}).

CONCLUSION

Different toxins with intracellular sites of action enter cells in a similar, but not identical manner. These toxins first bind to receptors at the cell surface. In all cases the next step appears to consist of endocytic uptake of the bound toxin. Diphtheria toxin rapidly enters the cytosol from the endosomes when these have aquired a low pH. The effect of the acidic conditions is to expose a hydrophobic region in diphtheria toxin B fragment which then inserts itself into the membrane and probably assists in the transfer of the A fragment. Also modeccin and Pseudomonas toxin appear to require low pH for entry, but we have so far no indication as to the effect of the acidic conditions on these toxins and the nature of the vesicles from which they enter. Possibly these toxins require some kind of modification before they are able to cross the lipid bilayer. Abrin, ricin and viscumin do not require low pH for entry. These toxins enter most rapidly when the medium is alkaline and the possibility should be considered that they cross the lipid bilayer from vesicles having pH above neutrality. It is likely that the activity of chimeric toxins will depend upon the extent to which the new ligand is able to direct the conjugate to the appropriate vesicular compartment.

REFERENCES

Alving, C.R., Iglewski, B.H., Urban, K.A., Moss, J., Richards, R.L. and Sadoff, J.C. 1980, Binding of diphtheria toxin to phospholipids in liposomes, Proc. Natl. Acad. Sci. USA, 77: 1986.

Boquet, P. and Pappenheimer, A.M. Jr. 1976, Interaction of diphtheria toxin with mammalian cell membranes, J. Biol. Chem., 251: 5770.

Boquet, P., Silverman, M.S., Pappenheimer, A.M., Jr. and Vernon, W.B. 1976, Binding of Triton X-100 to diphtheria toxin, crossreacting material 45, and their fragments, Proc. Natl. Acad. Sci. USA, 73:4449.

Brodsky, W.A., Sadoff, J.C., Durham, J.H., Ehrenspeck, G., Schachner, M. and Iglewski, B.H. 1979, Effects of Pseudomonas toxin A, diphtheria toxin, and cholera toxin on electrical characteristics of turtle bladder, Proc. Natl. Acad. Sci. USA, 76: 3562.

Chang, T.M. and Neville, D.M., Jr. 1978, Demonstration of diphtheria toxin receptors on surface membranes from both toxin-sensitive and toxin resistant species, J. Biol. Chem., 253: 6866.

Chung, D.W. and Collier, R.J. 1977, Infect. Immun., 16: 832.

Donovan, J.J., Simon, M.I., Draper, R.K. and Montal, M. 1981, Diphtheria toxin forms transmembrane channels in planar lipid bilayers, Proc. Natl. Acad. Sci. USA, 78: 172.

Dorland, R.B., Middlebrook, J.L. and Leppla, S.H. 1981, Effect of ammonium chloride on receptor-mediated uptake of diphtheria toxin by Vero cells, Exp. Cell Res., 134: 319.

Draper, R.K. and Simon, M.I. 1980, The entry of diphtheria toxin into the mammalian cell cytoplasm: Evidence for lysosomal involvement, J. Cell Biol., 87: 849.

Duncan, J.L. and Groman, N.B. 1969, Activity of diphtheria toxin. II. Early events in the intoxication of HeLa cells, J. Bacteriol., 98: 963.

Eiklid, K., Olsnes, S. and Pihl, A. 1980, Entry of lethal doses of abrin, ricin and modeccin into the cytosol of HeLa cells, Exp. Cell Res., 126:321.

Eiklid, K. and Olsnes, S. 1980, Interaction of Shigella Shigae cytotoxin with receptors on sensitive and insensitive cells, J. Receptor Res., 1: 199.

FitzGerald, D.,Morris, R.E. and Saelinger, C.B. 1980, Receptor-mediated internalization of Pseudomonas toxin by mouse fibro-blasts, Cell 21: 867.

Funatsu, G., Yoshitake, S. and Funatsu, M. 1978, Primary structure of Ile chain of ricin D, Agric. Biol. Chem. 42: 501.

Funatsu, G., Kimura, M. and Funatsu, M. 1979, Primary structure of Ala chain of ricin D, Agric. Bio. Chem. 43: 2221.

Gonatas, N.K., Stieber, A., Kim, S.U., Graham, D.I. and Avrameas, S. 1975, Internalization of neuronal plasma membrane ricin receptors into the Golgi apparatus, Exp. Cell Res., 94: 426.

Gonatas, N.K., Kim, S.U., Stieber, A. and Avrameas, S. 1977, Internalization of lectins in neuronal GERL, J. Cell. Biol., 73: 1.

Kagan, B.L., Finkelstein, A. and Colombini, M. 1981, Diphtheria toxin fragment forms large pores in phospholipid bilayⁿr membranes, Proc. Natl. Acad. Sci. USA, 78: 4950.

Kayser, G., Lambotte, P., Falmagne, P., Capiau, C., Zanen, J. and Ruysschaert, J.M. 1981, A CNBR peptide located in the middle region of diphtheria toxin fragment B includes conductance change in lipid bilayers, Biochem. Biophys. Res. Commun. 99: 358.

Kessel, M. and Klink, F. 1980, Archebacterial elongation factor is ADP-ribosylated by diphtheria toxin, Nature, 287: 250.

Kim, K. and Groman, N.B. 1965, Mode of inhibition of diphtheria toxin by ammonium chloride, J. Bacteriol., 90: 1557.

Lambotte, P., Falmagne, P., Capiau, C., Zanen, J., Ruysschaert, J.M. and Dirkx, J. 1980, Primary structure of diphtheria toxin fragment B: Structural similarities with lipid binding domains, J. Cell Biol., 87: 837.

Leppla, S.H., Martin, O.C. and Muehl, L.A. 1978, The exotoxin of P.aeruginosa: A proenzyme having an unusual mode of activation, Biochem. Biophys. Res. Commun., 81: 532.

Leppla, S.H., Dorland, R.B. and Middlebrook J.L. 1980, Inhibition of diphtheria toxin degradation and cytotoxic action by chloroquine, J. Biol. Chem., 255: 2247.

Lory, S. and Collier, R.J. 1980, Diphtheria toxin: Nucleotide binding and toxin heterogeneity, Proc. Natl. Acad. Sci. USA, 77: 267.

Lory, S., Carroll, S.F. and Collier, R.J. 1980a, Ligand interactions of diphtheria toxin, II. Relationships between the NAD site and the P-site, J. Biol. Chem., 255: 12016.

Lory, S., Carroll, S.F., Bernard, P.D. and Collier, R.J. 1980b, Ligand interations of diphtheria toxin. I. Binding and hydro-lysis of NAD, J. Biol. Chem., 255: 12011.

Marnell, M.H., Stookey, M., and Draper, R.K. 1982, Monensin blocks the transport of diphtheria toxin to the cell cytoplasm, J. Cell Biol. 93: 57.

Mekada, E., Uchida, T. and Okada, Y. 1979, Modification of the cell
     surface with neuraminidase increaes the sensitivities of cells
     to diphtheria toxin and Pseudomonas aeruginosa exotoxin,
     Exp. Cell Res., 123: 137.
Mekada, E., Uchida, T. and Okada, Y. 1981, Methylamine stimulates
     the action of ricin toxin but inhibits that of diphtheria toxin,
     J. Biol. Chem., 256: 1225.
Middlebrook, J.L., Dorland, R.B. and Leppla, S.H. 1978, Association
     of diphtheria toxin with Vero cells. Demonstration of a
     receptor, J. Biol. Chem., 253: 7325.
Middlebrook, J.L. 1981, Effect of energy inhibitors on cell surface
     diphtheria toxin receptor numbers, J. Biol. Chem., 256: 7898.
Moehring, T.J. and Crispell, J.P. 1974, Biochem. Biophys. Res.
     Commun., 60: 1446.
Moynihan, M.R. and Pappenheimer, A.M., Jr. 1981, Kinetics of
     adenosinediphosphoribosylation of elongation factor 2 in
     cells exposed to diphtheria toxin, Infect. Immun., 32: 575.
Nicolson, G.L. 1974, Ultrastructural analysis of toxin binding and
     entry into mammalian cells, Nature, 251: 628.
Nicolson, G.L., Lacorbiere, M. and Ekhart, W. 1975a, Qualitative
     and quantitative interactions of lectins with untreated and
     neuraminidase-treated normal, wild-type and temperature-
     sensitive polyoma-transformed fibroblasts, Biochemistry,
     14: 172.
Nicolson, G.L., Lacorbiere, M. and Hunter, T.R. 1975b, Mechanism
     of cell entry and toxicity of an affinity-purified lectin
     from Ricinus communis and its differential effects on normal
     and virus-transformed fibroblasts, Cancer Res., 35: 144.
Olsnes, S. and Abraham, A.K. 1979, Elongation-factor-2-induced
     sensitization of ribosomes to modeccin. Evidence for specific
     binding of elongation factor 2 to ribosomes in the absence of
     nucleotides, Eur. J. Biochem., 93: 447.
Olsnes, S. and Pihl, A. 1982, in "The molecular action of toxins
     and viruses", P. Cohen and S. van Heyningen, eds. Elsevier/
     North Holland, Amsterdam.
Olsnes, S., Reisbig, R. and Eiklid, K. 1981, Subunit structure of
     Shigella cytotoxin, J. Biol. Chem., 256: 8732.
Olsnes, S., Sandvig, K., Eiklid, K. and Pihl, A. 1978, Properties
     and action mechanism of the toxic lectin modeccin: Interaction
     with cell lines resistant to modeccin, abrin and ricin,
     J. Supramol. Struct., 9: 15.
Olsnes, S., Sandvig, K., Refsnes, K. and Pihl, A. 1976, Rates of
     different steps involved in the inhibition of protein
     synthesis by the toxic lectins abrin and ricin, J. Biol. Chem.,
     257: 3985.
Pappenheimer, A.M., Jr. 1977, Diphtheria toxin, Ann. Rev. Biochem.,
     46: 69.
Proia, R.L., Eidels, L. and Hart, D.A. 1981, Diphtheria toxin:
     Receptor Interactions. Characterization of the receptor
     interaction with the nucleotide-free toxin, the nucleotide-

bound toxin, and the B fragment of the toxin, J. Biol. Chem., 256: 4991.

Proia, R.L., Hart, D.A. and Eidels, L. 1979, Interaction of diphtheria toxin with phosphorylated molecules, Infect. Immun., 26: 942.

Proia, R.L.,Wray, S.K., Hart, D.A., and Eidels, L. 1980, Character-ization and affinity labelling of the cationic phosphate-binding (nucleotide binding) peptide located in the receptor-binding region of the B fragment of diphtheria toxin, J. Biol. Chem., 255: 12025.

Ray, B. and Wu, H.C. 1981, Enhancement of cytotoxicities of ricin and Pseudomonas toxin in Chinese hamster ovary cells by nigericin, Mol. Cell. Biol., 1: 552.

Reisbig, R., Olsnes, S., and Eiklid, K. 1981, The cytotoxic activity of Shigella toxin. Evidence for catalytic inactivation of the 60S ribosomal subunit, J. Biol. Chem., 256: 8739.

Rosen, S.W. and Hughes, R.C. 1977, Effects of neuraminidase on lectin binding by wild-type and ricin-resistant strains of hamster fibroblasts, Biochemistry, 16: 4908.

Sanai, Y., Morihara, K., Tsuzuki, H., Homma, J.Y., and Kato, I. 1980, Proteolytic cleavage of exotoxin A from Pseudomonas aeruginosa. Formation of an ADP ribosyltransferase active fragment by the action of Pseudomonas elastase, FEBS letters, 120:131.

Sandvig, K., Olsnes, S. and Pihl, A. 1976, Kinetics of the binding of the toxic lectins abrin and ricin to surface receptors of human cells, J. Biol. Chem., 251: 3977.

Sandvig, K., Olsnes, S. and Pihl, A. 1978, Binding, uptake and degradation of the toxic proteins abrin and ricin by toxin-resistant cell variants, Eur. J. Biochem., 82: 13.

Sandvig, K.,Olsnes, S. 1979, Effect of temperature on the uptake, excretion and degradation of abrin and ricin by HeLa cells, Exp. Cell Res., 121: 15.

Sandvig, K., Olsnes, S. and Pihl, A. 1979, Inhibitory effect of ammonium chloride and chloroquine on the entry of the toxic lectin modeccin into HeLa cells, Biochem. Biophys. Res.Commun., 90:648.

Sandvig, K. and Olsnes, S. 1981, Rapid entry of nicked diphtheria toxin into cells at low pH. Characterization of the entry process and effects of low pH on the toxin molecule, J. Biol. Chem., 256: 9068.

Sandvig, K. and Olsnes, S. 1982, Entry of the toxic proteins abrin, modeccin, ricin, and diphtheria toxin into cells. II. Effect of pH, metabolic inhibitors, and ionophores and evidence for toxin penetration from endocytic vesicles, J. Biol. Chem., 257: 7504.

Van Ness, B.G., Howard, J.B. and Bodley, J.W. 1980, ADP-ribosy-lation of elongation factor 2 by diphtheria toxin. NMR spectra and proposed structures of ribosyl-diphthamide and its hydro-lysis products, J. Biol. Chem., 255: 10710.

Vasil, M.L., Kabat, D. and Iglewski, B.H. 1977, Structure-activity
    relationships of an exotoxin of Pseudomonas aeruginosa,
    Infect. Immun., 16: 353.
Villafranca, J.E. and Robertus, J.D. 1981, J. Biol. Chem., 256: 554.
Yamaizumi, M., Mekada, E., Uchida, T. and Okada, Y. 1978, One mole-
    cule of diphtheria toxin fragment A introduced into a cell can
    kill the cell, Cell, 15: 245.
Youle, R.J. and Neville, D.M., Jr. 1979, Receptor-mediated transport
    of the hybrid protein ricin-diphtheria toxin fragment A with
    subsequent ADP-ribosylation of intracellular elongation factor
    II, J. Biol. Chem., 254: 11089.

ROLE OF THE B-CHAIN IN THE CYTOTOXIC ACTION OF

ANTIBODY-RICIN AND ANTIBODY-ABRIN CONJUGATES

Deirdre McIntosh and Philip Thorpe

Institute of Cancer Research
Chester Beatty Laboratories
Fulham Road, London SW3 6JB, U.K. and
Imperial Cancer Research Fund
Lincoln's Inn Fields
London WC2A 3PX, U.K.

INTRODUCTION

Attempts have been made in several laboratories to construct cell type-specific cytotoxic agents by linking antibody molecules to the highly potent toxins, abrin and ricin (for a review of the mode of action of abrin and ricin, see chapter by S. Olsnes, this volume). Two main strategies have been adopted. The first is to link the holotoxin directly to the antibody. Conjugates of this type always appear to exert a powerful cytotoxic effect upon cells with the appropriate antigens but suffer from a lack of complete specificity because they can also bind to non-target cells by the galactose-binding sites on the toxin B-chain (reviewed by Thorpe et al., 1982a). The other approach is to link the antibody by a disulfide bond to the isolated toxin A-chain. Conjugates of this second type, although free from the problem of non-specific binding, show great variability in cytotoxic potency, some being as effective as the native toxin and others being weakly or non-cytotoxic (reviewed by Olsnes and Pihl, 1982a; Martinez et al., 1982; Thorpe et al., 1982a). The consistent and generally superior cytotoxic performance of the intact toxin conjugates appears to be attributable to an ability of the B-chain to facilitate delivery of the A-chain moiety to the cytosol.

In this chapter we describe ways of maintaining the high specificity of cytotoxic effect of the A-chain conjugates whilst preserving the entry-promotion function of the B-chain. This can be achieved by the separate addition of ricin B-chain to cells

105

coated with an antibody-ricin A-chain conjugate or by blocking the galactose-binding sites in intact toxin conjugates. Lastly we consider the possible mechanisms by which the B-chain could assist A-chain entry.

SYNERGISM BETWEEN RICIN B-CHAIN
AND ANTIBODY-RICIN A-CHAIN CONJUGATES

Recently, McIntosh and her colleagues (1983) reported that free ricin B-chain markedly enhanced the cytotoxic action of ricin A-chain conjugates to a monoclonal antibody, 11/160, against rat fibrosarcoma cells. The conjugate was prepared using the SPDP reagent according to the method described by Thorpe and Ross (1982). The A-chain released from the conjugate by reduction with dithiothreitol was fully able to inhibit protein synthesis in rabbit reticulocyte lysates and the capacity of the antibody moiety to bind to a benzepyrene-induced fibrosarcoma cell line, HSNtc, was demonstrated by indirect radioimmunoassay to be unaltered by the conjugation procedure.

The conjugate used alone was entirely devoid of cytotoxic effect upon HSNtc cells in tissue culture. No reduction in the capacity of the cells to incorporate (^{3}H)leucine into cellular protein was observed even when the conjugate was applied for 24 h at 10^{-7}M, a concentration in excess of that needed for saturation of the cell surface antigens (Fig. 1a).

It had been reported by Youle and Neville (1982) that the cytotoxic action of a highly effective A-chain conjugate, anti-Thyl.1-ricin A, was accelerated and enhanced in the presence of free B-chain. This suggested that the addition of free B-chain to fibrosarcoma cells coated with 11/160-ricin A might likewise facilitate the expression of toxicity.

Fig. 1b shows the potent cytotoxic effect obtained when HSNtc cells were incubated first with 11/160-ricin A and then for 1 h with ricin B-chain at 10^{-7}M. The concentration of the A-chain conjugate now needed to reduce the (^{3}H)leucine incorporation of the cells by 50% (i.e. the ID_{50}) was 3.2×10^{-12}M, representing more than a 100,000-fold potentiation of toxicity. The cytotoxic effect was specific since exposure of the cells to a conjugate made from a monoclonal antibody (M10/76) of irrelevant specificity followed by B-chain was without effect. Even greater enhancement in toxicity ($ID_{50} = 10^{-13}$M) was obtained when HSNtc cells were pre-incubated with ricin B-chain for 1 h prior to adding the 11/160-ricin A conjugate (Table 1).

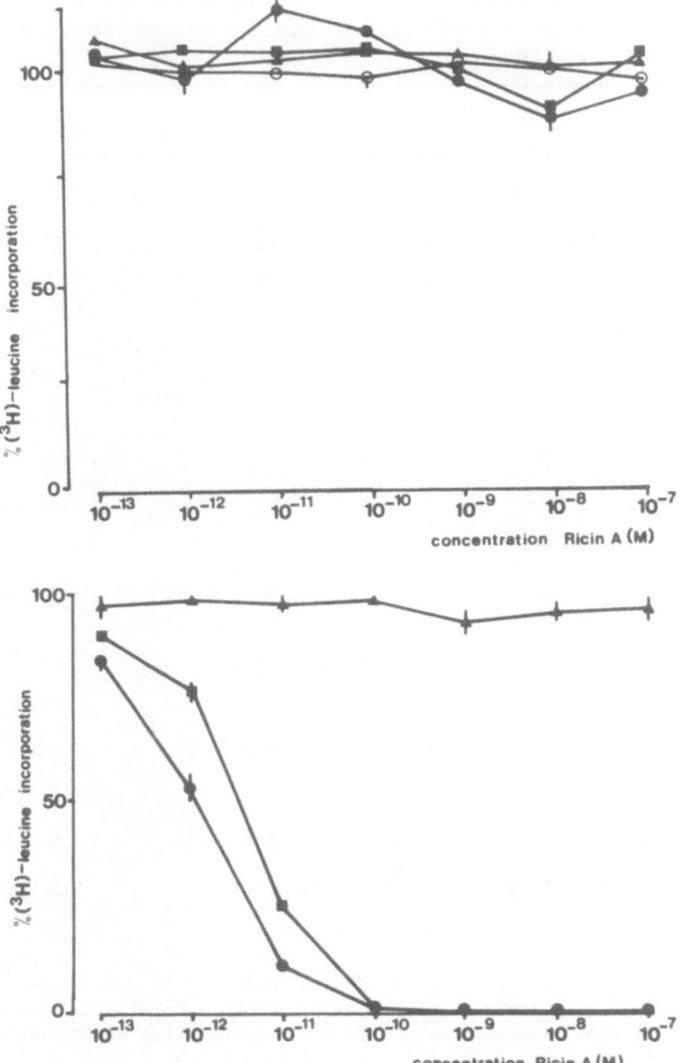

Fig. 1. Potentiation of toxicity of 11/160-ricin A to HSNtc
fibrosarcoma cells by adding free ricin B-chain. In
(a) the cells were incubated for 1 h with 11/160-ricin
(■), ricin A (●), 11/160 (▲) or with 11/160 plus
ricin A (○) before washing and setting up in tissue
culture. In (b) the cells were treated with 11/160-
ricin A (■), M10/76-ricin A (▲) for 1 h followed by
a 1 h incubation with ricin B-chain at 10^{-7}M. (●)
Treatment with ricin for 1 h. The capacity of the cells
to incorporate (^{3}H)leucine was measured 27 h later.

Table 1. The effects of ricin, its isolated A and B
          chains and ricin A chain conjugates on HSNtc
          cells

Treatment 1	Treatment 2	ID_{50}(M ricin A)
Ricin, 1 h	-	1.1×10^{-12}
Ricin + lactose, 1 h	-	2.4×10^{-10}
Ricin, 1 h	6 x lactose wash	1.1×10^{-11}
Ricin B, 1 h	-	10^{-7}
Ricin A, 1 h	-	10^{-7}
Ricin A, 24 h	-	5×10^{-8}
11/160-ricin-A, 1 h	-	no effect at 10^{-7}
11/160-ricin-A, 24 h	-	no effect at 10^{-7}
11/160-ricin-A, 1 h	ricin B chain, 1 h	3.2×10^{-12}
11/160-ricin-A, 1 h	ricin B chain + lactose, 1 h	5.3×10^{-12}
M10/76-ricin A	ricin B chain, 1 h	10^{-7}
Ricin A, 1 h	ricin B, 1 h	10^{-7}
Ricin A, 1 h	ricin B + lactose, 1 h	10^{-7}
Ricin B, 1 h	ricin A, 1 h	2.6×10^{-11}
Ricin B, 1 h	ricin A + lactose, 1 h	1.5×10^{-9}
Ricin B + lactose, 1 h	ricin A chain	4.5×10^{-9}
Ricin B	M10/76-ricin-A, 1 h	2.5×10^{-10}
Ricin B	11/160-ricin-A, 1 h	10^{-13}

ID_{50} is the dose required to inhibit the incorporation of (3H)leucine
into protein by 50%. HSNtc cells were plated out at 2×10^5 ml^{-1} for
24 h prior to a 1 h incubation at 37°C with materials listed in Treatment
1. Cells were then washed three times and incubated for 24 h in DMEM and
pulsed with (3H)leucine as in Fig. 1. All determinations were made in
triplicate.

     The mechanism whereby 11/160-ricin A was converted into a
potent cytotoxic agent by the addition of B-chain was investigated.
It was proposed that the B-chain either exerted its effect independently by binding to the cell surface and in some way signalling the
cell to internalize the conjugate or by spontaneously reassociating
with the A-chain moiety of the conjugate to form a pseudo-holotoxin

conjugate. To distinguish between these possibilities, an experiment
was performed in which the B-chain was applied to conjugate-coated
cells in the presence of 100 mM lactose to inhibit the binding of
the B-chain to galactose receptors on the cell surface. As Table 1
shows, the inclusion of lactose did not significantly diminish the
potentiation of toxicity afforded by the B-chain. The same concentration of lactose reduced the toxicity of native ricin 220-fold,
confirming that these conditions provide an effective blockade of
B-chain binding to cell surface receptors. It was concluded that
reannealing of B-chain with the A-chain component of the conjugate
at the cell surface was the likely mechanism of conjugate activation.

There is evidence that the affinity of free ricin A and B-
chains for each other is high. Thus, Houston (1982) reported that
at low concentration purified ricin A and B-chains efficiently reassociate to form a molecule as toxic as native ricin to leukemia
cells and that it makes no difference whether B-chain is added to
the cells first, followed a short time later by A-chain, or whether
the two chains are premixed and added. Similar results are presented
in Table 1.

It is unclear whether antibody-ricin A associated non-covalently
with ricin B-chain is itself cytotoxic or whether the SH group on
the ricin B-chain eventually reforms a disulfide bond with the A-
chain by thiol-disulfide exchange. Lappi et al. (1978) have claimed
that the interchain disulfide bond in ricin has to be intact for
toxicity which casts doubt on the effectiveness of the non-covalent
complex. If the complex formed between antibody-A-chain and B-chain
does break down to release ricin, this must occur inside the cell
because lactose would otherwise have inhibited toxicity.

Vitetta et al. (1983) have reported that an anti-Ig-ricin A
conjugate and an anti-Ig-ricin B conjugate act synergistically to
kill immunoglobulin-bearing target cells. Since the SH-group of
the B-chain had been used to form the conjugate, the likelihood of
spontaneous thiol-disulfide interchange in the complex to reform
native ricin is remote: the complex is therefore either itself the
toxic entity or cellular reduction systems come into play to split
the conjugates and allow the free A and B-chains to reunite.

The synergistic action of free or conjugated B-chain with antibody-A chain offers exciting possibilities for therapy. It has
immediate application in bone marrow transplantation procedures in
which conjugates are used in-vitro to destroy malignant cells in
marrow autografts or T-lymphocytes in marrow allografts in the
treatment of leukemia and other diseases. In the longer term, the
in-vitro administration of two reagents which individually are non-
cytotoxic but which together culminate in a specific cytotoxic
effect has implications for the treatment of tumors and their metastatic deposits.

BLOCKADE OF THE GALACTOSE-BINDING SITES IN INTACT TOXIN CONJUGATES

Competitive Antagonism with Galactose or Lactose

An easy and very effective way to abrogate the non-specific toxicity of antibody-ricin and antibody-abrin conjugates to cells in-vitro is to inhibit the galactose-binding properties of the toxin moiety with excess galactose or lactose. The specific component of toxicity is antagonized little, if at all, by these sugars. This was first demonstrated for ricin conjugates by Youle and Neville (1980) and for abrin conjugates by Thorpe et al. (1981). It has been put to practical use in preventing ricin conjugates from binding to and killing hematopoietic progenitors in rodent (Thorpe et al., 1982b; Mason et al., 1982) bone marrow grafts when purging the grafts of leukemic cells or T-lymphocytes.

Chemical Modification of Amino Acids responsible for the Galactose-

Recognition Properties of the Toxin

Experiments in which conjugates have been prepared from toxins that have been modified chemically to reduce their powers of galactose recognition have produced contradictory results. Youle et al. (1981), in agreement with an earlier study by Sandvig et al. (1978), found that the O-acetylation of one or two tyrosine residues in ricin weakened its binding to Sepharose (a β-galactosyl matrix) and reduced its toxicity to fibroblasts 10-fold. Similar treatment of a conjugate of ricin and the fibroblast-binding molecule, mannose-6-phosphate, produced the same 10-fold drop in specific toxicity and lead the authors to propose that the ricin B-chain portion of the conjugate needs to interact with a galactose-containing receptor for toxicity.

Thorpe and Ross (1982) modified lysine residues in abrin with the SPDP reagent and fractionated the dithiopyridyl-abrin thus formed into components that bound or did not bind to Sepharose. These components differed 100-fold in toxicity to Thy1.1-expressing T-lymphocytes in accordance with their galactose-binding capability but, when coupled to monoclonal anti-Thy1.1 antibodies, they were equally potent ($ID_{50} = 6 \times 10^{-10}M$).

The different outcomes of these two sets of experiments could reflect differences in the requirements for A-chain entry via the M-6-P and Thy1.1 uptake pathways. Alternatively, the abrin conjugates could be disrupted intracellularly to release the free toxin. Blocking agents with better stability and specificity (e.g. the phenylazide derivatives of galactose and lactose described by

Houston, 1982, 1983) are needed to determine whether or not the galactose-binding sites are needed for the cytotoxic activity of intact toxin conjugates.

Obstruction of the Galactose-Binding Sites of Ricin by its Linkage to Antibody

A very effective means of blocking the non-specific cell binding properties of antibody-intact ricin conjugates has been described by Thorpe et al. (1983a, b) and appears to derive from obstruction of the galactose-binding sites of the toxin by the antibody moiety itself.

A conjugate of monoclonal anti-Thy1.1 antibody and ricin was prepared in three steps. The N-hydroxysuccinimidyl ester of iodo-acetic acid was used to introduce iodoacetyl groups onto lysine residues in the toxin. Next, the immunoglobulin was thiolated by treatment with the SPDP reagent followed by dithiothreitol. On mixing, the derivatized proteins react to form a conjugate in which the linkage is a thio-ether bond. When chromatographed on a Sepharose column, most (80%) of the conjugate passed through the column and, of this, 24% also passed through an asialofetuin-Sepharose column. No diminution in Sepharose or asialofetuin binding was observed with unconjugated iodoacetylated ricin showing that the blockade was not due to direct modification of lysine residues important for integrity of the galactose-binding sites but that it was a consequence of linking the modified toxin to the antibody. The most plausible explanation is that the iodoacetylating reagent preferentially reacts with a hyperreactive lysine residue (or with one of congregation of lysine residues) close to the galactose-binding sites in the toxin, with the result that the conjugates tend to form with the sites orientated towards the antibody.

Fluorescence-activated cell sorter (FACS) analyses revealed that the component of anti-Thy1.1-ricin that adhered to the Sepharose column showed marked non-specific binding to Thy1.2-expressing CBA thymocytes and EL4 cells whereas the 'blocked' conjugates that passed through the Sepharose and asialofetuin columns showed little or no tendency to bind non-specifically to these cells. All three fractions of conjugate bound to Thy1.2-expressing AKR thymocytes and AKR-A lymphoma cells to an extent indistinguishable from native antibody.

The 'blocked' and 'non-blocked' ricin conjugates were extremely powerful cytotoxic agents for AKR-A cells in tissue culture. A 1 h period of exposure of the cells to the conjugates at 2 to 5 x 10^{-12}M

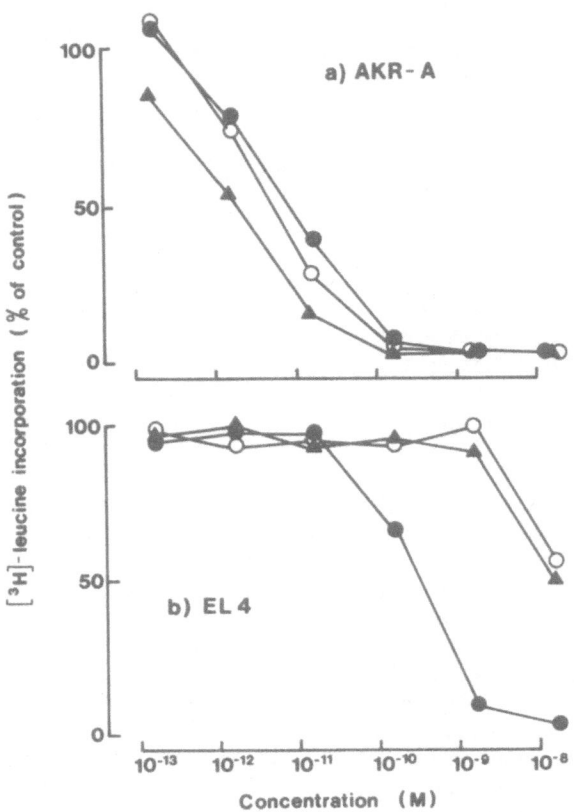

Fig. 2.   Specific toxic effects of anti-Thy1.1-ricin conjugates
          with blocked galactose-binding capacity. AKR-A (a) and
          EL4 (b) cells were incubated with the conjugates for 1 h
          at 37°C, washed, and their capacity to incorporate (^{3}H)
          leucine was measured 23 h later. The conjugates tested
          were i) not retained by asialofetuin-Sepharose (O),
          ii) not retained by Sepharose (▲), or iii) retained by
          Sepharose (●).

sufficed to reduce their rate of (^{3}H)leucine incorporation by 50%
(Fig. 2a). The conjugates were approximately 10 times as potent as
ricin used under the same experimental conditions. When tested for
non-specific toxicity to EL4 cells, which are as sensitive to native
ricin as AKR-A cells, the 'blocked' ricin conjugates were found to
be toxic only at 1.5×10^{-8}M (Fig. 2b), representing a differential
of 3000 to 7500-fold between the potencies of the specific and non-
specific cytotoxic effects. As predicted from its column-binding
properties, the 'non-blocked' conjugate substantially retained non-

specific toxicity to EL4 cells and this diminished the differential
between its specific and non-specific effects to only 60-fold.

The non-specific toxicity of the conjugate that retained
Sepharose binding capacity was mediated through the recognition of
galactose residues on the surface of the EL4 cells and anti-Thy1.1-
coated AKR-A cells; this was evident from the 100-fold drop in
toxicity when the conjugate was applied to these cells in the
presence of 50 mM lactose (results not shown). By contrast, the
weak non-specific toxicity of the conjugate that passed through the
asialofetuin-column was antagonized little, if at all, by lactose
and could possibly have been due to fluid-phase pinocytosis of the
conjugate and proteolytic cleavage to release active ricin.

A conjugate with diminished galactose-binding capacity was
also prepared from the W3/25 monoclonal antibody which recognizes
an antigen upon helper T-lymphocytes in the rat (Williams et al.,
1977; White et al., 1978). It displayed impressive potency and
selectivity in its cytotoxic effect upon W3/25 antigen-exposing rat
leukemia cells (Fig. 3). The conjugate was as toxic as the non-
blocked form and 10-fold more so even than ricin. This is an important result because W3/25 antibody disulfide linked to isolated
ricin A-chain is not toxic even when its endocytosis is induced with
rabbit anti-mouse immunoglobulin antiserum (Thorpe and Ross, 1982;

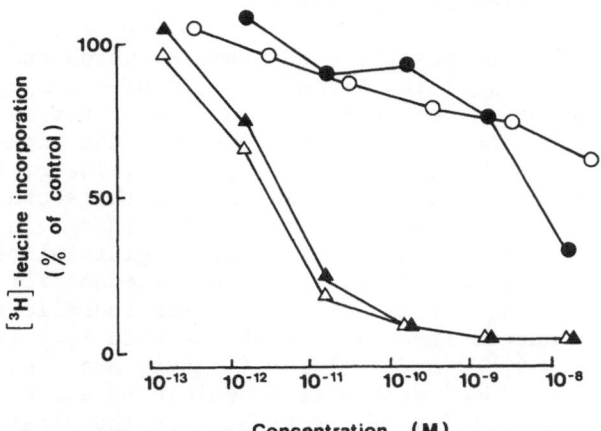

Concentration (M)

Fig. 3.   Toxic effect of W3/25 antibody conjugates upon PVG rat
          leukemic cells. The cells were incubated with the con-
          jugates for 1 h at 37°C, washed, and their capacity to
          incorporate (^{3}H)leucine measured 23 h later. The con-
          jugates tested were i) W3/25-ricin A-chain (O), ii)
          W3/25-intact ricin that passed through a Sepharose
          column (▲), iii) W3/25-intact ricin that was retained
          by a Sepharose column (Δ), or iv) anti-glycophorin-
          ricin that passed through a Sepharose column (●).

Mason et al., 1982). Although not conclusive since the conjugate could be cleaved intracellularly to re-expose the galactose-binding sites, this result provides suggestive evidence that the property of the B-chain that is needed for trans-membrane transport of the A-chain is independent of the recognition of galactose.

ROLE OF THE B-CHAIN IN A-CHAIN ENTRY

Ricin B-chain consistently improves the efficiency with which the A-chain portion of a conjugate is delivered to the cytosol. This was clearly seen above in the enhancement in toxicity gained when free B-chain was allowed to reassociate with A-chain conjugates and in the contrast between the regular and powerful cytotoxic action of intact ricin conjugates and the variable and generally weaker toxicity of A-chain conjugates. It is still unclear whether the B-chain assists the entry of the A-chain moiety by interacting with a galactose-containing macromolecule with the result that the conjugate kills the cell by the same route as the native toxin. On balance, the evidence from studies with intact ricin and abrin conjugates whose galactose-binding capacity has been attenuated support the opposite view that the role of the B-chain in A-chain entry is independent of galactose recognition.

We envisage two possible mechanisms by which the B-chain could facilitate the entry of the A-chain moiety of a conjugate without the recognition of galactose receptors. The first is that hydro-phobic domains on the B-chain, once brought into close proximity to the cell surface by antibody-antigen binding, insert into the lipid core of the membrane and in some way assist the A-chain to traverse it. Reduction of the disulfide interchain linkage in the toxin by cytosolic thiol-containing molecules (e.g. glutathione) could then release the A-chain inside the cell. A precedent for this mechanism is set by nicked diphtheria toxin for which there is compelling evi-dence for spontaneous insertion of the B-chain into liposomal (Donovan et al., 1982) and cellular (Sandvig and Olsnes, 1980, 1981; Draper and Simon, 1980) membranes at acidic pH and appearance of the ADPR transferase activity of the A-chain on the other side of the membrane (Donovan et al., 1982). Unlike diphtheria toxin, in which only the B-chain has hydrophobic segments, both the A and B chains of ricin have hydrophobic domains (Funatsu et al., 1978; Kimura and Funatsu, 1981). Ricin and both its separated subunits have been demonstrated by Ishida et al. (1983) to insert into viral membranes that contain GM_1: insertion of the toxin and of the free B-chain is promoted by binding to the ganglioside whereas that of the A-chain is not receptor-mediated. It can be inferred from the findings of Ishida et al. (1983) and of a similar study with liposomes by Beugnier et al. (1982) that isolated ricin A-chain applied to eukaryotic cells also inserts spontaneously into the plasma membrane but is unable thereafter to escape and kill the cell. Thus it could be that the function of the B-chain in ricin or ricin conjugates is to enter the membrane and, possibly by associating with other intra-membranal B-chain molecules, form a channel from which the A-chain can unfold on the far side of the membrane. In those A-chain conjugates that are effective without the B-chain, the molecule to which the conjugate attaches may itself aid A-chain entry.

Some sort of driving force appears to be needed for trans-membrane transport of the A-chain. This was suggested by the finding of Sandvig and Olsnes (1982b) that a combination of inhibitors of oxidative phosphorylation and glycolysis prevents abrin and ricin toxicity and that this cannot be explained solely by cessation of endocytosis since these agents individually did not prevent endocytosis but did inhibit toxicity. What the driving force is remains to be discovered but contenders are the membrane potential and ion fluxes through the membrane. Ca^{2+} ions have been reported by Sandvig and Olsnes (1982a) to be involved in the entry of ricin and abrin A-chains into the cytosol, possibly as an influx together with the toxins. It is conceivable that abrin and ricin conjugates might likewise be dependent for their toxicity on Ca^{2+} ion entry.

A second possible mechanism by which the B-chain could facilitate the entry of the A-chain moiety of a conjugate is by interacting with an unidentified membrane receptor that directs the conjugate to an intracellular compartment equipped with a transport system to convey macromolecules to the cytosol. The natural cytological role of such a transport system could be to internalize important macromolecules (e.g. polypeptide hormones or growth factors) destined for the cytosol. As pointed out by Olsnes and Pihl (1982b) cells may have evolved intracellular mechanisms for transporting molecules to avoid gross leakage of ions and other small molecules which, on the basis of studies on virus entry (Fernandez-Puentes and Carrasco, 1980) might be expected if the macromolecules were to enter at the cell surface. As yet there is little direct evidence that endocytosis of abrin and ricin conjugates is a pre-requisite for toxicity. Ricin A-chain conjugates made from anti-Thyl antibodies persist for long periods on cell surfaces (Thorpe et al., unpublished results) and are at least as toxic as conjugates made from epidermal growth factor (Cawley et al., 1980), asialofetuin (Cawley et al., 1981) and transferrin (Raso and Basala, this volume) which, like their parent ligands, are most probably endocytosed rapidly through coated pits. The best indication that endocytosis is needed for toxicity is that NH_4Cl enhances the potency of ricin A-chain conjugates (Casellas et al., 1982; Myers et al., 1983) and of intact ricin conjugates (Thorpe et al., unpublished results): this presumably is attributable to an alteration in intracellular fate of the conjugate as a consequence of elevating endosomal and lysosomal pH.

REFERENCES

Beugnier, N., Falmagne, P., Zanen, J. and Jansen, F.K., 1982,
    Interaction of ricin and its two chains with model mem-
    branes, Archives Internationales de Physiologie et
    Biochimie, 90:93.
Cawley, D.B., Herschman, H.R., Gilliland, D.G. and Collier, R.J.,
    1980, Epidermal growth factor - toxin A chain conjugates:
    EGF-ricin A is a potent toxin while EGF-diphtheria fragment
    A is non-toxic, Cell, 22:563.
Cawley, D.B., Simpson, D.L. and Herschman, H.R., 1981, Asialoglyco-
    protein receptor mediates the toxic effects of an asiolo-
    fetuin-diphtheria toxin fragment A conjugate on cultured
    rat hepatocytes, Proc. Natl. Acad. Sci. U.S.A., 78:3383.
Casellas, P., Brown, J.P., Gros, O., Gros, P., Hellstrom, I.,
    Jansen, F.K., Poncelet, P., Roncucci, R., Vidal, H. and
    Hellstrom, K.E., 1982, Human melanoma cells can be killed
    in-vitro by an immunotoxin specific for melanoma-associated
    antigen p97, Int. J. Cancer, 30:437.
Donovan, J., Simon, M. and Montal, M., 1982, Diphtheria toxin frag-
    ment A crosses lipid membranes at acid pH, Biophys. J., 37:
    256a.
Draper, R.K. and Simon, M.I., 1980, The entry of diphtheria toxin
    into the mammalian cell cytoplasm: evidence for lysosomal
    involvement, J. Cell Biol., 87:849.
Fernandez-Puentes, C. and Carrasco, L., 1980, Viral infection per-
    meabilizes mammalian cells to protein toxins, Cell, 20:769.
Funatsu, G., Yoshitake, S. and Funatsu, M., 1978, Primary structure
    of Ile chain of ricin D, Agric. Biol. Chem., 42:501.
Houston, L.L., 1982, Transport of ricin A after prior treatment of
    mouse leukemia cells with ricin B chain, J. Biol. Chem.,
    257:1532.
Houston, L.L., 1983, Inactivation of ricin using 4-azidophenyl-β-
    galactopyranoside and 4-diazophenyl-β-D-galactopyranoside,
    J. Biol. Chem., 258:7208.
Ishida, B., Cawley, D.B., Reve, K. and Wisnieski, B.J., 1983,
    Lipid-protein interactions during ricin toxin insertion
    into membranes, J. Biol. Chem., 258:5933.
Kimura, M. and Funatsu, G., 1981, Amino acid sequences of two
    cyanogen bromide fragments CB11 and CB111, and the complete
    sequence of Ala chain of ricin D, Agric. Biol. Chem., 45:277.
Lappi, D.A., Kapmeyer, W., Beglau, J.M. and Kaplan, N.O., 1978,
    The disulfide bond connecting the chains of ricin, Proc.
    Natl. Acad. Sci. U.S.A., 75:1096.
Martinez, O., Kimura, J., Gottfried, T.D., Zeicher, M. and Wofsy, L.,
    1982, Variance in cytotoxic effectiveness of antibody-toxin
    A hybrids, Cancer Surveys, 1:373.
Mason, D.W., Thorpe, P.E. and Ross, W.C.J., 1982, Elimination of
    leukemic cells from rodent bone marrow in-vitro with anti-

body-ricin conjugates: implications for autologous marrow
    transplantation in man, Cancer Surveys, 1:389.
McIntosh, D.P., Edwards, D.C., Cumber, A.J., Parnell, G.D., Dean,
    C.J., Ross, W.C.J. and Forrester, J.A., 1983, Ricin B chain
    converts a non-cytotoxic antibody-ricin A chain conjugate
    into a potent and specific cytotoxic agent, FEBS Lett.,
    164:17.
Myers, C.D., Thorpe, P.E., Ross, W.C.J., Cumber, A.J., Katz, F.E.
    and Greaves, M.F., 1984, An immunotoxin with therapeutic
    potential in T cell leukemia - WTI-ricin A, Blood (in press).
Olsnes, S. and Pihl, A., 1982a, Chimeric toxins, Pharmac. Ther.,
    15:355.
Olsnes, S. and Pihl, A., 1982b, Cytotoxic proteins with intracellular
    site of action: mechanism of action and anti-cancer properties,
    Cancer Surveys, 1:467.
Sandvig, K., Olsnes, S. and Pihl, A., 1978, Chemical modifications
    of the toxic lectins abrin and ricin, Eur. J. Biochem., 84:323.
Sandvig, K. and Olsnes, S., 1980, Diphtheria toxin entry into cells
    is facilitated by low pH, J. Cell Biol., 87:828.
Sandvig, K. and Olsnes, S., 1981, Rapid entry of nicked diphtheria
    toxin into cells at low pH. Characterization of the entry
    process and effects of low pH on the toxin molecule, J. Biol.
    Chem., 256:9068.
Sandvig, K. and Olsnes, S., 1982a, Entry of the toxic proteins abrin,
    modeccin, ricin and diphtheria toxin into cells. II. Effect
    of pH, metabolic inhibitors and ionophores and evidence for
    toxin penetration from endocytotic vesicles, J. Biol. Chem.,
    257:7504.
Sandvig, K. and Olsnes, S., 1982b, Entry of the toxic proteins,
    abrin, modeccin, ricin and diphtheria toxin into cells.
    I. Requirement for calcium, J. Biol. Chem., 257:7495.
Thorpe, P.E., Cumber, A.J., Williams, N., Edwards, D.C., Ross, W.C.J.
    and Davies, A.J.S., 1981, Abrogation of the non-specific
    toxicity of abrin conjugated to anti-lymphocyte globulin,
    Clin. Exp. Immunol., 43:195.
Thorpe, P.E. and Ross, W.C.J., 1982, The preparation and cytotoxic
    properties of antibody-toxin conjugates, Immunol. Reviews,
    62:119.
Thorpe, P.E., Edwards, D.C., Ross, W.C.J. and Davies, A.J.S., 1982a,
    Monoclonal antibody-toxin conjugates: aiming the magic bullet,
    in: "Monoclonal Antibodies in Clinical Medicine", J. Fabre and
    and A. McMichael, eds., Academic Press, London.
Thorpe, P.E., Mason, D.W., Brown, A.N.F., Simmonds, S.J., Ross,
    W.C.J., Cumber, A.J. and Forrester, J.A., 1982b, Selective
    killing of malignant cells in a leukemic rat bone marrow with
    an antibody-ricin conjugate, Nature, 297:594.
Thorpe, P.E., Brown, A.N.F., Foxwell, B., Myers, C., Ross, W.C.J.,
    Cumber, A.J. and Forrester, J.A., 1983a, Blockade of the
    galactose-binding site of ricin by its linkage to antibody,
    in: "Monoclonal Antibodies in Cancer", B.D. Boss, R.E.

Langman, I.S. Trowbridge and R. Dulbecco, eds., Academic
    Press, New York (in press).
Thorpe, P.E., Ross, W.C.J., Brown, A.N.F., Myers, C., Cumber, A.J.,
    Foxwell, B. and Forrester, J.A., 1983b, Blockade of the
    galactose binding sites of ricin by its linkage to antibody:
    specific cytotoxic effects of the conjugates, Eur. J. Biochem.
    (submitted for publication).
Vitetta, E.S., Cushley, W. and Uhr, J.W., 1983, Synergy of ricin A
    chain - containing immunotoxins and ricin B chain - containing
    immunotoxins in the in-vitro killing of neoplastic human B
    cells, Proc. Natl. Acad. Sci. U.S.A., 80:6332.
White, R.A.H., Mason, D.W., Williams, A.F., Galfre, G. and Milstein,
    C., 1978, T-lymphocyte heterogeneity in the rat: separation
    of functional subpopulations using a monoclonal antibody,
    J. Exp. Med., 148:664.
Williams, A.F., Galfre, G. and Milstein, C., 1977, Analysis of cell
    surfaces by xenogeneic myeloma-hybrid antibodies: Different-
    iation antigens of rat lymphocytes, Cell, 12:663.
Youle, R.J. and Neville, D.M., 1980, Anti-Thy1.2 monoclonal anti-
    body linked to ricin is a potent cell-type-specific toxin,
    Proc. Natl. Acad. Sci. U.S.A., 77:5483.
Youle, R.J., Murray, G.J. and Neville, D.M., Jr., 1981, Studies on
    the galactose-binding site of ricin and on the hybrid toxin
    Man-6-P-ricin, Cell, 23:551.
Youle, R.J. and Neville, D.M., 1982, Kinetics of protein synthesis
    inactivation by ricin-anti-Thy1.1 monoclonal antibody hybrids:
    role of the ricin B subunit demonstrated by reconstitution,
    J. Biol. Chem., 257:1598.

MONOCLONAL ANTIBODIES AS CELL TARGETED CARRIERS OF COVALENTLY

AND NON-COVALENTLY ATTACHED TOXINS

Vic Raso and Marylu Basala

Department of Pathology

Harvard Medical School and the
Dana-Farber Cancer Institute

INTRODUCTION

The strategy of using cell specific monoclonal antibodies to
modify the indiscriminate binding characteristics of lethal toxins
has stimulated considerable interest because of its therapeutic
potential. Intact plant or bacterial toxins linked to antibodies
may acquire some cell selectivity but because toxin binding sites
are retained, non-specific toxicity can still occur. Since the
binding region of such toxins is distinct from the enzymatically
active toxic portion, these components can be cleaved and isolated
(Olsnes and Pihl, 1973). Thus, purified toxic subunits can be
coupled to antibodies to produce highly selective cytotoxic agents
which are devoid of residual toxin binding.

A disulfide bond connects the toxic A and lectin B chains of
ricin and this same covalent linkage has been used to construct
antibody-ricin A chain conjugates. By analogy to the action of
ricin, it was suspected that reductive cleavage of A chain from its
carrier would be required for conjugate toxicity after entry into
the cell (Olsnes et al., 1976). The disulfide bond of ricin, unlike
the synthetic linkage of antibody-A chain conjugates, may be shielded
by subunit interactions since it is not easily split by mercaptoethanol (Lappi et al., 1978). This distinction has led to speculation
that disulfide linked cytotoxins might be cleaved in vivo by thiols
before localizing on target cells, and thereby rendered ineffective
as therapeutic agents.

Conjugate analogs of the disulfide linked molecule can be made
using a reduction stable thioether bond since antibody amino groups

119

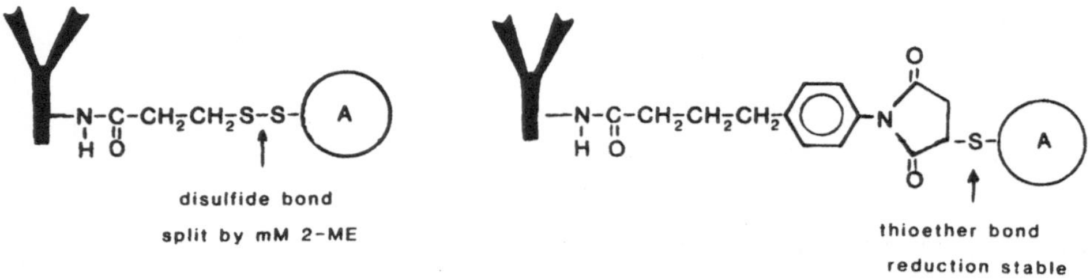

Fig. 1.   Chemical linkage of antibody-ricin A chain
          conjugates.

and the intrinsic sulfhydryl of A chain are points of attachment in
both instances (Fig. 1).   Previous work has shown that thioether
linked conjugates are either inactive or required at 100-fold higher
doses to produce cytotoxicity comparable to disulfide bonded forms
(Masuho et al., 1982; Jansen et al., 1982).   In this study, these
conjugates were used not only to explore the importance of linkage
type for cytotoxic potency in vitro but also to measure their serum
levels after i.v. injection into rabbits.   A comparison of elimination rates for the reduction stable and sensitive forms could indicate whether or not disulfide linkage is too labile for in-vivo use.

Antibody-A chain conjugates are slower acting than whole toxins,
presenting a possible drawback if lengthy exposure times are required
to ensure killing of all cells in vitro (Raso et al., 1982; Youle and
Neville, 1982).   The importance of this is more pronounced for in-
vivo situations where conjugate molecules can be rapidly eliminated
or degraded.   Such pharmacological problems might be overcome by a
system of non-covalent toxin delivery since this would allow the
flexibility needed to maximize exposure time.

Previous studies using conventional affinity purified antibodies
directed against ricin A chain and human IgG on the surface of lymphoma and leukemia cells showed that hybrid antibodies containing
both specificities could mediate delivery of toxin into cells (Raso
and Griffin, 1981).   These hybrids were constructed by reducing the
disulfide bond connecting the equivalent antibody halves and allowing
the Fab' portions of both specificities to mix and randomly reassociate (Nisonoff and Rivers, 1961).   This process yields heterodimeric
hybrid molecules having a single site directed against each determinant as well as homodimeric molecules which are not hybrids and
therefore useless for delivery.

To improve these reagents, murine hybridoma cell lines were

generated to produce homogeneous monoclonal anti-ricin A chain and
anti-ricin B chain antibodies. A unidirectional method for linking
anti-ricin antibody to a second antibody or alternate carrier was
developed to yield strictly heterodimeric hybrids having two sites
for each determinant. Potent carrier specific cytotoxicity was
achieved using such hybrids to carry either purified ricin A chain
or whole ricin in the presence of lactose to receptor bearing cells.
The effectiveness of these monoclonal hybrid delivery reagents was
much greater than earlier hybrid preparations (Raso and Griffin,
1981; Raso, 1981) since cell kill was extensive and toxicity was
enhanced 1000-fold over undelivered A chain or blocked toxins.

COVALENTLY LINKED CONJUGATES

   A murine monoclonal IgG directed against human K light chain
determinants present on normal and malignant B lymphocytes was used
as a carrier molecule. Methods for disulfide attachment of purified
ricin A chain using SPDP reagent have been described (Raso et al.,
1982). Thioether linkage was accomplished by reacting 10 mg of
purified monoclonal antibody in 3 ml of 0.1 M sodium phosphate buffer
pH 6.6, with 20 μl of a 0.2 M solution of succinimidyl 4-(p-
maleimidophenyl)butyrate in dimethyl formamide. After stirring for
3 h at room temperature, the reaction mixture was applied to a
Sephedex G-25 column equilibrated with PBS, pH 6.6, and the substituted protein was collected. This was mixed with a 2.5 molar excess
of freshly reduced A chain and allowed to react overnight. The conjugate was isolated by passage through Sephacryl S-300 sizing column.
Disulfide and thioether conjugate species having predominately a 1:1

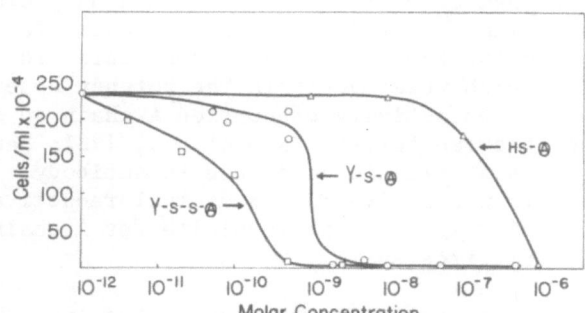

Fig. 2.   Dose response curves for K-directed conjugates
          and free ricin A chain. Daudi cells plated at
          30x10⁴ cells/ml received the designated con-
          centrations of disulfide linked conjugate (□—□),
          thioether linked conjugate (O—O) or free ricin A
          chain (Δ—Δ). Viable cells were counted after 4
          days.

molar ratio of antibody and ricin A chain were used throughout this
study. SDS polyacrylamide gel electrophoretic analysis showed that
free ricin A chain was released from the disulfide but not the
thioether conjugate when incubated for 1 h in 10 mM mercaptoethanol
at 37°C.

Daudi Burkitts lymphoma cells express substantial amounts of
K light chain on their surface and provide a good target for cytotoxicity experiments with K-directed agents. When disulfide conjugate, thioether conjugate or free ricin A chain was added to
cultures of proliferating Daudi cells, each produced inhibition of
cell growth and cell death, displaying ID_{50} levels of $1 \times 10^{-10}M$,
$1 \times 10^{-9}M$ and $5 \times 10^{-7}M$ respectively (Fig. 2). Both the disulfide and
thioether analogs produced good carrier-mediated cytotoxicity, being
5000-fold and 500-fold more potent respectively than A chain alone.

The 10-fold differential between thioether and disulfide conjugate activity was not as large as seen for other systems where the
thioether may be 100-fold or less active (Masuho et al., 1982;
Jansen et al., 1982). For example, A chain disulfide linked to monoclonal antibody directed against the common acute lymphoblastic
leukemia antigen had an $ID_{50}=2 \times 10^{-10}M$ for Nalm-1 target cells (Raso
et al., 1982) while an $ID_{50} > 5 \times 10^{-7}M$ was found for the thioether
analogue. Thus, the relative potency of the two analogues seems to
be influenced by the nature of the target antigen on the cell surface.

The basis for reduced potency of the thioether analogue was
examined further by comparing the A chain activity and cell binding
properties of the two conjugates. The inhibitory effects of antibody-A chain conjugates or free ricin A chain on protein synthesis
were evaluated using a cell free rabbit reticulocyte lysate system
(Fig. 3). The conjugate analogue had comparable ribosome inactivating capacities which were one-fifth the potency observed for free
A chain. The reduced activity of coupled A chain on isolated ribosomes has been observed before (Raso et al., 1982; Jansen et al.,
1982) and suggests steric hindrance due to antibody. The fact that
the thioether form was active indicates that reductive cleavage from
the antibody carrier is not a prerequisite for A chain action on
ribosomes in a cell free system.

The differential cytotoxic potency of the two conjugates was
not attributable to disparities in binding to K determinants on
Daudi cells (Fig. 4). At $10^{-7}M$ concentrations, each saturated the
available sites on the cell membrane as detected by fluoresceinated
anti-A chain or anti-mouse IgG probes. Furthermore, when using
titration methods to find the concentrations required to give maximal immunofluorescence on cells analyzed by flow cytofluorimetry
(Raso, 1981), the conjugates displayed identical affinity characteristics.

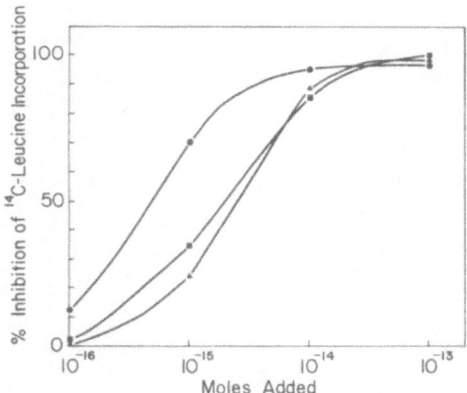

Fig. 3. Ribosome inactivation by free ricin A chain
         and antibody-A chain conjugates. The inhibition
         of protein synthesis by free A chain (●—●),
         thioether-linked antibody-A chain (▲—▲) and
         disulfide-linked antibody-A chain (■-■) was
         measured using a rabbit reticulocyte lysate
         system which contained all the necessary
         components for synthesizing ^{14}C-labeled protein.

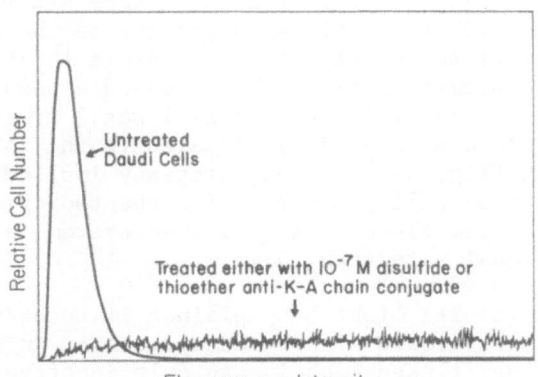

Fig. 4. Cell surface localization of K-directed con-
         jugates. Daudi cells were treated with PBS
         or 10^{-7}M thioether or disulfide-linked anti-
         body-A chain conjugates, washed, and exposed
         to a fluorescein labeled rabbit anti-A chain
         reagent. Surface immunofluorescence was
         analyzed by flow cytofluorimetry.

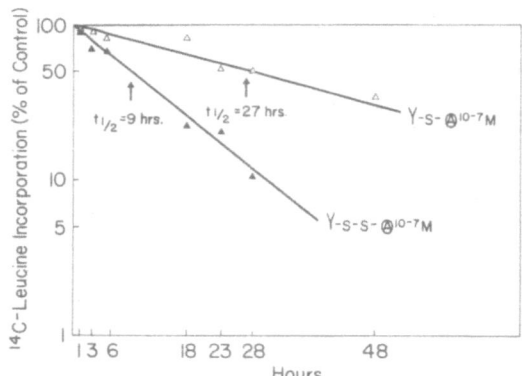

Fig. 5.   Kinetics of inhibition of cellular protein
          synthesis by anti-K-ricin A chain conjugates.
          Daudi cells in leucine free media were treated
          with 10^{-7}M disulfide (▲–▲) or thioether (△–△)
          linked anti-K-ricin A chain cytotoxin for
          various lengths of time. The cells were pulsed
          with ^{14}C-L-leucine, harvested, and assayed for
          incorporation of radioactivity.

     Since neither intrinsic A chain activity nor binding affinity
explained the reduced cytotoxic potency of the thioether analogue,
the kinetics of inactivation of cellular protein synthesis was evaluated (Youle and Neville, 1982).   Daudi cells were treated with an
excess of either conjugate so that cell surface K sites were saturated and equal amounts of A chain were present on the membrane (Fig.
4).   The capacity of these cells to incorporate ^{14}C-L-leucine into
protein decreased progressively with increased exposure time to
cytotoxin (Fig. 5).   The rate of cell kill was 3 times faster for
the disulfide conjugate, $t_{1/2}$=9 h, compared to the thioether conjugate, $t_{1/2}$=27 h (Fig. 5).   This discrepancy implied differences
in internalization or cell processing for the two cytotoxins based
on their linkage type, since in a cell free system they inactivated
ribosomes at an equal rate (Fig. 3).

     The kinetic results might be explained if both disulfide and
thioether conjugates arrive within the cytosol as intact molecules
(Fig. 6).   Thioether-linked A chain can only inactivate ribosomes
directly while disulfide-linked A chain might be cleaved from its
antibody carrier allowing the more potent action of free A chain
(Fig. 3). While this scheme fits the results found using K-directed
antibody-A chain cytotoxins, systems in which the differential between cytotoxicity is 100 to 1000-fold may be better explained by
an inability of thioether conjugate to reach ribosomes.   Perhaps

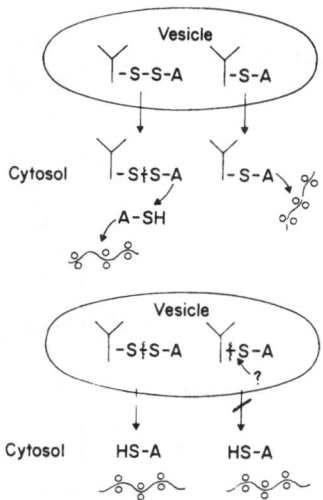

Fig. 6.   Hypothetical pathways for thioether and
          disulfide linked cytotoxins.

separation of A chain from antibody must precede passage from vesicles to cytosol in these instances so that the non-cleavable thioether analogue is virtually inactive.

Pharmacokinetic studies were undertaken in rabbits to assess
the in-vivo stability of i.v. administered thioether and disulfide
conjugates. The possibility that normal levels of circulating thiol
would be sufficient to cleave disulfide-coupled A chain could be
evaluated since the thioether linkage is not susceptible to reduction
(Fig. 1). For comparative purposes, the component parts of these
conjugates, mouse monoclonal antibody and free ricin A chain, were
also injected into separate animals to measure their rates of clearance from the circulation.

Three separate radioimmunoassays were devised to measure A
chain, antibody and intact antibody-A chain conjugate. The scheme
for assaying the latter (Fig. 7) utilized a flexible polyvinyl
chloride microtitre plate with affinity purified rabbit anti-A chain
bound to its inner surface (Klinman et al., 1976). Subsequent treatment with BSA prevented any non-specific attachment of protein.
Intact conjugate adhered to these wells via its A chain component
and was detected with an[125]I-labeled goat anti-mouse IgG reagent
which bound to the carrier portion of the conjugate. This assay
exclusively measured intact conjugate molecules. Mouse antibody

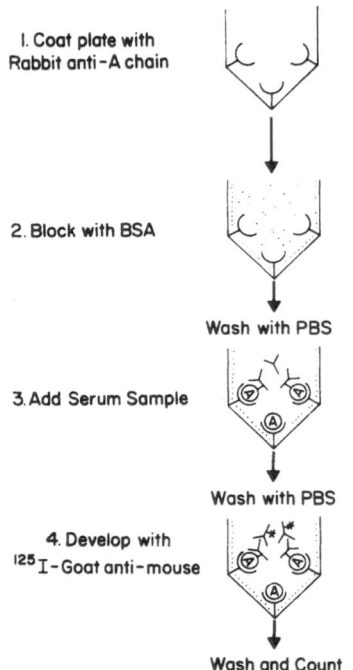

1. Coat plate with
Rabbit anti-A chain

2. Block with BSA

Wash with PBS

3. Add Serum Sample

Wash with PBS

4. Develop with
^{125}I-Goat anti-mouse

Wash and Count

Fig. 7.   RIA protocol for measuring serum levels of
intact antibody-ricin A chain conjugates.

alone could not adhere to the plate and was washed away while any
uncoupled A chain bound to the plate escaped recognition by the
^{125}I-labeled probe. A standard calibration curve for this assay,
showing the increase in bound radioactivity as a function of added
thioether conjugate is shown in Fig. 8.

Radioimmunoassays for free A chain and mouse antibody were
variations of this coated plate technique. The assay for ricin A
chain also utilized affinity purified rabbit anti-A chain attached
to the plate but an^{125}I-anti-ricin A chain reagent was used to
detect the amount bound to treated wells. Affinity purified rabbit
anti-mouse IgG was substituted to coat plates for the antibody
assay and an ^{125}I-labeled goat anti-mouse IgG probe quantified the
amount of specifically bound antibody.

In-vitro recovery experiments from whole rabbit blood were performed to ensure that the assay performed well with biological specimens and to test the stability of disulfide and thioether conjugates.

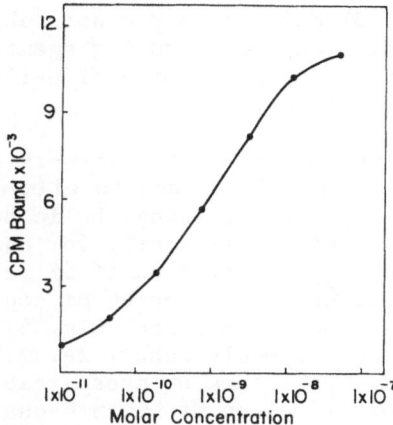

Fig. 8. Standard calibration curve for the coated plate RIA used to measure intact conjugate. Microtitre wells coated with affinity purified rabbit anti-A chain received known concentrations of thioether linked antibody-A chain conjugate. Bound cytotoxin was detected with ^{125}I labeled goat anti-mouse IgG.

Table 1. Recovery of anti-K-ricin A chain conjugates from whole rabbit blood in-vitro

Addition (M)		Time of Incubation at 37°C (h)	Recovered (M)	% Recovery
Disulfide	2×10^{-8}	0	2.2×10^{-8}	110
Disulfide	2×10^{-8}	24	1.3×10^{-8}	65
Thioether	2×10^{-8}	0	1.6×10^{-8}	80
Thioether	2×10^{-8}	24	1.8×10^{-8}	90

Table 1 shows that intact conjugates were recovered in good yield
following addition to freshly drawn heparinized whole blood and
incubation in blood at 37°C for 24 h did not substantially degrade
either form. This indicated that reducing agents in rabbit blood
did not cause extensive cleavage of disulfide-linked antibody-A
chain in an in-vitro situation.

Although the conjugates used for this work were directed
against human K chain, it was important to eliminate the possibility
of cross-reactivity with rabbit immunoglobulin before initiating
pharmacology studies. Conjugate toxicity for human K-bearing Daudi
cells was undiminished by the inclusion of 5% rabbit serum in the
test culture media. Furthermore, binding of conjugate to Daudi
cells as measured by immunofluorescence (Fig. 4) was not reduced if
performed in the presence of whole rabbit serum but was totally
blocked by human serum. This lack of cross-reactivity eliminated
the possibility that distribution of human K-chain specific conjugates in-vivo would be upset by binding to circulating rabbit immunoglobulins.

For in-vivo studies, separate rabbits received a dose of 1 mg/kg
of monoclonal anti-K antibody, purified ricin A chain or either of
the two conjugates in a small volume of PBS as a rapid i.v. bolus
injection. Blood samples were removed from the opposite ear at several time points starting at 5 min post administration. Serum was
collected and stored frozen and all specimens as well as serum from
a pre-bleed were then assayed.

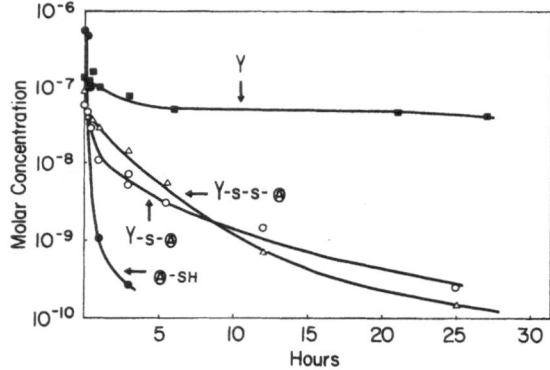

Fig. 9. Serum clearance curves for antibody, ricin-A
         chain, and antibody-A chain cytotoxins. Serum
         levels of antibody (■-■), A chain (●-●) thio-
         ether (O-O) or disulfide (△-△) linked conju-
         gates were measured at various times post
         administration using specific radioimmuno-
         assays.

The concentration of mouse monoclonal antibody in rabbit serum was 1.3×10^{-7}M 5 min post injection. Within a few hours, it fell to one-half this level but then declined very slowly (Fig. 9). Measurements taken for up to 4 days showed that the serum half-life of the slow phase was 40 h, very close to values determined by radiolabel methods (Spiegelberg and Weigle, 1965). In contrast, ricin A chain was only fleetingly present in the serum of rabbits. The 5×10^{-7}M concentration measured at 5 min was reduced to 10^{-9}M by 60 min.

Clearance from the circulation of both disulfide and thioether conjugates was substantially faster than unmodified antibody but not as rapid as free A chain (Fig. 9). The 5 min level for the disulfide form was 9.4×10^{-8}M while that for the thioether analogue was 5.6×10^{-8}M. These concentrations fell to 5×10^{-9}M by 5.5 and 3 h respectively. This indicates that the animal has a mechanism for eliminating such conjugates since they were stable in rabbit blood for up to 24 h in-vitro (Table 1). There was no large difference in the time course for clearance of disulfide versus thioether analogues suggesting that reductive cleavage of A chain from antibody is not a crucial factor in the elimination process.

In additional experiments, rabbits were injected with disulfide conjugate and monitored for intact antibody-A chain cytotoxin as well as cleaved A chain. Measurements from the two respective radioimmunoassays showed that A chain levels were totally accounted for by surviving intact conjugate molecules, indicating that no circulating free A chain was present. The mechanism by which disulfide and thioether antibody-A chain conjugates are removed from the serum is unknown but the relative stability of unmodified antibody in-vivo suggests that A chain may govern this elimination process. If this study in the rabbit is representative, it appears that antibody-A chain cytotoxins would have to localize within tumors fairly rapidly to be therapeutically effective.

HYBRID DELIVERY SYSTEMS

Monoclonal antibodies directed against ricin A and B subunits can be used to non-covalently bind either whole toxin or its toxic A subunit. When these antibodies are connected by a covalent bond to a cell-directed antibody or cell receptor ligand, a hybrid molecule with dual binding specificity is formed which can carry toxin into and kill target cells (Fig. 10).

Mice were immunized with whole ricin toxoid prepared by formaldehyde treatment as previously described (Pappenheimer et al., 1974). The animals were boosted one month later and after 3 days their spleens were removed and processed for somatic cell fusion (Kohler and Milstein, 1975). Solid phase radioimmunoassays (Klinnman et al., 1976) were used to select clones which secreted anti-ricin antibodies (Table 2). Anti-A clones produced antibody

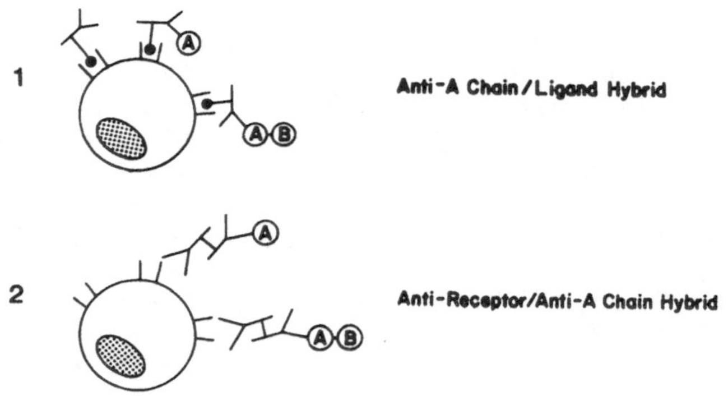

Fig. 10. Modes of hybrid delivery of ricin A chain.

which bound to polyvinyl microtitre plates coated either with whole
ricin (A plus B chains) or isolated A chain and were detected with
^{125}I-labeled anti-mouse IgG reagent. Antibody from B chain clones
bound to whole ricin coated plates but not the A chain plate. Moreover, anti-A chain but not anti-B chain binding was blocked by
addition of excess free A chain. After characterization these individual clones were injected into mice to produce ascites containing
monoclonal antibody. Salt fractionation and S-300 chromatography
was used to isolate these antibodies. The carrier monoclonal antibody directed against human K-immunoglobulin light chain has been
previously described (Raso, 1981) Commercially available human transferrin was also used as a cell specific carrier in these studies.

Table 2. Detection and characterization of antibodies
produced by hybridoma clones

Addition	Ricin Plate (A-B)(cpm)	A Chain Plate(cpm)
Media alone	313	318
Anti-A chain A1H7	3,267	3,119
+ A chain block	760	959
Anti-B chain (B1H8	2,291	380
+ A chain block	2,377	350

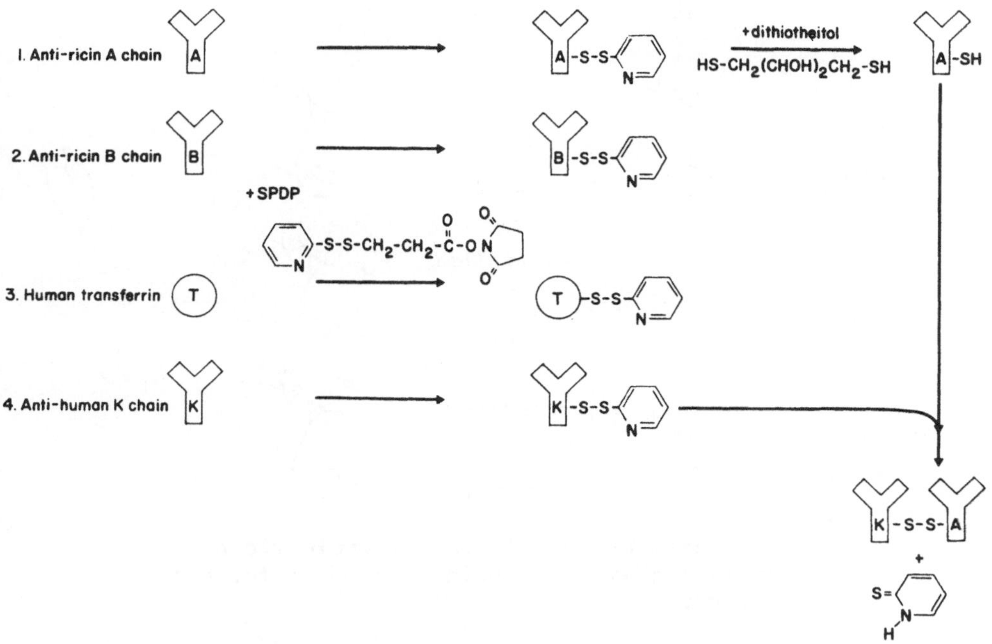

Fig. 11. Schemes for the construction of disulfide
         linked hybrid reagents.

Transferrin or monoclonal antibodies directed against human K
chain, ricin A chain or ricin B chain, were each reacted with a 6-
fold molar excess of SPDP reagent. Uncoupled reagent was removed
by passage over a Sephadex G-25 column equilibrated with 0.1 M
sodium acetate 0.1 M NaCl, pH 4.5. Approximately 3 moles of SPDP
were attached per mole of protein in each case (Fig. 11).

To produce any desired hybrid antibody combination, one SPDP
substituted protein (e.g. anti-ricin A chain in Fig. 11) was reduced
with 0.4 mM dithiothreitol for 1 h and isolated by passage over a
G-25 column equilibrated in PBS. The thiolated antibody was mixed
with a second SPDP substituted protein (e.g. anti-K chain in Fig. 11)
and allowed to react overnight. The product, a disulfide linked
heterodimeric hybrid antibody (MW 300,000) with two binding sites
for ricin A chain plus two sites directed against a cell surface
determinant, was separated from unreacted antibody (MW 150,000) by
chromatography on Sephacryl S-300.

Fig. 12 shows four different means by which anti-ricin A chain

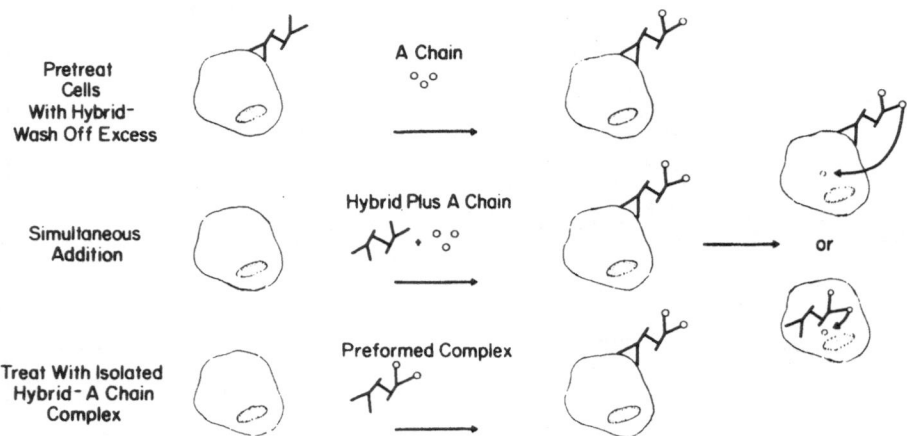

Fig. 12. Alternatives for binding of whole ricin or
ricin A chain to hybrid antibodies for cell
delivery.

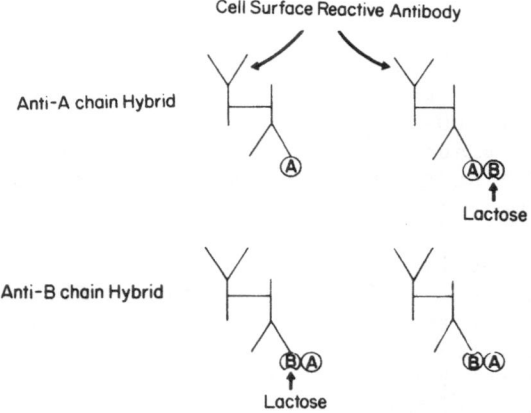

Fig. 13. Methods for delivering ricin A chain to
cells by hybrid antibodies.

and anti-ricin B chain antibodies may be used in combination with a second cell-reactive carrier antibody or ligand. Anti-A chain antibodies used with a carrier can deliver either free ricin A chain, which has no intrinsic affinity for cells, or whole ricin in the presence of lactose to block the binding site of its B chain. Similarly anti-B chain antibodies can deliver whole ricin in the presence on lactose or without it if antibody attaches to the sugar binding site precluding interaction with cell surface glycoproteins.

Fig. 13 illustrates different ways to load the toxic moiety (in this case ricin A chain) on to hybrid for delivery to cells. In the first instance, hybrid is added directly to cells in the absence of A chain and allowed to react via the carrier antibody. Washing removes any unattached hybrid and the surface localized anti-A chain sites are filled by addition of free ricin A chain. The second involves exposing cells simultaneously to hybrid and A chain so that interactions may occur continuously between toxin, hybrid and cell. Finally, hybrid and toxin can be preincubated to saturate all anti-toxin binding sites. Performed hybrid-toxin can be added to cells following removal of unbound A chain (MW 30,000) by passage of the complex (MW 360,000) over a molecular sizing column. Once the complex is situated on the cell surface by any of these protocols, either dissociated toxin or the entire hybrid-toxin complex can be enzymatically destroy ribosomes, shut down de novo protein synthesis and kill the cell.

Daudi Burketts lymphoma cells having K light chain determinants on their membrane, were treated simultaneously with 3×10^{-9}M anti-K chain/anti-ricin A chain hybrid and 6×10^{-8}M ricin A chain. Whereas untreated Daudi cells proliferated in 4 days, hybrid plus A chain treated cells decreased in number (Table 3). This cytotoxicity was totally blocked by K determinants on human gamma globulin (HGG) which fill hybrid sites preventing attachment of the complex to cells. Heterologous bovine gamma globulin (BGG) provided no protection since it is non-reactive with human K chain directed antibody sites. Furthermore, addition of excess K chain directed monoclonal antibody blocked cytotoxicity by preoccupying K determinants on the cell surface, preventing interaction with hybrid.

In another experiment Daudi cells were pretreated with the anti-K chain/anti-ricin A chain hybrid, washed and incubated at 30×10^4 cell/ml with varying concentrations of whole ricin in the presence of 10 mM lactose. Four days later these cells were counted and compared to controls which were incubated with ricin plus lactose but received no pretreatment with hybrid (Fig. 14). Cytotoxic effects were observed only in the presence of hybrid. A similar experiment using an anti-K chain/anti-ricin B chain hybrid gave parallel results.

In order to show that carriers other than antibody could be

Table 3. Cytotoxicity anti-K chain/anti ricin A
hybrid and ricin A chain.

Addition	Cells/ml x 10^{-4} at 4 days
Daudi alone	202
+ Hybrid + A chain	7
+ Hybrid + A chain + HGG	219
+ Hybrid + A chain + BGG	6
+ Hybrid + A chain + anti-K	166
Daudi alone	200
+ HGG	188
+ BGG	242
+ Anti-K	170

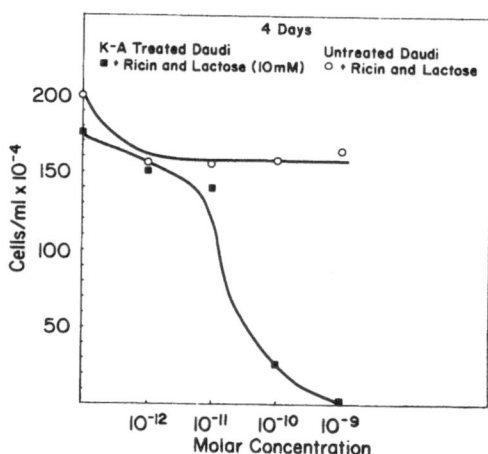

Fig. 14. Hybrid-mediated ricin toxicity. Anti-K/anti-
A chain hybrid-treated (■-■) and untreated
(O-O) Daudi cells were cultured in medium
containing 10 mM lactose and the designated
concentrations of ricin. Viable cells were
counted after 4 days.

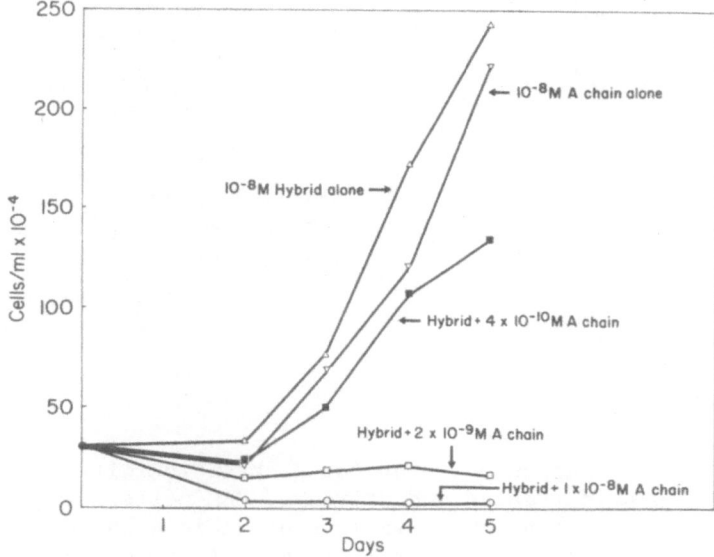

Fig. 15. Combined effects of transferrin/anti-ricin A
         chain plus ricin A chain on the growth of CEM
         cells. Various concentrations of ricin A
         chain were added to CEM cells cultured in the
         presence of 10^{-8}M transferrin/anti-ricin A
         chain hybrid, and viable cells were counted
         daily. Control cultures received either
         hybrid or 10^{-8}M A chain.

used to direct anri-ricin antibodies to specified cells, human transferrin was disulfide-linked to monoclonal anti-ricin A chain (Fig.
10 and 11). This transferrin/anti-ricin A chain hybrid was added at
10^{-8}M to human leukemic CEM cells which possess transferrin receptor
on their surface. As increasing concentrations of ricin A chain
were included in the culture, progressive cytotoxicity was produced
(Fig. 15). Cells exposed to transferrin/anti-ricin A chain hybrid
alone or ricin A chain alone grew normally. The dose response
curves of A chain toxicity for cells plus hybrid and cells alone
(Fig. 16) indicate that 1000-fold increased potency is achieved via
hybrid-mediated delivery. Table 4 indicates that cytotoxicity is
completely blocked by excess native transferrin. Attachment and
penetration of the transferrin/anti-A chain hybrid plus A chain complex was prevented by ligand bound to transferrin receptors.

Several additional possibilities exist for modification of this

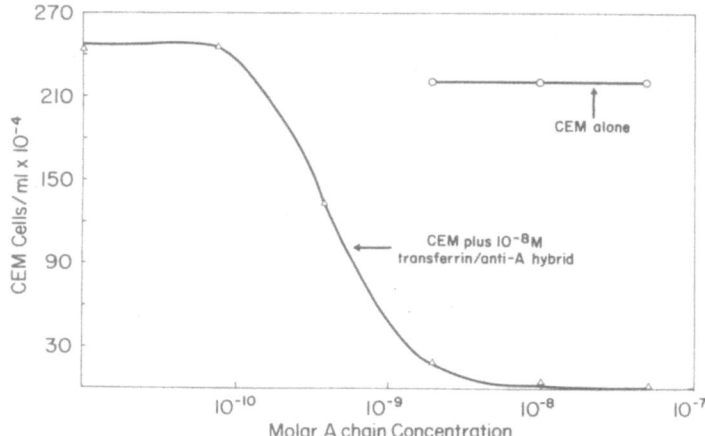

Fig. 16. Dose response curves for A chain toxicity on
         hybrid treated and untreated CEM cells. The
         designated concentrations of ricin A chain
         were added to transferrin/anti-A chain treated
         (△-△) or untreated (O-O) CEM cells. Cultures
         were assayed on day 4 for viable cells.

Table 4. Blockage of toxin delivery using native human
         transferrin

Additions	Cells/ml x 10^{-4}	
	0 h	96 h
CEM cells alone	30	270
+ A chain 10^{-8}M	30	300
+ TF-hybrid 10^{-7}M	30	270
+ A chain + TF-hybrid	30	17
Transferrin + A + TF-hybrid	30	228
Tranferrin + TF-hybrid	30	249

hybrid approach. These include 1) non-covalent transport of toxins or drugs other than ricin, 2) varying choice of cell targets by using antibodies or ligands with different cell binding specificities, 3) using reagents other than SPDP to link cell-directed carrier to toxin binding antibody, and 4) substituting $F(ab')_2$ (100,000 MW) or Fab (50,000MW) antibody fragments for whole antibody to produce tetravalent hybrids of 200,000 MW or bivalent hybrids of 100,000 MW respectively.

These hybrid delivery systems offer a number of potential advantages for therapeutic applications. Infusion of innocuous hybrid molecules may precede introduction of the toxic moiety. Thus, maximal specific localization of hybrid on target tissue may be achieved over a period of time before toxin is administered and sequestered by cell-bound hybrid molecules. This could be advantageous for treating solid or poorly vascularized tumors. Multiple application of toxin or toxic fragments to hybrid coated targets might be feasible. The hybrid may alternatively be preloaded with toxin, the complex isolated and administered as a single agent. Perhaps hybrid-bound A chain will have different pharmacological qualities than covalently linked A chain. Different molecular weight hybrids can be constructed using whole antibodies or fragments and these would be expected to possess distinct pharmacological properties. Binding valency may be similarly varied. This delivery system renders non-cytotoxic toxin subunits or fragments such as ricin A chain lethal to cells. Moreover, inactivated, modified or binding site-blocked whole toxins (e.g. ricin plus lactose or affinity labeled ricin) can be used in a selective manner to kill specified cells when delivery is mediated by hybrids.

ACKNOWLEDGEMENTS

This work was supported in part by Grant CA29039 from the National Cancer Institute.

REFERENCES

Jansen, F.K., Blythman, H.E., Carriere, D., Casellas, P., Gros, O., Gros, P., Laurent, J.C., Paolucci, F., Pau, B., Poncelet, P., Richer, G., Vidal, H. and Voisin, G.A., 1982, Immunotoxins: hybrid molecules combining high specificity and potent cyto-toxicity, Immunol. Rev., 62:185.
Klinnman, N.R., Pickard, A.R., Sigal, N.H., Gearhart, P.J., Metcalf, E.S., and Pierce, S.K., 1976, Assessing B cell diversification by antigen receptor and precursor cell analysis, Ann. Immunol., 127C:489.
Kohler, G. and Milstein, C., 1975, Continuous cultures of fused cells secreting antibody of predefined specificty, Nature, 256:495.
Lappi, D.A., Kapmeyer, W., Beglaw, J.M. and Kaplan, N.O., 1978,

The bond connecting the chains of ricin, Proc. Natl. Acad. Sci. USA, 75:1096.

Masuho, Y., Kishida, K., Saito, M., Umemoto, N. and Hara, T., 1982, Importance of antigen-binding valency and the nature of the cross-linking bond in ricin A-chain conjugates with antibody, J. Biochem., 91:1583.

Nisonoff, A. and Rivers, M.M., 1961, Recombination of a mixture of univalent antibody fragments of different specificity, Arch. Biochem. Biophys., 93:460.

Olsnes, S., and Pihl, A., 1973, Different biological properties of the two constituent peptide chains of ricin. A toxic protein inhibiting protein synthesis, Biochemistry, 12:3121.

Olsnes, S., Sandvig, K., Refsnes, K. and Pihl, A., 1976, Rates of different steps involved in the inhibition of protein synthesis by the toxic lectins abrin and ricin, J. Biol. Chem., 257:3985.

Pappenheimer, A.J. Jr., Olsnes, S. and Harper, A.A., 1974, Lectins from Abrus precatorius and Ricinus communis. I. Immunochemical relationships between toxins and agglutinins, J. Immunol., 113:835.

Raso, V., 1981, Antibody mediated delivery of toxic molecules to antigen bearing target cells, Immunol. Rev., 62:93.

Raso, V., and Griffin, T., 1981, Hybrid antibodies with dual specificity for the delivery of ricin to immunoglobulin-bearing target cells, Cancer Res., 41:2073.

Raso, V., Ritz, J., Basala, M. and Schlossman, S.F., 1982, Monoclonal antibody-ricin A chain conjugate selectivity cytotoxic for cells bearing the common acute lymphoblastic leukemia antigen, Cancer Res., 42:457.

Spiegelberg, H.L. and Weigle, W.O., 1965, The catabolism of homologous and heterologous 7S gamma globulin fragments, J. Exp. Med., 121:323.

Youle, R.J. and Neville, D.M. Jr., 1982, Kinetics of protein synthesis inactivation by ricin-anti-thy 1.1 monoclonal antibody hybrids. Role of the ricin B subunit demonstrated by reconstitution, J. Biol. Chem., 257:1598.

LONG-TERM KINETICS OF ANTIBODY-TOXIN CONJUGATES

SHOW RESISTANT CELLS

Richard J. Youle and David M. Neville, Jr.

National Institute of Mental Health
Laboratory of Neurochemistry
Section on Biophysical Chemistry
9000 Rockville Pike
Bethesda, Maryland 20205

ABSTRACT

A monoclonal anti-$Thy_{1.1}$-ricin A chain conjugate (OX-7-A chain) inactivated protein synthesis in $Thy_{1.1}$ positive cells with first order kinetics for 15 h. After this period, the surviving cells began growing again. The maximal rate of protein synthesis inactivation was one log inactivation each 10 h and was reached when the cell surface antigen was saturated with conjugate. Therefore, the maximal inactivation of one addition of conjugate over the 15 h lifetime would leave 3% survivors. If the ricin B chain was added to cells treated with OX-7-A chain, the maximal rate of protein synthesis inactivation was increased three-fold, but the length of time the conjugate was effective (15 h) was not changed. Extrapolation shows that the maximum cell killing with OX-7-A chain plus B chain would leave only 0.003% survivors.

We tested the effect of a second addition of conjugate after the first had reached its maximal inhibition of protein synthesis. OX-7-A was incubated at subsaturating concentrations with cells for 30 h and the survivors were then treated with a second addition of OX-7-A at the same concentration as the first. The survivors were inhibited by the second incubation only 5-20% as much as the initial cell population was inhibited by the first addition. Identical results were found for second additions of OX-7-A plus B and for OX-7-ricin plus lactose. Therefore, repeated additions of conjugate do not give repeated inhibition of cell growth. This result is consistent with a model that the target cell population was heterogenous in sensitivity to antibody-toxin conjugates and that the differences between cells were stable for 46 h.

INTRODUCTION

Cell-type-specific toxins have many applications in science and medicine. Several laboratories have made such reagents by linking monoclonal antibodies which bind cell surface antigens to toxic proteins[1-3]. Though often these antibody-toxin conjugates are selectively toxic for antigen bearing cells in vitro, when examined, the rate of cell killing compared to native toxins is slow[4-5]. We found the maximal rate of protein synthesis inhibition by a mono-clonal anti-$Thy_{1.1}$ receptors were saturated with conjugate. This maximal rate was 1 log inactivation each 10 h. When the B chain of ricin was added, the maximal rate was increased three-fold[4]. The length of time conjugates remained toxic in vitro was not determined, thus, the maximum number of cells which could be killed by these methods remained unknown.

In vivo anti-tumor effects of antibody-toxin conjugates have been disappointing. This is probably due in part to the slow killing rate of conjugates combined with a short (30 min) half-life in vivo[5]. Several groups have given repeated injections of antibody-toxin conjugates to tumor bearing animals to help circumvent the first half-life[5.7]. In this report we examine long term kinetics of several types of antibody-toxin conjugates. We also test the effect of repeated doses of conjugate on target cells in-vitro.

RESULT AND DISCUSSION

$Thy_{1.1}$ AKR cells were incubated with subsaturating concentra-tions of an anti-$Thy_{1.1}$ monoclonal antibody linked to ricin A chain (OX-7-A chain) and periodically examined for protein synthesis rate (Fig. 1). After 20 h incubation of cells with ricin the rate of protein synthesis has been shown to represent the proportion of surviving cells[8]. The protein synthesis rate of cells incubated with 25 ng/ml or approximately 1.4×10^{-10} M OX-7-A chain decreased for the first 10-20 h, then began increasing at the rate of control cells. Previous work with OX-7-A shows first order inactivation for the first 10 h with no lag time before inactivation begins[4]. The maximal rate of inactivation reached at saturating concentra-tions was 1 log inactivated each 10 h. If saturating concentrations of OX-7-A follow the same long-term kinetics as subsaturating concentrations, then 3% viable cells would remain after a single exposure to a saturating concentration of conjugate.

When ricin B chain is added to OX-7-A chain the rate of protein synthesis inactivation increased at least 5-fold at sub-saturating concentrations of OX-7-A and 3-fold at saturating concentrations[4]. The long-term kinetics of OX-7-A plus B (Fig. 2) are very similar to OX-7-A chain alone (Fig. 1). Though the 1000

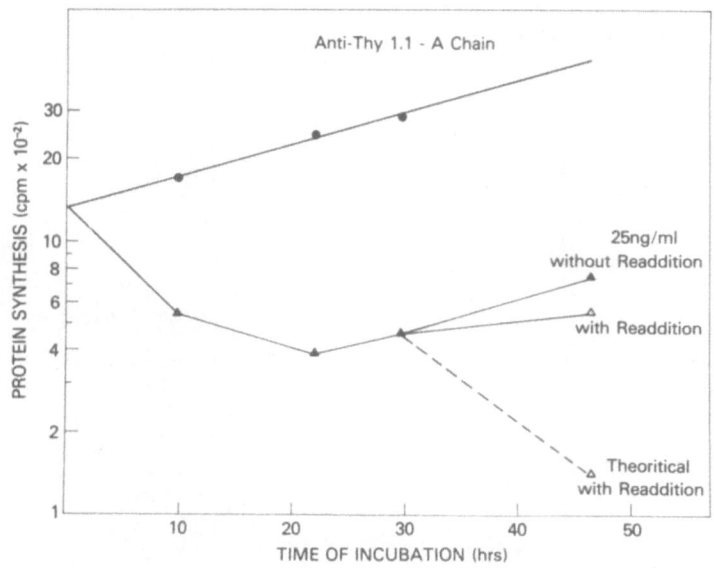

Fig. 1. Long-term kinetics of OX-7-A chain. AKR-SL2
cells, 2×10^6 cells in 2 ml R PMI media plus
10% fetal calf serum (media), were incubated
with 25 ng/ml OX-7-A chain for 3 h at 37°C.
Cells were then pelleted and resuspended in
2 ml fresh media, dispensed in 100 μl aliquots
into 96 well plates. After further incubation
at 37°C and 5% CO_2 the cells were pulsed with
0.1 μCi ^{14}C leucine per well for 1.5 h. A
second addition of OX-7-A chain after 29 h was
made directly to the appropriate wells to
yield 25 ng/ml OX-7-A and remained with the
cells for the remainder of the assay. Cells
were harvested onto glass fibre filters with
a Titertek cell harvester, washed, dried and
counted. Time of incubation plotted includes
the 3 h incubation with OX-7-A and one-half of
the 1.5 h pulse with labeled leucine. Triplicate wells were averaged to yield data
points. OX-7-A was synthesized as previously
reported[4]. Control cells with no addition,
(●——●).Cells incubated with 25 ng/ml OX-7-A
at 0 time,(▲——▲).Cells incubated with 25
ng/ml OX-7-A at 0 time and again after 29 hrs,
(△——△).Theoretical curve for the second
addition if it inhibited synthesis as much as
the first,△---△

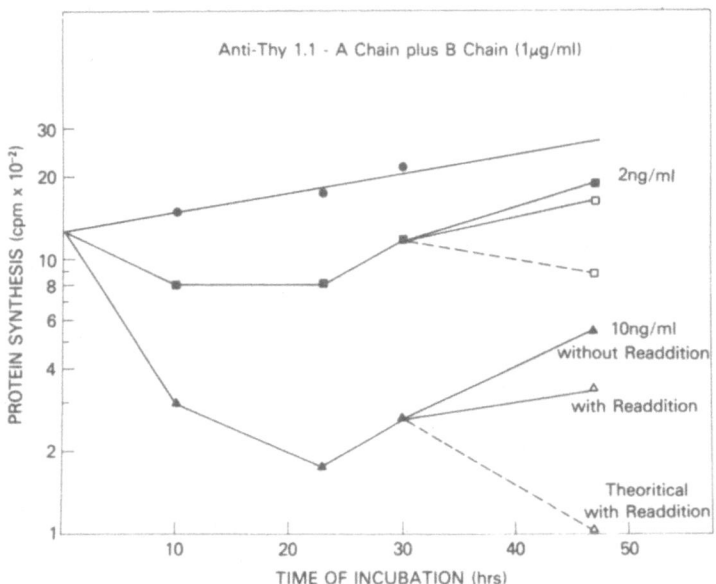

Fig. 2.   Long-term kinetics of OX-7-A chain plus
          ricin B chain. Cells were treated as in
          Fig. 1 except the OX-7-A chain incuba-
          tions included ng/ml ricin B chain.
          Ricin B chain was purified and assayed
          as previously reported[4]. Control cells
          with no addition,(●——●).Cells incubated
          at 0 time with ng/ml OX-7-A plus ng/ml
          ricin B chain (■——■) or 10 ng/ml
          OX-7-A plus 1 ng/ml ricin B chain(▲——▲)
          Cells incubated at 0 time and again after
          30 h with 2 ng/ml OX-7-A chain plus 1
          ng/ml ricin B chain (□——□) or 10 ng/ml
          OX-7-A chain plus 1 ng/ml ricin B chain
          (△——△)   Theoretical curves for the
          second addition of it inhibited as much
          as the first 2 ng/ml OX-7-A chain (□---□)
          or 10 ng/ml OX-7-A chain (△---△).

ng/ml of B chain activates OX-7-A 3- fold, the length of time
inhibition of protein synthesis occurs is not changed. If the
maximal rate of OX-7-A plus B seen previously[4] has the same long-
term kinetics as submaximal doses seen here, then 0.003% of the
original cells will survive one exposure to a saturating concentration of conjugate. The fraction of surviving cells from OX-7-A
plus B ($3x10^{-5}$) would be 10^{-3} fold lower than OX-7-A alone
($3x10^{-2}$). For cancer therapy activation of OX-7-A would be highly
advantageous.

We also examined the long-term kinetics of OX-7-Ricin in the presence of 50 mM lactose (Fig. 3). Cells were exposed to OX-7-ricin plus 50 mM lactose for 3 h, washed and incubated in 10 mM lactose for the remaining time. This conjugate had the same kinetics as OX-7-A chain and OX-7-A plus B chain. Based on concentration, OX-7-Ricin was more than 2.5 times more effective than OX-7-A plus B and 8 times

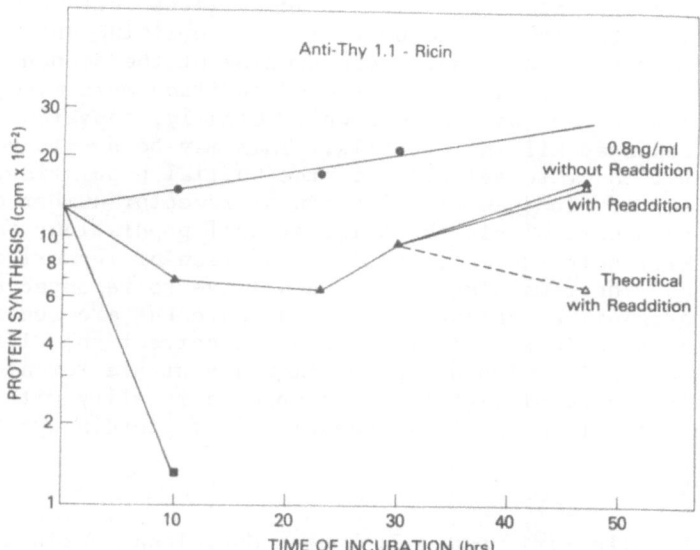

Fig. 3. Long-term kinetics of OX-7-ricin plus lactose. Cells were treated as in Fig. 1 with OX-7-ricin plus 50 mM lactose for 3 h, washed and further incubated in media plus 10 mM lactose. The second addition of OX-7-ricin included lactose to yield 50 mM. OX-7-ricin was synthesized and purified as previously reported[4]. Control cells with no addition, (●————●) Cells incubated at 0 time with 4 ng/ml OX-7-ricin plus 50 mM lactose (■————■) or 0.8 ng/ml OX-7-ricin (▲————▲) Cells incubated at 0 time and again after 30 h with 0.8 ng/ml OX-7-ricin plus 50 mM lactose, (△————△) Theorectical result for the second addition of 0.8 ng/ml OX-7-ricin if it inhibited as much as the first addition, (△---△).

more effective than OX-7-A. In previous experiments with different
batches of conjugates assayed differently the OX-7-A chain was as
toxic as OX-7-Ricin[4]. All three types of conjugate had the same
length of protein synthesis inhibition at low rates. The long-term
kinetics of ricin alone[8] are also almost identical to the results
with conjugates shown here.

    Since we could see maximal effects of one dose of conjugate we
asked how effective repeated doses would be in-vitro. Long after
the first addition of OX-7-A chain, OX-7-A plus B, or OX-7-ricin was
no longer toxic (30 h), we made a second addition of OX-7-A,
OX-7-A plus B, or OX-7-ricin at the same concentration as the first.
In all cases the second addition was much less toxic than the first
(Fig. 1-3). The maximal effect seen for a second addition was 20%
of the first addition (Fig. 2). All conjugates were much below
saturation of the Thy[1.1] receptors so the remaining antibody from
the first addition could not block binding of the second addition.
If higher concentrations of the second addition were made, more
inhibition was seen (data not shown). Clearly, repeated doses do
not give repeated killing of cells. This may be due to receptor
down regulation or to selection of the initial population. In the
latter case, if Gaussian distribution of receptor number or ricin
sensitivity occurs within the original cell population, the initial
dose may kill only the most sensitive portion of the original
population. The remaining are not sensitive to repeated doses at
the same original concentration of conjugate but are sensitive to
higher concentrations[1]. If this model is correct the differences
in cell sensitivity within a population are stable for at least
46 h. This may be similar to the stable variability within a Hela
cell population to Shiga toxin inhibition of protein synthesis[9].

REFERENCES

1. D.M. Neville, Jr., and R.J. Youle, Monoclonal antibody-ricin A
    chain hybrids: Kinetic analysis of cell killing for tumor
    therapy, Immunol Rev. 62:75 (1982).
2. S. Olsnes, and A. Pihl, Chimeric toxins, Pharmac. Ther.
    15:355 (1982).
3. E.S. Vitetta, K.A. Krolick, M Miyama-Inaba, W. Cushley and
    J.W. Uhr, Immunotoxins: A new approach to cancer therapy,
    Science 219:644 (1983).
4. R.J. Youle, and D.M. Neville Jr., Kinetics of protein
    synthesis inactivation by ricin-anti-Thy[1.1] monoclonal
    antibody hybrids, J. Biol. Chem. 257:1598 (1982).
5. F.K. Jansen, H.E. Blythman, D. Carriere, P. Casellas, O. Gros,
    P. Gros, J.C. Laurent, F. Paolucci, B. Pau, P. Poncelet,
    G. Richer, H. Vidal, and G.A. Voisin, Immunotoxins: Hybrid
    molecules combining high specificity and potent cytotoxicity,
    Immunol. Rev. 62:185 (1982).

6. I.S. Trowbridge, and D.L. Domingo, Anti-transferrin receptor monoclonal antibody and toxin conjugates affect growth of human tumor cells, <u>Nature</u> 294:171 (1981).

7. K.A. Krolick, J.W. Uhr, S. Slavin and E.S. Vitetta, In-vivo therapy of a murine B cell tumor (BLC_1) using antibody-ricin A chain immunotoxins, <u>J. Exp. Med</u>. 155:1797 (1982).

8. K. Eiklid, S. Olsnes and A. Pihl, Entry of lethal doses of abrin, ricin, and modeccin into the cytosol of Hela Cells, <u>Exp.Cell. Res</u>. 126:321 (1980).

9. R. Reisbig, S. Olsnes and K. Eiklid, The cytotoxic activity of Shiga toxin. Evidence for catalytic inactivation of the 60S ribosomal subunit, <u>J. Biol. Chem</u>. 256:8739 (1981).

SIGNIFICANCE OF THE KINETICS OF IMMUNOTOXIN CYTOTOXICITY

F.K. Jansen, H.E. Blythman, B. Bourrie, D. Carriere,
P. Casellas, D. Dussossoy, O. Gros, J.C. Laurent,
M.C. Liance, P. Poncelet, G. Richer and H. Vidal

Centre de Recherches Clin-Midy
Section Immunologie
Rue du Prof. Joseph Blayac
34082 Montpellier, France

INTRODUCTION

Since 1980, immuno-A-toxins (I-A-Ts) which are conjugates between an antibody and the A-chain of a polypeptide toxin, e.g., diphtheria toxin or ricin, have been described by several laboratories as cytotoxic agents with high specificity for their target cells.[1-15] However, there was great variation from high to low potency depending on the antigen-immunotoxin (IT) system used. The advantage of the I-A-Ts is their low non-specific toxicity in-vivo (LD/50 in mice 20mg/kg),[16] which will allow their utilization in human patients in the near future.

Another approach is represented by Immuno-AB-toxins (I-AB-Ts) which are conjugates between whole toxins such as ricin or diphtheria toxin.[17-19] Although these conjugates show in general high potency, they retain a considerable non-specific toxicity which does not allow their utilization in human patients in-vivo. The non-specific toxicity is, however, inhibited in-vitro, by high lactose concentrations.

The diminished potency of some of the I-A-Ts corresponds to slow kinetics and seems to be due to the lacking helper function which is attributed to the B-chain of the whole toxin, which enhances the translocation of the A-chain into the cytosol.[20,21] It can therefore be expected, that the potency of I-A-Ts may be considerably augmented, if the underlying mechanisms of kinetics are understood and corrected. This was tried by increasing the kinetics of

147

our I-A-Ts with some small molecular weight drugs.[16,22,23] Thereby
we were able to obtain highly potent in-vitro cytotoxicity by which
target cells were destroyed with less than 40 IT molecules bound per
cell (at IC/50) or by which the number of surviving target cells was
reduced to less than 1/10⁴. As a consequence of the influence of
kinetics on in-vivo potency we found that one of our I-A-Ts with
inherent rapid kinetics was able to induce 50% long-term survival
in a mouse leukemia model, while an I-A-Ts with slow kinetics was
much less effective. These results suggest that rapid kinetics are
a major condition for high in-vitro and in-vivo potency of I-A-Ts.

The significance and implication of the kinetics of I-A-Ts
cytoxicity are described in the following chapters.

SIGNIFICANCE, PHASES AND VALUES OF CYTOTOXIC KINETICS

The cytotoxicity of immunotoxins is a time-dependent process
and the kinetics of this process, either slow or rapid, are of biological importance. Tumour cell destruction is in balance with tumour
cell proliferation. The time needed by cytotoxic molecules to inhibit or destroy cells should be much shorter than the proliferation
time of these cells. If not, the cytotoxic activity of ITs will not
be able to reduce the tumour cell number, because there would be a
steady state between cell destruction and cell proliferation.

Another interference of the kinetics of cytotoxicity with the
target cell can be expected by cellular defense mechanisms, since
cells seem to be capable of inhibiting or degrading polypeptide
toxins by lysosomic enzymes or by other mechanisms. If, therefore,
the kinetics of cytotoxicity, which include the translocation of
the toxic A-chain into the cytosol, are not much faster than IT
neutralization, the cytotoxic potential will not become apparent.
With regard to cell death, several successive phases may be distinguished, which are defined by the different methods used. Each phase
is governed by its own kinetics which are rapid for early, and slow
for late phases of cell death.

Phases of cell death

An analogy to the phases of cell death by a toxin might be
found in the classical literature where the philosopher Plato
described the different phases of the death of Socrates, who was
forced by the people of Athens to drink the toxin containing hemlock.
The sentence of death pronounced by the judges of Socrates was irre-
versable and could correspond to the first phase of death (Fig. 1).
The second phase would be represented by the intoxication when
Socrates drank the hemlock extract. The third phase is characterized by the first signs of paralysis, which began in the legs and
slowly spread through the whole body. The last phase of death is
the eventual loss of all physiological functions leading to the
degradation of the organs.

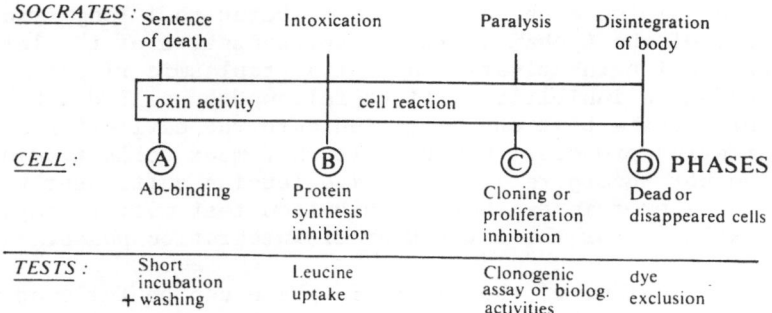

Fig. 1.   Different phases of cell death and the corres-
           ponding test systems.

Cell death by intoxication with an immunotoxin may be regarded
in a very similar way. The incubation of target cells with IT resembles the sentence of death, phase A, since the binding of IT to
its target cells will ultimately lead to cell death. The intoxication phase, phase B, corresponds to the first interaction of the
toxin with its subcellular targets inducing the inhibition of protein synthesis at the ribosomal level. Both phases, A and B, depend
only on IT, the A phase on its antibody binding and the B phase on
its internalization and toxic activity. The subsequent phases, C
and D, represent the reaction of the cell to the intoxication, i.e.,
the inhibition of protein synthesis and its consequences. Phase C,
the phase of paralysis, may be defined as the loss of biological
functions such as the responses to mitogens or, as the ultimate aim
of tumour therapy, the paralysis of cell proliferation. Phase D, or
the disintegration phase, represents the cell degradation or the
complete disappearance of the tumour cell.

Characteristics of different phases of the kinetics of cytotoxicity

Each phase corresponds to a particular test system. Phase A
can be measured by the binding of IT to its target, followed by the
washing out of excess IT and the waiting for cell destruction. This
phase is characterized by the antibody-binding kinetics, which may
be achieved within 10-15 min finally leading to cell death. Phase
A may be of major importance for the clinical efficiency of IT, if
a short contact for binding is sufficient to induce the eventual
destruction of the target cell. In this case, a short half life of
ITs in-vivo would not represent a crucial limitation.

Phase B, characterized by the translocation of the toxin into
the cytosol and the inhibition of protein synthesis, can be followed
by the radioactive leucine uptake of living cells. This phase lasts
several hours or even days depending on the IT used. Since ribosomes
are rapidly inactivated (1500 ribosomes/minute with 1 A-chain[24] the

long duration of the B phase may be attributed to a long translocation process of the A-chain. It is representative of the lethal hit
of the toxin which inactivates the intracytoplasmic ribosomes. Since
protein synthesis inhibition will be followed by cell death, this
method would be the best choice to indicate the earliest stages of
the intoxication process. However, fresh tumour cells from clinical
material do not incorporate radioactive leucine sufficiently. In
this case, the protein synthesis inhibition test must be replaced
by a dye exclusion of the later D or disintegration phase.

Phase C kinetics of the paralysis phase can be followed up by
assays depending on cell function, especially cell proliferation.
The cloning assay is, however, not suitable for kinetic measurements,
because it measures cell proliferation amplified by new prolifera-
tions until a colony can be observed after about 10 days. Thus the
proliferation time of the initial target cell, in which we are interested is not measurable. Other test systems, which directly
measure the time of single cell mitoses would be more suitable, for
this phase.

Phase D, the disintegration phase, can easily be studied by
dye uptake of cells, indicating holes in the cell membrane or by
following the disappearance of target cells. These kinetics are the
longest, lasting several days and characterize the final consequence
of ITs i.e., tumour cell destruction. It is essential for IT efficiency that B phase kinetics are shorter than C phase kinetics and
this means that protein synthesis must be inhibited before cell
proliferation starts. Thereby ITs cannot be diluted in daughter
cells thus becoming ineffective.

Measurement of kinetics

In an example of B phase kinetics it is shown (Fig. 2) how
kinetics can be measured. The leucine uptake of cells is plotted
on the ordinate on a log scale versus a linear time scale on the
abscissa. Ricin, as well as four other ITs on the same CEM cells
(Fig. 2), show a linear decrease of leucin uptake i.e., first order
kinetics. Below a certain threshold concentration, kinetics are
highly concentration dependent. Above this limit high concentrations can no longer increase kinetics, being therefore at saturation.
As seen with several immunotoxins the saturation dose corresponds
well with the saturation of all accessible antigens on target cells[25]
Only when extremely high antigen densities per cell are present
(TNP for TNP-IT or receptors for ricin : 10^7/cell) saturation kinetics are obtained with doses, which do not saturate all antigens.
Such ITs indicate that the translocation of the A-chain does not
depend only on the number of IT molecule per cell.

Kinetics can be calculated in two ways, either by indicating
the rate constants of the kinetic curve[21] or by measuring the time

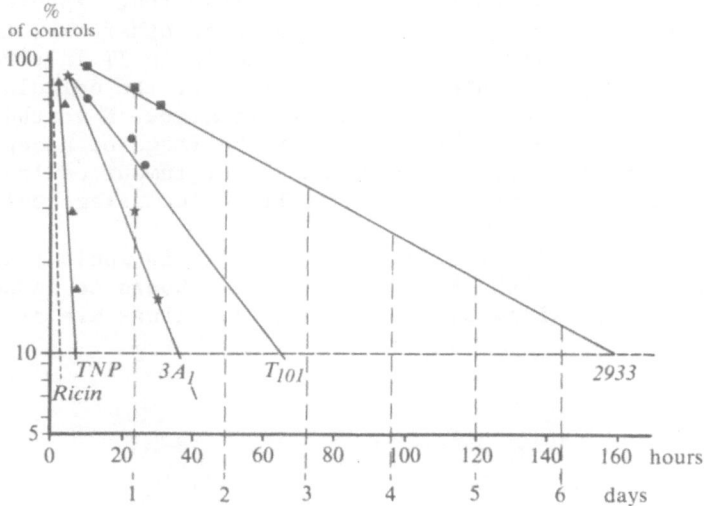

Fig. 2. Variability of the B phase kinetics of different
         I-A-Ts on the same cell line. Leucine uptake
         inhibition was measured at different time inter-
         vals with the 2933 IT (MAb: 2933 against a leuko-
         cyte antigen, Hybritech) with the T 101 IT (MAb T
         101 against the T65 antigen, Hybritech) with the
         3 Al IT (MAb against a T cell antigen, NIH) and
         with the TNP IT (MAB against the DNP hapten, Clin
         Midy). T10 values are defined as the time needed
         to reduce leucine uptake 10% of the control value
         and are obtained by extrapolation of the linear
         kinetic curves.

needed to inhibit protein synthesis to 50% or 10% of the control
values, called T50 or T10 value. The T50 value is easier to measure
but the T10 value has the advantage of enabling a rapid estimate of
the time needed to kill all tumour cells in a clinical situation.
Under the assumption that first order kinetics are maintained the
T10 value has only to be multiplied with the exponent of the number
of target cells (for 10^9 cells: T10 in hours x 9).

Theoretic limitation of kinetic values

     With the help of the T10 values, the minimum time necessary to
kill the last tumour cell in a patient could be calculated. First
order kinetics, however, will probably not be maintained to kill
the 10^{12} tumour cells expected in last stage patients. This was
suggested by highly sensitive clonogenic assays showing that target
cells can be reduced to about $1/10^3$ in a short period and that

further reduction needed a longer time than that expected by first
order kinetics.   Therefore, and for several other reasons i.e., a
different accessibility of all tumour cells to IT in-vivo or a
heteregenous population of target cell etc., the calculated minimum
time is certainly considerably underestimated.   Nevertheless, these
theoretical considerations will limit the range of acceptable T10
values for the therapy of large numbers of tumour cells in the range
from 10^9 (clinical onset) to 10^{12} cells (final stage patients).

In order to kill the last target cell, it would take less than
half a day with a B-phase T10 value of 1.5 hours corresponding to
the one of ricin (Fig. 3).   When such T10 values are arbitrarily

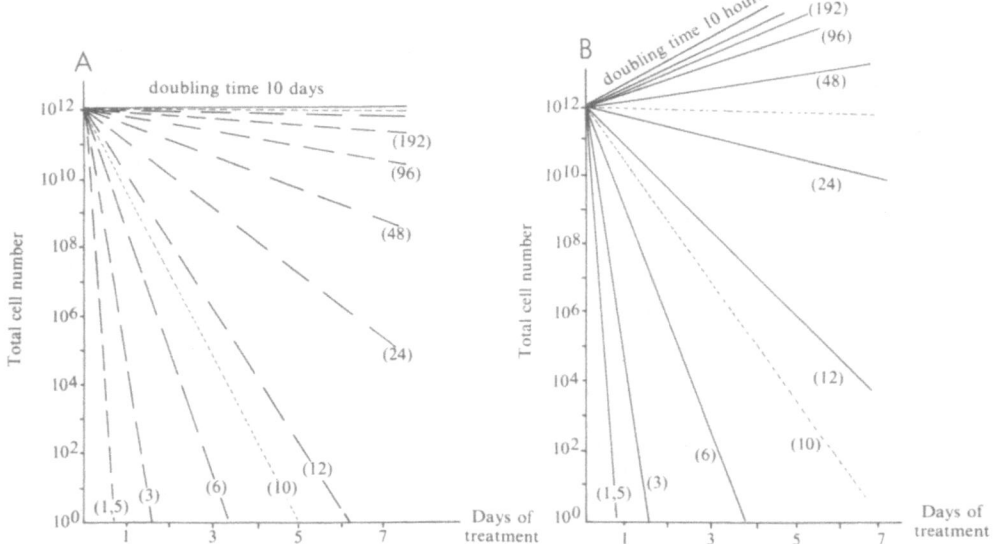

Fig. 3. Theoretical kinetics of protein synthesis inhibition
(B-phase) from a variety of ITs with increasing kinetics.   Under the simplified assumption that 10^{12} cells
(a tumour mass in a final stage patient) can be killed
by IT with first order kinetics, it is calculated after
how many days the last cell will be killed.   The T10
values of ITs were arbitrarily defined as multiples of
the T10 value of ricin (1.5 hours), by multiplying them
repetitively with the factor 2 up to 192.   The interference with the proliferation of slowly growing cells
(A: a doubling time of 10 days) and of rapidly growing
cells (B: a doubling time of 10 hours) is taken into
account in the formula $N = N_o e \uparrow (\alpha - \beta) t$ in which N_o
is the starting cell number,   the proliferation con-
stant,   the constant of destruction and t the time
interval.

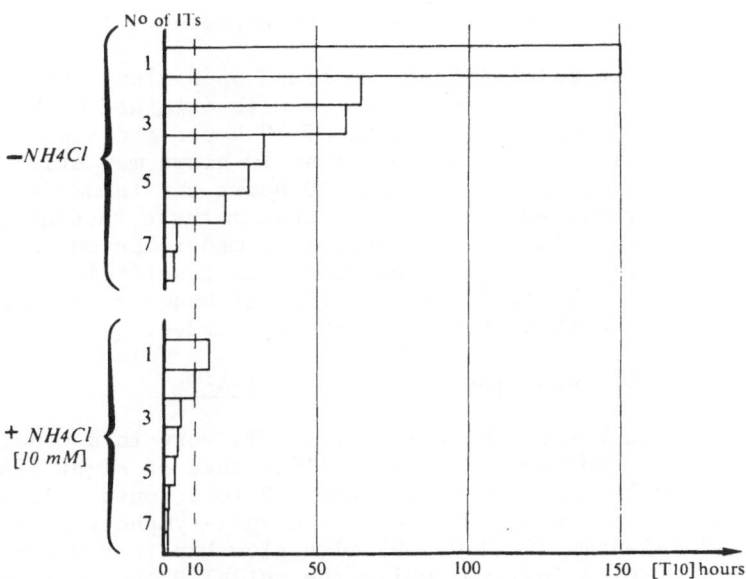

Fig. 4.   Leucine inhibition or B phase kinetics from different I-A-Ts. T10 values (the time needed to
reduce leucine uptake to 10% of the control value)
were measured for eight I-A-Ts. They are arranged
for decreasing values without any activation
(upper part) or with activation due to the presence of 10 mM ammonium chloride (lower part). The
following ITs and cell lines were used: 1. anti-
human leukocyte IT on CEM cells (Mab T2933, Hybritech); 2. anti-human melanoma IT on SK Mel 28
cells (Mab P 96.5, Hellström); 3. anti-human T
cell IT on CEM cells, (Mab T 101, Hybritech); 4.
anti-human T cell IT on CEM cells, (Mab 3A1, NIH);
5. anti-human melanoma IT on SK Mel 28 cells (Mab
B5B1, Clin Midy); 6. anti-mouse T cell IT on WEHI
7 cells (Mab, IgM, Olac); 7. anti-mouse T cell IT
on WEHI 7 cells (Mab AT15E); 8. anti-hapten IT on
TNP cells (Mab F9 D2, Clin Midy).

increased by a factor of 2, the time to kill the last tumour cell
rapidly increases. With values of more than 10 hours, it would take
about a week to kill the last tumour cell out of 10^{12} cells, which
is for the time being, a limiting factor for all treatments with immunotoxins because of the probable induction of neutralizing antibodies in the patients. With long kinetics corresponding to T10
values of more than 50 hours, as found with several I-A-Ts (Fig.4),

no sufficient therapeutic effects can be expected.

Although these calculations are based on a slowly growing tumour with a doubling time of 10 days, the results obtained with rapidly growing tumours with a doubling time of 10 hours are similar for rapid kinetics. T10 values of less than 10 hours may still be acceptable, while with values of more than 50 hours the tumour progression is not even stopped. By taking into account these theoretical considerations and the other limiting factors mentioned one would expect that T10 values of more than 10 hours may be insufficient in clinical situations with 10^9 to 10^{12} tumour cells, although they show some effect in animals with a much lower tumour burden.

Experimentally obtained kinetic values of I-A-Ts

The measured B phase kinetics of I-A-Ts vary to a considerable degree from rapid to slow (Fig. 4). T10 values of eight I-A-Ts differ randomly by a factor of 50 from 150 to 3 hours. In accordance with the theoretical considerations, the anti-lymphocyte IT (T10=60h) and the anti-melanoma IT (T10=65h) with slow kinetics did not give satisfactory results in-vivo, while the anti-TNP and anti-Thy 1,2 IT with rapid kinetics gave much better results (see later).

ACCELERATION OF IT KINETICS IN-VITRO

Following the idea that the B-chain of ricin has two functions, one to bind the whole toxin to the cell membrane and the other to enhance the translocation of the A-chain, several laboratories tried to introduce the helper function of the B-chain in ITs. With the whole toxin bound to antibodies, the non-specific binding of the B-chain must be inhibited by high lactose concentrations, restricting this approach to in-vitro treatments. If the non-specific binding can be blocked by steric inhibition through the coupled antibody, as in the new approach of Thorpe et al,[27] there is the danger that ricin may be separated from the antibody in-vivo and induce high non-specific toxicity. The use of the free B-chain in addition to I-A-Ts (McIntosh and Thorpe, this book) seems[21] to need a high excess of B-chain in order to be effective and may then be limited by B-chain toxicity.

If this B-chain is itself carried by an antibody,[28] it may be recombined with the A-chain within the target cell to reconstitute active ricin, thus leading to specific killing. In several approaches we failed to obtain a significant activation of I-A-Ts with B-chains coupled to antibodies. Fearing that the recombination of A and B chains to ricin might regenerate non-specific toxicity, we blocked this with high asialofetuin concentrations and then observed no activation. It might therefore be that recombined ricin again needs the galactose receptor to become toxic, which is in agreement with the theory of Youle and Neville.[26] Our own approach to the I-A-Ts activation was quite different.

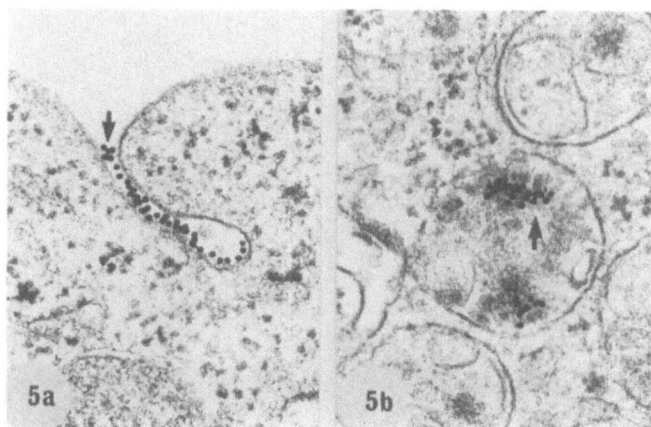

Fig. 5. Endocytosis of IT coated colloidal gold part-
         icles in target cells without activators. The
         T101 IT was absorbed on colloidal gold part-
         icles and incubated with its target cells (CEM)
         for about 1 hour at 4°C to be bound, and then
         for different time periods to be internalized.
         The gold particles are localized at the inner
         membrane of small endocytotic vesicles (a) or
         clustered in secondary lysosomes (b) after
         different time periods.

Endocytosis modified by lysosomotropic amines

It was found with the electron microscope that ITs adsorbed on
gold particles enter through small endocytosis vesicles (Fig.5a)
leading them to secondary lysosomes very rapidly, where they can be
destroyed. About 50% of the particles reach lysosomes of a normal
diameter of about 500 nm (Fig. 5b) within about 30 min. We therefore tried to inhibit the enzymatic function of secondary lysosomes
with the help of lysosomotropic amines, such as NH_4Cl, methylamine,
chloroquine and others.[22,23] According to the literature these
drugs increase the pH in secondary lysosomes and thereby inhibit
lysosomal enzyme activity. Ammonium chloride did not definitely
inhibit the entry of ITs into secondary lysosomes but retarded it,
so that 50% of IT gold particles needed about 2 hours instead of
30 min to be localized in secondary lysosomes. During this time,
such particles are seen in small endosomes (Fig. 6a) and in enlarged
endocytotic vesicles of a diameter between 500 and 1500 nm (Fig 6b)
without acid phosphatase activity, before they enter secondary lysosomes containing acid phosphatase activity.[29]

Acceleration of I-A-T kinetics

In the presence of NH_4Cl at a concentration of 10mM the kinetics

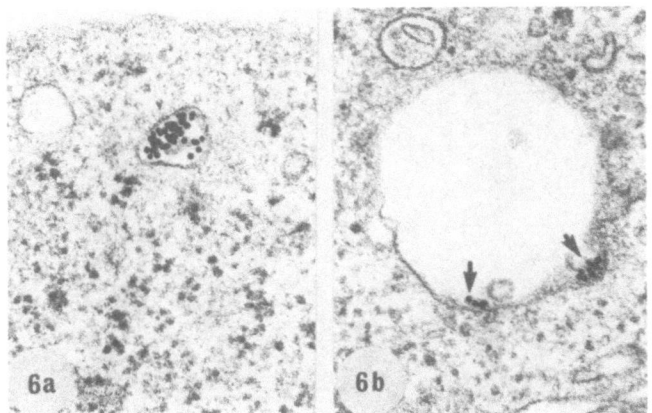

Fig. 6.   Endocytosis of IT coated gold particles in tar-
          get cells in the presence of an activator for IT
          (10 mM ammonium chloride). The T101 IT is ad-
          sorbed on gold particles and incubated for 1 hour
          with CEM target cells at 4°. For different time
          periods during incubation at 37° 10 mM ammonium
          chloride is present in the incubation medium.
          Gold particles are localized at the inner mem-
          brane of normal small endocytotic vesicles (a) or
          at the membrane of enlarged endocytotic vesicles
          without acid phosphatase activity (b).

of ITs are in general accelerated to a considerable extent.[16,22,23,30] The slower the kinetics without NH_4Cl the more they can be accelerated reaching levels similar to those of ricin. Long kin-etics of 65 hours thus become accelerated to about 5 hours (Fig. 4). It might therefore be, that the prolonged presence of IT in prelyso-somal endosomes induced by ammonium chloride, was one important con-dition to give the A-chain a better chance to cross the cell mem-brane.

Increase of specific activity

The same conditions which accelerated IT kinetics also increased the specificity of I-A-Ts in all our IT models. In the case of the anti-human T-cell IT, the specificity increase was particularly im-pressive. Without potentiation, and with an incubation period of 24 hours, 50% protein synthesis inhibition was reached in concentra-tions of 10^9 and showed a flat slope on the dose response curve (Fig. 7,-NH_4Cl). In the presence of ammonium chloride, 50% inhib-ition was obtained with $10^{-13}M$ concentrations, while the non-specific A-chain activity was only slightly augmented, inducing a considerable

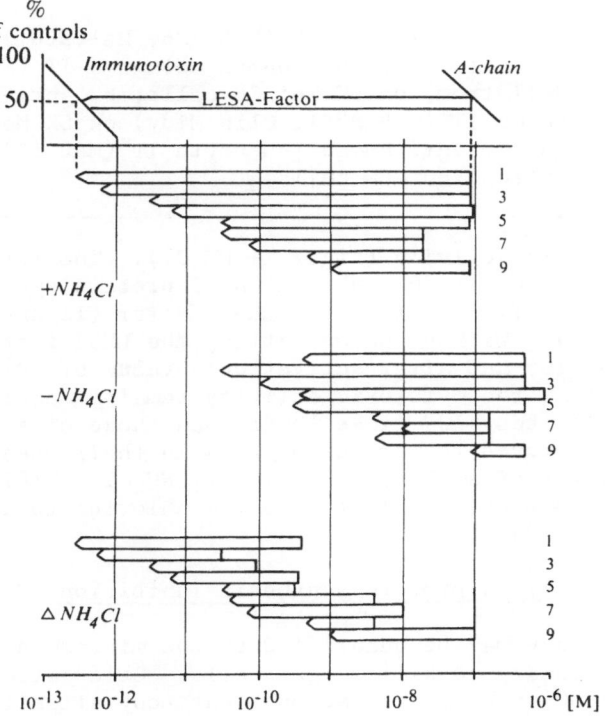

Fig. 7. Dose range of specific activity called LESA
factor (ligand enhanced specific activity factor)
of a variety of I-A-Ts. The LESA factor is the
ratio of the concentrations of IT versus A-chain
leading to 50% protein synthesis inhibition on
the same cell and represented as an arrow. The
body of the arrow covers all concentrations of
IT, in which only specific activity is found. At
lower concentrations than the head of the arrow
there is no activity and at higher concentrations
than its tail there is only non-specific activity.
The LESA factors of nine I-A-Ts in the presence of
ammonium chloride are grouped in the upper part,
the same I-A-Ts on the same cells but without NH$_4$Cl
in the middle part and the increase of specific
activity by the use of NH$_4$Cl in the lower part of
the graph. The following IT's and cell lines were
used: 1. anti-human T cell IT (MAB T101, Hybritech)
on CEM cells; 2. anti-human T Cell IT (MAB 3A1,
NIH) on CEM cells; 3. anti-microglobulin IT (MAB
3E10, Clin Midy) on CEM cells; 4. anti-calla IT
(MAB 10T5a, Immunotech) on RAJI cells; 5. anti-
hapten IT (MAB F9D2, Clin Midy) on CEM cells; 6.

(continued)

Fig. 7 continued
anti-human melanoma IT (MAB 2G6, Hellstom) on SK
Mel 28 cells; 7. anti-human melanoma IT (MAB P96,
5, Hellström) on SK Mel 28 cells; 8. anti-human
melanoma IT (MAB B5B1, Clin Midy) on SK Mel 28
cells; 9. anti-human leukocyte IT (MAB T2933,
Hybritech) on CEM cells.

---

increase in specific activity (Fig.7, + NH_4Cl). The ratio of the
concentrations leading to 50% inhibition of protein synthesis (IC50)
for A-chain versus IT, is called the LESA factor (ligand enhanced
specific activity). Without potentiation, the LESA factor reaches
a value of about 10^2 and after activation a value of 10^5 (Fig. 7).
This indicates that the cytotoxic activity remains specific in concentrations of IT about 10^5 times lower than those of A-chain. All
other ITs were improved to various degrees in their specific act-
tivity, by a factor of at least 10 (Fig. 7, NH_4Cl). This means that
there is a different potential of ammonium chloride to activate ITs
in the different models.

Augmented efficiency in protein synthesis inhibition

IT's efficiency may be quantified by the minimum number of bound
IT molecules necessary to kill target cells. Calculations were performed at the IC50, taking into account antibody affinity and antigen
quantity per cell. With the human T cell IT only 40 bound molecules
were needed in the presence of NH_4Cl, and about 20 000 without it
(Fig. 8). With the human melanoma IT, many more bound molecules were
needed, about 300 000 in the absence and 200 000 in the presence of
NH_4Cl. In the mouse system, with the anti-Thy 1,2 (IgM) IT, similarly high amounts were necessary (Fig. 8). Even higher quantities
were required with the anti-TNP on the human leukemia cell line CEM
showing that other factors besides the number of IT molecules per
cell are also to be considered for IT activity.

Increased potency in colony formation assay

The method of leucine uptake is not sensitive enough to prove
whether or not the last target cell has been killed. This aspect of
IT efficiency was measured in a colony formation assay. With the
human T cell line CEM, a cloning efficiency of 60% and a total of
10^6 colonies, could be taken into account in controls as calculated
from dilutions;[31] without activation, the human T cell IT (MAb T101)
was only slightly active after 24 hours of incubation, reducing
colony numbers to 1/10 of the initial quantity. The activator 10mM
ammonium chloride did not inhibit colony formation by itself, but
in combination with the IT, colony numbers were reduced to 10^{-4}
(Fig. 9). High amounts of non-target cells in the range of 10^{10}
cells/L did not influence the activity of this IT.[31] This will be
of importance for the treatment of leukemia cells in human bone
marrow.

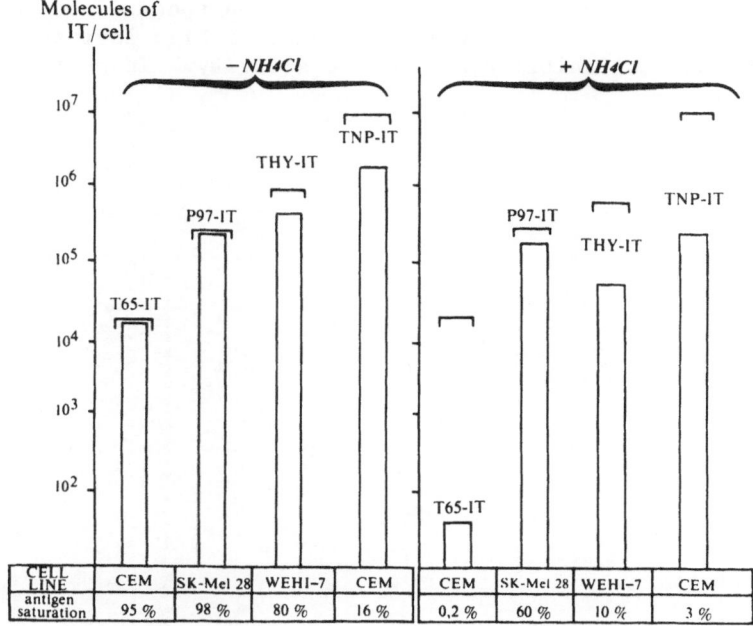

Fig. 8. Number of IT molecules bound per target cell
at IC 50. The affinity constants (methods of
Scatchard) and the maximum antibody binding
per cell were determined with radioactive antibodies. With the help of antibody affinity and
the antigen density/cell (maximum accessible
antigen) the amount of IT molecules bound per
target cell was calculated for the concentration
of IT leading to 50% inhibition of leucine uptake. The corresponding quantity of IT molecules
bound per target cell is shown for ITs without
any activation (left part) and for IT with activation by 10 mM ammonium chloride (right part).
The percent of saturation of the accessible antigen is indicated at the bottom.

Colonies formed by cells escaping IT treatment were much more
resistant to a new IT treatment and had less antigen. This might
be explained by selection of cells with low antigen density. One
possibility to kill such escaping cells could be a cocktail of ITs
directed against different antigens.

Comparison between activated I-A-Ts and I-AB-Ts

Immuno-AB-toxins are in general active in lower concentrations

than I-A-Ts, but have the disadvantage of non-specific toxicity.
Therefore I-AB-Ts are used in the presence of high lactose concentrations in order to inhibit non-specific activity. In contrast, adju-

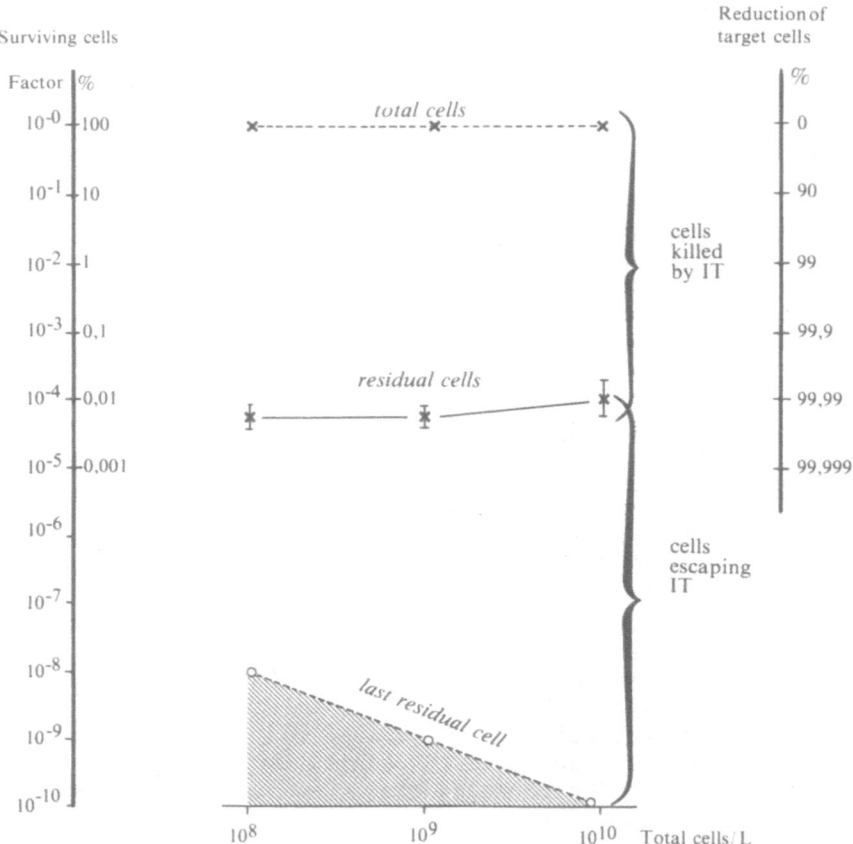

Fig. 9. Residual surviving clonogenic cells after incubation with IT. Cells were incubated in high
concentrations (10^8 - 10^{10}/l) with the T101 IT
at 10^{-8}M for 20 hours and then cloned in soft
agar in different dilutions. With a cloning
efficiency of 60% up to 10^{10} clonogenic cells
survived in the controls. (x-----x) IT reduces
this amount by several orders of magnitude
(ordinate) in relation to the initial number of
target cells (abscissa). The area between total
cells (x----x) and residual cells (x———x) after
IT treatment indicates the cells killed by IT.
The area between residual cells and the last
residual cell (o----o) shows the quantity of
cells escaping the IT treatment.

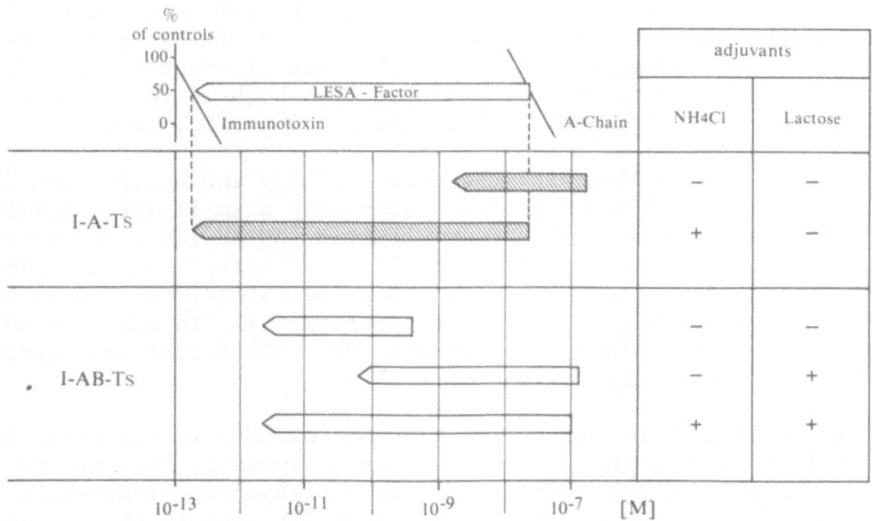

Fig. 10. Comparison of the concentration range of specificity (LESA factor) of I-A-Ts with respect
to I-AB-Ts. Antibody toxin conjugates were
assembled between the T101 antibody (Hybritech)
and the A-chain of the toxin ricin, I-A-T (from
our laboratory) or with the whole toxin ricin,
I-AB-T, (from Leonard and Royston, La Jolla,
Calif.) Both conjugates were tested in parallel
on the same CEM cell line after incubation for
18 hours at 37°. LESA factors (ligand enhanced
specific activity factors), are represented as
arrows and indicate the concentration range in
which these ITs are specifically active. At
lower concentrations than indicated by the head
of the arrow, there is no activity. At higher
concentrations than determined by the tail of
the arrow, there is non-specific activity. The
ITs were used in the presence or absence of adjuvants: 10 mM NH_4Cl or 100 mM lactose.

vants like lysosomotropic amines are not generally used with I-A-Ts,
although they can increase their specific activity considerably. We
therefore compared immuno-AB-toxin (from Dr. Royston, La Jolla,
California), also assembled with the T101 antibody, with our Immuno-
A-toxin.[32] The range of concentrations in which ITs become only
specifically active is characterized by the LESA factor and represented by an arrow (Fig. 10). In lower concentrations than those
indicated by the arrow there is no cytotoxicity and in higher concentrations there is only non-specific cytotoxicity.

The range of specific activity of our anti-T-cell I-A-T without ammonium chloride is narrow (small LESA factor of 10^2) and limited to high concentrations. But in the presence of ammonium chloride the specific activity becomes extended from high to very low concentration of 10^{-13} (large LESA factor of 10^5) (Fig. 10).

The I-AB-T studied with the same antibody and on the same CEM cell without the adjuvant lactose shows only a small specificity range (LESA factor of 10^2) but now limited to the low concentration range. In the presence of lactose the specificity range is larger (LESA factor of 10^3) and begins in high concentrations. It does not, however, reach very low concentrations. If, in addition to lactose, ammonium chloride is present, the LESA factor now becomes extended to low concentrations of 10^{-12}M.

There is only a small difference in the LESA factor when the I-AB-T is tested in the presence of both adjuvants, lactose and ammonium chloride, and compared to the corresponding I-A-T with ammonium chloride alone. Therefore, both ITs seem to show a similar efficiency in-vitro in the presence of adjuvants. The advantage of I-A-Ts becomes more apparent when they are injected with their adjuvant in-vivo. The adjuvants are immediately diluted. I-A-Ts then lose their activity, but the I-AB-Ts lose the protection by lactose and become highly non specifically toxic. This may be of major interest, if human bone marrow treated with ITs in-vitro is reinjected into the patient. Although the whole I-A-T content may be injected with the marrow, I-AB-Ts must be carefully eliminated beforehand, which implies manipulation of the marrow with the danger of bacterial contamination and loss of stem cells.

FACTORS DETERMINING CYTOTOXIC KINETICS

IT kinetics are determined by several factors : the phases of cell death, the antigen density, the antigen antibody system and the cell type of target cells.

Phases of kinetics and their relationships

Kinetics differ widely depending in the phases of cell death studied. Phase A kinetics, or antibody binding kinetics, are obtained in minutes, phase B or translocation kinetics in hours, and finally phase D, or cell disintegration kinetics, are only completed within days (Fig. 11). Due to such large differences, kinetics must be classified according to their phases, and can only be compared within the same phase and with the same method.

During phase A, the time needed to bind 90% of the human T-cell IT is in the order of 10 minutes. Saturation of all accessible antigens, about 20000 per cell, on CEM cells is reached within 15 minutes.[33] Ammonium chloride has no influence on these kinetics.

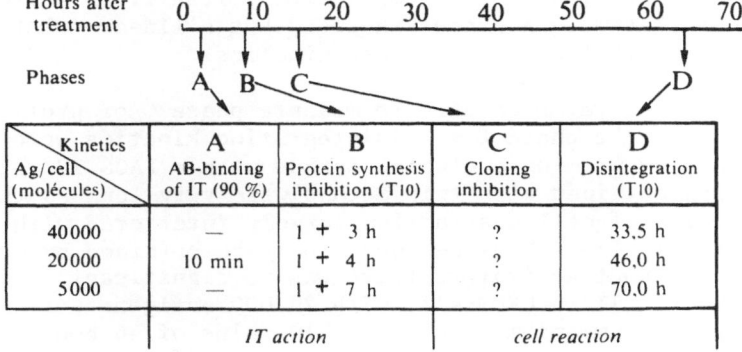

Fig. 11. Kinetics of the different phases of cell death
induced by I-A-Ts in relation to antigen density. Phase kinetics were studed with the T101
IT on stable cell clones derived from a heterogeneous CEM cell line. They had been selected
for their different antigen density. Binding
kinetics were determined with radioactive antibodies and expressed as the time needed to bind
90% of antibodies. (phase A). The kinetics of
protein synthesis inhibition were examined by
the leucine uptake inhibition on intact cells
and defined as the lag time (one hour in this
model) plus a T10 value, corresponding to the
time needed to reduce leucine uptake to 10%.
(phase B). No reliable kinetic test was available for the proliferation phase (phase C). The
kinetics of the cell disintegration (phase D)
were determined with the ethidium bromide exclusion test by cytofluorimetry. Since during
these long kinetics cell proliferation interfered with cell destruction, the T10 values were
extrapolated from the results during the first
24 hours.

It would be expected that these values would vary with the affinity
of different antibodies and with the accessibility of the antigens.

With the same IT on CEM cells with about 20 000 antigens per
cell the translocation kinetics of phase B were much longer, with
T10 values being of the order of 60 hours. In the presence of activators such as NH_4Cl, T10 values were reduced to 4 hours, plus an
additional lag time of about 1 hour (Fig. 11). This value seems to
depend essentially on the translocation time of the A chain into
the cytoplasm, and not so much on the time needed for endocytosis

or for the inhibition of ribosome activity itself. B phase kinetics are not correlated to A phase kinetics, but a linear relationship can be found with the later D phase kinetics.

We have not yet been able to measure phase C or proliferation kinetics. But the phase D or disintegration kinetics were measured by the ethidium bromide exclusion method with a FACS IV cell sorter.[33] The kinetics obtained with this method were very long so that first order kinetics of cell destruction largely interfered with cell proliferation. The T10 values could only be obtained by extrapolation. Without activators there was no significant activity of the anti-T-cell IT on CEM cells with 20 000 antigens per cell. In the presence of ammonium chloride a T10 value of 46 hours was obtained, which is about 9 times longer than the leucine uptake inhibition in the B phase (Fig. 11). This indicates that cell disintegration is a late consequence of IT activity, and that target cells survive a long time after inhibition of their protein synthesis without measurable damage to their cell membrane.[33]

Antigen density

In order to measure the influence of antigen density on the cytotoxicity induced by ITs, mutants from the same cell line were obtained. The initial CEM cell line, heterogeneous for antigen expression, was cloned with the FACS, according to different antigen densities. Subclones of 44000, 20000 and 5000 antigens per cell were obtained. The B phase kinetics measured with these subclones showed a clear correlation between kinetics and antigen density. With 44000 antigens, T10 values were in the order of 3 hours in the presence of activators, with 20000 antigens about 4 hours, and with 5000 about 7 hours.[31,33] A similar correlation was obtained between antigen density and D phase kinetics, 33, 5, 46 and 76 hours respectively which was about 9 times longer than the B phase kinetics.[33]

Antigen-antibody system

In order to evaluate the influence of different kinds of antigens, the influence of antigen density and antibody affinity had to be discarded. Therefore kinetics were established at non saturation concentrations, in which only 10000 IT molecules were bound per cell. The corresponding IT concentrations were determined with a quantitative cytofluorometric method. These kinetics, called antigen related kinetics, showed a great variety of T10 values without activators, ranging from approximately 300 hours for the leucocyte IT (Mab 29.33, Hybritech) to 150 hours for a betamicroglobulin IT (Mab 3E10, Clin Midy) and 80 hours for the T-cell IT (Ab T101, Hybritech). In the presence of activators, the values were 80, 25 and 5 hours respectively, showing that the antigen antibody system had a great influence on the kinetics of ITs.[34]

Cell type

A preliminary result on the influence of the cell type could be obtained with an anti-DNP IT on three different cell lines modified with the hapten TNP : the human T leukemia CEM, the mouse leukemia WEHI-7 and the human melanoma cell line SK-mel 28. The corresponding T10 values of the B phase kinetics were 1.5, 3.5 and 6 hours respectively. This may indicate an influence of the cell type. One cannot, however, exclude the possibility that the hapten modified proteins are of a different nature and of different densities in the three cell types studies.

PHARMACOKINETICS OF I-A-Ts

Besides cytotoxicity kinetics, serum half-life of IT in-vivo appears to be a predominant factor in IT efficiency in-vivo. The disulphide linkage between A chain and antibody was always suspected to be labile under in-vivo conditions, i.e., rapidly cleaved in the blood or by the passage through organs like the liver. The half-life of IT and the stability of the disulphide bridge was therefore tested first in-vitro in the presence of whole blood or plasma, and then in-vivo after injection into rabbits.

In-vitro stability

An anti-DNP-IT was mixed with heparinized plasma or heparinized whole blood and incubated for 24 hours at 37°. Thereafter IT activity in the supernatant was measured against its target cells, TNP-CEM cells, by the leucine uptake inhibition test. No significant decrease in IT activity could be found after a 24-hour incubation period with plasma or whole blood. This indicates that the instability of IT cannot be attributed to thiols or enzymes in the serum, or to an interaction with blood cells.[35]

In-vivo half-life

The presence of ITs in the serum after i.v. injection in rabbits was followed up with radioimmunoassays for 24 hours. The antibody was removed from the serum with polyclonal anti-Ig antibodies adsorbed on plastic and revealed with radioactive anti-Ig antibodies. Free or bound A chain was immobilized with polyclonal anti-A chain antibodies, adsorbed on plastic and detected with radioactive anti-A chain antibodies. The presence of intact IT, i.e., A-chain attached to an Ig molecule, could be demonstrated with immobilized anti-A chain anitbodies binding to the hybrid molecules and with radioactive anti-Ig antibodies used to detect them.

After i.v. injection, free antibody decreases slowly in the plasma (Fig. 12). The half-life is in the order of 36 hours. When A chain or whole IT were injected, the pharmacokinetic profiles were

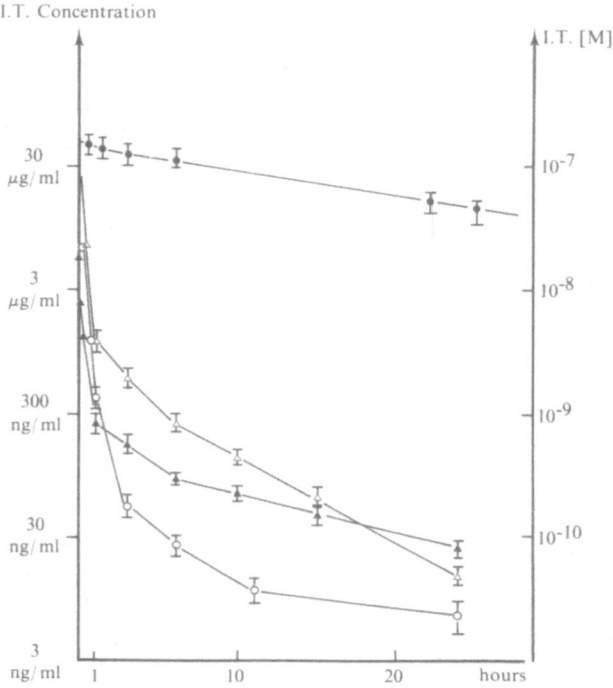

Fig. 12. Pharmacokinetics of I-A-Ts with a disulphide or
         with a thioether bridge. The test substances
         were injected in low doses i.v. into rabbits.
         Blood samples were collected after different
         time periods and assayed for the presence of the
         injected materials. With the help of a double
         determinant radioimmunoassay, consisting of ad-
         sorbed antibodies against A chain and radioactive
         anti A chain antibodies, the A chain concentra-
         tion in the serum was determined (O——O). Sim-
         ilarly with adsorbed anti-mouse γ-globulin anti-
         bodies and radioactive anti-mouse γ-globulin
         antibodies, T101 antibodies (●——●) were detected.
         Immunotoxins with a disulphide bridge (△——△) or
         with a thiother bridge (▲——▲) were studied with
         adsorbed antibodies against A chain and radio-
         active antibodies against mouse γ-globulin.[25]

faster. No clear cut change of the slopes of the plasma concentration versus time curves indicates a first distribution phase and a
second metabolic phase. A chain is cleared more rapidly than free
Ig and the whole IT has an intermediate clearance. This suggests
that the A chain induces the rapid disappearance of I-A-T. Since

the decrease of IT concentration during the distribution phase is greater than expected in view of its dilution in the extravascular fluid, other mechanisms have to be considered such as adsorbtion or degradation of IT during its passage through organs.

The half-life during the second phase may be more important for in-vivo activity since IT in the blood is in equilibrium with the extravascular fluid. The computerized data indicate a half-life of 15 hours for the antibody and 4.5 hours for whole IT. Half-life for free A chain is difficult to estimate.[25]

In-vivo stability of the disulphide bridge

The suspected instability of the disulphide bridge in-vivo was studied by comparing the half-life of a disulphide conjugate with a thioether conjugate, which represents a chemically stable bond between A chain and antibody. The half-life of the thioether conjugate in the rabbit (9.5 hours) was two times longer than the one with a disulphide bridge (4.5 hours).[25] Although this difference in half-life seems to be of interest, the thioether linkage does not represent a great advantage over the disulphide bond, because thioether conjugates are in general much less active than the corresponding disulphide conjugates (a thioether anti-Thy 1,2 IT is about 100 times less active than its disulphide homologue).

In-vivo half-life of IT activity

The radioimmunoassay detecting the IT molecule by its antigenicity and the integrity of the linkage does not reflect its activity, i.e., in a therapeutic situation circulating antigens could adsorb its antibody activity or the Ab or A chain moieties may be partially degraded. Measurements of IT activity are therefore a necessary complementary test, which can even be more sensitive with some ITs than the radioimmunoassay. With the anti-Thy 1,2 in C57 Bl/Ka (BL 1.1) congenic mice, which do not bear the antigen, the half-life of activity in the serum was measured with an in-vitro test on T2 mouse leukemia cells.[35] The calculated half-life of the second phase (5 hours) was similar to the corresponding half-life determined by the radioimmunoassay. It is however, more important to know the remaining IT activity in molar concentrations, which can be compared to in-vitro concentrations. The IT concentration is thought to be in equilibrium with the extravascular fluid during the second or metabolic phase of pharmacokinetics and can be calculated from the intravascular concentrations. With the anti-human T-cell IT injected into mice at doses of 1/5 of the LD50 it was found that a molar concentration of 10^{-8} (with respect to A chain) was maintained during at least 16 hours. For the T101 IT this is a concentration, which would saturate all antigens of 10^7 target cells. This corresponds to about the quantity of Ichikawa cells (a human T-cell leukemia bearing the T-65 antigen) injected i.p. in the nude mice (see below). Therefore there is a sufficiently high concentration of IT over a period of time long enough to be active on easily accessible leukemia cells. If, however, the localization of ITs on solid tumours in-vivo needs several days to be accomplished, as indicated by imaging studies,[36] a half-life of 5 hours may be insufficient for IT activity.

CONSEQUENCES OF IT KINETICS IN-VIVO

In order to avoid the problems of accessibility of tumour cells in-vivo we excluded solid tumour systems at the beginning of our studies. The therapeutic value of experiments with i.p. transplanted tumour cells and an i.p. treatment with IT is therefore difficult to extrapolate to spontaneous tumours in humans. One advantage in this particular situation is, however, the accessibility of transplanted and rapidly treated tumour cells, so that the half-life of ITs might not be a limiting factor. Since the extravascular T101 IT concentration is sufficient to saturate the antigens on Ichikawa cells over a long period (16 hours). the influence of the kinetics of cytotoxicity should be predominant in this model system.

Low efficiency with a slow I-A-T

In a recent experiment, the anti-human T-cell (Mab:T101, Hybritech) with slow B phase kinetics (T10=60 hours), was injected i.p. into nude mice bearing 25×10^6 Ichikawa cells. Treatment was performed on 8 consecutive days with doses of 1.5 micrograms A chain (bound to Ab) per mouse per day. This corresponds to about 1/50 of the LD50 (single dose). Control mice and those treated with the antibody mixed with non-chemically bound A chain died within 43 days. The IT treated group showed no long-term survivors but the mean survival time was prolonged by up to 2 weeks (Fig. 13). This effect was obtained with a minimum quantity of injected tumour cells leading to 100% tumour take.[35]

Higher efficiency with a rapid I-A-T

An anti-Thy 1,2 IT assembled with a rat IgG1 antibody and presenting fast cytotoxicity kinetics (B phase T-10 value of 4 hours) gave encouraging results. Congenic B57 BL/Ka (B1 1.1) mice were injected i.p. with T2 cells (a mouse T cell leukemia) and treated on day 1 with 100 micrograms and on days 2 and 3 with 50 micrograms of anti-Thy 1,2 IT corresponding to 1/2 of the LD50, single dose.

With a limited number of 5×10^5 leukemia cells control animals died within 50 days but in the IT or antibody treated groups, 90 or 80% respectively of the animals survived for more than 150 days. By increasing the tumour cell number 10 times in a second experiment (5×10^6 cells), a great difference appeared between the antibody and the IT treated groups (Fig. 14). While in the controls the mean

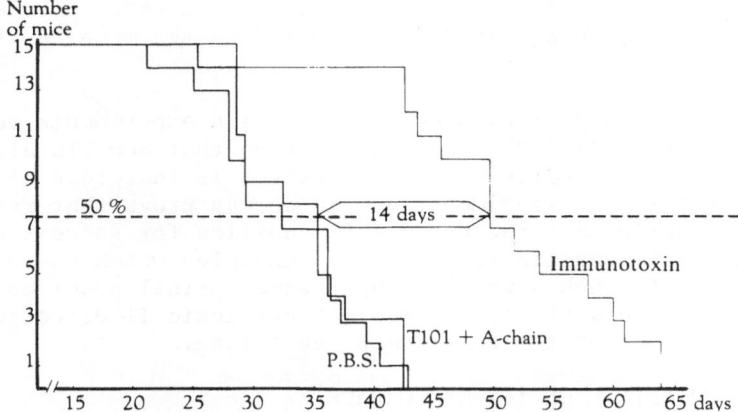

Fig. 13. Survival of nude mice after transplantation of
of a human leukemia cell line and treatment with
IT. Groups of 15 nude BALB/C mice were injected
with 25×10^6 Ichikawa cells i.p. (kindly provided
by Dr. Shaw Watanabe) and treated during 8 days
with daily i.p. injections of $1.5 \mu g$ (A chain content) of the T101 IT. Control groups received
equivalent amounts of a mixture of unmodified
antibodies with unbound A chain or only PBS.

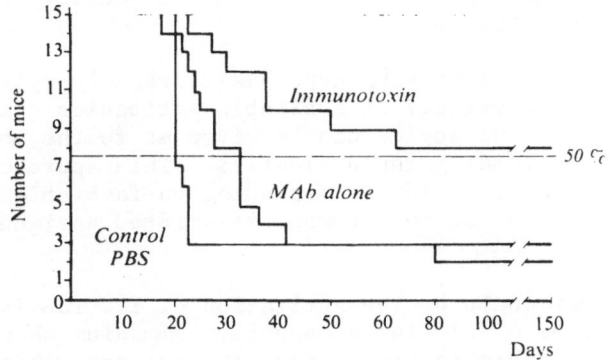

Fig. 14. Survival of mice after transplantation of a
mouse leukemia cell line and treatment with IT.
Groups of 15 C57Bl/Ka (BL 1.1.) mice were injected with 5×10^6 T 2 cells i.p. and treated
with an anti-Thy 1.2 IT (MAB AT15E, IgG from the
the rat) in doses of $100 \mu g$ the first day and 50
μg the two following days. Control groups were
treated with equivalent amounts of unmodified
antibodies or with PBS alone. Survival was observed over a period of 150 days.

survival time was increased from 20 days to 32 days after antibody treatment alone, more than 50% of the IT treated mice survived more than 150 days.[35]

There are only a few successful in-vivo experiments reported in the literature.[2,7,37,38] The confirmation that new ITs also inhibit tumours in-vivo in different model systems is therefore of interest. Although these two experiments can in no way prove that rapid cytotoxicity kinetics are the essential condition for success of ITs in-vivo, they nevertheless represent two examples which would be in accordance with such a working hypothesis. Final proof could only be obtained with a slowly and rapidly cytotoxic IT directed against the same epitope on the same tumour cell line.

CLINICAL POTENTIAL OF I-A-Ts TO DATE

In-vivo treatment of cancer patients

The low non-specific toxicity of I-A-Ts with an LD50 of about 20 mg/Kg in mice allows their utilization in the human patient today.[16] Toxicology studies in two other species, the rat and the monkey, confirm the low toxicity of conjugates assembled with the A chain of ricin. Therefore the word immunotoxin, evoking highly toxic substances, is misleading with respect to I-A-Ts, using the A chain only. However, the efficiency of tumour treatment with ITs cannot be taken for granted, since only for some animal experiments performed under favorable conditions, was a positive effect of ITs reported in the literature.

According to our working hypothesis, rapidly cytotoxic ITs could be selected from a variety of available antibodies against different antigens. To date no advice can be given as to the most favorable conditions for obtaining rapid kinetics. This approach based on a general screening of rapid ITs depending on favorable antigens will, however, be limited, because tumour associated antigens are not very frequent.

Another way could be the activation of ITs in-vivo with lyso-somotropic amines or similar compounds. Ammonium chloride does not seem to be applicable in-vivo, since the necessary high concentration of 10^{-2} would probably not be obtained over a sufficiently long time. Therefore, other similarly acting small molecular weight drugs are presently under study for their utilization in-vivo.

Ex-vivo treatment of bone marrow from leukemia patients

While the activation of IT kinetics is still a problem in-vivo, lysosomotropic amines make some ITs extremely potent in-vitro. They may therefore now be used in eradicating contaminating leukemia cells in autologous bone marrow[39] or T-cells in allogenic bone marrow

grafts.[40] Some tumours, such as leukemias, are sensitive to supralethal doses of radiotherapy or chemotherapy, which induces bone
marrow toxicity. Such problems may be overcome by autologous or
allogeneic bone marrow grafts. However, in autologous grafts contaminating leukemia cells must be eliminated to prevent recurrence of
leukemia. In allogenic grafts T-cells should be eliminated in order
to avoid graft-versus-host reactions. Treatment of bone marrow has
so far been effected with monoclonal antibodies alone or in the
presence of complement.[39,41] ITs may have several advantages over
these methods. The efficacy of I-A-Ts[42,43] as well as of I-AB-Ts[44]
were demonstrated in animal experiments.

    As described in a previous paragraph, the anti-T cell IT in con-
centrations of 10^{-8} is capable of reducing T-leukemia cells to less
than 10^{-4} of their initial quantity in whole bone marrow (Fig. 9).
An additional IT i.e. an anti-beta microglobulin IT recognizing the
cells by another antigen increases[31] the cell reduction to more
than 10^{-6}. About 10^{10} nucleated cells are present in one litre of
bone marrow. Contamination by leukemia cells in remission marrows
may vary from undetectable quantities to a few percent of nucleated
cells. A contamination of 10% would correspond to 10^9 tumour cells.
The anti-T-IT could reduce these tumour cells to less than 10^5.
Although 99·99% of leukemia cells can thus be destroyed, some cells
would still survive. But it might well be that such a reduction is
sufficient for the normal defence mechanisms of the body to eradicate
the residual tumour cells. If not, the efficiency of ITs will be
considerably increased by the use of a cocktail of ITs against different antigens.

    It is essential that the therapeutic IT concentrations for an
ex-vivo treatment do not harm bone marrow stem cells. Therefore it
was verified that the presence of ammonium chloride (10mM) does not
diminish the formation of CFUc and BFUe colonies. The toxicity of
IT was measured with different IT concentrations from 10^{-6} to 10^{-9}M.
After an incubation of 4 hours no significant reduction of colony
formation could be found.[45] Only after an incubation of 8 hours
the highest IT dose of 10^{-6}M inhibited BFUc colonies to 61% and
BFUe colonies to 29%, while the concentration of 10^{-7}M remained non
toxic.[46] The toxic doses for human bone marrow are therefore about
100 times higher than the optimal therapeutic doses for leukemia
cell elimination of 10^{-8}M.

    An advantage of the I-A-Ts is the low non-specific toxicity.
The total quantity of IT corresponding to a 10^8M concentration in
one litre of bone marrow is about 2 mg (0·04 mg/kg for a patient of
50 kg, calculated as Ig plus A chain content). Since this dose
represent less than 1/100 of the LD50 in the mouse, there should be
no problem in reinfusing this quantity with the marrow into the
patient. The ammonium chloride content of the treated bone marrow
would be immediately diluted and excreted and thus lose its poten-

tiating activity in-vivo. Thus I-A-Ts crossreacting with other organs will become inactive after infusion.

The great practical advantage of I-A-Ts is their easy manipulation, since there is no need to isolate or concentrate nucleated bone marrow cells prior treatment or to wash them thereafter. It is sufficient to inject IT into the bag containing whole marrow, to incubate it for the time needed and to reinfuse it with the IT into the patient. In contrast to non-specific toxicity due to antibodies plus complement, an I-A-T can be standardized. Since it is much easier to handle, a loss of bone marrow cells during the manipulations can be avoided. When comparing the efficiency of one IT with a cocktail of complement-fixing antibodies, an activated I-A-T is at least equal.[47] A cocktail of I-A-Ts, available in the future, should then be more potent than a cocktail of complement fixing antibodies.

CONCLUSIONS

One aspect of immunotoxins, their kinetics of cytotoxicity, has not been sufficiently studied in the past. It now seems to be established for the in-vitro activity of ITs and suggested for their in-vivo activity, that only rapidly cytotoxic ITs are efficient in tumour cell destruction. The reason for the importance of rapid kinetics is the fact that IT is in competition with as yet unknown cellular defense mechanisms. If the cell proliferation or the neutralization of an IT by the cell is more rapid than the cytotoxic activity of the IT, target cells will escape. Only with a rapidly cytotoxic IT (an anti-Thy 1,2 IT with a B phase T10 value of 5 hours) did we obtain good in-vivo results with more than 50% of longterm surviving mice after the transplantation of a mouse leukemia. This confirms that well selected ITs can be efficient in-vivo.

Different classes of kinetics may be defined, which correspond to different phases (A-B-C-D) of cell death and can be measured with the appropriate test systems as T50 or T10 value. In general, cytotoxic kinetics depend on antigen density, the antigen-antibody system, and the cell type.

Theoretical considerations suggest that ITs with B phase kinetics (protein synthesis inhibition) characterized by T10 values of more than 10 hours may be less promising for a large tumour burden in-vivo. Since experimentally obtained T10 values vary from 4 to 150 hours, only some of these ITs will be useful for in-vivo treatment. Kinetics of cytotoxicity can, however, be artificially accelerated by B chains of ricin as well as by lysosomotropic amines, which may inhibit IT neutralization during the endocytotic process. Thus, extremely high efficiency can be obtained with certain ITs in the presence of ammonium chloride. With only about 40 IT molecules bound per cell, target cells will be killed at IC50% concentrations.

In a colony formation assay target cells are reduced to less than one in 10^4 cells.

The in-vivo effect of IT not only depends on the kinetics of cytotoxicity but also on the in-vivo half-life. Other yet insufficiently studied conditions such as soluble antigens in the serum or neutralizing antibodies will also interfere. ITs show a half-life of about 5 hours. Therefore a high IT concentration of $10^{-8}M$ can be maintained for 16 hours in the mouse. Since this optimal therapeutical dose is maintained long enough, the pharmacokinetic behaviour of ITs is not a limiting factor for easily accessible tumours such as leukemias. They will however, become much more important with less accessible solid tumours.

The first promising clinical use of I-A-Ts is the ex-vivo treatment of leukemic bone marrow of patients subjected to autologous or allogenic bone marrow transplantations, because rapid kinetics can easily be obtained in-vitro but not yet in-vivo. Here ITs seem to be superior to antibodies plus complement (Fig. 15). The ex-vivo treatment of autologous bone marrow aims to reduce contaminating leukemia cells, which might lead to recurrence of leukemia, to a minimum before the marrow is reinjected to the patient. A reduction of leukemia cells to less than one in 10^4 cells can be obtained with I-A-Ts due to the acceleration of kinetics with ammonium chloride. Another hopeful use of activated I-A-Ts will be the destruction of T cells in human bone marrow. This will reduce the graft-versus-host reaction after transplantation of an allogenic bone marrow graft.[40,48] Although the ex-vivo treatment with activated I-A-Ts

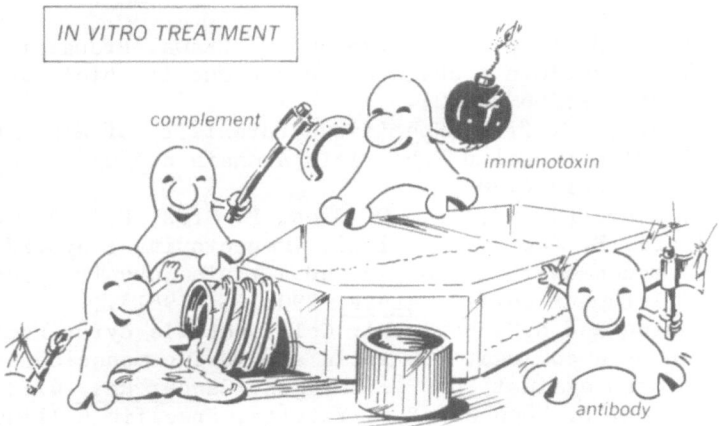

Fig. 15. A simplified view of the treatment of human
          bone marrow. ITs are in competition (for
          simplicity and efficiency) with antibodies,
          or antibodies plus complement.

can now be undertaken, the in-vivo acceleration of the kinetics of ITs is still under study. Rapid cytotoxic kinetics of ITs will, however, be one of the essential conditions for the final break-through in the treatment of cancer patients with ITs in the future.

ACKNOWLEGMENTS

We would like to thank Dr. G.A. Voisin for stimulating discuss-ions and advice, all our technicians for their skillful work, O. Thurneyssen for verification of the literature and Miss S. Dusfour and Miss A. Garcia for secretarial assistance.

REFERENCES

1. D.G. Gilliland, Z. Steplewski, R.J. Collier, K.F. Mitchell, T.H. Chang and H. Koprowski, Antibody-directed cytotoxic agents : Use of monoclonal antibody to direct the action of toxic A chains to colorectal carcinoma cells, Biochem. J. 77:4539 (1980).

2. F.K. Jansen, H.E. Blythman, D. Carriere, P. Casellas, J. Diaz, P. Gros, J.R. Hennequin, F. Paolucci, B. Pau, P. Poncelet, G. Richer, S.L. Salhi, H. Vidal and G.A. Voisin, High specific cytotoxicity of antibody-toxin hybrid molecules (immunotoxins) for target cells, Immunol. Lett. 2:97 (1980).

3. K.A. Krolick, C. Villemez, P. Isakson, J.W. Uhr and E.S. Vitetta, Selctive killing of normal or neoplastic B cells by antibodies coupled to the A chain of ricin, Proc. Natl. Acad. Sci USA, 77:5419 (1980).

4. Y. Masuho and T. Hara, Target-cell cytotoxicity of a hybrid of Fab' of immunoglobulin and A chain of ricin, Gann, 71:759 (1980).

5. H. Miyazaki, M. Beppu, T. Terao and T. Osawa, Preparation of antibody (IgG)-ricin A chain conjugate and its biologic act-ivity, Gann, 71:766 (1980).

6. V. Raso and T. Griffin, Specific cytotoxicity of a human immunoglobulin-directed Fab' ricin A-chain conjugate, J. Immunol. 125:2610 (1980).

7. H.E. Blythman, P. Casellas, O. Gros, P. Gros, F.K. Jansen, F. Paolucci, B. Pau and H. Vidal, Immunotoxins : Hybrid mole-cules of monoclonal antibodies and a toxin subunit specifi-cally kill tumour cells, Nature. 290:145 (1981).

8. L.L. Houston and R.C. Nowinski, Cell-specific cytotoxicity expressed by a conjugate of ricin and murine monoclonal anti-body directed against Thy 1.1 antigen, Cancer Res. 41:3913(1981).

9. K.A. Krolick, D. Yuan and E.S. Vitetta, Specific killing of a human breast carcinoma cell line by a monoclonal antibody coupled to the A chain of ricin, Cancer Immunol. Immunother, 12:39 (1981).

10. T.N. Oeltmann and J.T. Forbes, Inhibition of mouse spleen cell function by diphtheria toxin fragment A coupled to anti-mouse

Thy 1.2 and by ricin A-chain coupled to anti mouse IgM. Arch.
Biochem. Biophys. 209:362 (1981).

11. I.S. Trowbridge and D.L. Domingo, Anti-transferrin receptor
monoclonal antibody and toxin-antibody conjugates affect
growth of human tumour cells, Nature, 294 (1981).

12. T.W. Griffin, L.R. Haynes and J.A. Demartino, Selective cytotoxicity of a ricin A-chain-anti-carcinoembryonic antigen
antibody conjugate for a human colon adenocarcinoma cell line,
JNCI, 69:799 (1982).

13. V. Raso, J. Ritz, M. Basala and S.F. Schlossman, Monoclonal
antibody-ricin A-chain conjugate selectively cytotoxic for
cells bearing the common acute lymphoblastic leukemia antigen,
Cancer Research, 42:457 (1982).

14. M.I. Bernhard, K.A. Foon, T.N. Oeltmann, M.E. Key, K.M. Hwang,
G.C.Clarke, W.L. Christensen, L.C. Hoyer, M.G. Hanna, Jr. and
R.K. Oldham, Guinea pig line 10 hepatocarcinoma model :
Characterization of monoclonal antibody and in-vivo effect of
unconjugated antibody and antibody conjugated to diphtheria
toxin A chain, Cancer Res. 43:4420 (1983).

15. G. Moller, ed., "Antibody Carriers of Drugs and Toxins in
Tumor Therapy", Immunological Reviews vol. 62, Munksgaard,
Copenhagen (1982).

16. F.K. Jansen, H.E. Blythman, D. Carriere, P. Casellas, O. Gros,
P. Gros, J.C. Laurent, F. Paolucci, B. Pau, P. Poncelet, G.
Richer, H. Vidal and G.A. Voisin, Immunotoxins : Hybrid mole-
cules combining high specificity and potent cytotoxicity, in:
"Antibody Carriers of Drugs and Toxins in Tumor Therapy",
Immunological Reviews, G. Moller, ed., vol.62, Munksgaard,
Copenhagen (1982).

17. F.L. Moolten and S.R. Cooperband, Selective destruction of
target cells by diphtheria toxin conjugated to antibody
directed against antigens on the cells, Science, 169:68 (1970).

18. P.E. Thorpe, W.C.J. Ross, A.J. Cumber, C.A. Hinson, D.C.
Edwards, A.J.S. Davies, Toxicity of diphtheria toxin for
lymphoblasted cells is increased by conjugation to anti-
lymphocytic globulin, Nature, 271:752 (1978).

19. J.R. Youle and D.M. Neville, Jr., Anti-Thy 1.2 monoclonal
antibody linked to ricin is a potent cell-type-specific
toxin, Proc. Natl. Acad. Sci. USA, 77:5483 (1980).

20. F.K. Jansen, H. Blythman, D. Carriere, P. Casellas, P. Gros,
F. Paolucci, B. Pau, P. Poncelet, G. Richer and H. Vidal,
Replacement of the B chain of ricin with specific conventional
or monoclonal antibodies, in: "Receptor-mediated binding and
internalization of toxins and hormones", J.L. Middlebrook and
L.D. Kohn, eds., Academic Press, New York (1981).

21. R.J. Youle and D.M. Neville, Kinetics of protein synthesis
inactivation by ricin-anti-Thy 1.1 monoclonal antibody hybrids:
Role of the ricin B subunit demonstrated by reconstitution,
J. Biol. Chem. 257:1598 (1982).

22. P. Casellas, J.P. Brown, O. Gros, P. Gros, I Hellstrom,

F.K. Jansen, P. Poncelet, R. Roncucci, H. Vidal and K.E. Hellstrom, Human melanoma cells can be killed in-vitro by an immunotoxin specific for melanoma-associated antigen p97, Int. J. Cancer, 30:437 (1982).

23. P. Casellas, H.E. Blythman, P. Gros, G. Richer and F.K. Jansen, High efficiency of immunotoxin potentiated by lysosomotropic amines to specifically kill tumor cells in:"Protides of the Biological Fluids", H. Peeters, ed., Pergamon Press, Oxford (1982).

24. S. Olnes and A. Pihl, Toxic lectins and related proteins, in: "The molecular actions of toxins and viruses", Ph. Cohen and S. van Heyningen, eds., Elsevier/North Holland, Amsterdam (1980).

25. P. Casellas and B. Bourrie, in preparation.

26. R.J. Youle, G.J. Murray and D.M. Neville, Jr., Studies on the galactose-binding site of ricin and the hybrid toxin man6P-ricin, Cell, 23:551 (1981).

27. P. Thorpe, A. Brown, B. Foxwell, C. Myers, W. Ross, A. Cumber and T. Forrester, Blockade of the galactose-binding site of ricin by its linkage to antibody, in: "Monoclonal Antibodies and Cancer", Proceedings of the 4th Armand Hammer Cancer Conference, La Jolla, R.E. Langman, I.S. Trowbridge and R. Dulbecco, eds., Academic Press, New York (1983).

28. E.S. Vitetta, W. Cushley and J.W. Uhr, Synergy of ricin A chain-containing immunotoxins and ricin B chain-containing immunotoxins in in-vitro killing of neoplastic human B cells, Proc. Natl. Acad. Sci. USA, 80:6332 (1983).

29. D. Carriere, submitted.

30. P. Casellas, submitted.

31. P. Casellas, in preparation.

32. O. Gros, P. Casellas, J. Leonard, I. Royston and F.K. Jansen, Immunotoxins with whole ricin or A chain : comparison of efficiency and specificity, in: "Bacterial Protein Toxins", Proceedings of the Seillac Workshop Conference, J.E. Alouf, F. Fehrenbach, H. Freer and J. Jeljaszewicz, eds., Academic Press, London (1983).

33. P. Poncelet, in preparation.

34. P. Casellas, D. Carriere, O. Gros, J.C. Laurent, P. Poncelet and F.K. Jansen, Properties of antibody-ricin A-chain conjugates (immunotoxins) in specific cell killing, in: "Bacterial Protein Toxins", Proceedings of the Seillac Workshop Conference, J.E. Alouf, F. Fehrenbach, H Freer and J. Jeljaszewicz, eds., Academic Press, London (1983).

35. H. Blythman, in preparation.

36. M.I. Bernhard, K.M. Hwang, K.A. Foon, A.M. Keenan, R.M. Kessler, J.M. Frincke, D.J. Tallam, M.G. Hanna, Jr., Leona Peters and R.K. Oldham, Localization of ^{111}In-and ^{125}I-labeled monoclonal antibody in guinea pigs bearing line 10 hepatocarcinoma tumors, Cancer Res., 43:4429 (1983).

37. K.A. Krolick, J.W. Uhr, S. Slavin and E.S. Vitetta, In-vivo

therapy of a murine B cell tumor (BCL1) using antibody-ricin A-chain immunotoxins, J. Exp. Med. 155:1797 (1982).

38. M. Seto, N. Umemoto, M. Saito, Y. Masuho, T. Hara and T. Takahashi, Monoclonal Anti-MM46 antibody ricin A chain conjugate : In-vitro and in-vivo antitumor activity, Cancer Res. 42:5209 (1982).

39. H. Kaiser, R. Levy, C. Browall, C.I. Civin, D.J. Fuller, S.H. Hsu, B.G. Leventhal, R.A. Miller, E.S. Milvenan, G.W. Santos and M.D. Wharam, Autologous bone marrow transplantation in T-cell malignancies : A case report involving in-vitro treatment of marrow with a pan-T-cell monoclonal antibody, J. Biol. Resp. Modif., 1:233 (1983).

40. D.A. Vallera, R.J. Youle, D.M. Neville, Jr., and J.H. Kersey, Bone marrow transplantation across major histocompatibility barriers. Protection of mice from lethal graft-versus-host disease by pretreatment of donor cells with monoclonal anti-Thy 1.2 coupled to the toxin ricin, J. Exp. Med. 155:949 (1982).

41. R.C. Bast, J. Ritz, J.M. Lipton, M. Feeney, S.E. Sallan, D.G. Nathan and S.F. Schlossman, Elimination of leukemic cells from human bone marrow using monoclonal antibody and complement, Cancer Res., 43:1389 (1983).

42. K.A. Krolick, J.W. Uhr and E.S. Vitetta, Selective killing of leukaemia cells by antibody-toxin conjugates : implications for autologous bone marrow transplantation, Nature. 295:604 (1982).

43. M. Muirhead, P.J. Martin, B. Torok-Storb, J.W. Uhr and E.S. Vitetta, Use of an antibody-ricin A-chain conjugate to delete neoplastic B cells from human bone marrow, Blood. 62:322 (1983).

44. P.E. Thorpe, D.W. Mason, A.N.F. Brown, S.J. Simmonds, W.C.J. Ross, A.J. Cumber and J.A. Forrester, Selective killing of malignant cells in a leukemic rat bone marrow using an antibody-ricin conjugate, Nature. 297:594 (1982).

45. N.C. Gorin, L. Douay, F.K. Jansen, G.A. Voisin, C. Baillou, A. Najman and G. Duhamel, Etude de l'immunotoxine T101 : Absence de toxicite envers les proge-niteurs hematopoietique CFUc et BFUe : application possible a l'auto-greffe de moelle osseuse, Hematologie 25:204 (1983).

46. N.C. Gorin, L. Douay, F.K. Jansen, G.A. Voisin, C. Baillou, A. Nahman, and G. Duhamel, Study of immunotoxin T101 : Absence of cytotoxicity against hemopoietic progenitor stem cells. Application to in-vitro therapy prior to autologous bone marrow transplantation, in: "Minimal Residual Disease In Acute Leukemia", B. Lowenberg, A. Hagenbeeck and M. Nijhoff, eds., The Hague Netherlands Publishers (1984).

47. R.C. Bast, Jr., P. Maver, C. Lipton, J. Rit, J. Nadler, L. Sallan, S. Nathan, D.G. and S.F. Schlossman, Elimination of malignant clonogenic cells from human bone marrow using multiple monoclonal antibodies and complement. Proceedings of the Annual Meeting of the American Association for Cancer Research, Vol. 24. (1983).

48. D. Vallera, R.C. Ash, E.D. Zanjani, J.M. Kersey, T.W. Lebien, P.C.L. Beverley, D.M. Neville and R.J. Youle, Anti-T-cell reagents for human bone marrow transplantation : Ricin linked to three monoclonal antibodies, <u>Science,</u> in press.

THE USE OF RICIN A CHAIN-CONTAINING IMMUNOTOXINS

TO KILL NEOPLASTIC B CELLS

Ellen S. Vitetta and Jonathan W. Uhr

Department of Microbiology, University of Texas
Southwestern Medical School, Dallas, Texas 75235

INTRODUCTION

Paul Ehrlich first discussed the potential use of antibodies
as carriers of pharmacologic agents. During the last 10 years there
have been many attempts to apply this concept to the elimination of
neoplastic and other target cells using antibodies coupled to toxic
agents. A cell-binding antibody conjugated to a plant or bacterial
toxin has been termed an "immunotoxin". One such toxin, ricin, like
most toxic proteins produced by bacteria and plants, has a toxic
polypeptide (A chain) attached to a cell binding polypeptide (B
chain) (Olsnes and Pihl, 1973). The B chain is a lectin that binds
to galactose-containing glycoproteins or glycolipids on the cell
surface. By mechanisms that are not yet well understood, the A
chain of the cell-bound ricin gains access to the cell cytoplasm
presumably by receptor-mediated endocytosis and penetration of the
membrane of the endocytic vesicle (Olsnes and Pihl, in press). There
is evidence that the B chain can also facilitate the translocation
of the A chain through the membrane of the endocytic vesicle, possibly by forming a pore (Jansen, et al., 1982; Neville and Youle,
1982; Thorpe and Ross, 1982; Houston, 1982). In the cytoplasm, the
A chain of ricin inhibits protein synthesis by enzymatically inactivating the EF-2 binding portion of the 60S ribosomal subunit. It
is thought that one molecule of A chain in the cytoplasm of a susceptible cell can kill it (Olsnes and Pihl, 1981).

The A and B chains of ricin can be separated, purified and
covalently linked to antibodies derivatized with the thiol-containing
cross-linker, SPDP. In the case of A chain-containing immunotoxins,
the antibody portion substitutes for the lectin portion (B chain)
thus allowing the specific targeting of the toxic A chain to the
relevant target cells.

179

THE MURINE BCL₁ MODEL

This disease bears a close resemblance to the prolymphocytic
form of chronic lymphocytic leukemia in the human i.e. splenomegaly
and severe leukemia (Slavin and Strober, 1977; Muirhead, et al.,
1981). Injection of one BCL₁ cell into a normal BALB/c mouse
results in leukemia in approximately one-half of the recipients 12
weeks later (Vitetta, et al., 1982). Tumor-bearing mice usually
survive for 3-4 months after receiving 10^5 - 10^6 tumor cells. The
BCL₁ tumor cells bear large amounts of cell surface IgMλ and traces
of IgDλ, both of which have the same idiotype.

ELIMINATION OF BCL₁ CELLS FROM BONE MARROW

In initial experiments, immunotoxins containing anti-idiotypic
antibody directed against the tumor-derived Ig were incubated with
populations of BCL₁ tumor cells and control cells. The specific
immunotoxin decreased protein synthesis in the populations containing
tumor cells by 70-80%; the percentage of tumor cells in these populations was also 70-80%. Control immunotoxins containing irrelevant
antibodies had no effect on BCL₁ cells nor did specific immunotoxins
have an effect on normal splenocytes, on T cell tumors, or on another
B cell tumor bearing a different idiotype (Fig. 1). Anti-idiotype
antibody by itself did not affect protein synthesis in BCL₁ cells.
These results indicate that immunotoxin-mediated killing of neoplastic B cells in a mixed population is specific (Krolick, et al.,
1980).

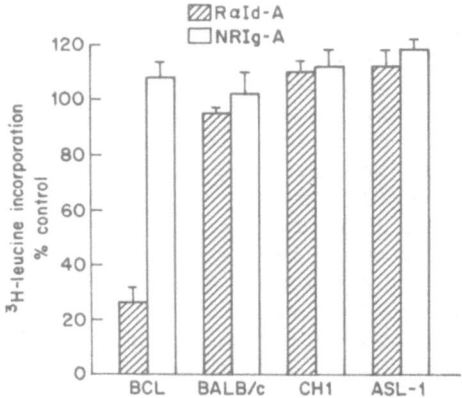

Fig. 1. Effect of anti-Id-A chain on protein synthesis in BCL₁
cells, normal BALB/c splenocytes, ASL-1 tumor cells,
and CH1 tumor cells. Anti-Id-A chain was used at
0.2 μg/ml. The CH1 cells express IgMλ on their surface but lack the BCL₁ idiotype. Hatched bars, anti-
Id-A chain; empty bars, NRIg-A chain.

Similar studies were performed using a tumor-infiltrated bone marrow (Krolick, et al., 1982a) (containing 15% BCL_1 cells) because of the clinical implications of removing tumor cells from marrow. In addition, it was possible to evaluate the nonspecific killing of stem cells by adoptively transferring the treated cells into lethally irradiated recipients. In these studies, anti-Ig immunotoxin was used since the only requirement for the specificity of the immuno-toxin was that it kill all the tumor cells but not the stem cells. Thus, it was possible to use a polyvalent antibody against Ig rather than an anti-idiotypic antibody. The results of these experiments (Fig. 2) indicate that 1) the hematopoietic system of all the ani-mals was reconstituted because all lethally irradiated mice survived. 2) 15 of 20 mice treated with tumor-reactive immunotoxin did not develop tumor over a period of 25 weeks of observation. Of the 5 animals that relapsed, all had idiotype positive cells that were susceptible to the in-vitro lethal effect of anti-Ig containing immunotoxins. Hence, no evidence was obtained for the emergence of an immunotoxin-resistant variant. Rather, the results of immuno-toxin treatment in these studies was consistent with the survival of 1 cell per 1 x 10^6 cells injected. Results similar to ours have been obtained by Thorpe et al. (1982) using antibody-ricin conjug-ates in the presence of lactose to delete tumor cells from rat bone marrow. Furthermore, we have recently extended this approach to the removal of neoplastic B cells from human bone marrow and demon-strated that the tumor cells are killed but that the CFU_{GM} BFU_E are not (Muirhead, et al. 1983).

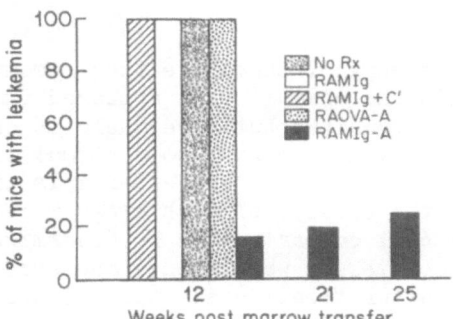

Fig. 2. Adoptive transfer into lethally irradiated recipients of BCL_1-containing bone marrow cells treated with rabbit antibody (Ab) to mouse Ig conjugated with A chain. Bone marrow cells containing 10 to 15 percent tumor cells were injected into groups of 20 mice at 10^6 marrow cells per mouse. Every two weeks after adoptive transfer the mice were examined for leukemia. At 25 weeks, all sur-viving mice were killed and 10^6 spleen cells were adopt-ively transferred into normal recipients. The spleen cells from one of the mice caused a tumor in these recip-ients 10 weeks later. Thus, this mouse is scored as leukemic at 25 weeks.

IN-VIVO THERAPY OF BCL$_1$

For these experiments (Krolick et al., 1982b), mice bearing
massive tumor burdens (20% of body weight or approximately 10^{10}
tumor cells) were employed. The strategy was to reduce the tumor
burden by at least 95% using nonspecific cytoreduction and to
eliminate the remaining tumor cells with immunotoxins directed
against either the idiotype or the δ chain of sIgD on the BCL$_1$
cells. The rationale for using anti-δ is that sIgD is present on
a large proportion of B cell tumors and, therefore, would present
a more practical reagent for clinical therapy. Furthermore, after
cytoreductive therapy, there are virtually no sIgD-positive normal
B cells or serum IgD to bind the immunotoxin. Normal B cells can
also be regenerated from sIgD$^-$ cells. In these experiments, non-
specific cytoreduction was accomplished with a combination of
splenectomy and fractionated total lymphoid irradiation (TLI).
Animals receiving no further treatment other than TLI and splenectomy
were dead within 8 weeks (Fig. 3). The injection of these cyto-
reduced mice with control immunotoxins or antibody alone did not
prolong their survival. In contrast, animals receiving anti-
immunotoxins appeared disease-free as judged by the absence of
detectable idiotype-positive cells 1 to 18 weeks later in 3 of
4 experiments. In one experiment, treated mice relapsed at 8-10
weeks after immunotoxin therapy. It should also be noted that 14
weeks after such immunotoxin treatment, mice in remission had normal
or above normal levels of sIgD-bearing B lymphocytes. Hence, stem
cells, pre-B cells or sIgD$^-$ lymphocytes had fully restored the virgin
B cell compartment.

These results suggest that 1) either remaining tumor cells had
been eradicated in the animals that appeared tumor-free or that some
viable tumor cells remained but were "held in check" by host resistance mechanisms. 2) Immunotoxin to a normal tissue component, in
this case sIgD, can be used to render animals disease-free and the
host can survive the effects of such cross-reactivity and can reconstitute the B cell compartment. To determine whether the animals
were disease-free, tissues were then transferred from disease-free
animals 25 weeks after treatments. All animals adoptively transferred tumor into normal mice indicating that the animals were not
tumor-free and suggesting that host resistance had developed.

The partial success of these experiments was probably due to
the fact that nonspecific cytoreduction was successful in reducing
the number of remaining tumor cells to a level which could be
effectively killed by a non-lethal dose of the immunotoxin. In
addition, the immunotoxins in this instance did not kill all the
remaining tumor cells yet prolonged remissions occurred. Presumably,
the remaining viable tumor cells did not produce progressive disease
because of a tumor-specific immune response.

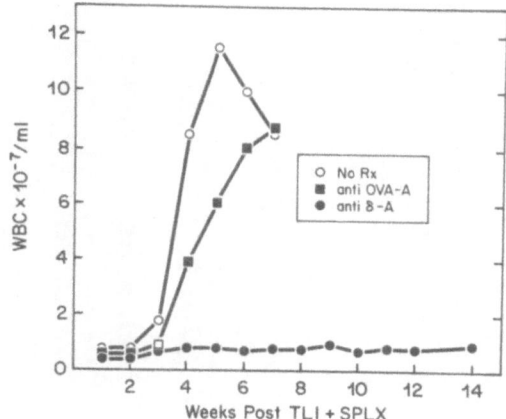

Fig. 3.   Effect of TLI, splenectomy, and administration of immunotoxin on leukemic relapse of BCL$_1$-bearing mice. After
nine doses of TLI and splenectomy, mice were injected
with two doses of 20 μg of anti-δ or control immunotoxin
or were not injected. There were nine mice per group.
Leukemic relapse was monitored by determining the number
of white cells in the blood of the treated mice. The
control mice were all dead at 7 weeks after TLI. The
rabbit anti-mouse δ-A chain-treated group was monitored
for a period of 14 weeks post-TLI, at which point the
experiment was terminated. O, no treatment; ■, anti-OVA-
A; ●, anti-δ-A.

USE OF B CHAIN-CONTAINING IMMUNOTOXINS TO POTENTIATE A CHAIN-

CONTAINING IMMUNOTOXINS

It is known that in many cases ricin conjugates are significantly more toxic than antibody-A chain conjugates (Jansen et al.,
1982; Neville and Youle, 1982; Thorpe and Ross, 1982; Houston, 1982).
In addition, free B chains can synergize in-vitro with A chain-
containing immunotoxins in specifically killing target cells (Neville
and Youle, 1982). It is postulated, therefore, that the greater
toxicity of ricin-containing immunotoxins as compared to A chain-
containing immunotoxins is due to the capacity of the B chain to
facilitate the entry of A chain into the cytoplasm (reviewed in
Neville and Youle, 1982). It would be desirable to develop a
strategy in which the putative transport role of the B chain could
be preserved while eliminating and minimizing its function as a
lectin. One approach would be to utilize two types of immunotoxins.
Tumor reactive antibodies could be conjugated to either ricin A
chain or ricin B chain. Affinity purification of the immunotoxins
on their respective antigens would be used to remove free A and B

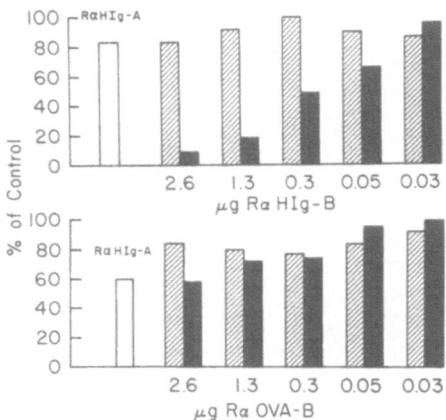

Fig. 4. The use of mixtures of A chain and B chain containing
         immunotoxins to kill Daudi cells in-vitro. Daudi cells
         were cultured with a nontoxic concentration of either
         RAHIg-A ⊔, or nontoxic doses of either RAHIg-B (upper
         panel) ◻ , or RAOVA-B (lower panel) ◻. The solid bars
         ∎, represent mixtures of the single dose of the R HIg-A
         plus different concentrations of either RAHIg-B (upper
         panel), or RAOVA-B (lower panel). Cells were treated
         with immunotoxin for 15 min at 4°C, washed and cultured
         for 16 h at 37°C in medium containing immunotoxin. Cells
         were labeled for 4 h with ^{3}H-leucine and harvested. The
         controls were not treated with immunotoxins but were in-
         cubated and labeled in the same manner.

chains. Using the two immunotoxins, the two subunits of the ricin
toxin could thereby be delivered independently to the same target
cell.

     As seen in the upper panel of Fig. 4 (Vitetta et al., 1983),
when Daudi cells were treated with either a low dose of rabbit anti-
human Ig-A chain (RAHIg-A) or a variety of doses of rabbit anti-
human Ig-B chain (RAHIg-B), no toxicity was observed. However,
when the RAHIg-A was mixed with various concentrations of RAHIg-B
there was significant cytotoxicity. It is of interest that this
treatment of the Daudi cells with the mixture of immunotoxins was
performed in medium lacking galactose. As shown in the bottom panel
of Fig. 4, when Daudi cells were treated with a low dose of RAHIg-A,
a variety of doses of rabbit anti-ovalbumin-B (RAOVA-B) or mixtures
of the two, no toxicity was observed except at the highest dose of
the RAOVA-B. These results indicate that the target cell specificity
of the antibody combining site of the immunotoxin is essential for
synergy.

The precise events that underlie the synergy are unclear. One possibility is that the two immunotoxins that bind to the same target cell are endocytosed together and are present in the same endosome. Therein, interchain disulfide bonds may be split and free ricin may be formed in the endocytic vesicle. The B chain would then facilitate translocation of the A chain into the cytoplasm with resultant cell death. These results represent a new strategy for utilizing the potential toxic property of the A chain and the pore-forming ability of the B chain in a manner which retains the specific toxicity conferred by the antibody.

ACKNOWLEDGEMENTS

We thank our colleagues Drs. Krolick, Villemez, Isakson and Cushley who collaborated with us on these studies, our technicians Ms. C. Bockhold, Mr. Y. Chinn, Ms. K. Gorman, Ms. R. Baylis, Mr. J. Hudson, Ms. L. Trahan and Mr. T. Tucker, and Ms. D. Tucker for secretarial assistance. These studies were supported by NIH grant CA-28149.

REFERENCES

Houston, L.L., 1982, Transport of ricin A chain after prior treatment of mouse leukemia cells with ricin B chain, J. Biol. Chem., 257:1532.

Jansen, F.K., Blythman, H.E., Carriere, D., Casellas, P., Gros, O., Gros, P., Laurent, J.C., Paolucci, F., Pau, B., Poncelet, P., Richer, G., Vidal, H. and Voison, G.A., 1982, Immunotoxins: Hybrid molecules combining high specificity and potent cytotoxicity, Immunol. Rev., 62:185.

Krolick, K.A., Uhr, J.W., Slavin, S. and Vitetta, E.S., 1982a, In-vivo therapy of a murine B cell tumor (BCL$_1$) using antibody-ricin A chain immunotoxins, J. Exp. Med., 155:1797.

Krolick, K.A., Uhr, J.W. and Vitetta, E.S., 1982b, Selective killing of leukemia cells by antibody-toxin conjugates; implications for autologous bone marrow transplantation in the treatment of cancer, Nature, 295:604.

Krolick, K.A., Villemez, C., Isakson, P., Uhr, J.W. and Vitetta, E.S., 1980, Selective killing of normal or neoplastic B cells by antibodies coupled to the A chain of ricin, Proc. Natl. Acad. Sci. USA, 77:5419.

Muirhead, M.J., Holbert, M.H., Uhr, J.W. and Vitetta, E.S., 1981, BCL$_1$, a murine model of prolymphocytic leukemia, Amer. J. Pathol., 105:306.

Muirhead, M.J., Martin, P.J., Torok-Storb, B., Uhr, J.W. and Vitetta, E.S., 1983, Use of an antibody-ricin A chain conjugate to delete neoplastic B cells from human bone marrow, Blood, 62:337.

Neville, D.M., Jr. and Youle, R.J., 1982, Monoclonal antibody ricin
     or ricin A chain hybrids: Kinetic analysis of cell killing
     for tumor therapy, Immunol. Rev., 62:75.
Olsnes, S. and Pihl, A., 1973, Different biological properties of
     the two constituent peptide chains of ricin. A toxic protein
     inhibiting protein synthesis, Biochemistry, 12:3121.
Olsnes, S. and Pihl, A., 1981, Chimeris toxins, in: "Pharmacology
     of Bacterial Toxins", J. Drews and F. Dornes, eds., Pergamon
     Press, New York.
Slavin, S. and Strober, S., 1977, Spontaneous murine B cell leukemia,
     Nature, 272:624.
Thorpe, P.E., Mason, D.W., Brown, A.N.F., Simmonds, S.J., Ross, W.C.J.
     Cumber, A.J. and Forrester, J.A., 1982, Selective killing of
     malignant cells in a leukemic rat bone marrow using an anti-
     body ricin conjugate, Nature, 297:594.
Thorpe, P.E. and Ross, W.C.J., 1982, The preparation and cytotoxic
     properties of antibody toxin conjugates, Immunol. Rev., 62:119.
Vitetta, E.S., Cushley, W. and Uhr, J.W., 1983, Synergy of ricin A
     chain-containing immunotoxins and ricin B chain-containing
     immunotoxins in in-vitro killing of neoplastic human B cells,
     Proc. Natl. Acad. Sci. USA, 80:6332.
Vitetta, E.S., Krolick, K.A. and Uhr, J.W., 1982, Neoplastic B cells
     as targets for antibody-ricin A chain, Immunol. Rev., 62:159.

SELECTIVE CYTOTOXICITY OF RICIN A CHAIN-ANTI CARCINOEMBRYONIC

ANTIBODY CONJUGATES TO HUMAN ADENOCARCINOMA CELLS

Thomas W. Griffin, Linda R. Haynes and
Larissa V. Levin

Division of Oncology, Department of Medicine
University of Massachusetts Medical School, Worcester
Massachusetts 01605

INTRODUCTION

Studies of immunotoxins as selective anti-tumor agents have
been performed primarily with hematologic tumor antigens, such as
mouse T cell differentiation antigens (1, 2) and immunoglobulin on
the surface of murine and human B cell leukemia cells (3-6). We
have recently explored this approach utilizing antibodies directed
against the carcinoembryonic antigen, CEA (7-9), a well described
tumor-associated antigen of adult human solid malignancies.

The carcinoembryonic antigen (CEA) was first described in
1965 by Gold and Friedman (10), as an antigen associated with neoplastic and embryonic gastrointestinal tissues. Although CEA is
not tumor specific, its concentration may be greatly increased in
a variety of epithelial malignancies (11). Malignancies which
demonstrate CEA expression detected by immunohistochemical techniques are presented in Table 1. Of particular interest, studies
in both murine models and in patients have suggested that radiolabeled antibody raised against carcinoembryonic antigen may
selectively localize in tumors bearing this antigen in-vivo (12,
13). Therefore, CEA may be an attractive target for clinically
relevant immunotoxins.

There are, however, potential difficulties with the use of
immunotoxins directed against the carcinoembryonic antigen. The
CEA is a shed antigen, and therefore free antigen either in the
circulation or in the microenvironment of the tumor may interfere
with immunotoxin localization. In addition, CEA is not tumor-
specific, and glycoproteins cross reactive with CEA are expressed

Table 1. Expression of CEA by human malignancies

Tumor type	% of tumors expressing CEA
Colonic Carcinoma	62
Gastric Carcinoma	62
Cervical Carcinoma	63
Ovarian Carcinoma	67
Lung Carcinoma	82

at low levels in normal adult tissues. Finally, there is the problem of antigenic heterogeneity, which has two aspects relevant to CEA. First, present knowledge indicates that the CEA is immunochemically complex, and the antigen may represent a family of glycoproteins which differ in molecular weight and carbohydrate content (14). Also, antigen expression, as determined by immunohistochemistry, may be heterogenous in a given tumor specimen, with some tumor cells expressing a large amount of antigen, and others entirely devoid of antigen expression.

We have prepared immunotoxins of ricin A chain linked to 8 different murine monoclonal antibodies and to affinity-purified polyclonal antibodies from three species: goat, rabbit and baboon. Several of these preparations will be described in detail. The general experimental design for our in-vitro characterization of these immunotoxins consists of four steps:

1. Ricin A chain is isolated from the parent toxin by affinity chromatography.

2. The antibody is conjugated to the isolated ricin A chain with a disulfide bond, mirroring the chemical linkage found in the parent toxin.

3. The integrity of A chain enzymatic activity and the binding capacity of the antibody is demonstrated in the conjugate.

4. Selectivity of conjugate cell killing is demonstrated by experiments with specific and control antibody, specific and control cell lines, and blocking experiments with unreacted antibody and free antigen.

These methods are described in more detail in the next section.

EXPERIMENTAL METHODS

1. Isolation of Ricin A Chain

Ricin A chain is isolated from ricin (castor bean toxin) obtained from Sigma Chemical Company, St. Louis, Missouri, and EY Labs, San Mateo, California. The toxin is reduced by the addition of mercaptoethanol to a final concentration of 5%, incubated at 20°C for 12 h, and then centrifuged at 3,000 rpm for 15 min. The supernatant is applied to a lactose-substituted agarose (Selection 12, Pierce Chemical Co., Rockford, Illinois). The A chain is eluted with phosphate-buffered saline (PBS; pH 7.4) containing 0.2% β-mercaptoethanol. The eluate is concentrated by Amicon ultrafiltration with a PM 10 membrane, and traces of contaminating ricin and B chain are removed by chromatography on a Sepharose 4B column, pretreated with 1 M acetic acid and washed extensively with elution buffer prior to the experiment. The eluted A chain peak tubes are pooled and concentrated again. The homogeneity of the A chain preparation is assessed by sodium dodecyl sulfate (SDS)-polyacrylamide electrophoresis. Cell toxicity of successfully purified A chain demonstrates an ID_{50} (median inhibitory dose) of $1.7 \times 10^{-7}M$ for LoVo cells. Immediately prior to conjugation, each batch of A chain is reduced with 100 mM dithiothreitol and desalted on Sephadex G-25.

2. LoVo, WiDr, and HT29, all lines of human adenocarcinoma cells, were purchased from the American Type Culture Collection, Rockville, Maryland, as was the murine melanoma cell line, Cloudman (S91). LoVo, WiDr and HT29 produce moderate amounts of CEA in culture.

3. Synthesis of Ricin A Chain Antibody Conjugates

Monoclonal and polyclonal antibodies, at a concentration of 1.8 mg/ml, are dialyzed against PBS for 16 h at 4°C with two changes of the buffer, and then reacted with a 10/.25-fold molar excess of the bifunctional coupling reagent SPDP (N-succinimidyl-3-) 2-pyridyldithio) proprionate; Pharmacia, Piscataway, NJ). The reaction is allowed to proceed for 30 min at room temperature. The progress of the reaction is monitored by release of pyridine-2-thione (judged by the increase in 343-nm absorbance). Reaction products are then separated by gel filtration on a Sephacryl S-300 column (1 cm x 120 cm). Synthesis of conjugates is confirmed by SDS-polyacrylamide electrophoresis.

4. Inhibition of Protein Synthesis in Cell-Free System

A rabbit reticulocyte lysate system (Bethesda Research Laboratories, Inc., Rockville, MD) is used for assaying the

inhibitory activity of ricin A chain, specific conjugates, and
control conjugates on protein synthesis. The components for
protein synthesis are incubated in the presence of (^{3}H) leucine
(New England Nuclear, Boston, MA) for 1.5 h at 30° in a final
volume of 30 μl with A chain or reduced conjugate. Protein is
precipitated with 25% trichloroacetic acid (TCA) and centrifuged.
The pellets are washed twice with 5% TCA and are then redissolved
and assayed for (^{3}H) leucine incorporation by scintillation
counting.

5. Radio-Antibody Assay for CEA

     Near-confluent LoVo and murine melanoma cells are dispersed
with trypsin-EDTA to a concentration of 6 x 10^5 cells/ml. Tubes
containing 1 ml of the cell suspension are pelleted by centrifugation at 200 rpm for 10 min and resuspended in PBS containing
2% fetal calf serum (FCS). Different dilutions of ^{111}In-labeled
diethylenetriamine penta-acetic acid-C (DTPA-C)antibody (15)
(0.75 μCi/μg) in a 0.1 ml volume are added to the tubes in triplicate. Control tubes contain labeled antibody in the absence of
cells. Cells are incubated at 23°C for 1 h with occasional mixing
and then washed three times with PBS containing 2% FCS. The tubes
are counted in a gamma counter.

6. Blockage of ^{111}In-DTPA Antibody Binding by Unlabeled Antibody

     In this experiment, cell number is the same as in radiolabeled
antibody binding assay. The cells are first preincubated with
100 μl of unlabeled antibody (1-100 μg) for 20 min at 23°C. ^{111}In-
DTPA-C antibody (3 μCi in 100 ml) is then added, and the tubes were
counted for 30 min. Washing and counting are performed as described
below.

7. Blockage of ^{111}In-DTPA-C Antibody Binding by C-19 Ricin A
    Chain Conjugate in LoVo Cells

     The procedure is identical to the blockage experiments with
unlabeled antibody, with the following modifications. The cell
number is 6 x 10^4/tube; samples are run in duplicate. Preincubation
with PBS, the conjugate (100-fold molar excess), and anti-CEA antibody (100 and 1000-fold excess over the radioactive antibody) proceeds for 30 min, and subsequent incubation with ^{111}In-DTPA-anti-
body (0.054 μCi/100 μl) lasts for 50 min.

8. Inhibition of Cellular Protein Synthesis

     LoVo and murine cells are removed from near-confluent T-
flasks with 0.25% trypsin and 0.02% EDTA. The cells are suspended
in leucine-free medium (Minimum Essential Medium, 10% FCS, 1%
antibiotics) and seeded into microtiter wells (7,000 cells in a

final volume of 200 μl per well). The cells are incubated with
PBS alone (control) or with specified additions of ricin A chain,
specific conjugate, control conjugates, or unreacted antibody for
46 h at 37°C in 5% CO_2. Thereafter, 0.1 μCi (^{14}C) leucine (New
England Nuclear, Boston) is added to each well. Following a 3 h
incubation the cells are collected on to glass fiber filters with
a Mash II cell harvester. Incorporation of (^{14}C) leucine into
cellular protein is measured by scintillation counting of the glass
fiber discs.

9. Toxicity Blocking by Free Antibody

LoVo cells are seeded into microtiter wells in the manner
described above. The cells are incubated with PBS (control) or with
specific or control antibody for 1 h at 37°C in 5% CO_2. The cells
are treated with PBS or with specific conjugate and incubated for
an additional 46 h, followed by a 3 h (^{14}C) leucine pulse as
described above.

10. Toxicity Blocking by Free Antibody

Specific conjugate is preincubated with either an equimolar
amount or a 10-fold excess of CEA antigen at 23°C for 1 h. LoVo
cells were seeded in the usual manner. The cells are treated with
PBS, CEA antigen, specific antibody, specific conjugate, or one of
the preincubated mixtures and incubated for 24 h. The cells are
then pulsed with (^{14}C) leucine, harvested, and assayed for protein
synthesis as previously described.

11. Radiolocalization Studies

CX-1 colorectal xenografts have been implanted in genetically
athymic mice in three different sites: a subcutaneous site: 3-4 mm^3
fragment of tumor tissue is implanted in the right flank of a nude
mouse midway between the axillary and inguinal areas; a tumored ear
site: a 2 mm^3 fragment of solid tumor is implanted subcutaneously
into the pinna of the mouse with a trocar; and the subrenal capsule
site: a 1 mm^3 fragment of tumor is implanted with a trocar under
the externalized renal capsule. 10 μg of radiolabeled antibody is
injected intravenously by tail vein, and at 24 h the degree of
specific localization determined by sacrifice and dissection.

RESULTS

1. Anti-CEA Immunotoxins produced with murine monoclonal antibodies

Monoclonal antibodies directed against CEA produce potent
immunotoxins. For example, the C-19 murine monoclonal antibody
conjugated to ricin A chain has been studied in detail in our

laboratory (8). This antibody, originally developed at the
Hoffman-LaRoche Laboratories in Basel, Switzerland, has been conjugated to ricin A chain by the SPDP linkage. Such conjugates
retain integrity of their antibody combining site, as demonstrated
by the ability to displace ^{111}In-DTPA-C-19 antibody bound to carcinoembryonic antigen positive cell lines. In addition, A chain
from conjugate reduced with dithiothreitol inhibited cell-free
protein synthesis in a reticulolyte lysate system at concentrations
similar to those of free A chain, demonstrating the retention of A
chain toxicity in the conjugate. Conjugates of ricin A chain with
C-19 antibody are potent cytotoxins for LoVo adenocarcinoma cells,
producing 50% inhibition of ^{14}C-leucine incorporation at concentrations of 5×10^{-10}M which is within one log of the ID_{50} of intact
ricin on this cell line (3×10^{-11}M) under similar experimental
conditions. C-19 conjugates were 570-fold as potent in producing
inhibition of (^{14}C) leucine incorporation in the CEA-bearing LoVo
cell line as A chain alone. Such toxicity could be blocked by preincubation of the conjugate with fluid-phase antigen, or of the
cells with unconjugated antibody. With 50 nM conjugate, almost
complete inhibition of (^{14}C) leucine incorporation was seen. The
conjugates were 270 times more toxic for LoVo cells than for a
control, CEA negative, murine melanoma cell line. Immunotoxins
with similar potency and specificity have also been produced with
other high affinity, anti-CEA murine monoclonal antibodies, such as
A-1 C (Fig. 1). Conjugates of specific anti-CEA antibody with ricin
A chain were greater than one hundred fold more toxic to the adenocarcinoma cell line than similar conjugates constructed with a monoclonal antibody (anti-prostatic acid phosphatase, New England Nuclear
Corporation) that did not bind to the adenocarcinoma target cells.

2. Anti-CEA immunotoxins produced with polyclonal antibodies

Anti-CEA immunotoxins have been produced with affinity-
purified polyclonal antibodies from three species: goat, rabbit
and baboon. The goat antibody has been affinity-purified to remove
cross-reaction with normal antigens, and is specific for CEA.
Compared to a control immunotoxin constructed with affinity-purified
goat anti-human IgG antibody, this anti-CEA conjugate showed a 40-
fold greater cytotoxicity towards LoVo and WiDr cell lines (7, 9).

We have also produced an affinity-isolated, specific immunotoxin with baboon anti-CEA antibody. This reagent has been
developed by Dr. Darrow Haagensen of Harvard University, and these
studies were done in collaboration with Dr. Haagensen (9). Of
particular interest is the fact that clinical trials of this primate
antibody in patients with colorectal cancer have demonstrated
limited toxicity and immunogenicity of this preparation (16).

The baboon antibody immunotoxin is a potent cytotoxin for
LoVo adenocarcinoma cells (ID_{50} 2×10^{-9}M) (9). At levels of

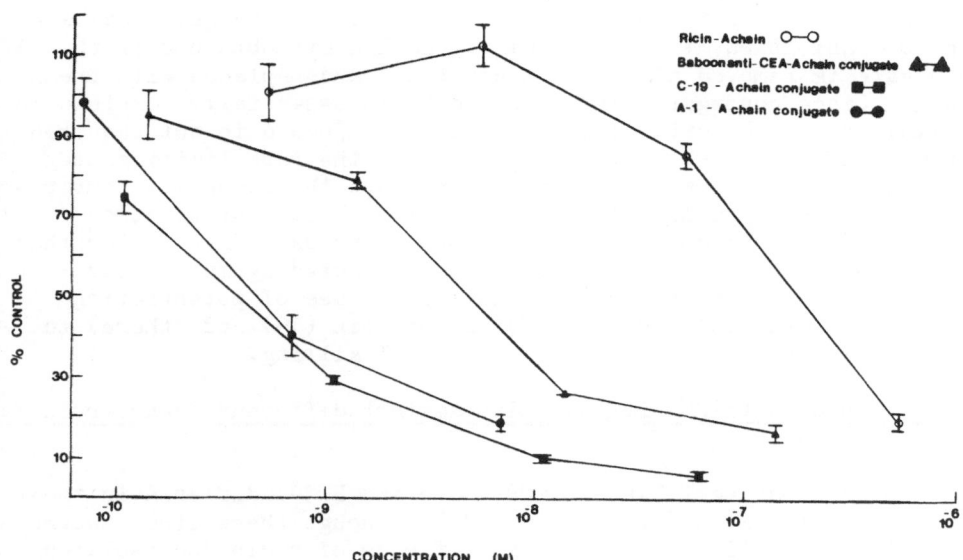

Fig. 1. Inhibition of ^{14}C-leucine incorporation in LoVo adenocarcinoma cells by specific conjugate: Inhibition of
^{14}C-leucine incorporation in LoVo cells by ricin A chain,
baboon conjugate (heteroantibody), C19 conjugate (monoclonal), and A1 conjugate (monoclonal). 1.5 x 10^4 cells/
microtiter well in leucine-free media and 10% fetal bovine
serum were incubated at 37°C in 5% CO_2 for 46 h, following
appropriate additions of the above compounds at designated
concentration. At 46 h the cells were treated with ^{14}C-
leucine (.1μCi/well) for 3 h, and harvested onto filter
discs with a Mash II cell harvester. Radioactivity incorporated was counted.

10^{-8}M, complete inhibition of protein synthesis is seen after a 24
h incubation, and cell death occurs in 2 to 4 days. These toxic
effects appear antibody-mediated, in that a non-covalent mixture
of antibody and A chain has limited cytotoxic effect, and complete
blockage of conjugate action is seen if its binding was prevented
by the presence of excess unreacted baboon antibody. Comparison
of the cytotoxicity of this baboon conjugate with murine monoclonal
conjugates is provided in Fig. 1.

3. Kinetics of immunotoxin action

The kinetics of inhibition of protein synthesis by anti-CEA
immunotoxins may be of great relevance to their ultimate utility
as anti-tumor reagents (17). At saturating concentrations of monoclonal anti-CEA immunotoxin A-1 (1 x 10^{-8}M), 50% inhibition of

[14]C-leucine incorporation is seen at 12 h after conjugate addition and 90% inhibition at 18 h. Similar results are obtained if the conjugate is removed after 1 h incubation, and replaced with fresh media. Since conjugate binding to cells is essentially complete in 30 min, and since continuous presence of conjugate is not required for maximal kinetics, it would appear that the rate limiting step in conjugate intoxication is processing into the endosomal compartment after endocytosis. The kinetics of cellular intoxication exhibited by these monoclonal anti-CEA immunotoxins are faster than those described with other immunotoxins directed against solid tumors. We are presently investigating the use of potentiating agents (ammonium chloride (18), viral protein (19) and others) to further improve the kinetics of tumor cell killing.

4. Action of anti-CEA immunotoxins against different adenocarcinoma cell lines

Three human cell lines (WiDr, LoVo and HT29) differed in their sensitivity to ricin immunotoxins (9), although these lines showed equivalent sensitivity to the toxic effects of ricin and isolated ricin A chain. We examined the possibility that the differing sensitivities of these cell lines may be related to differing amounts of CEA expressed on the cell membranes. The total content of CEA in homogenized cells was determined by radioimmunoassay. The cell line LoVo contained the greatest amount of CEA (10.5 micrograms/10^7 cells) and showed to have greatest sensitivity to the C19 immunotoxin. HT29, with the lowest amount of cellular CEA, was least sensitive to the immunotoxin.

While realizing the inherent limitations of Scatchard analysis of binding data when using intact cells, we attempted to approximate the number of CEA molecules on the cell surface using binding data with monoclonal anti-CEA antibody, radiolabeled by a previously described method (15). The results of equilibrium binding studies were expressed as Scatchard plots, and the maximum number of antibody molecules capable of binding to each cell was calculated utilizing a linear regression program. Again, a positive correlation between CEA expression and immunotoxin sensitivity was seen. The more sensitive LoVo cell line had a ten fold greater amount of surface CEA (10^5 sites/cell) compared to the less sensitive WiDr cells. These experiments again suggest the toxic effects of these conjugates are mediated through the CEA combining site of the antibody.

5. Immunotoxins directed against specific epitopes of the CEA antigen

Previous studies have suggested that monoclonal antibodies directed against two determinants of the same antigen may act synergistically to increase antigen modulation and endocytosis (20) and complement-dependent cytotoxicity (21). We have therefore examined the effects of several monoclonal antibodies directed against sterically distinct epitopes of the CEA antigen, for their ability to mediate ricin A chain effect on CEA-bearing target cells. Monoclonal antibodies CEL 007, CET 149, CEV 122, and CEJ 326 were generously provided by Hybritech, Inc. (San Diego, CA). Immunotoxins produced with each of these antibodies demonstrated ten to hundred-fold selective cytotoxic effect towards the LoVo adenocarcinoma line, as compared to control antibody. Mixtures of two different immuno-toxins (e.g. CET 149 and CEV 122) gave additive (but not synergistic) cytotoxic effect. This suggests that the use of multiple immuno-toxins in combination in the treatment of CEA-bearing tumours may produce increased cytotoxicity. This approach may be particularly important in view of the known heterogeneity of this antigen. Also, it may be possible to deliver the toxic moiety (ricin A chain) con-jugated to an antibody directed against one epitope, and a potent-iating substance (for example, ricin B chain) with an antibody directed against a second epitope.

6. The subrenal capsule model for in vivo studies of immunotoxins

The most important studies of immunotoxins as potential anti-tumor agents are in-vivo trials. To approach this in a systematic way, we have first examined the biodistribution of our carrier molecule (radiolabeled C-19 monoclonal anti-CEA antibody) in genetically athymic mice bearing xenografts of human colorectal carcinomas. We have used tumor fragments, rather than cell lines for these experiments, in order to mirror the cell to cell contact, spatial arrangements, and heterogeneity of the tissue of origin. Anti-CEA antibodies were radiolabeled to a specific activity of $0.2\,\mu Ci/\mu g$, either with iodine, by the chloramine T method, or with indium, by the DTPA cyclic anhydride method (15). Radioantibody ($10-20\,\mu g$) was administered to tumor bearing mice by tail vein in-jection. Three different xenograft implantation sites were compared: the flank, the pinna of the ear, and under the renal capsule.

All three sites gave sufficient localization for imaging (15). Also, autoradiography demonstrated penetration of the radiolabeled antibody into the substance of the tumor. We quantatively compared antibody-targeted localization in these three model systems (Table 2). Superior localization is seen with the subrenal capsule implantation site.

Similar superior localization of radiolabeled antibody in tumor fragments implanted under the renal capsule of the mouse, as opposed to subcutaneous implantation, has recently been described by Sands et al (22), utilizing a different model system which employs anti-body directed against murine T lymphocytes. The improved local-ization ratios of radiolabeled antibody in tumors implanted in this site appears related to the improved vascularity and blood flow of

Table 2. Tumor (CX-1) Uptake of Radiolabeled Antibody

	% injected dose/g tissue (range)
Subrenal	25.6 (20.4 - 35.4)
Ear	9.4 (8.4 - 10.4)
Subcutaneous	11.7 (8.1 - 14.3)

subrenal capsule xenografts, as compared to those implanted in the subcutaneous site.

Another potential advantage of the subrenal capsule implantation site is the ability to demonstrate specific localization of radiolabeled antibody in tumor explants from fresh surgical specimens. Tumor fragments of human colorectal cancer removed at surgery maintain tumor histology and antigen expression for up to four days after implantation. In studies done with Drs. Bogden, Hnatowich and Doherty, excellent specific localization of murine monoclonal anti-CEA antibody was demonstrated in fresh surgical explants of four colorectal cancers implanted under the renal capsule of normal, immunocompetent mice (23).

DISCUSSION

Disulfide-linked conjugate produced with polyclonal or monoclonal antibodies directed against the carcinoembryonic antigen and affinity-purified ricin A chain retain their ability to inactivate ribosomes in a cell free system, and to bind specifically to CEA-bearing adenocarcinoma cells. These conjugates were 50-1000 times as potent in producing inhibition of ^{14}C-leucine incorporation as A chain alone in CEA-bearing human colorectal adenocarcinoma cells. Blocking experiments with free CEA antigen and with unreacted antibody indicate that the cytotoxicity of the conjugate is mediated by the antibody combining site. Limited cytotoxicity is seen towards CEA-negative murine and human melanoma cell lines. Relative susceptibility of different cell lines to the cytotoxic effect of the conjugate appears to be correlated to their cellular content and surface expression of CEA.

The potency and kinetics of these anti-CEA immunoconjugates appear favorable to other ricin A immunoconjugates directed against solid tumors (18, 24, 25). This potency of anti-CEA immunotoxins may be due to antigenic modulation or some related form of endocytosis of the antigen-antibody complexes. Such a phenomenon has been described with goat polyclonal anti-CEA antibody (26). Most instances of antigenic modulation that have been observed with monoclonal antibodies have been associated with leukemia and lymphoma cells. Antigenic modulation or some related form of endocytosis of the antigen-antibody complexes may be required for the toxic effect of most immunoconjugates. This mechanism may be particularly important for those immunotoxins constructed with ricin A chain alone, without use of the B chain. The fact that carcinoembryonic antigen produces productive internalization of ricin A chain suggests that this antigen may demonstrate antigenic modulation. Also, the wide distribution of this antigen on human epithelial malignancies (27) suggests that immunotoxins may have a role in the treatment of solid tumors, as well as hematologic malignancies.

Despite the potential difficulties with the use of immunotoxins against CEA noted above, this approach may yet prove clinically feasible. Although free antigen in the circulation may interfere with immunotoxin localization, radiolabeled antibodies to CEA do localize to human colonic carcinomas in man and animals even in the presence of shed soluble antigen (12, 13). Also, we have demonstrated cell surface localization of radiolabeled monoclonal anti-CEA antibody in human colorectal xenografts implanted in genetically athymic mice. Finally, the problem of antigenic heterogeneity may be overcome by the use of mixtures of immunotoxins directed against different epitopes on CEA, or, alternately, by the combination of anti-CEA immunotoxins with immunotoxins directed against other tumor-associated antigens on colorectal cancer cells (19).

The potential uses of conjugates of this type await further study. Conjugates such as these may be useful in functional studies of CEA, by their use in isolation of CEA-negative mutants of a CEA-positive cell line. Also, the conjugates may be of use in in-vitro destruction of CEA-positive cells (e.g. from lung cancer) in bone marrow prior to autologous bone marrow transplantation. Their potential as clinical anti-cancer agents awaits the results of careful animal studies of in-vivo toxicity and anti-tumor effect.

ACKNOWLEDGEMENTS

Supported in part by CA 29160-01 from the National Cancer Institute. Dr. Griffin is a Junior Clinical Faculty Fellow of the American Cancer Society.

REFERENCES

1. F.K. Jansen, H.E. Blythman, D. Carriere, P. Casellas, J. Diaz,
    P. Gros, J.R. Hennequi, S. Paolucci, B. Pau and P. Poncelet.
    High specific cytotoxicity of antibody-toxin hybrid molecules
    (immunotoxins) for target cells, Immunol. Lett. 2:97 (1980).
2. R.J. Youle and D.M. Neville. Anti-thy 1,2 monoclonal antibody
    linked to ricin is a potent cell-type-specific toxin, Proc.
    Natl. Acad. Sci. USA. 77:4883 (1980).
3. V. Raso and T.W. Griffin. Specific cytotoxicity of a human
    immunoglobulin directed Fab'-ricin A chain conjugate,
    J. Immunol. 125:2610 (1980).
4. V. Raso and T.W. Griffin. Hybrid antibodies with dual specif-
    icity for the delivery of ricin to immunoglobulin-bearing
    target cells, Cancer Res. 41:2073 (1981).
5. K.A. Krolick, C. Villemez, P. Isakson, J.W. Uhr and E.S. Vitetta.
    Selective killing of normal and neoplastic B cells by anti-
    bodies coupled to the A chain of ricin, Proc. Natl. Acad. Sci.
    USA. 77:5419 (1980).
6. K.A. Krolick, J.W. Uhr, S. Slavid and E.S. Vitetta. In-vivo
    therapy of a murine B cell tumor (BCL) using antibody-ricin
    A chain immunotoxins, J. Exp. Med. 155:1797 (1982).
7. T.W. Griffin, L.R. Haynes and J.A. DeMartino. Selective cyto-
    toxicity of a ricin A chain anti-CEA antibody conjugate for
    a human colonic adenocarcinoma cell line, J. Natl. Cancer
    Inst. 69:799 (1982).
8. L.V. Levin, T.W. Griffin, L.R. Haynes and C.J. Sedor. Selective
    cytotoxicity for a colorectal carcinoma cell line by a mono-
    clonal anticarcinoembryonic antigen antibody coupled to the
    A chain of ricin, J. Biol. Resp. Modif. 1:149 (1982).
9. L.V. Levin, T.W. Griffin, L.R. Haynes and D. Haagenson.
    Conjugates of antibody directed against CEA disulfide linked
    to ricin A chain demonstrate selective cytotoxicity to human
    CEA bearing cells in culture, Proc. Am. Ass. Cancer Res. 24:
    222 (1983).
10. P. Gold and S.O. Freedman. Demonstration of tumor specific
    antigens in human colonic carcinomas by immunological toler-
    ance and absorption technique, J. Exp. Med. 122:439 (1965).
11. C.H.J. Ford, C.E. Newman, J.R. Johnson, C.S. Woodhouse, T.A.
    Reeder, G.F. Rowland and R.G. Simmonds. Localization and
    toxicity study of a vindesine-anti-CEA conjugate in patients
    with advanced cancer, Br. J. Cancer 47:35 (1983).
12. J.P. Mach, F. Buchegger, M. Forni, J. Ritschard, C. Berche,
    J.D. Lumbruso, M. Schreger, C. Giradet, R.S. Accolla and
    S. Carrel. Use of radiolabeled monoclonal anti-CEA antibodies
    for the detection of human carcinomas by external photo-
    scanning and tomoscintography, Immunol. Today 2:239 (1981).
13. D.M. Goldenberg, F. Leland, E. Kim, S. Bennett, F.J. Primus,
    J.R. van Nagell, N. Estes, P. DeSimone and P. Rayburn. Use
    of radiolabeled antibodies to carcinoembryonic antigen for

the detection and localisation of diverse cancers by external photoscanning, N. Engl. J. Med. 298:1384 (1978).

14. E. Alpert. The immunochemical complexity of CEA. A golden
dream or molecular nightmare, Cancer 42:1585 (1978).

15. D.J. Hnatowich, W.W. Layne, R.L. Childs, D. Lanteigne, M.A.
Davis, T.W. Griffin and P.W. Doherty. Radiolabeling antibody: a simple and efficient method, Science 222:613 (1983).

16. D.E. Haagensen, M.S. Huberman, A. Kaldany, S. Davis, K. George
and C. Moore. Phase I-II immunotherapy trial of non-human
primate (baboon) anti-CEA antibody for treatment of metastatic adenocarcinoma which secretes CEA, Proc. Am. Ass.
Cancer Res. 24:209 (1983).

17. R.J. Youle and D.M. Neville. Kinetics of protein synthesis
inactivation by ricin-anti-thy 1.1 monoclonal antibody
hybrids, J. Biol. Chem. 257:1598 (1982).

18. P. Cassellas, J.P. Brown, O. Gros, P. Gros, I. Hellstrom,
F.K. Jansen, P. Poncelet, R. Roncucci, H. Vidal and K.E.
Hellstrom. Human melanoma cells can be killed in-vitro by
an immunotoxin specific for melanoma-associated antigen,
Int. J. Cancer 30:437 (1982).

19. D.J R. Fitzgerald, R. Padmanabran, I. Pastan and M.C.
Willingham. Adenovirus-induced release of epidermal growth
factor and pseudomonas toxin into the cytosol of B cells
during receptor-mediated endocytosis, Cell 32:607 (1983).

20. K.S. Webb, J.L. Ware, S.F. Parks, W.H. Briner and P. Paulson.
Monoclonal antibodies to different epitopes on a prostate
tumor-associated antigen, Cancer Immunol. Ther. 14:155 (1983).

21. I. Hellstrom, J.P. Brown and K.E. Hellstrom. Monoclonal antibodies to two determinants of melanoma-antigen 97 act synergistically in complement-dependent cytotoxicity, J. Immunol.
127:159 (1981).

22. H. Sands, P.L. Jones, W. Neacy, L. Canin and B.M. Gallagher.
A comparison of the ability of radio-iodinated monoclonal
anti-rat Thy 1 (0x7) to image SL2 tumors located subcutaneously or in the subrenal capsule, Proc. 30th Ann. Meeting
Soc. Nucl. Med. J. Nucl. Med. 24:102 (1983).

23. A.E. Bogden, D.J. Hnatowich, P.W. Doherty and T.W. Griffin.
In-vivo localization of monoclonal antibody in fresh surgical
explants of human tumors: 3 day subrenal capsule assay, Proc.
Am. Ass. Cancer Res. 24:218 (1983).

24. D.G. Gilliland, P. Stephenski, R.J. Collier, K.F. Mitchell,
T.H. Chang and H. Koprowski. Antibody directed cytotoxic
agents: use of monoclonal antibody to direct the action of
toxic A chains to colorectal carcinoma cells, Proc. Natl.
Acad. Sci. USA. 77:4539 (1980).

25. K.A. Krolick, D. Yuan and E.S. Vitetta. Specific killing of
a breast carcinoma cell line by a monoclonal antibody coupled
to the A chain of ricin, Cancer Immunol. Immunotherapy 12:39
(1981).

26. K.L. Rosenthal, W.A.F. Tompkins and W.E. Rawls. Factors

affecting the expression of carcinoembryonic antigen at the
surface of cultured human colon carcinoma cells, <u>Cancer Res.</u>
40:4744 (1980).

27.   D.M. Goldenberg, R.M. Sharkey and F.F. Primas. Immunocytochemical detection of carcinoembryonic antigen in conventional
histopathology specimens, <u>Cancer</u> 42:1546 (1978).

DEVELOPMENT OF MONOCLONAL ANTIBODIES WITH

SPECIFICITY FOR HUMAN EPITHELIAL CELLS

Joyce Taylor-Papadimitriou and
Andrew B. Griffiths

Imperial Cancer Research Fund
Lincoln's Inn Fields
London WC2A 3PX, U.K.

INTRODUCTION

The technique developed by Kohler and Milstein (1975) for the directed production of monoclonal antibodies has had an enormous impact not only on the field of immunology but in every area of cellular and molecular biology, because such antibodies provide highly specific markers for molecules and cells. It is likely that monoclonal antibodies will also be of great importance in clinical medicine, in a number of areas, among which is the receptor-mediated targeting of drugs, toxins and isotopes. The potential of monoclonal antibodies as vehicles for drug targeting is obvious, and many laboratories are involved in attempts to produce antibodies which react specifically with components of tumour cell membranes not found in normal cells. The major solid tumours in man are carcinomas which develop from epithelial cells, and therefore the membrane of the malignant epithelial cell is of prime interest. It is probably fair to say that most of the antibodies reacting with malignant epithelial cells which have been produced so far, are tumour-associated rather than tumour-specific and show a positive reaction with some other normal cells, usually epithelial in origin. It is evident from this work that there are components of the cell membrane which are epithelial specific; whether there are components which are only found in carcinomas or in one particular kind of epithelial cell is not so evident. In the present article, we will briefly describe some of the general methods and approaches used in the production of monoclonal antibodies and discuss in more detail the results of applying these methods to the production of antibodies which show some specificity for human epithelial cells.

NATURAL MODELS OF PRODUCTION OF MONOCLONAL ANTIBODIES

The basic structure of the immunoglobulin molecule has been elucidated largely through the availability of single species of antibodies produced in large quantities by monoclonal malignant antibody-secreting B cells (myelomas). Using these naturally occurring monoclonal antibodies, it has been possible to elucidate the structure of the immunoglobulin molecule which has been found to have 2 light and 2 heavy chains, both with variable and constant regions. Immunoglobulins belonging to a specific class share the same constant region in their heavy chains and may have K or λ type light chains. The specificity of a particular immunoglobulin, is however determined by amino acid sequences in the variable region which are similar in the light and heavy chains and it is these sequences which bind to the specific antigenic determinant recognised by the antibody.

With few exceptions, such as the autoantibodies found in cold agglutin disease (Feizi, 1981a; see also later reference to these) the antigenic specificities of the naturally occurring monoclonal antibodies are not known. In spite of this, the antibody producing monoclonal malignant B cell has provided an invaluable model system for studying the general mechanisms underlying differentiation in the B cell, both in the embryo and in the adult. As a result of investigations of this system using recombinant DNA technology, the genetic rearrangements involved in the development of an antibody producing lymphocyte, and in the switch from IgM to IgG in the plasma cell are now largely understood (for review see Gottleib, 1980). The phenotypic changes and cellular interactions which are important for the proliferation and differentiation of activated B cells are also being elucidated using B cell lymphomas and leukemias (Kishimoto, 1983). Fig. 1 illustrates concepts which have come out of these experiments regarding the factors involved in the production of Ig-secreting clones of B cells. In the experimental systems used, an anti-idiotype antibody is used in place of antigen to stimulate the resting B cell, and the T cells are stimulated by lectins to produce factors which are crucial to the proliferation and differentiation of the B cell. Even though the cellular interactions involved in the stimulation of proliferation and differentiation in the B cell are complex, it is now possible to "immunize" spleen cells in-vitro (Pardue et al., 1983). However, these studies represent a very recent development and the directed production of monoclonal antibodies to particular antigens has largely been achieved in an animal model.

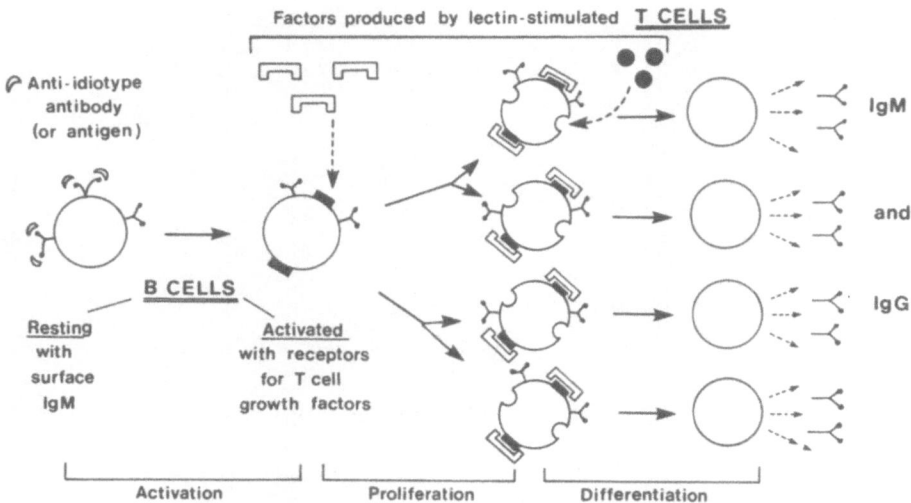

Fig. 1. Activation of B cells by anti-idiotype Ig and T cell
         factors. Based on experiments with malignant human B
         cells (Kishimoto, 1983).

PRODUCTION OF MOUSE AND RAT MONOCLONAL ANTIBODIES DIRECTED TO

SPECIFIC ANTIGENS

Principles of the Technique

     It is important to remember that even if a purified protein
(or other immunogenic large molecules) is used to immunize an animal,
many antibodies will be produced, each directed to a different epitope or determinant in the molecule. Therefore all antisera, even
those directed to a single molecular species of antigen are polyclonal, containing immunoglobulins with differences in both the C
and V regions of the heavy and light chains. At the moment there
is no obvious and easy way to fractionate out monoclonal antibodies
from such a mixture. The technique developed by Kohler and Milstein
(1975) depends on actually separating out the stimulated B cells
from the spleen of an immunized mouse, by immortalizing them through
hybridisation with a myeloma cell line and selecting for those
hybrids producing the required antibody. Fig. 2 gives an outline
for the basic technique for production of monoclonal antibodies as
developed by Kohler and Milstein.

· The success of the fusion technique comes from the use of a
mouse myeloma line deficient in the enzyme hypoxanthine guanine
phosphoribosyl transferase, thus allowing for the selection of

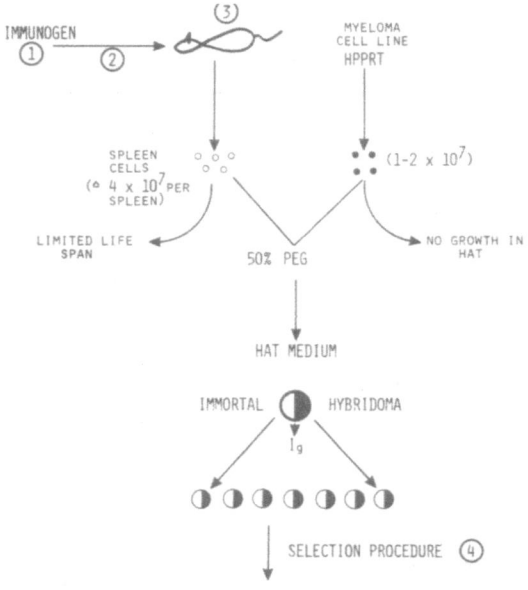

FACTORS INFLUENCING SPECIFICITY AND AFFINITY OF ANTIBODY

1. IMMUNOGEN SELECTS THE B CELLS WHICH ARE STIMULATED.
2. IMMUNISATION SCHEDULE MAY AFFECT THE AFFINITY OF THE AB PRODUCED.
3. IMMUNE RESPONSE OF THE MOUSE - STRONGER WITH SOME ANTIGENS THAN OTHERS.
4. SCREENING AGAINST SPECIFIC ANTIGENS DETERMINES SPECIFICTY AND MAY
    ALSO SELECT FOR AFFINITY.

Fig. 2.   Outline of procedure for production of monoclonal anti-
          bodies by in-vivo immunization and in-vitro selection.

hybrid clones in a medium containing hypoxanthine, thymidine and
an inhibitor of the endogenous synthesis of purines (methotrexate
or aminopterin). The parent spleen cells have a limited life span
in-vitro so that in the HAT medium only hybrid cells develop.
Originally the spleen cells and the myeloma cell line were fused
with Sendai virus, but since the use of polyethylene glycol was
introduced by Pontecorvo (1975) this method has superceded virus
fusion. The cell line originally used by Kohler and Milstein was
P3 from a Balb C mouse which produced some light chains of the K
class and some heavy chains of the IgG1 type. Now several other
derivatives of this and other mouse myeloma lines are available
which do not produce immunoglobulin chains and some have been
developed from inbred rats. A list of such lines is given in
Table 1 together with data on the production of immunoglobulin
chains.

     In directing the production of antibodies to specific antigens
as outlined in Fig. 2, the specificity and/or affinity of the anti-

Table 1. Myeloma cell lines for production of hybridomas

Full name of cell line	Abbreviated name	Animal, strain origin	Immunoglobin chains known to be produced	Other comments	Reference
P3-X63-Ag8 Parent cell line to:	P3	Mouse Balb/c	K γd IgG1 (K)		Kohler and Milstein, 1975
P3-NS1-1Ag4-1	NS1	Mouse Balb/c (via P3-X63-Ag8)	K	K intracellular but secreted by hybrids	Kohler and Milstein, 1976
X63-Ag8 653	NS2	Mouse Balb/c (via P3-X63-Ag8)	None		Kearney et al., 1979
SP2/0-Ag14	SP2	Mouse Balb/c	None	Low fusion efficiency	Schulman et al., 1978
(MPC-11) 45.6GT1.7		Mouse Balb/c	IgG2b (K)		Margulies et al., 1976
PuBu1-Ou		Mouse Balb/c	IgG2a (K)		Kohler and Milstein, 1976
210-RCY30AgI	Y3	Lou rat spontaneous myeloma	K		Galfre et al., 1979
L363		Human	λ on cell surface. None secreted		Bischoff et al., 1982
GM - 1500 6TGA1		Human B cell myeloma G.M. 1500	IgG2 (K)		Croce et al., 1980
GM - 1500 6TGA2					
SKO 007		Human Myeloma cell line. U266	IgE (λ)		Olsson and Kaplun, 1980
LICR LON/HMy2		Human	IgG		Edwards, 1981

bodies produced may be influenced at various stages. Thus, while the potential specificity of the antibodies depends mainly on the choice of immunogen and screening procedure, this potential is limited by the immune response of the mouse or rat. Obviously antibodies recognising determinants which are strongly antigenic in the animal will be produced more readily and in fact the first attempts to make tumour specific antibodies have yielded a group of antibodies reacting with similar determinants (see below) which are clearly highly immunogenic in the mouse.

The affinity of the antibodies produced can also be influenced by a variety of factors, some of which are ill-defined. Some immunization schedules, for example those involving several inject-ions spaced well apart (3 weeks) using low levels of antigens, tend to produce high affinity antibodies. Conversely two injections (10-21 days apart) of high level of antigens, followed by daily injections up to the fusion of the spleen have in our experience yielded many low affinity antibodies. The screening procedure will also determine to some degree the affinity of the antibodies selected, since a very sensitive assay, such as solid phase RIA will pick up low affinity antibodies, whereas precipitation assays, and possibly those using a fluorescent second antibody might select for the higher affinity ones.

Assays used in Selecting Antibodies

Most screening assays depend on the detection of binding of the tissue culture supernatant from the hybrid clones to a prepar-ation of the antigen by using a second antibody (goat or rabbit antimouse) labelled with radioactivity (RIA), fluorescein (indirect immunofluorescence) or an enzyme which subsequently converts a colourless substrate to coloured products (ELISA) (see Fig. 3). In dealing with membrane antigens, whole cells (live or fixed), membrane preparations, or tissue sections may be used as a source of antigen.

Cloning and Growth of Hybridomas

The number of hybridomas produced from one fusion can vary from a few hundred to a few thousand, depending on the number of activated B cells there are in the spleen. One reason why the technique is so successful is that there appears to be a preferential production of hybrids from the stimulated B cells, so that a large fraction (10-80%) of the hybridomas may actually be producing anti-bodies; there is some evidence that a higher proportion of hybrids formed from rat myelomas produce antibodies than is the case with the mouse (Milstein et al., 1980). It is therefore important to select as early as possible so as to reduce the number of hybridomas

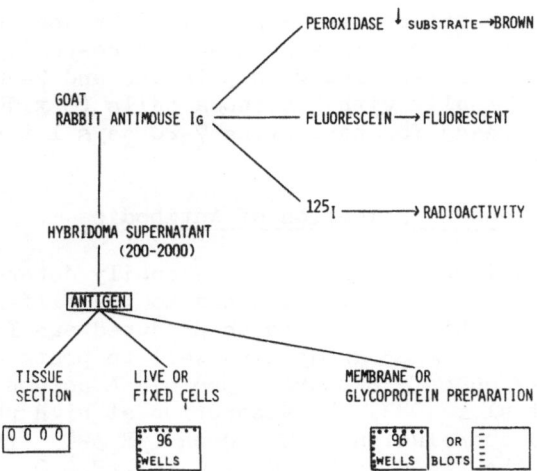

Fig. 3. Principles of screening methods used for selection of monoclonal antibodies.

which have to be cultured and then to clone these selected cells. Cloning serves to separate the antibody producing hybrid from any other viable hybrids found in the same original well, and also selects for a stable variant. Cloning by limited dilution, with spleen cell feeders, has been found in our laboratory to be suitable for most situations although cloning in agar can also be used.

In general, hybrid cells have been selected, cloned and grown up in medium containing a high level of foetal calf serum previously selected for its ability to support the growth of hybrid cells. Some investigators are now using chemically defined medium (Chang et al., 1983) which has advantages for large scale production for purification. It should be noted however that optimal conditions for growth are not necessarily optimal for antibody production and it is important to monitor antibody levels, which usually range from 3-30 μg/ml, while growing the cells. Higher levels can be produced by growing the cells as an ascites tumour (yield around 1 mg/ml, 1-2 ml of ascites per mouse). Not all hybridomas grow well as ascites and some have a tendency to form solid tumours. Although growing the cells in mice avoids the need for large scale tissue culture facilities, there are disadvantages to using ascites fluid for purifying antibodies, especially for use in patients, since other mouse Ig may be present and these may co-purify. For growth of hybridomas, mice are primed with pristane and 7-10 days later injected intraperitoneally with hybridoma cells (3×10^6 per mouse). Ascites fluid is ready for harvesting 7-20 days later.

Characterization and Purification of Antibodies

The class of immunoglobulin can be easily determined using Ochterlony plates and specific antisera to the different classes. The common classes which appear to be produced are IgM, IgG and some IgG2a and IgG2b. The IgG2 group bind well to protein A and can be effectively purified on a Sepharose-protein A column (Ey et al., 1978; Wagener et al., 1983), by adsorption at high pH (0.1 M citrate, pH 8.0) and elution with low pH (pH 3-4) after eluting impurities from the FCS with 0.1 M citrate, pH 6.0. Some immunoglobulins of the IgG1 class bind well enough to protein A to be effectively purified in this way; impurities can be eluted at pH 6.5, before eluting the monoclonal antibody at pH 6.0 or less. Demonstration of purity is best made using isoelectric-focusing.

PRODUCTION OF HUMAN MONOCLONAL ANTIBODIES

In principle, there should be advantages to using human rather than mouse monoclonal immunoglobulin molecules for administration to patients, since these homologous antibodies should not produce an immune response in the human host. Some mutant myeloma cell lines are now available (see Table 1) and human monoclonal antibodies can be produced by fusing them with antibody producing cells from humans. It is also possible to get stable hybrids by fusing mouse myeloma lines with human B cells (Levy and Dilley, 1978). However, the problem lies in there being no obvious way in which the specificity of the antibody produced by the human cell can be manipulated. Hybrids have been made using lymphocytes from regional lymph nodes in patients with various forms of cancer. Although most of the monoclonal antibodies produced so far have not shown specificity of tumour cells, this approach could yield useful information regarding the profile of those antigens which appear to be immunogenic in the autologous human host. Immortalisation of antibody producing cells (which can be circulating lymphocytes) can be achieved by transformation with EBV virus. This technique has been used effectively to obtain monoclonal antibodies to influenza virus, by using lymphocytes from patients injected with influenza vaccine (Crawford et al., 1983). As the process of induction of antibody formation becomes better understood (see Fig. 1) it may be possible to stimulate human spleen cells (taken at surgery or autopsy) completely in-vitro. At the moment, however, the monoclonal antibodies with a well-defined specificity (including those directed to epithelial cell membrane components) are those which have been developed in the rodent.

MOUSE MONOCLONAL ANTIBODIES TO EPITHELIAL MEMBRANE COMPONENTS

Most of the antibodies reacting with epithelial membrane components have been raised using cells cultured from carcinomas or membrane extracts prepared from them. This is partly because cell lines and tissue from malignant tumours are readily available, and also because many investigators were looking for antibodies reacting specifically with the cancer cells. Where normal cells and/or their membranes are available these may also be used. In particular the membrane from the terminally differentiated mammary epithelial cell which surrounds the milk fat globule and which is found in the cream fraction of milk, has been widely used to produce antibodies reacting specifically with the same epithelial cell membranes. We will discuss both of these approaches, and it will be seen that in both the normal and malignant epithelial cell membrane, certain antigenic determinants appear to be very immunogenic in the mouse.

Antibodies Raised Against Carcinoma Cells or their Membranes

Table 2 gives a representative but not comprehensive list of antibodies which have been produced against membrane antigens in some human carcinomas and which show specificity for epithelial cells. Where known, the size of the component(s) carrying the antigenic determinant and/or the chemical nature of the determinant is shown.

The carcinoembryonic antigen (CEA): This represents a special case where a semi-purified preparation of a glycoprotein has been used as immunogen. Carcinoembryonic antigen was originally defined by Gold and Freedman (1965) as a component (180-200k) found in perchloric acid extracts of colon carcinomas, which was not found in any normal adult tissues, but was found in foetal gut, liver and pancreas. CEA was initially recognised by a polyclonal antiserum, directed to colon carcinoma cells made specific by adsorption with normal colon and blood components. It is now clear that CEA is not a single antigen, or even a single molecular component, but rather a heterogeneous family of molecules (glycoproteins) bearing several antigenic determinants (von Kleist and Burtin, 1979). Even before the advent of monoclonal antibodies and finer methods for examining structural similarity between molecules, it was obvious that other components could be extracted from normal adult tissues, from adult feces (Kuroki et al., 1981) and from meconium (Primus et al., 1983a), which showed immunological cross reactivity with CEA. Of these the NCA (non-specific cross reacting antigen) (von Kleist et al., 1972) found in high levels in normal lung and with a MW of 60k, and the

Table 2. Characteristics of certain immunogens

Immunogen	Antibody	MW Antigen	Chemical nature of antigen	Spectrum of reactivity	Reference
Carcinoembryonic antigen (CEA)	NP Series	60K, 185K, 200K	Glycoproteins	Normal and malignant colonic epithelium. Other epithelia	Primus et al., 1983b
	V11 23e 37			Granulocytes	Accolia et al., 1980
	CEA-63-3E			Neutrophils	Wagener et al., 1983
CELL LINES					
MCF-7 (breast cancer)	MBr.1, MBr.2 MBr.3	?	Glycolipid	Normal and malignant breast epithelium Sweat Gland	Menard et al., 1983 Canevari et al., 1983
	24-17-1	95K	Glycoprotein	Normal and malignant breast Other tumors	Thompson et al., 1983
	24-17-2	100K	Glycoprotein	Normal and malignant breast Other tumors	
SW1116 (colon cancer)	19.9		Glycolipid[a]	Gastro-intestinal cancers	Steplewski et al., 1981 Magnani et al., 1982
Colo 38 (melanoma)	376.965	94K	Glycoprotein	Melanomas Carcinomas	Imai et al., 1982

Source	Antibody		Type	Reactivity	Reference
NCI H59 (Small cell carcinoma of lung)	30% of 85 Hybridomas		Glycolipid[b]	Lung Cancers	Cuttitta et al., 1981
				Breast Cancers	Magnani et al., 1982; Huang et al., 1983
SKRC-7 -6 -28 (Renal cancer cell lines)	S4	160K	Glycoprotein	Normal and malignant	Ueda et al., 1981
	S23 S25	120K —	Glycoprotein Glycolipid	Kidney epithelium	
HPAF (Pancreatic cancer)	DUPAN-1,2,3	?	?	Normal and malignant pancreas. Other tumors, and foetal tissues	Metzgar et al., 1982

EXTRACTS OF TUMOR CELL MEMBRANE

Source	Antibody		Type	Reactivity	Reference
Liver metastasis tumor	B72.3	220K	Glycoprotein	Some breast tumors	Colcher et al., 1981
Breast	B6.2	90K	Glycoprotein	Some breast tumors	
Colon adenoma	C14/1/46/10		Glycolipid[c] Glycoprotein	Malignant (and normal) colonic mucosa	Brown et al., 1983
Cultured cell line (Hep 2)	Cal	350 and 390K	Glycoprotein	Human cancers	Ashall et al., 1982
Ovarian mucinous carcinoma	Mov 1	Large	Mucin	Ovarian and other cancers	Ines-Colnaghi et al., 1983

[a] NeuNAcα2-Galβ1 → 3GlcNAcβ1 → Galβ1 → 3GlcNAcβ1
                                          1,4
                                          Fuc α

[b] Galβ1 → 4GlcNAcβ1 → Galβ1 → 4GlcNAcβ1
                                  1,3
                                  Fuc α

[c] Galβ1 → 4GlcNAc
           1,3   1,3
           Fuc α Fuc α

MA (meconium antigen) with a molecular weight of 185k are probably the best known. Now that monoclonal antibodies to CEA are available it is possible to examine their reactivity both with immunologically similar molecules and with tissue sections. Primus and colleagues have described a series of antibodies to CEA which fall into three classes according to their cross-reactivity with NCA and MA (Primus et al., 1983b). In this series, cross-reacting antibodies like NP-1 react most strongly and most consistently with tumours, but also show a stronger reaction with normal colonic mucosa (and also with neutrophils in blood vessels and in extravascular tissue). Antibodies of the NP-4 type show only a weak staining reaction with normal colon, but show a more heterogeneous reaction with tumours. Thus, 30% of primary colon tumours do not react with NP-4 and most of the liver and nodal metastases from NP-4 positive tumours show a negative reaction with the antibody (Primus et al., 1983c). Similarly antibodies raised by Wagener et al. (1983) show cross-reactions with liver tissue and granulocytes, although those developed by Accolia and colleagues (1980) appear to be more tumour specific.

These observations with CEA focus on two problems common to most of the studies on tumour associated antigens, namely, their presence in normal tissues and the heterogeneity of their expression in tumours. In the case of CEA where the reaction of the CEA-specific monoclonals with normal tissues is reported to be weak, and many such antibodies are available, the use of a cocktail may overcome the heterogeneity problem. Such a cocktail may also be more useful for the detection of the tumour associated antigens in serum (Buchegger et al., 1982) or for localisation of tumours in vivo.

Whole cells: Most of the monoclonal antibodies which have been raised against epithelial membrane antigens using whole cells have been obtained by injecting mice with cultured cell lines (Table 2), since it is technically difficult to obtain homogeneous preparations of intact cancer cells from primary carcinomas or their solid metastases. Metastatic cells growing in suspension in serous effusions (found in the pleural or abdominal cavities) can be harvested more readily and we have produced antibodies against metastatic breast cancer cells from a pleural effusion (Burchell, Hurst and Taylor-Papadimitriou, manuscript in preparation). However, the use of cultured cell lines has received much more attention, and some useful antibodies have been obtained in this way, using cell lines derived from the major solid tumours (see Table 2).

Early studies on the specificity of these antibodies produced using cell lines and tissue extracts suggested that they were tumour specific. The use of a more sensitive assay, i.e. immunoperoxidase staining of frozen or fixed tissue sections has shown this generally not to be the case. Most of them react with some other normal

epithelial cells albeit less strongly and sometimes other carcinomas.
Some of the antigenic determinants recognised by antibodies raised
against cultured carcinoma cells are found on membrane glycoproteins
such as the histocompatibility antigens (Koprowski et al., 1979) and
the transferrin receptor (Newman et al., 1982) which are present on
a wide range of cell types. However, many of the antibodies which
show some degree of specificity for tumours react with carbohydrate
determinants which may be in glycolipids and/or glycoproteins.
Although it is not possible to make a conclusive statement, an examination of the literature and our own experience suggests that more
antibodies to epitopes on membrane glycolipids are produced when
whole cells (as compared to membrane extracts) are used as immunogens. The glycolipid antigens which have been well characterised
are the sialylated Le^a group recognised by the 19.9 antibody
(Magnani et al., 1982) and the lacto-N-fucopentaose III sequence
which is recognised by many antibodies raised against a variety of
cell types (Brockhaus et al., 1982; Huang et al., 1983). Although
the oligosaccharide sequence may be characterised from a membrane
glycolipid, it may also be present on a large molecular weight
glycoprotein. For example in the case of the sialylated Le^a antigen
and the MBr1 glycolipid, the antigens are detected in the serum of
cancer patients on large molecular weight glycoproteins (Koprowski
et al., 1981). The membrane glycoproteins recognised by antibodies
raised against whole cells showing some tumour specificity are in
the 90-160K range, but in most cases it is not clear whether the
epitope recognised by the antibody is a peptide or oligosaccharide
sequence.

Membrane extracts: Extracts of the extranuclear cell membranes
have been used effectively to produce antibodies which show a high
degree of specificity for carcinomas, both of the kind from which
the membrane preparation was derived and for other carcinomas.
Table 2 lists some of the better known antibodies which have been
produced in mice injected with membrane extracts from primary and
metastatic tumours and from cell lines. Membrane extracts from
benign tumours (e.g. prostate) have also been used and antibodies
obtained which react specifically with the epithelial cell type used
to immunize (Frankel et al., 1982).

While some of the antigenic determinants recognised by this
group of antibodies are defined only by the MW of the component
carrying the determinant some have been characterised and found to
be carbohydrates. The oligosaccharide determinants are often found
on large mucin like glycoproteins (MW > 200K); antibody C14/1/46/10,
which was raised against membranes prepared from a human colonic
adenoma has been shown to react with the difucosylated N-acetyl
lactosamine sequence.

Antibodies to Membrane Components of Differentiated Cells:
As indicated earlier, the human milk fat globule (HMFG) membrane,

Table 3. Monoclonal antibodies to components of the human
         milk fat globule

Antibody	Antigenic determinant	Reference
HMFG-1 HMFG-2	Carbohydrate determinants of large molecular weight (>300 component)	Taylor-Papadimitriou et al., 1981 Arklie et al., 1981 Burchell et al., 1983
M Series M18, M39	I (MA) determinant	Foster et al., 1982; Gooi et al., 1982
Mam Series Mam 3	Large molecular weight component	Hilkens (personal communication)
BLMRL-HMFG Mc3	46K	Ceriani et al., 1982

which can be extracted by churning, or by solvent-extraction of the
cream fraction of human milk provides a ready source of a membrane
representative of a functionally differentiated epithelial cell.
Antisera to such preparations have been used in the past to identify
normal mammary epithelial cells in culture (Ceriani et al., 1977)
and to demonstrate the presence of differentiation antigens on
mammary cancers (Heyderman et al., 1979). More recently, at least
three groups have prepared monoclonal antibodies to HMFG and these
are listed in Table 3. Of all the components present in this membrane (around 15 bands on polyacrylamide gel), the most immunogenic
appears to be a large molecular weight component (> 300K) which is
glycoprotein in nature and consists of at least 50% carbohydrate
(Shimizu and Yamauchi, 1982). The antibodies, HMFG-1 and 2 developed by us (Taylor-Papadimitriou et al., 1981) appear to be against
this component (Burchell et al., 1983); M8, M18 and M39 may also be
directed to it. The actual determinants recognised by M18 and M39
have been shown to be of the type recognised by anti-I antibodies
(Gooi et al., 1983). These antigens are the precursors of the
blood group antigens which are formed by adding sugars to terminal
galactose (Gal) and N-acetyl glucosamine (GlcNAc) moieties (see
Fig. 5 and below).

The antibodies reacting with the large molecular weight component of the HMFG tend to be epithelial specific, and some, like
the HMFG-2 antibody, react preferentially with tumours. They are
therefore proving to be useful in the localisation of tumours and
in in-vitro diagnosis (Epenetos et al., 1982a and b; Gatter et al.,
1982). An antibody to a smaller molecular weight component of the
milk fat globule, as well as the HMFG-2 antibody also detect anti-

gens in the sera of breast cancer patients (Ceriani et al., 1982).

Carbohydrate Antigens in Epithelial Cell Membranes: While there are certainly antibodies available which show a high degree of specificity for epithelial cells and which react with antigenic determinants forming part of a polypeptide chain in a glycoprotein, these antigens have as yet been less well characterised than those which are carboyhdrate in nature. The components carrying the peptide determinants are usually characterised only by their molecular weight and their presence or absence in various cells or tissues. There are some exceptions such as the transferrin (Newman et al., 1982) receptor and the placental alkaline phosphatase (McLaughlin et al., 1982) although neither of these proteins can be considered epithelial specific. Pragmatically some of these antibodies, such as B6.2 used by Colcher et al. (1983) to locate breast tumours in nude mice may prove to be very useful and important. However since so many of the antibodies produced against malignant or differentiated epithelial cell membranes are proving to be against oligosaccharide sequences in glycolipids and glycoproteins, it is appropriate to put this recent information into a wider context.

For some time there has been considerable interest in blood group related carbohydrate antigens and in the changes in their expression which have been observed in differentiation, both in the embryo and in the adult (Kapadia et al., 1981; Feizi, 1981b) and in malignancy (Oh-Uti, 1949; Springer et al., 1975; Feizi et al., 1979; Javadpour, 1983). These studies have been given a new impetus by the availability of monoclonal antibodies to defined oligosaccharide sequences which were previously defined by lectins (Newman et al., 1979) or the naturally occurring monoclonal antibodies to the I antigens (Feizi, 1981a).

Many of the epithelial specific antibodies react with carbohydrate determinants found on large molecular weight glycoproteins with a high carbohydrate content (see Tables 2 and 3). These mucin-like molecules are commonly found on the surface of epithelial cells, and probably play an important role in their barrier function. The most widely studied have been the gastric mucins, where the structure of the carbohydrate side chains is best understood (Hounsell and Feizi, 1982). Fig. 4 shows a diagrammatic representation of the structure of a mucin and Fig. 5 shows how the various blood group activities (H, A, B, Le[a] and Le[b]) are conferred by adding monosaccharides to the Type 1 or Type 2 backbone structure. In gastric mucins, the I antigenic determinants are masked by the addition of sugars to produce the H, A, B and Le[a] and Le[b] antigens except in those individuals who lack the glycosyl transferase which adds the fucose to produce the H antigen (non-secretors). However, in some tumours of the gastrointestinal tract, I or i antigens are clearly detectable, and there may be aberrant expression of other blood group antigens (Feizi et al., 1979). Fig. 5 also shows the

relation of the naturally occurring blood group substances to some
of the determinants which have been found to react with antibodies
raised against human tumours and epithelial cells such as the
difucosylated antigen of Brown et al. (1983) and the lacto-N-fucopentaose III sequence, against which at least 25 monoclonal antibodies have been produced. As mentioned above, this latter sequence
appears to be very immunogenic in the mouse, and antibodies to it
have been obtained using carcinomas of the lung, stomach and colon
as well as melanomas (Brockhaus et al., 1982; Huang et al., 1983).
The sequence was first identified as being the epitope recognised
by antibodies to the stage specific embryonic antigen of the mouse
(SSEA-1) which identifies mouse embryos at a certain stage of
differentiation (Gooi et al., 1981; Shevinsky et al., 1982). Antibodies raised against SSEA-1 react also with some human tissues,
and tumours (Fox et al., 1983).

        Clearly, the changes in the pattern of carbohydrate determin-
ants seen in the large mucin-like glycoproteins (and glycolipids)
of malignant and/or differentiated epithelial cells must reflect
changes in the profile of glycosyl transferases. These observations therefore correlate well with results coming from other
studies which indicate a changed level in the activity of some
transferases of malignant cells, particularly those associated with
the addition of sialic acid (Bernacki and Kim, 1977; Katapodis et
al., 1982).

USE OF ANTIBODIES HMFG-1 AND 2 TO STUDY DIFFENENTIATION AND

MALIGNANCY IN THE HUMAN MAMMARY GLAND

        Antibodies like HMFG-1 and 2, which are directed to epithelial
cells at a certain stage of differentiation have a great potential

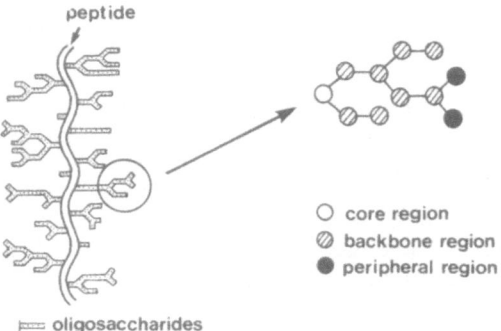

Fig. 4.   Schematic representation of a portion of a mucin glyco-
          protein molecule.

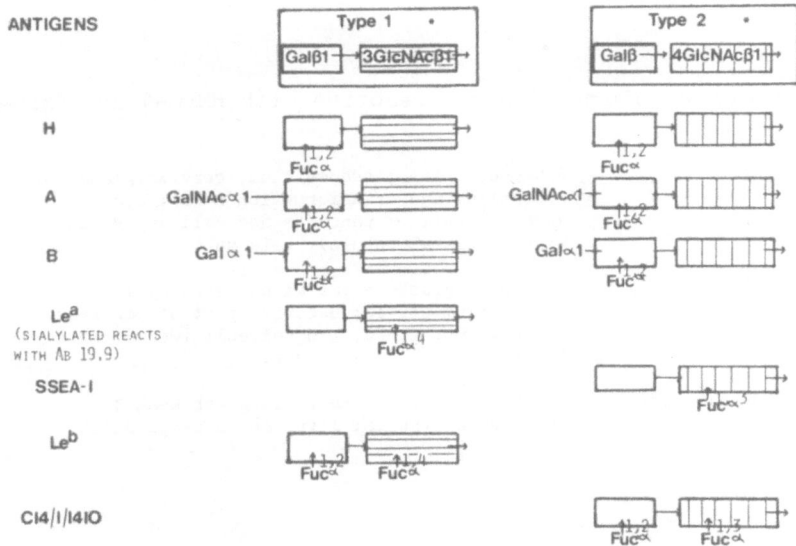

Fig. 5. Monosaccharides which confer H, A, B, Lewis[a], Lewis[b] and
         SSEA-1 antigenic activities when added to the periphery of
         type 1 and type 2 backbone sequences. Straight backbone
         = i specificity, Branched chain = I specificity. Mono-
         clonal antibodies M18 and M39 show I specificity.

for studying differentiation and its relation to malignancy.
Although the sequence of sugars in the determinants recognised by
antibodies HMFG-1 and 2 is not yet known, the oligosaccharide
nature of the determinants has been established (Burchell et al.,
1983) since peanut and wheat germ lectins can completely block
binding of both antibodies to the HMFG; the sialic acid binding
lectin (Limulus polyhemus agglutinin) also effectively blocks the
binding of HMFG-2 (but not of HMFG-1) to HMFG. In the milk fat
globule, both determinants are carried on the large molecular
weight mucin (Taylor-Papadimitriou et al., 1983b; Burchell et al.,
1983).

    The results of staining tissues with the antibodies in an
immunoperoxidase reaction show that both of the epitopes recognised
by HMFG-1 and 2 are expressed in the lactating breast. Although
neither epitope is strongly expressed in the resting breast other
normal tissues do express the antigens and these are shown in
Table 4. These normal tissues are in general epithelia associated
with a secretory function. Since major organs are negative, the
antibodies, particularly HMFG-2 are proving to be useful for in-vivo
localisation of tumours (Epenetos et al., 1982b).

Table 4. Normal tissues reactive with HMFG-1 and HMFG-2

HMFG-1	Sebaceous gland; endometrium, cervix (glandular epithelium and transformation zone); ovary; exocrine part of pancreas and salivary gland; kidney (distal tubule); bile duct.
HMFG-2	Sebaceous gland; sweat gland; epididymis endometrium cervix; exocrine part of pancreas and salivary gland; lung alveoli (weak); kidney (distal tubule); bile duct.
Negative for both HMFG-1 and HMFG-2	Gastro-intestinal tract; lung (or weak); squamous epithelia; liver and non-epithelial tissue.

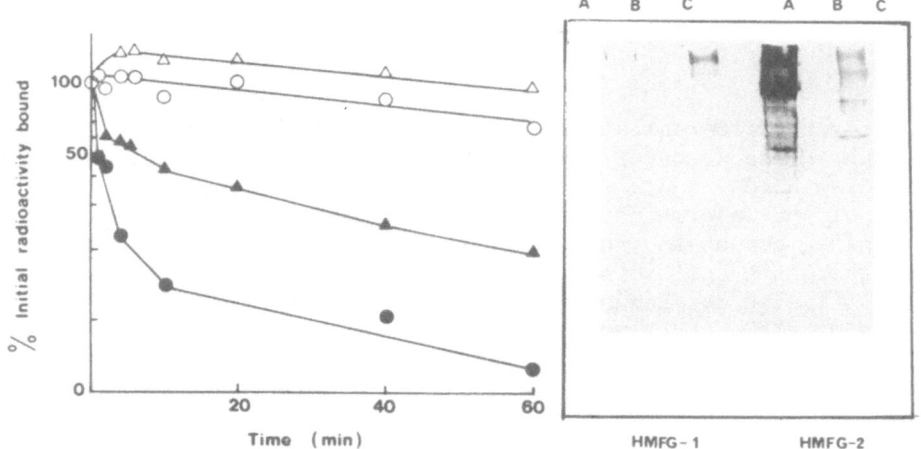

Fig. 6. Dissociation of ^{125}I-labelled HMFG-1 and HMFG-2 from T47D and milk cells and blots.

In view of the weak reaction of the antibodies with the non-lactating breast, the positive reaction observed with most primary and secondary breast cancers (Arklie et al., 1981) is interesting. We have examined in more detail the expression of the antigenic determinants recognised by HMFG-1 and 2 on normal and malignant cells in culture and in tissue sections, to see if their expression relates to a differentiation stage which is expressed in the malignant cells.

The first observation of interest is that the antibody HMFG-1 reacts with lactating gland or milk cells, at a lower concentration than antibody HMFG-2, whereas the reverse is true for breast cancer cell lines and many metastatic breast cancers. Studies with ^{125}I-labelled antibodies show that the large mucin found in the HMFG and on a proportion of the epithelial cells cultured from human milk has fewer sites for HMFG-2 than for HMFG-1, but both show a high affinity for their respective sites (Burchell et al., 1983). The kinetics of binding of the labelled antibodies to T47D cells (breast cancer cell line) however, show that the higher binding capacity of HMFG-2 to these cells is due to differences in the affinities of the antibodies. The dissociation of ^{125}I-labelled HMFG-1 and 2 from T47D and milk cells shown in Fig. 6 clearly demonstrates heterogeneity in the affinity of the sites on T47D for either antibody, but a reasonable proportion of HMFG-2 is still bound after 1 hour.

When immunoblots of gel separated lysates of the two cell types are stained with the antibodies, it can be seen that HMFG-1 reacts with a high molecular weight material in milk cells, while HMFG-2 also reacts with lower molecular weight components in both cell types (see Fig. 6). Indirect immunofluorescent staining of cultured cells and tissue sections, indicates that HMFG-2 staining is usually cell associated and possibly intracytoplasmic, while HMFG-1 can stain extracellular material (Taylor-Papadimitriou et al., 1983b). The possibility arises therefore that the smaller molecular weight components with antigenic sites binding HMFG-2 are precursors of the large mucin which has more HMFG-1 sites and is secreted at the lumenal surface of the secretory cells. The expression of large amounts of the HMFG-1 antigen by cultured milk cells has been shown to be associated with a decreased growth rate (Burchell et al., 1983) and a high extracellular expression of this antigen by primary ductal carcinomas appears to be related to a good prognosis (Wilkinson and Howell, personal communication). A high level of expression of the HMFG-1 antigen on a large molecular weight material as it is seen in the HMFG, is therefore probably indicative of the normal differentiation process of the potentially secretory mammary epithelial cells. A high expression of a determinant(s) reacting with HMFG-2 on the other hand, which is (are) consistently found in metastatic breast cancers, and can be found on smaller molecular weight components, may reflect an aberrant processing of the carbohydrate side chain.

Fig. 7 represents a speculative version of the relationship between the components expressing the HMFG-1 and 2 determinants in the normal and malignant cell. It is likely that the antigenic site(s) on tumour cell components binding HMFG-2 with a reasonable affinity are not identical to the site on normal HMFG or milk cell components, since they do not bind the antibody as strongly as the normal site. However, whether the smaller molecular weight components do indeed contain the same core protein as the large mucin, is not known, and Fig. 7 will remain speculative until this can be demonstrated. It is of interest to note that many of the glycoproteins carrying antigens with some specificity for epithelial tumours and melanomas are in the MW range 90-160K (Table 2). Possibly these glycoproteins are similar to the molecules carrying the HMFG-2 determinant. Such a possibility could be tested by looking for cross-reactivity of the antibodies using precipitation techniques and immunoblotting.

The above observations show that a carbohydrate determinant recognised by a single monoclonal antibody may be expressed on more than one membrane component and, moreover, the affinity of the antibody for the antigenic site may vary from cell to cell. This is an important point to consider when using antibodies like HMFG-2 for in-vivo localisation of tumours, since the time at which antibody is maximally concentrated in the tumour may vary from one patient to the next and several scans may be necessary.

ANTIBODIES TO INTERMEDIATE FILAMENTS OF EPITHELIAL CELLS

The intermediate filaments of epithelial cells are formed from a complex group of proteins, the cytokeratins which are coded for by at least 20 genes in the human and which are found only in epithelial cells (Franke et al., 1978). Antibodies to the cytokeratins have been developed which are specific for either a wide or narrow range of epithelial cell types. Thus the LE61 antibody, which reacts with all simple, non-stratifying normal epithelia (Lane, 1982) and carcinomas derived from them (Gatter et al., 1982) is a very useful reagent for characterisation of epithelial cells and tumour diagnosis in-vitro. This antibody also distinguishes lumenal (potentially secretory) cells in the human mammary gland from basal or myoepithelial cells (Lane, 1982; Taylor-Papadimitriou et al., 1983a). Other antibodies such as the K92 antibody reacts only with keratinizing epithelium (Gatter et al., 1982). The anti-keratin antibodies have therefore great potential as tissue specific markers. The problem in applying them to the targeting of isotopes and/or drugs in-vivo lies in the fact that the antigens they react with are generally intracellular. It is hard to see how this problem can be overcome in the case of drug-targeting where an intact cell is the target. However, it is conceivable that extracellular keratins exist within a solid tumour and are accessible to radio-

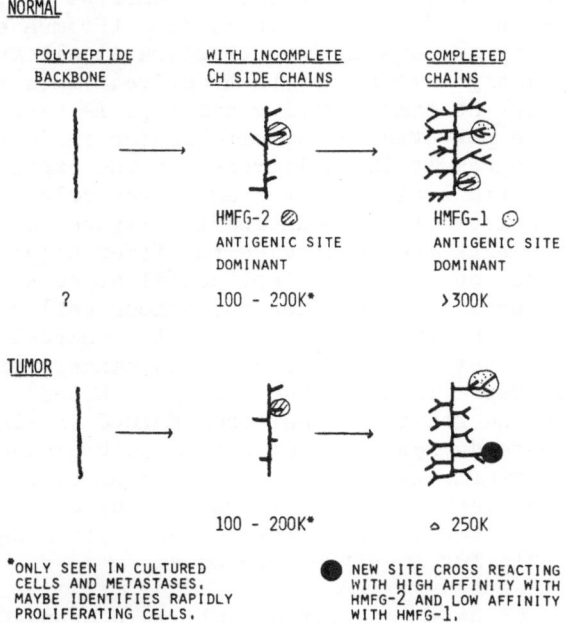

NORMAL

POLYPEPTIDE   WITH INCOMPLETE   COMPLETED
BACKBONE   CH SIDE CHAINS   CHAINS

                   HMFG-2 ⊘   HMFG-1 ○
                   ANTIGENIC SITE   ANTIGENIC SITE
                   DOMINANT   DOMINANT

       ?   100 - 200K*   >300K

TUMOR

                   100 - 200K*   ∘ 250K

*ONLY SEEN IN CULTURED   ● NEW SITE CROSS REACTING
 CELLS AND METASTASES.   WITH HIGH AFFINITY WITH
 MAYBE IDENTIFIES RAPIDLY   HMFG-2 AND LOW AFFINITY
 PROLIFERATING CELLS.   WITH HMFG-1.

Fig. 7. Speculative precursor relationship between antigenic
         sites recognised by antibodies HMFG-1 and 2.

activity labelled antibody. Presumably such extracellular material
reacting say with LE61 would not be found in normal tissues, and
tumour-detection may be feasible.

APPLICATIONS OF MONOCLONAL ANTIBODIES TO CELL IDENTIFICATION

IN VITRO AND IN VIVO

     The use of monoclonal antibodies for identifying cells in-vitro
and studying their differentiation has been mentioned. Their application to targeting of isotopes and drugs will be dealt with more
fully in other chapters of this book. Here we will briefly discuss
the potential problems referred to earlier, namely 1) the antigenic
heterogeneity in tumour cell populations and 2) the difficulty in
finding an antibody to a completely tumour specific antigen.

     The experience of most investigations is that even with a
sensitive technique like the immunoperoxidase staining of sections,
each antibody shows heterogeneity in its staining pattern of tumours.

Where a strong specificity for a tumour is claimed as with some of
the breast cancer directed antibodies (Colcher et al., 1981; Kufe
et al., 1983) or the CEA specific antibodies (Primus et al., 1983a),
some of the primary tumours show no reaction at all and metastases
from positive primary tumours may be negative. This heterogeneity
between tumours of the same histological type is seen with most
antibodies, and is potentially exploitable for indication of prognosis and may even result in an increase in the expression of an
antigen on metastatic lesions (metastases from colon carcinomas
originally unreactive with HMFG antibodies may become reactive).
It is less marked with the antibodies to differentiation antigens
which are expressed on the normal epithelial at some stage of
differentiation, presumably because the tumour cell consistently
expresses the differentiation phenotype. Thus HMFG-2 reacted
positively with 59 out of 60 primary breast cancers and with 49 out
of 50 metastatic deposits from them (in lymph nodes). M8 which
reacts more with the resting breast than HMFG-2 is also consistent
in reacting to some degree or other with most breast cancers. The
antigens on the intermediate filaments can also be considered to be
differentiation antigens and that recognised by antibody LE61 is
highly conserved in carcinomas. This makes their reagents extremely
useful and reliable for in-vitro diagnosis.

    The problem of heterogeneity of antigenic expression within
tumours (Horan-Hand et al., 1983) is found with most antibodies,
whether they were raised against tumour cells or differentiated
cell membranes. In our experience this heterogeneity can be more
marked in fixed sections, particularly where the antigenic determinant can be carried on a component with different chemical
properties (e.g. glycolipid and glycoprotein); more cells appeared
to stain with HMFG antibodies in frozen unfixed sections than in
sections of formalin-fixed paraffin embedded tissue. However, even
in frozen sections, heterogeneity of antigen expression is observed
and this is clearly a major problem to face in drug targeting.
Within this problem is that of antigenic drift where metastatic
cancer cells may lose antigens expressed by the primary tumour.

    Successful targeting of isotopes for tumour localisation and
scanning is obviously going to be easier to develop using the antibodies now available, than is the targeting of toxic drugs. Antigens
of the kind recognised by HMFG antibodies which are expressed at high
levels by most breast and ovarian carcinomas and even by some other
metastatic carcinomas can be applied rather effectively to tumour
localisation. Cocktails of the more tumour specific antibodies may
be required to achieve the same result. If such cocktails become
available and the antibodies are truly tumour specific (as are
possibly those against placental alkaline phosphatase) then it may
indeed be possible to effectively use them to target toxic drugs.
Until this aim is realised much useful information can be gained
from a more detailed study of the nature of the antigens so far
identified as being epithelial specific, and their expression in

cells at different stages of differentiation. Such studies may lead
us to a more directed approach to the production of antibodies with
the level of specificity required for drug targeting.

APPENDIX

METHOD FOR PRODUCTION OF HYBRID ANTIBODY PRODUCING CELLS

     The following method is employed in the author's laboratory;
the crucial step is the addition of the polyethylene glycol and
detailed instructions are given for its preparation and addition.
Four days after completion of the immunisation programme the mouse
spleen is teased out under sterile conditions to form a cell suspension and washed once in sterile RPMI before re-suspension in a
small volume (2-5 ml). 3×10^7 myeloma cells (NS2) are washed twice
in RPMI and mixed with the spleen cells in a universal container,
and centrifuged (1000 rpm, bench centrifuge, 150 x g) for 5 min.
The almost dry cell pellet is suspended in 1 ml of PEG (50% weight
by volume, sterilised by autoclave) at 37°C over a period of 1 min,
mixing gently by taking up into the pipette. After 45 seconds at
37°C the mixture is diluted as follows: (a) 1 ml warm RPMI is added
over a period of 2 min; (b) 5 ml warm RPMI is added over 2 min, and
(c) 10 ml warm RPMI is added slowly. The mixture is then centrifuged
as above before re-suspending the cell pellet in RPMI + 20% serum
and plating out into 24 well plates, containing spleen feeders
(1 spleen to 4 x 24 well plates) to a final volume of 1 ml. Plating
out may be into 16 x 24 well plates equally, or the cells may be
diluted (half of the cells into 8 plates, quarter into 8 plates,
one eighth into 8 plates and one sixteenth into 8 plates). Assays
may then be restricted to plates with approximately 1 hybridoma per
well.

     The following day a further 1 ml of 2 x HMT is added. (HMT -
RPMI + 20% FCS containing hypoxanthine 10^{-4}M, thymidine 1.6×10^{-5}
and methotrexate 10^{-5}M). Clones will appear at 7-12 days. If
dilution has been performed plates with many clones per well may
be discarded, and wells with only one clone apparent marked. After
marking of single clones, a medium change removing 1 ml and replacing
approximately 1 ml HMT will usually stimulate cell growth. Assay of
supernatants from clones of suitable size may be performed 48-72
hours after changing medium. Positive wells should be re-cloned by
dilution in 96 well plates (one each at $\frac{1}{2}$, 1 and 3 cells/well and
$\frac{1}{2}$ plate each of 10 and 100 cells containing spleen feeders, 1 spleen
to 10 x 96 well plates). If a positive well was thought to contain
more than 1 clone it is especially important that this cloning be
done early before the antibody producing hybridoma becomes overgrown
by other clones in the well. Cells from early passages of positive

original wells may be stored in liquid nitrogen in case of failure
of cloning. Clones will appear after 7-10 days and single clones
should be marked before changing medium. Assay after a further 48-
72 hours enables the positive single clones from the greatest
dilution to be selected and transferred to 24 well plates and
weaned off feeders before transfer to tissue culture flasks. Methotrexate is withdrawn between the second and third week and H + T
slowed between the third and fourth. Passage of cells should always
be with a portion of "conditioned" media since hybridomas show considerable variation in their growth and resilience.

SCREENING ASSAYS

Solid Phase RIA or ELISA

      Using membrane extracts or protein preparation: Wells in Limbro
plates can be coated with membrane extracts or proteins either after
pre-treatment with 0.25% glutaraldehyde or merely after washing with
buffer. The advantage of using glutaraldehyde treated plates is
that duplicate wells tend to be more reproducible and a definite
amount of antigen can be bound. The following is a method used in
the author's laboratory for the preparation of plates which may be
used for RIA or ELISA: 96 well PVC Limbro V bottom plates are filled
with 0.25% glutaraldehyde in PBSA and left to stand for 5 min.
Wells are drained and blotted dry. The protein solution ($25 \mu l$) of
the desired concentration in PBSA (e.g. 20 ng of solvent extracted
HMFG, up to $1 \mu g$ of a membrane extract of solid tissue) is added
and dried overnight at $37^{o}C$. These plates may then be stored
indefinitely at $4^{o}C$. Before assay the wells are rinsed with PBSA
containing 0.1% BSA + 0.05% azide and blotted dry. The supernatant
to be screened is added, $25 \mu l$ per well usually in duplicate, and
incubated overnight at room temperature in humidified air (wet box).

      Using fixed cells: The base of the wells in flat bottomed 96
well tissue culture plates, which have better optical properties for
ELISA, is coated with poly-L-lysine (0.1 mg/ml in PBSA) and left at
room temperature for 1 hour. The poly-L-lysine is then flicked out,
and the plate blotted dry. The cells (taken from suspension or from
monolayer culture by agitation with glass beads) are washed twice
in PBSA and added at 2.5×10^{5} per well in 50 μl PBSA. The plates
are centrifuged (1000 rpm.175 x g for 8 min) then left at room temperature for 15 min. Glutaraldehyde is then added to almost fill
the wells and the plates left for a further 15 min. The wells are
flicked out and washed three times with PBSA and blotted dry. For
storage they may be filled with PBSA containing $200 \mu g/ml$ gelatin,
and 0.02% azide and covered with clingfilm; at $4^{o}C$ they keep for up
to 4 months. To use, the plates are washed with PBSA three times,
filled with 0.1% phenyl hydrazine, and left for 1 hour. After
three washes with PBSA and one with PBSA containing 0.2% tween 20,
the plates are blotted dry, supernatants added to duplicate wells

and the wells incubated for 2 hours before continuing the assay as
described below.

For RIA the plates are washed three times with PBSA containing
0.1% BSA and 0.05% azide. The second, radiolabelled (^{125}I), antibody is added, e.g. $30\,\mu$l rabbit anti-mouse diluted in the above
solution to give 100,000-200,000 cpm. The plates are incubated for
4 hours before washing 4 times as above and then cut up and counted
on a scintillator counter. When using the hard plates used for
fixed cell assays, the cells are removed in an alkaline solution
for counting.

For ELISA the plates are washed three times with PBSA and once
with PBSA containing 0.2% tween 20, then blotted dry. $25\,\mu$l and
peroxidase-conjugated goat anti-mouse antiserum (1:20 in 200 mM tris
pH 7.6 and 10% normal goat serum using GAm from Central Laboratory
of the Netherlands Red Cross Blood Transfusion Service) is added
and the plates incubated for 30 min. The plates are again washed
as above and blotted dry before adding $25\,\mu$l of peroxidase-anti-
peroxidase antibody (a mouse monoclonal antibody, Central Laboratory
of the Netherlands Red Cross Blood Transfusion Service diluted
1:100 in 200 mM tris pH 7.6 and 10% normal goat serum). After
incubation for 30 min, the plates are washed as before and $100\,\mu$l
of substrate is added. Substrate is made up fresh just before use
by adding 20 mg of D-phenylenediamine and $5\,\mu$l of H_2O_2 (100 vols)
to 10 ml of buffer (22.5 g Na_2HPO_4 5.6 g citric acid made up to
1 litre and brought to pH 6.0 with conc. HCl). The plates are
incubated in the dark (by wrapping in foil), before being read
immediately at 450 nm on the titertek multiskan machine, or the
reaction may be stopped by adding $50\,\mu$l 0.1 M NaF and read later,
or photographed as required.

Live Cell Assays

The cells expressing the antigen are suspended (without the
use of trypsin) in RPMI containing 0.2% BSA and 0.02% azide, $25\,\mu$l
of the suspension is added to each well in Limbro 96 well v-bottomed
PVC plates (larger cell numbers increase the background but also the
reproducibility of the duplicates as a more stable pellet is formed
on centrifuging). The hybridoma supernatant is added to the wells
(usually $50\,\mu$l in duplicate) and the plate is incubated at 4°C for
1 hour. Washing solution (RPMI + 0.2% BSA and 0.02% azide) at 4°C
is added to fill 75% of the well before centrifuging the plate at
1000 rpm, 175 x g, for 5 min. The supernatant is removed by careful
suction so as not to disturb the cell pellet, and the wash repeated
three times. A second radiolabelled antibody (e.g. ^{125}I-labelled
rabbit anti-mouse, diluted in washing solution to 200,000 cpm) is
added and incubated at 4°C for 1 hour, then the initial wash and
four subsequent washes carried out as above. The cells may then
be transferred in a small amount of washing solution to $400\,\mu$l
Sarsted tubes and counted in a scintillation counter.

spread the antibody and avoid bubbles or scratching of the cell
monolayer). Incubation and washing are as above, the coverslip
being floated off with PBSA. The final wash is with distilled
water and a new coverslip mounted over the sample area with Gelvatol,
an aqueous mounting medium (Monsanto Polymers and Petrochemicals Co.).
The sides of the dish are clipped off to avoid obstruction of the
objective lens and the plate viewed and scored for fluorescence in
ultraviolet epi illumination. Positive and negative samples may be
identified by this means under low power objectives; high power
objectives may give information as to distribution of the antigen
in the cell.

INDIRECT IMMUNO-FLUORESCENT STAINING OF CELLS IN SUSPENSION

     This technique has little place in the screening of large
numbers of wells as it is labour-intensive and time-consuming. It
is, however, a useful technique required for separating positive
from negative cells using the fluorescent-activated cell-sorter
(FACS). It can also be used to examine cells which occur in suspension in serous effusions, e.g. in the pleural or peritoneal
cavities of patients with advanced malignant disease.

     When dealing with cells in serous effusions the cells are
first washed with PBSA and the contaminating red cells removed by
centrifuging for 20 min at 1000 rpm over a Lymphoprep/PBSA gradient
(Nyegaard & Co., Oslo, Norway). The cells at the interface having
been washed twice in PBSA and re-suspended in RPMI containing 10%
FCS + 0.05% azide may then be counted in a haemocytometer (ignoring
the few remaining red cells). If cultured cells are used trypsin

Indirect Immune Fluorescence on Monolayers of Fixed Cells

     This technique (Lane and Lane, 1981) has proved a reliable
method of screening for antibodies against antigens stable to
methanol acetone fixation. Confluent cultures of cells expressing
antigens of interest, in 60 mm Nunc tissue culture dishes are
washed free of medium with PBSA and the cells fixed in situ with
methanol acetone (equal volumes) in one or two 5 min changes. The
dishes are dried thoroughly by evaporation. On the underside of
the dish a grid the size of a coverslip is marked off with a water-
proof pen and the supernatant to be tested systematically loaded on
the fixed dried cell layer inside the dish using a clean pipette
tip for each sample for 1-3 μl aliquots. Taking care not to scratch
the cell layer, over 100 samples may be tested within the area of
a 22 mm x 40 mm coverslip.

     After loading, the plate is incubated at room temperature for
20 min; the samples are then washed off with PBSA and a second antibody coupled to fluorescein isothyiocyanate (e.g. FITC-goat anti-

mouse from Cappel Laboratories 1:50) added to cover the marked area (as little as 50 μl may be used if a coverslip is used with care to is avoided in the preparation of cell suspension as it may change the cell membrane. Aliquots of cells (2 x 105) in universal containers are incubated for 30 min with antibody supernatant, washed with medium (20 ml of RPMI and 10% FCS) at 4°C and cell pellets resuspended in 0.5 ml of second antibody (e.g. FITC GAM diluted 1:40 with medium or PBSA containing 10% serum and 0.02% azide). Incubation is for 30 min at 4°C and the cells are then washed as above. The cells are mounted on a slide by re-suspending in 1 drop of Gelvatol and transferring this drop, by capillary action, in a pasteur pipette (to avoid bubbles), dropping it on to a slide and placing a coverslip over it. Slides may be examined immediately in ultraviolet epi-illumination. When staining cells for FACS the azide is omitted and solutions are sterile.

IMMUNE PEROXIDASE STAINING OF TISSUE SECTIONS

There is much variation in the staining properties of antibodies dependent on the method of fixation. Indeed, some antibodies will only react with frozen tissue, in which case, to form a permanent record, slides may be fixed, e.g. in neutral buffered formalin, after the peroxidase substrate reaction. Generally speaking, methanol-fixed tissue is preferred for immuno-histochemical techniques although the requirements for each individual antibody have to be worked out. There are advantages to antibodies that react well with formalin fixed tissue in that there may be much tissue available going back over a number of years because of its widespread routine use. The following represents the author's experience of the antibodies HMFG-1 and HMFG-2 which react well with formalin fixed tissue: formalin fixed and paraffin embedded sections are first de-waxed in xylene and alcohol in the usual way. Sections are immersed in a solution containing 1 part 3% hydrogen peroxide to 5 parts methanol for 15 min to remove endogenous peroxidase activity and then washed thoroughly two or three times in PBSA. The slides are then drained and incubated with the antibody, neat or diluted up to 1:100, for 40 min at room temperature. The antibody is rinsed off with PBSA as before and a second peroxidase conjugated antibody (e.g. rabbit anti-mouse DAKO diluted 1:50 with PBSA containing 10% normal rabbit serum) added. The slides are incubated for 40 min as before and the antibody rinsed off with PBSA thoroughly 2-3 times. The freshly made up peroxidase substrate diamino benzidine (DAB, 10 mg in 10 ml of 0.03% hydrogen peroxide in PBSA) is added and the peroxidase reaction allowed to continue for 5 min. The DAB is rinsed off with distilled water and the slides counterstained with a progressive haemotoxylin and mounted in Gelvatol.

When frozen unfixed sections are used they are first air-dried thoroughly and the hydrogen peroxide methanol blocking step omitted (the methanol fixes the tissue). The procedure is otherwise as

above from that point until the completion of the peroxidase
reaction when fixation before mounting helps preserve the slide.
Great care has to be exercised in handling frozen sections as they
are prone to wash off the slide. The use of glycerine albumin pretreated slides and the air-drying being performed overnight in a
vacuum dessicator has helped to overcome these problems.

THE IMMUNO-BLOTTING TECHNIQUE

This technique can be useful in the detection of antigens
resistant to protein denaturation by SDS treatment and gives information as to the molecular weight of the substance on which the
antigen is carried. Western (protein) blots can be performed
essentially as described by Towbin et al. (1979). Proteins are
electrophoretically transferred overnight in the cold, from polyacrylamide gels to nitrocellulose paper (Schleicher and Schuell)
at 50 volts, 0.29 amps. Detection of antigenic determinants is
carried out by incubating the blot with monoclonal culture supernatant for 30 min at $4^\circ C$. After extensive washing (5 times, 10 min
each wash with agitation) with PBS at $4^\circ C$, the paper is incubated
for a further 30 min with peroxidase-conjugated rabbit antimouse
immunoglobulin (DAKO) diluted 1:40 in PBS and containing 4% human
serum. After further extensive washing the colour is developed by
agitating the blots for 5-10 min in a solution of 4-chloronaphthol
(a saturated solution in ethanol, diluted 1:100 with buffer and
filtered before use) containing 0.03% hydrogen peroxide. A radioactive second antibody may be used in place of peroxidase conjugate,
and the blot then processed for autoradiography.

REFERENCES

Accolia, R.S., Carrel, S. and Mach, J.-P., 1980, Monoclonal anti-
        bodies specific for CEA and produced by two hybrid cell lines,
        Proc. Natl. Acad. Sci. USA, 77:563.
Arklie, J., Taylor-Papadimitriou, J., Bodmer, W.F., Egan, M. and
        Millis, R., 1981, Differentiation antigens expressed by epi-
        thelial cells in the lactating breast are also detectable in
        breast cancers, Int. J. Cancer, 28:23.
Ashall, F., Bramwell, M.E. and Harris, H., 1982, A new marker for
        human cancer cells. 1. The Ca antigen and the Cal antibody,
        Lancet, 2:1.
Bernacki, R.J. and Kim, U., 1977, Concomitant elevations in serum
        sialyltransferase activity and sialic acid in rats with
        metastasizing mammary tumours, Science, 195:577.
Bischoff, R., Eisert, R.M., Schedel, I., Vienken, J. and Zimmerman,
        U., 1982, Human hybridoma cells produced by electro fusion,
        FEBS Lett., 147:64.
Brown, A., Feizi, T., Gooi, H.C., Embleton, M.J., Picard, J.K. and
        Baldwin, R.W., 1983, A monoclonal antibody against human

colonic adenoma recognizes difucosylated Type-2-blood-group chains, Bioscience Rep., 3:163.

Brockhaus, M., Magnani, J.L., Herlyn, M., Blaszczyt, M., Steplewski, Z., Koprowski, H., Einsberg, V., 1982, Monoclonal antibodies directed against the sugar sequence of lacto-N-fucopentaose III are obtained from mice immunised with human tumours, Arch. Biochem. Biophys., 217:647.

Buchegger, F., Phan, M., Rivier, D., Carrel, S., Accolia, R.S. and Mach, J.-P., 1982, Monoclonal antibodies against carcino-embryonic antigen (CEA) used in a solid-phase enzyme immuno-assay. First clinical results, J. Immunol. Methods, 49:129.

Burchell, J., Durbin, H. and Taylor-Papadimitriou, J., 1983, Complexity of expression of antigenic determinants, recognised by monoclonal antibodies HMFG-1 and HMFG-2, in normal and malignant human mammary epithelial cells, J. Immunol., in press.

Canevari, S., Fossati, G., Balsari, A., Sonnino, S. and Colnaghi, M.I., 1983, Immunochemical analysis of the determinant recognised by a monoclonal antibody (MBr1) which specifically binds to human mammary epithelial cells, Cancer Res., 43:1301.

Ceriani, R.L., Thompson, K.E., Peterson, J.A. and Abraham, S., 1977, Surface differentiation antigens on human mammary epithelial cells carried on the human milk fat globule, Proc. Natl. Acad. Sci. USA, 74:582.

Ceriani, R.L., Sasaki, M., Sussman, H., Wara, W.M. and Blank, E.W., 1982, Circulating human mammary epithelial antigens in breast cancer, Proc. Natl. Acad. Sci. USA, 79:5420.

Chang, T.H., Steplewski, Z. and Koprowski, H., 1980, Production of monoclonal antibodies in serum free medium, J. Immunol. Meth., 39:369.

Cleveland, W.L., Wood, I. and Erlanger, B.F., 1983, Routine large-scale production of monoclonal antibodies in a protein-free culture medium, J. Immunol. Meth., 56:221.

Colcher, D., Horan Hand, P., Nuti, M. and Schlom, J., 1981, A spectrum of monoclonal antibodies reactive with human mammary tumour cells, Proc. Natl. Acad. Sci. USA, 78:3299.

Colcher, D., Zalutsky, M., Kaplan, W., Kufe, D., Austin, F. and Schlom, J., 1983, Radiolocalization of human mammary tumours in athymic mice by a monoclonal antibody, Cancer Res., 43:736.

Croce, C.M., Linnenbach, A., Hall, W., Steplewski, Z. and Koprowski, H., 1980, Production of human hybridomas secreting antibodies to measles virus, Nature, 288:488.

Crawford, D.H., Callard, R.E., Muggeridge, M.I., Mitchell, D.M., Zander, E.D. and Beverley, P.C.L., 1983, Production of human monoclonal antibodies to X31 influenza virus nuclear protein, J. Gen. Virol., 64:697.

Cuttitta, F., Rosen, S., Gazdar, A.F. and Minna, J.D., 1981, Mono-clonal antibodies that demonstrate specificity for several types of human lung cancer, Proc. Natl. Acad. Sci. USA, 78:4591.

Edwards, P.A.W., 1981, Some properties and applications of mono-
    clonal antibodies, Biochem. J., 200:1.
Epenetos, A.A., Canti, G., Taylor-Papadimitriou, J., Curling, M.
    and Bodmer, W.F., 1982a, Use of two epithelial-specific
    monoclonal antibodies for diagnosis of malignancy in serous
    effusions, Lancet, 2:1004.
Epenetos, A.A. Britton, K.E., Mather, S., Shepherd, J., Granowska,
    M., Taylor-Papadimitriou, J., Nimmon, C.C., Durbin, H.,
    Hawkins, L.R., Malpas, J.S. and Bodmer, W.F., 1982b,
    Targeting of iodine-123-labelled tumour-associated mono-
    clonal antibodies to ovarian, breast and gastrointestinal
    tumours, Lancet, 2:999.
Ey, P.L., Prowse, S.J. and Jenkin, C.R., 1978, Isolation of pure
    IgG_1, IgG_{2a} and IgG_{2b} immunoglobulins from mouse serum using
    protein A-Sepharose, Immunochem., 15:429.
Feizi, T., 1981a, The blood group Ii system: a carbohydrate antigen
    system defined by naturally monoclonal or alogoclonal auto-
    antibodies of man, Immunol. Commun., 10:127.
Feizi, T., 1981b, Carbohydrate differentiation antigens, Trends
    Biochem. Sci., 6:333.
Feizi, T., Picard, J., Kapadia, A. and Slavin, G., 1979, Changes in
    the expression of the major blood group antigens ABH and
    their precursor antigens Ii in human gastric cancer tissues,
    in: "Protides of Biological Fluids", H. Peeters, ed.,
    Pergamon Press, Oxford, New York.
Foster, C.S., Edwards, P.A.W., Dinsdale, E.A. and Neville, A.M.,
    1982, Monoclonal antibodies to the human mammary gland.
    1. Distribution of determinants in non-neoplastic mammary
    and extra mammary tissues, Virch. Arch. Path. Anat. Physiol.
    Kin. Med., 394:279.
Fox, G., Damjanov, I., Knowles, B.B. and Solter, D., 1983, Immuno-
    histochemical localisation of the mouse stage specific
    embryonic antigen I in human tissues and tumours, Cancer
    Res., 43:669.
Franke, W.W., Weber, K., Osborn, M., Schmid, E. and Freudenstein, C.,
    1978, Antibody to prekeratin, Exp. Cell Res., 116:429.
Frankel, A.E., Rouse, R.V. and Herzenberg, L.A., 1982, Human pros-
    tate specific and shared differentiation antigens defined by
    monoclonal antibodies, Proc. Natl. Acad. Sci. USA, 903:907.
Galfre, G., Milstein, C. and Wright, B., 1979, Rat x rat hybrid
    myelomas and a monoclonal anti-Fd portion of mouse IgG,
    Nature, 277:131.
Gatter, K.C., Abdulaziz, Z., Beverley, P., Corvalan, J.R.F., Ford, C.,
    Lane, E.B., Mota, M., Nash, J.R.G., Pulford, K., Stein, H.,
    Taylor-Papadimitriou, J., Woodhouse, C. and Mason, D.Y.,
    1982, Use of monoclonal antibodies for the histopathological
    diagnosis of human malignancy, J. Clin. Pathol., 35:1253.
Gooi, H.C., Feizi, T., Kapadia, A., Knowles, B.B., Solter, D. and
    Evans, M.J., 1981, Stage specific embryonic antigen SSEA-1
    involves al-3 fucosylated Type 2 blood group chains, Nature,
    292:156.

Gooi, H.C., Uemura, K., Edwards, P.A.W., Foster, C.S., Pickering, N.
     and Feizi, T., 1983, Two mouse hybridoma antibodies against
     human milk fat globules recognise the I(MA) antigenic deter-
     minant: Galβ1-$\rightarrow$4GlcNAcβ1-$\rightarrow$6, Carbohydrate Res., in press.
Gold, P. and Freedman, S.O., 1965, Specific carcinoembryonic anti-
     gens of the human digestive system, J. Exp. Med., 121:439.
Gottleib, P.D., 1980, Immunoglobulin genes, Molecular Immunol., 17:
     1423.
Horan-Hand, P., Nuti, M., Colcher, D. and Schlom, J., 1983,
     Definition of antigenic heterogeneity and modulation among
     human mammary carcinoma cell populations using monoclonal
     antibodies to tumour-associated antigens, Cancer Res., 43:728.
Hounsell, E.F. and Feizi, T., 1982, Gastrointestinal mucins.
     Structures and antigenicities of their carbohydrate chains
     in health and disease, Med. Biol., 60:227.
Heyderman, E., Stule, K. and Ormerod, M.G., 1979, A new antigen on
     the epithelial membrane: its immunoperoxidase localisation
     in normal and neoplastic tissue, J. Clin. Pathol., 32:35.
Huang, L.C., Brockhaus, M., Magnani, J.L., Cuttitta, F., Rosen, S.,
     Minna, J.D. and Ginsburg, V., 1983, Many monoclonal anti-
     bodies with an apparent specificity for certain lung cancers
     are directed against a sugar sequence found in Lacto-N-fuco-
     pentaose III, Arch. Biochem. Biophys., 220:318.
Imai, K., Wilson, B.S., Bigotti, A., Natali, P.G. and Ferrone, S.,
     1982, A 94,000-Dalton glycoprotein expressed by human melanoma
     and carcinoma cells, J. Natl. Canc. Inst., 68:761.
Ines-Colnaghi, M., 1983, Antibodies to ovarian cancer, in: "Mono-
     clonal Antibodies and Cancer", . Dulbecco and . Langman,
     eds., Academic Press, New York.
Javadpour, N., 1983, Immunocytochemical localisation of various
     markers in cancer cells and tumours, Urology, XXI:7.
Katapodis, N., Hirshaut, J., Geller, N.L. and Stock, C.C., 1982,
     Lipid-associated sialic acid test for the detection of human
     cancer, Cancer Res., 42:5270.
Kapadia, A., Feizi, T. and Evans, M.J., 1981, Changes in the ex-
     pression and polarization of blood group I and i antigens in
     post-implantation embryos and teratocarcinomas of mouse
     associated with cell differentiation, Exp. Cell Res., 131:185.
Kearney, J.F., Radbruck, A., Liesegang, B. and Rajewskyl, M., 1979,
     A new mouse myeloma cell line that has lost immunoglobulin
     expression but permits the construction of antibody-secreting
     hybrid cell lines, J. Immunol., 123:1548.
Kohler, G. and Milstein, C., 1975, Continuous culture of fused cells
     secreting antibody of predefined specificity, Nature, 256:495.
Kohler, G. and Milstein, C., 1976, Derivation of specific antibody
     producing tissue culture and tumour lines by cell fusion,
     Eur. J. Immunol., 6:511.
Koprowski, H., Steplewski, Z., Mitchell, K., Herlyn, D., Herlyn, M.
     and Fuhrer, P., 1979, Colorectal carcinoma antigens detected
     by hybridoma antibodies, Somatic Cell Genet., 5:957.

Koprowski, H., Herlyn, M., Steplewski, Z. and Sears, H.F., 1981,
    Specific antigen in serum of patients with colon carcinoma,
    Science, 212:53.
Kishimoto, T., 1983, Human neoplastic B cells: monoclonal models
    of B-cell differentiation, Immunology Today, 4:117.
Kufe, D.W., Nadler, L., Sargent, L., Shapiro, H., Hand, P., Austin,
    F., Colcher, D. and Schlom, J., 1983, Biological behavior
    of human breast carcinoma-associated antigens expressed
    during cellular proliferation, Cancer Res., 43:851.
Kuroki, M., Koga, Y. and Matsuoka, Y., 1981, Purification and
    characterization of carcinoembryonic antigen-related anti-
    gens in normal adult feces, Cancer Res., 41:713.
Lane, E.B., 1982, Monoclonal antibodies provide specific intra-
    molecular markers for the study of epithelial tonofilament
    organisation, J. Cell Biol., 92:665.
Lane, D.P. and Lane, E.B., 1981, A rapid antibody assay system for
    screening hybridoma cultures, J. Immunol. Meth., 47:303.
Levy, R. and Dilley, J., 1978, Rescue of immunoglobulin secretion
    from human neoplastic lymphoid cells by somatic cell hybrid-
    isation, Proc. Natl. Acad. Sci. USA, 75:2411.
Magnani, J.L., Nilsson, B., Brockhaus, M., Zop, D., Steplewski, Z.,
    Koprowski, H. and Ginsburg, V., 1982, A monoclonal antibody-
    defined antigen associated with gastrointestinal cancer is a
    ganglioside containing sialylated lacto-N-fucopentaose II,
    J. Biol. Chem., 257:14369.
Marguiles, D.H., Kuehl, W.M. and Scharff, M.D., 1976, Somatic cell
    hybridisation of mouse myeloma cells, Cell, 8:405.
McLaughlin, P.J., Cheng, M.H., Slade, M.B. and Johnson, P.M., 1982,
    Expression on cultured human tumour cells of placental
    trophoblast membrane antigens and placental alkaline phos-
    phatase defined by monoclonal antibodies, Int. J. Cancer,
    30:21.
Menard, S., Tagliabue, E., Canevari, S., Fossati, G. and Colnaghi,
    M.I., 1983, Generation of monoclonal antibodies reacting
    with normal and cancer cells of human breast, Cancer Res.,
    43:1295.
Metzgar, R.S., Gaillard, M.T., Levine, S.J., Tuck, F.L., Bossen, E.H.
    and Borowitz, M.J., 1982, Antigens of human pancreatic adeno-
    carcinoma cells defined by murine monoclonal antibodies,
    Cancer Res., 42:601.
Milstein, C., Clark, M.R., Galfre, G. and Cuello, A.C., 1980, Mono-
    clonal antibodies from hybrid myelomas, in: "Immunology 80,
    Progress in Immunology IV", I.M. Fougereau and J. Dausset,
    eds., Academic Press, London, New York.
Newman, R.A., Klein, P.J. and Rudland, P.S., 1979, Binding of peanut
    lectin to breast epithelium, human carcinomas, and a cultured
    rat mammary stem cell: Use of the lectin as a marker of
    mammary defferentiation, J. Natl. Canc. Inst., 63:1339.
Newman, R., Schneider, C., Sutherland, R., Vodinelich, L. and
    Greaves, M., 1982, The transferrin receptor, Trends Biochem.
    Sci., 7:397.

Nuti, A., Teramoto, Y.A., Mariani-Constantini, R., Horan-Hand, P., Colcher, D. and Schlom, J., 1982, A monoclonal antibody (B72.3) defines patterns of distribution of a novel tumour-associated antigen in human mammary carcinoma cell populations, Int. J. Cancer, 29:539.

Oh-Uti, K., 1949, Polysaccharides and glycidamin in the tissue of gastric cancer, Tohuku J. Exp. Med., 51:297.

Olsson, L. and Kaplan, H.S., 1980, Human-human hybridomas producing monoclonal antibodies of pre-defined antigenic specificity, Proc. Natl. Acad. Sci. USA, 77:5429.

Pardue, R.L., Brady, R.C., Perry, G.W. and Dedman, J.R., 1983, Production of monoclonal antibodies against calmodulin by in-vitro immunization of spleen cells, J. Cell Biol., 96:1149.

Pontecorvo, G., 1975, Production of mammalian somatic cell hybrids by means of polyethylene glycol treatment, Somatic Cell Genet., 1:397.

Primus, F.J., Freeman, J.W. and Goldenberg, D.M., 1983a, Immunological heterogeneity of CEA: Purification from meconium of an antigen related to CEA, Cancer Res., 43:679.

Primus, F.J., Newell, K.D., Blue, A. and Goldenberg, D.M., 1983b, Immunological heterogeneity of CEA: Antigenic determinants on CEA distinguished by monoclonal antibodies, Cancer Res., 43:686.

Primus, F.J., Kuhrs, W.J. and Goldenberg, D.M., 1983c, Immunological heterogeneity of CEA: Immunohistochemical detection of CEA determinants in colonic tumours with monoclonal antibodies, Cancer Res., 43:693.

Shevinsky, L.H., Knowles, B.B., Damjanow, I. and Solter, D., 1982, Monoclonal antibody to murine embryo defines a stage specific embryonic antigen expressed on mouse embryos and human teratocarcinoma cells, Cell, 30:697.

Shimizu, M. and Yamauchi, K., 1982, Isolation and characterisation of mucin-like glycoprotein in human milk fat globule membrane, J. Biochem., 91:515.

Shulman, M., Wilde, C.D. and Kohler, C., 1978, A better cell line for making hybridomas secreting specific antibodies, Nature, 276:269.

Springer, G.F., Desai, P.R. and Banatwala, I., 1975, Blood group MN antigens and precursors in normal and malignant human breast glandular tissue, J. Natl. Cancer Inst., 54:335.

Steplewski, Z., Chang, T.H., Herlyn, M. and Koprowski, H., 1981, Release of monoclonal antibody-defined antigens by human colorectal carcinoma and melanoma cells, Cancer Res., 41:2723.

Taylor-Papadimitriou J., Peterson, J.A., Arklie, J., Burchell, J., Ceriani, R.L. and Bodmer, W.F., 1981, Monoclonal antibodies to epithelium-specific components of the human milk fat globule membrane: production and reaction with cells in culture, Int. J. Cancer, 28:17.

Taylor-Papadimitriou, J., Lane, E.B. and Chang, S.E., 1983a, Cell lineages and interactions in neoplastic expression in the human breast, in: "Understanding Breast Cancer: Clinical and

Laboratory Concepts", P. Furmanski and M. Dekker, eds., in press.

Taylor-Papadimitriou, J., Burchell, J. and Chang, S.E., 1983b, Use of antibodies to membrane antigens in the study of differentiation and malignancy in the human breast, in: "Monoclonal Antibodies and Cancer", . Dulbecco and . Langman, eds., Academic Press, New York.

Thompson, C.H., Jones, S.L., Whitehead, R.H. and McKenzie, I.F.C., 1983, A human breast tissue-associated antigen detected by a monoclonal antibody, J. Natl. Cancer Inst., 70:409.

Towbin, H., Staehelin, T. and Gordon, J., 1979, Electrophoretic transfer of proteins from polyacrylamide gels to nitrocellulose sheets: Procedure and some applications, Proc. Natl. Acad. Sci. USA, 76:4350.

Ueda, R., Ogata, S.-I., Morrissey, D.M., Finstad, C.L., Szkudlarek, J., Whitmore, W.F., Oettgen, H.F., Lloyd, K.O. and Old, L.J., 1981, Cell surface antigens of human renal cancer defined by mouse monoclonal antibodies: Identification of tissue-specific kidney glycoproteins, Proc. Natl. Acad. Sci. USA, 78:5122.

von Kleist, S. and Burtin, P., 1979, Antigens cross-reacting with CEA, in: "Immunodiagnosis of Cancer", R.B. Herberman and C.R. McIntyre, eds., Marcel Dekker, Inc., New York.

von Kleist, S., Chavanel, G. and Burtin, P., 1972, Identification of an antigen from normal human tissue that cross reacts with the carcinoembryonic antigen, Proc. Natl. Acad. Sci. USA, 69:2492.

Wagener, C., Yang, Y.H.J., Crawford, F.G. and Shively, J.E., 1983, Monoclonal antibodies for carcinoembryonic antigen and related antigens as a model system: A systematic approach for the determination of epitope specificities of monoclonal antibodies, J. Immunol., 130:2308.

ATTEMPTED THERAPY OF A HUMAN XENOGRAFT COLONIC CANCER

HT29 USING ^{131}I-LABELLED MONOCLONAL ANTIBODIES

A.A. Epenetos*

Imperial Cancer Research Fund
44 Lincoln's Inn Fields
London, WC2A 3PX, U.K.

INTRODUCTION

The demonstration that ^{125}I- and ^{123}I-labelled monoclonal antibody AUA1 can successfully localise in xenografted tumours (HT29) in contradistinction to negative control antibody M236 (Epenetos et al, 1982) led to this study of ^{131}I-labelled antibodies for the therapy of xenografts.

METHODS

Antibodies: Monoclonal antibody AUA1, negative control monoclonal antibody M236, and the iodination procedures have been described before (Epenetos et al, 1982). Briefly, the AUA1 antibody is specific to an epithelial proliferating antigen which is strongly expressed on malignant gastrointestinal tissue, normal gastrointestinal tissue and weakly on other normal epithelial tissues. M236 is specific for cells of the T lymphocyte lineage and does not react with cells of epithelial origin. Purified antibodies were iodinated with ^{131}I using the iodogen method (at 1 mole I/mole IgG). Free iodine was removed by gel filtration using a G50 Sephadex column. The reactivity of antibodies before and after iodination was tested by an enzyme linked immunosorbent assay (Lansdorp et al, 1980), using whole cells as targets for the antibodies.

Tumour Model: HT29 cells (Fogh et al, 1977) were implanted

*Present address: Royal Postgraduate Medical School, Hammersmith Hospital, Du Cane Road, London, W12 OHS, U.K.

$(5 \times 10^6 - 10^7$ cells) intramuscularly in the thighs of immunodeficient (nude) mice or rats. Tumour size was determined by measuring the maximum diameter of each tumour at different intervals after implantation, beginning at day 7. Animals were killed at time intervals for the determination of tumour weights. It was found that the maximum diameter of tumours (in cm) correlated well with the weight of the tumours measured in g, e.g. 1 cm diameter tumour weighed 1 g, a 2 cm diameter tumour weighed 2 g etc.

Experimental design: Several studies were conducted in an attempt to define the cytotoxic effect on tumour growth, maximum dosage of ^{131}I-labelled antibody that could be given, whether fractional doses could be given advantageously and to compare the responses in two species of animals, i.e. mice and rats.

As mentioned earlier, a good correlation between increases in tumour size and weight was found, therefore tumour growth inhibition was assessed by measuring changes in the diameters of tumours with time. Antibodies were administered intraperitoneally at a time when tumours measured 1 cm in diameter unless otherwise stated. Studies in mice were conducted in triplicate and in rats in duplicate.

RESULTS

Study 1: 10 μg of unlabelled antibodies AUA1 and M236 were administered to mice which were compared with a control group

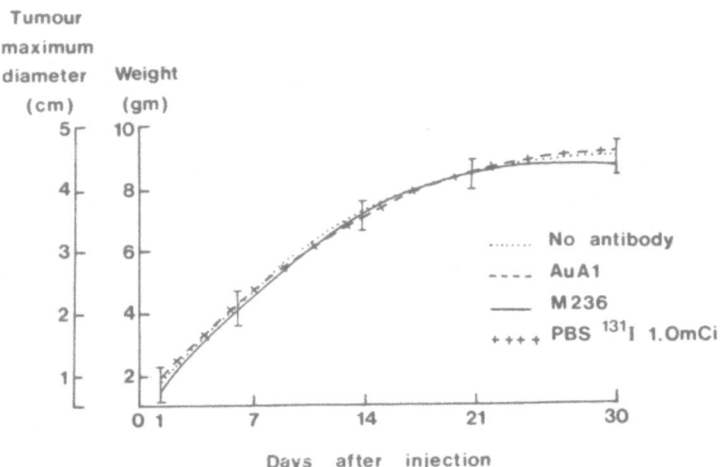

Fig. 1. The effect of unlabelled antibody on tumour growth. For other details see the text.

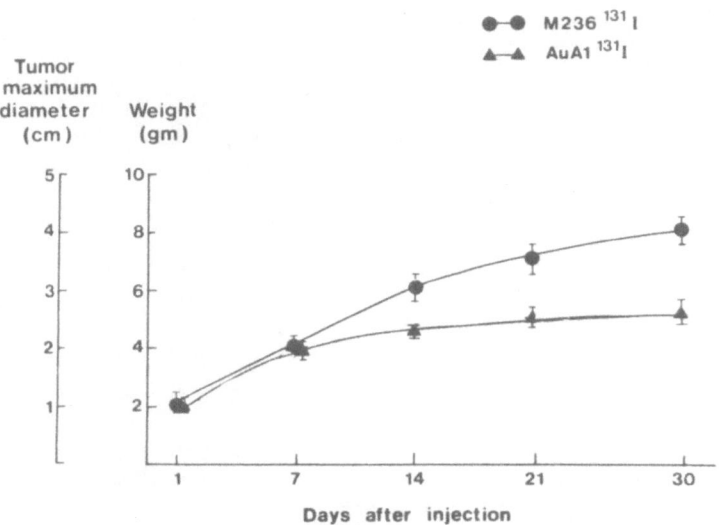

Fig. 2. The effect of labelled antibody on tumour growth. For other details see the text.

receiving no antibody. Fig. 1 demonstrates that unlabelled antibody has no effect on tumour growth. It was thus decided that in subsequent studies only labelled specific antibody would be compared with labelled control antibody.

Study 2a: 10 μg AUA1 labelled with 1 mCi of ^{131}I was compared to 10 μg M236 labelled with 1 mCi [131]I. Fig. 2 shows the findings. It can be seen that the rate of tumour size increase was half of that observed in the control group at 30 days after injection.

Study 2b: In this study fractionation of the same dosage as in Study 2a was attempted. Mice received on three occasions at weekly intervals 3.5 μg of AUA1 and M236 labelled with 0.35 mCi of [131]I. There was no difference in tumour size in the two groups. Tumours measured 8 cm in diameter by 30 days.

Study 3: Following evidence from Study 2a of the therapeutic efficacy of 1 mCi of administered dose resulting in a partial response and in no improvement from fractionation of that dose (Study 2b), 20 μg of AUA1 and M236 labelled with 2 mCi of ^{131}I were given. All mice in both groups were dead by day 15 with no difference in tumour size between the two groups. Mice developed jaundice and post-mortem examination demonstrated areas of haemorrhage. It was concluded that the cause of death was due to hepatic damage and bone marrow toxicity resulting from the higher dose of 2 mCi of [131]I.

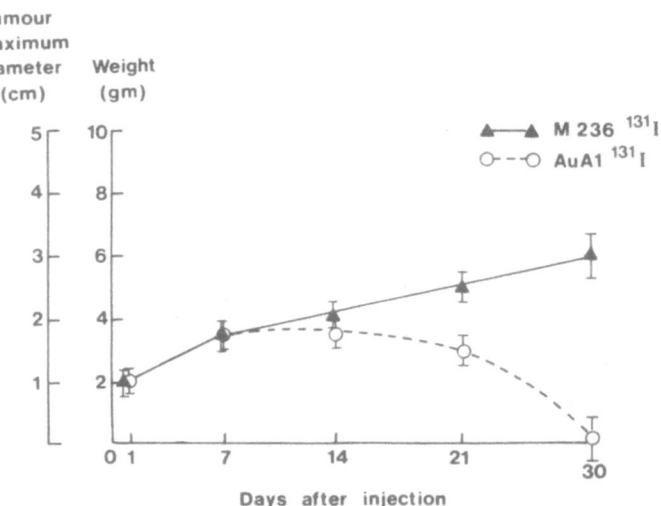

Fig. 3.   The effect of labelled antibody on tumour growth.
For other details see the text.

Study 4: In an attempt to overcome the problem of radiation
toxicity, fractionation of the dose given in Study 3 was used.
Mice were injected with 10 μg of AUA1 labelled with 1 mCi of [131]I
at day 0 and again given the same dosage at day 7. Mice in the
control group were dead by day 21 whilst those in the AUA1 group
were dead by day 30. There was no difference in tumour size
between the two groups but in both groups evidence of tumour growth
inhibition was seen similar to that achieved in Study 2a.

The following studies involved isotopically labelled antibodies
used in immunodeficient rats bearing HT29 tumours.

Study 5: Rats bearing intramuscular HT29 tumours, 1 cm in
diameter, were given intraperitoneally 10 μg AUA1 labelled with
1 mCi [131]I. Results are shown in Fig. 3. As it can be seen, by
30 days there was disappearance of tumours from the AUA1 group but
there was tumour diameter progression from 1 cm to 3 cm in the M236
group. The AUA1 group remained tumour-free for a further 60 days.
Rats were then killed and post-mortem examination showed no evidence
of tumour. The M236 group (at a time when tumour size had reached
3 cm in diameter) was given 25 μg AUA1 labelled with 1.0 mCi of
[131]I. However, by 30 days there was tumour progression to 5 cm in
diameter (Fig. 4). The rats were killed and post-mortem examination
of the tumours showed large areas of necrosis and haemorrhage
together with viable HT29 cells.

Study 6: In view of the findings in Study 5 of no response of

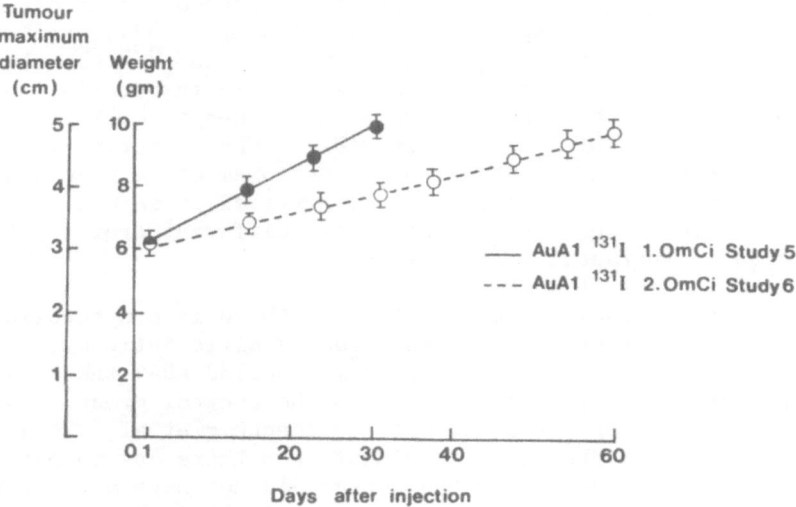

Fig. 4. The effect of labelled antibody on tumour growth. For other details see the text.

the 3 cm tumour to 1.0 mCi of ^{131}I-labelled AUA1, the dosage was doubled. Rats with HT29 tumours (3 cm in diameter) received 50 μg AUA1 labelled with 2.0 mCi [131]I. There was tumour progression but with a rate slower than that of Study 5 (Fig. 4).

Study 7: Fractionation of dosage was examined in this study. Rats bearing HT29 tumour (1 cm diameter) received intraperitoneally on three occasions at weekly intervals 3.5 μg of AUA1 and M236 labelled with 0.35 mCi of [131]I. At 30 days (from the time of first injection) there was tumour regression in the AUA1 group whereas in the M236 group tumour growth remained static at 1 cm diameter.

Immunoperoxidase staining: Formalin-fixed paraffin wax-embedded sections of HT29 tumours from Study 2 were examined following an immunoperoxidase reaction (Epenetos et al, 1982; Lansdorp et al, 1980) with freshly added specific antibody AUA1. It was of interest to note that HT29 tumours in mice that had received ^{131}I-labelled AUA1 were negative to the freshly added AUA1 while tumours of mice that had received ^{131}I-labelled M236 produced a positive reaction.

DISCUSSION

The HT29 human colonic carcinoma xenograft model which is CEA producing (Fogh et al, 1977), resistant to several conventional chemotherapy agents (Warenius and Bleehen, 1982) and to external X-irradiation (Belamy, A., personal communication) closely resembles the behaviour of human colonic cancer. When xenografted, it grows as a mucin-secreting poorly differentiated adenocarcinoma capable sometimes of metastasising (Epenetos and Bodmer, 1983). Following the evidence (Epenetos et al, 1982) that ^{123}I- and ^{125}I-labelled AUA1 antibody can consistently localise in HT29 tumours in immuno-deficient mice and rats, studies were undertaken to evaluate the anti-tumour effects of ^{131}I-labelled AUA1. The criteria for response were kept simple and involved measurement of size and weight of tumour and, in some studies, survival of animals. Two species of animals were used to study the different growth patterns in response to therapeutic manoeuvres.

The present studies demonstrate that there is a cytotoxic effect on tumour growth by ^{131}I-labelled specific antibody. In mice, at a single dose of 1 mCi of ^{131}I-labelled AUA1 there was 50% tumour growth inhibition as compared to the control group. This is consistent with other published data (Goldenberg et al, 1981). However, in none of the mice experiments was there any tumour regression. Fractionation of the dosage did not have a positive influence on tumour regression and as seen in Study 2b there was no difference in tumour size in the two groups whilst in Study 2a when a single dose was given, there was a significant advantage in the AUA1 group. Toxicity in mice was reached when 2 mCi of ^{131}I-labelled antibody was given. Fractionation of this dosage probably reduced or delayed toxicity as demonstrated by the pro-longed survival of mice in Study 4 as compared to control.

It was interesting to note that a more pronounced cytotoxic effect on tumours was achieved when immunodeficient rats were used. As seen from Study 5, there was complete tumour regression following a single injection of 1 mCi ^{131}I-labelled AUA1 while there was tumour progression in the control group. This effect was repro-duced when fractionation of dosage was used but it was also noted that there was a cytotoxic effect in the control group. It is thought that fractionation causing prolongation of circulating radioactivity was responsible for the tumour inhibition in the control group. However, other workers (Ettinger et al, 1982) have suggested that smaller fractional doses of radioimmunoglobulin could achieve potentially more radiation at the target. When larger than 1 cm diameter tumours were used in rats, complete regression of tumour was not achieved though there was delay in tumour growth when 5 mCi of ^{131}I-labelled antibody was given. No toxicity to the animals was noted at this dosage.

The marked difference in the responses to the same amount of radioimmunoglobulin between immunodeficient mice and rats is interesting. Nude mice, although athymic, do possess mechanisms for controlling development, growth and metastasis of tumours. It has been found (Stutman, 1978) that nude mice have the same incidence of spontaneous neoplasms as normal mice. Furthermore, human xenografts in nude mice rarely metastasise (Sharkey and

Fogh, 1979) although in our experience we have occasionally observed distant metastases from HT29 tumours (Epenetos and Bodmer, 1983) and from an ovarian cancer xenograft (unpublished data). Nude mice have been shown to have high levels of NK cell activity (Herberman, 1978) and enhancement of humoral responses (Parrot and de Sousa, 1974).

The nude rat (Festing et al, 1978) shares many properties with the nude mouse including the acceptance of xenografts. However, one major difference between the two species is the much higher level of resistance to tumour growth seen in the rat (Colston et al, 1981). Furthermore, spontaneous regression of xenografted tumours has been observed (Colston et al, 1981) including cell lines of colonic, pancreatic and lung cancer origins. In our experience, however, when HT29 tumours were established in rats, no spontaneous tumour regression was observed. It is thought that non-T-cell mediated mechanisms that control the growth of xenogeneic tumours appear to be even more important in the athymic rat (Colston et al, 1981). It may be, thus, concluded that ^{131}I-labelled monoclonal antibodies have an antitumour effect more markedly observed in less immunodeficient hosts.

Toxicity is a major limitation of this approach. Toxicity may be reduced by decreasing the amount of administered immunoglobulin and increasing the specific activity (isotope:antibody ratios). This approach is supported by the immunoperoxidase findings demonstrating that HT29 binding sites may be saturated by the administered amounts of AUA1 antibody. The immunoperoxidase findings may also support the possibility that following the in-vivo administration of ^{131}I-labelled AUA1 there was structural change or "antigenic modulation" at the cell surface of HT29 cells. This implies that for maximal cytotoxic benefit a single dose may be more suitable than fractional doses. However, as the isotope: antibody ratio increases there is loss of antibody reactivity.

^{131}I probably is not the ideal cytotoxic radionuclide. Firstly it has undesirable γ-rays and secondly dehalogenation probably occurs in-vivo leading to capture of ^{131}I by other normal tissues (Scheinberg and Strand, 1982) causing increased total body irradiation. With the increasing use of chelated monoclonal antibodies (Scheinberg et al, 1982) the repertoire of suitable cytotoxic radionuclides has become wider and the successful application of effective and yet selective therapy of neoplasia is now within reach.

REFERENCES

Colston, M.J., Fledsteel, A.H. and Dawson, P.J., 1981, Growth and regression of human tumour cell lines in congenitally allogeneic (rnu/nu) rats, J. Natl. Cancer Cancer Inst., 66:843.

Epenetos, A.A., Nimmon, C.C., Arklie, J., Elliot, A.T., Hawkins, L.A., Knowles, R.W., Britton, K.E. and Bodmer, W.F., 1982, Detection of human cancer in an animal model using radio-labelled tumour-associated monoclonal antibodies, Br. J. Cancer, 46:1.

Epenetos, A.A. and Bodmer, W.F., 1983, Analysis of a human xeno-graft metastasis, In press.

Ettinger, D.A., Order, S.E., Wharan, M.D., Parner, M.K., Klein, J.L. and Leichner, P.P., 1982, Phase I-II study of isotopic immunoglobulin therapy for primary liver cancer, Cancer Treat. Rep., 66:289.

Festing, M.F., May, D., Connors, T.A., Lovell, D. and Sparrow, S., 1978, An athymic mutation in the rat, Nature, 274:365.

Fogh, J., Fogh, J.M. and Orfeo, T., 1977, 127 cultured human tumour cell lines producing tumours in nude mice, J. Natl. Cancer Inst., 59:221.

Goldenberg, D.M., Gaffar, S.A., Bennett, S.J. and Beach, J.L., 1981, Experimental radioimmunotherapy of a xenografted human colonic tumour (GW-39) producing carcinoembryonic antigen, Cancer Res., 41:4354.

Herberman, R.B., 1978, Natural cell-mediated cytotoxicity in nude mice, in: "The nude mouse in experimental and clinical research", Fogh, J., Giovanella, B.C., eds., Academic Press, New York.

Lansdorp, P.M., Astaldi, G.C.B., Oosterhof, F., Janssen, M.C. and Zeijlemaker, W.P., 1980, Immunoperoxidase procedures to detect monoclonal antibodies against cell surface antigens. Quantitation of binding and staining of individual cells, J. Immunol. Meth., 39:393.

Parrot, D.M. and de Sousa, M.A., 1974, B cell stimulation in nude (nu/nu) mice, in: "Proceedings of the first international workshop on nude mice", Rygaard, J., Povlsen, R., eds., Gustav Fischer Verlag, Stuttgaart.

Scheinberg, D.A., Strand, M. and Gansow, O.A., 1982, Tumour imaging with radioactive metal chelate conjugates to monoclonal antibodies, Nature, 215:1511.

Scheinberg, D.A. and Strand, M., 1982, Leukemic cell targeting and therapy by monoclonal antibody in a mouse model system, Cancer Res., 42:44.

Sharkey, F.E. and Fogh, J., 1979, Metastasis of human tumours in athymic nude mice, Int. J. Cancer, 24:733.

Stutman, O., 1978, Spontaneous, viral and chemically induced tumours in the nude mouse, in: "The nude mouse in experi-mental and clinical research", Fogh, J., Giovanella, B.C., eds., Academic Press, New York.

Warenius, H.M. and Bleehen, N.M., 1982, In-vivo and in-vitro clonogenic assays in a human tumour xenograft with a high plating efficiency, Br. J. Cancer, 46:45.

FATE OF LIPOSOMES IN VIVO: CONTROL LEADING TO TARGETING

Gregory Gregoriadis, Judith Senior,
Barbara Wolff and Christopher Kirby

Division of Clinical Sciences
Clinical Research Centre
Harrow, Middlesex HA1 3UJ, U.K.

INTRODUCTION

It is well established (Gregoriadis, 1981) that following
intravenous injection, liposomes and entrapped drugs are sooner or
later taken up by the reticuloendothelial system (RES). Rate of
uptake in particular tissues of the RES depends on vesicle size,
surface charge and lipid composition, amount of liposomal lipid
given, animal species and physiological state. For instance, a
large vesicle size and/or a negative surface charge promote rapid
uptake by tissues. Further, depending on the lipid composition of
injected liposomes, plasma high density lipoproteins (HDL) will
remove phospholipid molecules, destabilize vesicle structure and
lead to entrapped drug release into the circulation. Thus, tissues
will take up vesicles in various stages of destabilization and with
a portion of their drug contents lost.

Targeting of liposomes to accessible alternative cells in the
body require that a number of conditions must be satisfied. Among
these, the following appear at present especially important: Liposomes must (a) retain quantitatively both targeting ligand and
entrapped drug en route to their destination; (b) circulate in the
blood for periods of time long enough so that they can seek and
interact with their target efficiently; (c) be endowed with target
selectivity by means of ligands attached onto their surface and
capable of binding to respective receptors on target cells. Recent
work in this laboratory (Gregoriadis, 1983; Senior and Gregoriadis,
1982a, b; Kirby and Gregoriadis, 1980; Kirby et al., 1980a; Wolff
and Gregoriadis, 1984) suggests that such liposomal properties are
feasible and can operate in-vivo.

RETENTION OF DRUGS BY LIPOSOMES IN BLOOD

The problem of massive drug loss from liposomes upon exposure
to blood encountered in the early days of liposome research has now
been resolved. Following the understanding (Krupp et al., 1976;
Scherphof et al., 1978; Kirby et al., 1980a, b; Allen, 1981) of the
role of plasma HDL in vesicle destabilization, it was reasoned
(Gregoriadis and Davis, 1979; Kirby et al., 1980a) that packing of
the liposomal bilayer with excess cholesterol and/or the inclusion
of phospholipids with a high liquid-crystalline transition temperature (Tc), may prevent HDL from removing phospholipids. In recent
experiments (Gregoriadis and Senior, 1980; Senior and Gregoriadis,
1982a, b) using quenched carboxyfluorescein as a water soluble
marker we were able to show that cholesterol equimolar to total
phospholipid, prevents solute loss in the presence of blood, plasma
or serum to an extent dependant on the nature of the phospholipid
used. A striking example of this is shown in Fig. 1 where an
increasing proportion of sphingomyelin (SM) relative to egg phosphatidylcholine (PC) in cholesterol-rich liposomes is parallelled
by increasing solute retention. Retention (measured as CF latency)
in the presence of blood plasma at $37^{\circ}C$ reaches 100% values when
at least 77% of the total phospholipid is SM. Even more stable
liposomes (100% CF latency is maintained for at least 48 h) are
obtained with equimolar distearoyl phosphatidylcholine (DSPC) and
cholesterol (Fig. 2).

Experiments (Kirby et al., 1980a, b; Kirby and Gregoriadis,
1980; Senior et al., 1983) with liposomes containing radiolabelled
phospholipids have shown repeatedly that such manipulations of
lipid composition do indeed reduce or abolish altogether loss of
phospholipid to HDL in the presence of plasma. Fig. 3 for example,
shows that increasing presence of SM in liposomes (which, as discussed already parallels increasing CF latency retention) is associated with decreasing elution of the radiolabelled lipid together
with HDL. Additional support for this role of HDL in removing
phospholipids has come from work in which cholesterol-free PC liposomes were exposed to plasma from lipoprotein-deficient mice
(Senior et al., 1983). There was virtually no transfer of 3H-PC to
the HDL (Fig. 4). Furthermore, when purified HDL, low and intermediate density (LDL + IDL) or very low density (VLDL) mouse lipoproteins were added to lipoprotein-deficient mouse plasma in
increasing quantities to cover physiological levels of the lipoproteins, only HDL (Fig. 5) had a destabilizing effect on liposomes.

It seems that removal of phospholipids by plasma HDL leads to
the formation of "pores" on the lipid bilayer through which entrapped
solutes leak. This became evident from experiments (Kirby and
Gregoriadis, 1981) in which leakage of solutes in the presence of
plasma was inversely related to their molecular weight. It also
appeared that the number and/or size of pores was smaller the

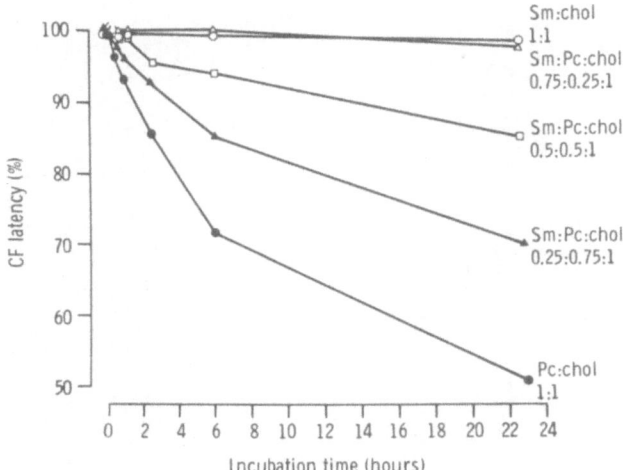

Fig. 1. The effect of phospholipid composition of liposomes on
         their permeability in plasma. Small unilamellar liposomes
         containing quenched CF and composed of SM (O), 77% SM, 23%
         PC (Δ), 47% SM, 53% PC (□), 23% SM, 77% PC (▲) and PC (●)
         were incubated in the presence of mouse plasma at 37°C.
         All liposomal preparations contained cholesterol, equi-
         molar to total phospholipid. CF latency values at time
         intervals are % of total CF present (from Senior and
         Gregoriadis, 1982b).

greater the content in cholesterol of liposomes, with those preparations exhibiting a phospholipid:cholesterol molar ratio of 1
being only slightly leaky. The notion of pore formation has been
questioned by others (Scherphof et al., 1984) on the basis that the
permanent presence of pores with hydrophobic surfaces would be
incompatible with the presence of a polar environment and the pores
would therefore tend to close by means of bilayer shrinkage. However, because shrinkage is unlikely to occur with the small liposomes used in our studies, the authors suggest that liposomes
disintegrate instead and that our findings probably concern a small
population of larger liposomes that is likely to exist in any preparation of small liposomes. We have previously suggested (Senior
and Gregoriadis, 1984) that the problem of maintenance of pores in
a polar environment can be circumvented if one accepts the possibility of pore surface interaction with the hydrophobic portions of
plasma components which could line the pore surface, with their
polar moieties exposed in surrounding water. In this way there
will be no need for bilayers to shrink or disintegrate, events
which would have been in any case incompatible with the observed
continuous leakage of solute over many hours of incubation.

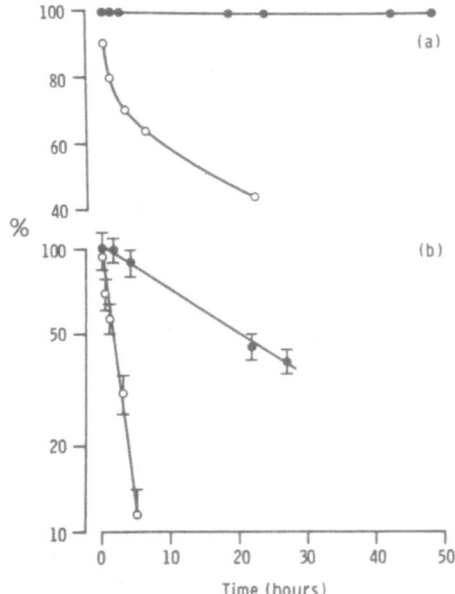

Fig. 2.   Stability of DSPC-liposomes in mouse plasma and clearance
           from the circulation. Small unilamellar liposomes contain-
           ing quenched CF and composed of DSPC (O) or equimolar DSPC
           and cholesterol (●) were incubated in the presence of
           mouse plasma at 37°C (a) or injected intravenously into
           4 mice (b). Values at time intervals are % latent CF of
           total CF present or % (mean ± S.D.) latent CF of injected
           latent CF per total mouse plasma (from Senior and
           Gregoriadis, 1982b).

        Whilst CF and other polar solutes (Kirby and Gregoriadis, 1981;
Wolff and Gregoriadis, 1984) are retained quantitatively by cholesterol-rich liposomes in the presence of plasma, other solutes of
lipophilic nature do not behave similarly (Kirby and Gregoriadis,
1983). Again, the presence of cholesterol and the correct choice
of phospholipid influence solute retention but this is less pronounced for lipophilic drugs. According to Table 1, for instance,
even cholesterol-rich SM liposomes retain only a fraction of entrapped vincristine or melphalan.

RETENTION OF DRUGS BY CIRCULATING LIPOSOMES AND LIPOSOME CLEARANCE

        The types of liposomes that retain their drug contents quantit-
ively in the presence of plasma in-vitro at 37°C, also do so in the
bloodstream after injection (Kirby et al., 1980; Kirby and

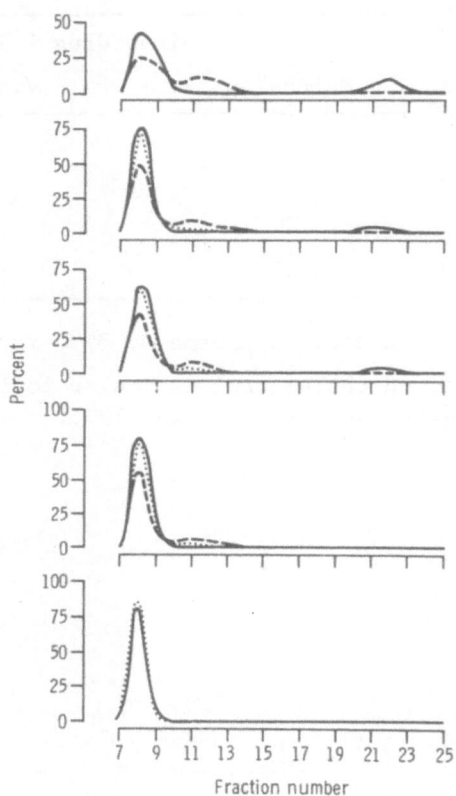

Fig. 3. Liposomal phospholipid transfer to HDL in the presence of
plasma. Liposomes of compositions described in Fig. 1,
radiolabelled with ^{3}H-PC and/or ^{14}C-SM and containing
quenched CF were mixed with 9 volumes of mouse plasma and
incubated at 37°C for 4 h. Subsequently, samples were
passed through Ultrogel AcA 34 columns and fractions
obtained assayed for CF (solid line), ^{3}H (broken line)
or ^{14}C (dotted line). Applied preparations from top to
bottom were: PC; 23% SM, 77% PC; 47% SM, 53% PC; 77% SM,
23% PC; SM liposomes. All preparations had cholesterol
equimolar to the total phospholipid. Liposomes, HDL and
free CF were eluted in fractions 7-9, 9-13 and 21-13
respectively. Values are % of ^{3}H, ^{14}C or CF applied to
the columns recovered per fraction. Recoveries were
greater than 89% of applied.

Table 1. Retention of lipophilic drugs by liposomes in plasma[a]

| Liposomes[b] | Retained drug (%) | |
	Melphalan	Vincristine
PC	67.6	48.9
DMPC	90.0	71.9
SM	85.5	82.5

[a]Liposomes were incubated in plasma at 37°C for 10 min.

[b]Liposomes contained cholesterol equimolar to the phospholipid. Modified from Kirby and Gregoriadis, 1983.

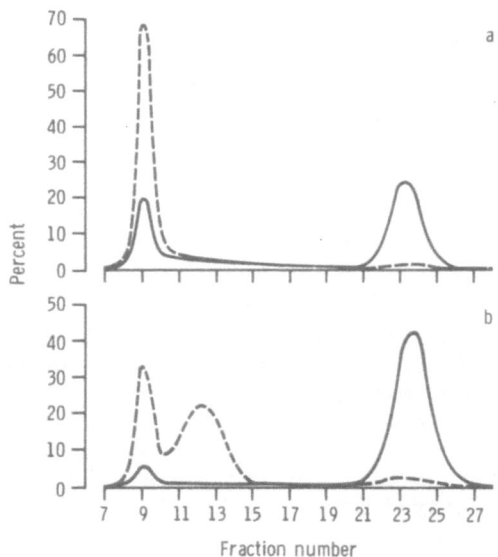

Fig. 4. Elution profiles of cholesterol-free PC liposomes after incubation with normal or lipoprotein-deficient plasma. Blood plasma from mice treated with 4-aminopyrazol (3, 4d) pyrimidine (a) or normal mice (b) was mixed with ^{3}H-PC liposomes containing CF and incubated at 37°C. Thirty min later, samples were passed through Ultrogel AcA 34 columns. ^{3}H radioactivity (broken line) and CF (solid line) were assayed in the fractions obtained. Values are % of ^{3}H or CF applied to the columns recovered per fraction. Recoveries were greater than 82% (^{3}H) or were 100% (CF) of applied (from Senior et al., 1983).

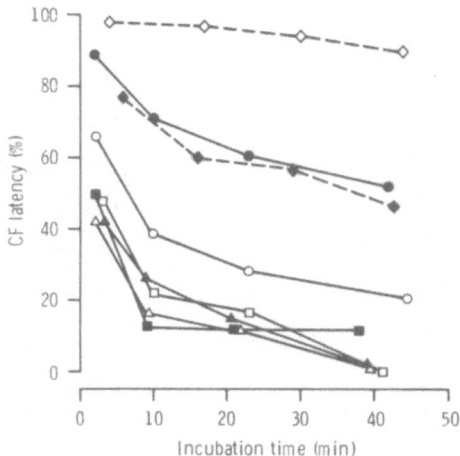

Fig. 5. Liposomal stability in lipoprotein-deficient plasma
supplemented with HDL. CF-containing cholesterol-free
PC liposomes were mixed phosphate-buffered saline (◇)
or plasma from normal (□) or lipoprotein-deficient (●)
mice. In other experiments liposomes were mixed as above
with lipoprotein-deficient plasma into which 188 (○), 375
(▲), 563 (△) or 750 (■) μg protein HDL had been previously
added or with phosphate buffered saline containing 1500 μg
protein HDL (◆). CF latency assayed at time intervals in
the mixtures incubated at 37°C was expressed as % of CF
latency in the original liposomal preparation. Concentration of HDL in normal mouse plasma was estimated as
1.5 mg protein per ml assuming 100% recovery of the lipoprotein following its purification from plasma (from
Senior et al., 1983).

Gregoriadis, 1980, 1983; Senior and Gregoriadis, 1982a, b). This
has been shown by the measurement of both a radioactive lipid label
and of latent CF and also by the monitoring of lipid label transfer
(if any) to HDL. For very stable liposomes the two markers are
cleared from the circulation at the same rate and there is virtually
no transfer of phospholipid to HDL (Senior and Gregoriadis, 1982a
and Fig. 3). Another indication that liposomes retain their
structural integrity in the blood is the persistence of highly
quenched CF in blood samples taken at time intervals after injection
(Gregoriadis and Senior, 1980; Senior and Gregoriadis, 1982a, b).

Recently we observed (Senior and Gregoriadis, 1982b) a direct
relationship between solute (CF) retention by liposomes in the
presence of blood and the rate of their clearance from the circulation. This is demonstrated in Figs. 6 and 7 where clearance rates

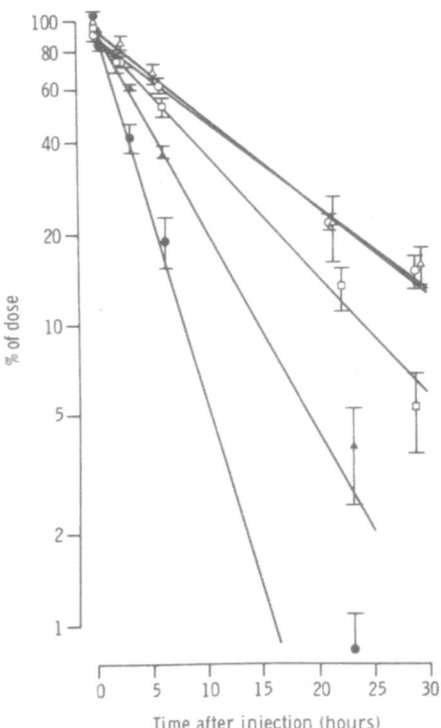

Fig. 6. The effect of phospholipid composition of liposomes on
         their clearance from the circulation. Liposomes composed
         as described in Fig. 1 were injected intravenously into
         mice. Latent CF values (mean ± S.D.; 3-6 mice) at time
         intervals are % of injected latent CF per total mouse
         plasma. Symbols as in Fig. 1 (from Senior and Gregoriadis,
         1982b).

(CF and lipid markers respectively) of 5 preparations of liposomes,
shown to be increasingly stable in plasma at 37°C (see Fig. 1),
exhibit increasingly longer half-lives after intravenous injection
ranging from about 2 h for PC cholesterol-rich liposomes to about
11 h for cholesterol-rich SM liposomes (CF marker). A similar
relationship is shown in the case of two preparations of distearoyl
phosphatidylcholine, one without and the other with (equimolar)
cholesterol. In the latter case half-life after intravenous injection was 20 h (see Fig. 1). It has been suggested (Patel et al.,
1984) that the reason for the slower clearance of cholesterol-rich
liposomes is that the sterol inhibits their uptake by the liver.
This, however, is not supported by their findings (Patel et al.,
1984) or by observations made with a number of cholesterol-rich

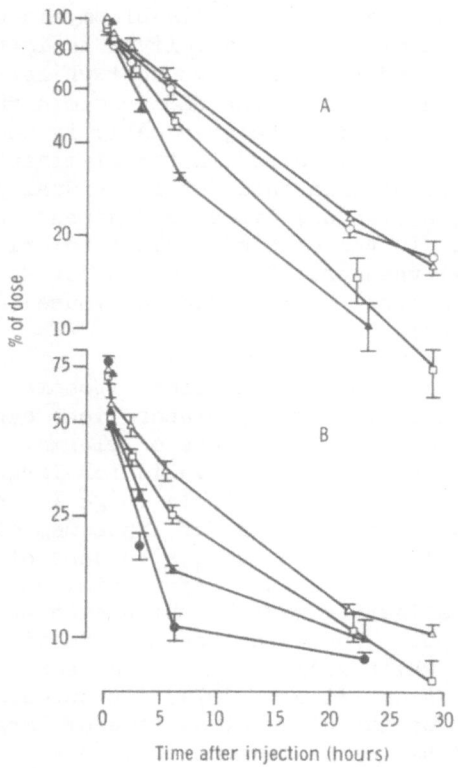

Fig. 7.   The effect of phospholipid composition of liposomes on
          their clearance from the circulation. Blood plasma from
          mice in experiment described in Fig. 6 was assayed for
          ^{14}C-SM (A) and/or ^{3}H-PC (B) radioactivity. Symbols as
          Figs. 1 and 6.

preparations of differing phospholipid composition (1:1 phospho-
lipid:cholesterol molar ratio) exhibiting a wide range of half-lives
of 0.1 h to 20 h (Gregoriadis and Senior, 1980; Senior and Gregoriadis, 1982a, b; Hwang et al., 1980; Figs. 1 and 6). A more likely
explanation for the direct relationship between vesicle clearance
rate and vesicle permeability in the presence of plasma may be that
the greater the permeability the less packed the bilayers are and
the easier for plasma proteins (such as opsonins) responsible for
the elimination of foreign matter from the circulation to adsorb or
insert themselves onto the vesicles and so to render the latter
recognizable for the RES. Longer half-lives, however, does not
mean that liposomes escape the RES to any significant extent to
interact with other intravascular tissues or leave the vascular
compartment. Our recent work (J. Senior and G. Gregoriadis,

unpublished) in which the fate of ^{111}In-bleomycin-containing small
unilamellar liposomes of varying half-lives in intravenously injected
mice was followed for up to 72 h shows that whilst uptake by the
hepatic and splenic moieties of the RES is diminished, much of the
remainder of the dose accumulates eventually in the bone marrow.
Regardless of the factors involved in the elimination of liposomes
from the blood, rates of clearance for individual preparations of
small unilamellar vesicles are linear and depend on the amount of
administered lipid. In the case of cholesterol-rich DSPC liposomes,
for instance, half-lives were 12, 7 and 3 h for 2, 0.6 and 0.2 mg
respectively administered phospholipid per mouse (J. Senior and
G. Gregoriadis, unpublished).

It therefore seems that in choosing liposomes which can exhibit
long half-lives, one must look for preparations capable of retaining
their entrapped solute in the presence of plasma. Unfortunately,
however, although this rule is also valid for larger vesicles, half-
lives achieved even for the most stable large liposomes are only
modestly long (about 60 and 20 min for liposomes of about 0.2 and
0.4 m diameter respectively injected at a dose of 0.6 mg phospholipid per mouse; J. Senior and G. Gregoriadis, unpublished). We
are at present investigating ways by which such short half-lives
could be improved upon. These include modification of the liposomal
surface with agents which would render it more hydrophilic, a state
which apparently (van Oss et al., 1974) discourages opsonin adsorption. Prolongation of the circulation time of large liposomes would
be a considerable achievement towards their use in drug targeting
in-vivo. Because of their size, large liposomes can carry much more
drug per lipid unit weight and are, in addition, easier to prepare.
Indeed, in a recent simple method developed in this laboratory
(Kirby and Gregoriadis, 1984a, b, c), liposomes prepared under
defined conditions in the absence of organic solvents, sonication
or detergents entrap drugs, proteins or nucleic acids in amounts of
up to 60-70% of the original materials. Such liposomes (named DRV
because of the dehydration-rehydration steps in their formation)
and REV which also entrap large amounts of solutes are obviously
most suitable for the rapid delivery of drugs to the liver and
spleen but would also be ideal for targeting to alternative tissues
if ways could be found to delay their removal by the RES.

TARGETING OF LIPOSOMES

It would appear from work already described that small unilamellar liposomes composed of equimolar DSPC and cholesterol com-
bine both great stability in plasma and prolonged survival in the
circulation and thus qualify as a highly appropriate preparation
for drug targeting within the vascular system. Furthermore, because
of their size such liposomes would also be suitable for targeting
by other routes of administration, for instance, after subcutaneous

injection. Thus, small liposomes havc been shown to enter the
periphery quantitatively and in intact form following injection
into the foot pads of rats (Tumer et al., 1983).

   Preparation of targeted liposomes, however, requires that these
are coupled with molecules such as glycoproteins and antibodies which
are expected to mediate vesicle association with respective ligands
on the surface of target cells (Gregoriadis, 1981; Toonen and
Crommelin, 1983). The question therefore arises as to whether reagents used for coupling as well as targeting molecules per se will
in some way alter liposomal stability and clearance patterns in ways
detrimental to targeting. For instance, the presence of immunoglobulin G on the surface of liposomes can promote their interception by
the RES via the Fc receptors (Gregoriadis et al., 1977).

   Initial attempts to target liposomes some 10 years ago
(Gregoriadis, 1974; Gregoriadis and Neerunjun, 1975) employed
liposomes sonicated in the presence of protein which, in the final
preparation, was partly exposed on the vesicle's surface. Amino
groups of the protein and of the targeting molecule, in this instance
asialofetuin which binds selectively to the galactose receptors in
the liver (Morell et al., 1971), were coupled via gluteraldehyde
(Gregoriadis, 1974). Alternatively, liposomes were sonicated in
the presence of anti-target cell IgG which again appeared partially
on the vesicle's surface. Such antibody-bearing liposomes were
shown to interact selectively with a variety of cell targets
(Gregoriadis and Neerunjun, 1975; Gregoriadis et al., 1977). The
same procedure was applied later (Gregoriadis et al., 1981) in the
preparation of (affinity chromatography-purified) antibody-bearing
small unilamellar liposomes. In an alternative procedure, anti-
IgG IgG was incorporated into liposomes as above and then interacted
with anti-target purified IgG which retained its ability to interact with the target (Gregoriadis et al., 1981). These latter studies
demonstrated that antibody-bearing liposomes can interact extensively
and selectively with circulating antigens in the blood in-vivo after
intravenous injection (Gregoriadis et al., 1981). Further, in-vitro
work with lymphocytes coated with antigen via $CrCl_3$ and exposed to
liposomes bearing the respective purified antibody revealed the
significance of antibody and antigen concentrations on the liposomal
and cellular surfaces respectively (Gregoriadis and Meehan, 1981).

   More recently we have attempted targeting of small unilamellar
liposomes by coupling targeting macromolecules on the liposomal
surface via heterobifunctional reagents (Gregoriadis and Senior,
1984; Wolff et al., 1984). Liposomes used were composed of DSPC
cholesterol and dipalmitoyl phosphatidylethanolamine (DPPE) which
provides the amino groups needed for coupling. At the same time
we established that the presence of up to 50% DPPE (of total phospholipid) in the liposome structure did not alter liposomal stability
significantly upon incubation at $37^{\circ}C$ in plasma for up to at least
50 h (Table 2).

Table 2. Effect of DPPE content of liposomes or their
          stability in plasma

DPPE	Incubation time (h)					
Content (%)	0.5	1	3	7	24	50
0	96.4	96.1	94.5	93.5	93.0	92.2
6	98.9	98.7	97.9	97.5	96.2	93.0
12	98.4	98.1	97.5	96.8	94.9	91.5
24	99.1	98.6	98.0	97.7	96.9	85.9[a]
50	99.8	99.6	99.5	99.1	98.2	89.5[a]
63	99.3	99.2	98.9	98.5	91.0	42.2
75	96.5	96.2	95.4	95.1	80.6	20.6

Liposomes composed of DSPC, increasing amounts of DPPE
(0-75% of total phospholipid) and cholesterol (equimolar
to total phospholipid) and containing CF were incubated
with mouse plasma at 37°C. Values represent CF latency as
% of the latency in the original preparations (> 98%).
Similar values were obtained for incubation of up to 6 h
in the presence of heparinised whole mouse blood (from
Wolff and Gregoriadis, 1984).

[a]Incubation was carried out for 68 h.

Maintenance of stability in the presence of plasma was also
observed when the DPPE component of liposomes was modified with the
heterobifunctional reagent N-succinimidyl 3-(2-pyridyldithio)propio-
nate (SPDP) to give PDP-DPPE. However, when liposomal PDP-DPPE was
coupled to the targeting ligand, stability (in plasma) in terms of
entrapped solute and targeting ligand retention by the carrier was
reduced, albeit modestly (see later).

TARGETING OF LIPOSOMES TO THE GALACTOSE RECEPTOR IN VIVO

     Targeting of liposomes to cells in-vivo was first demonstrated
by using heterogeneous preparations of multilamellar liposomes with
asialofetuin adsorbed on their surface (Gregoriadis and Neerunjun,
1975). More recent studies have been carried out with small unilamellar liposomes incorporating lactosylceramide (Szoka and Mayhew,

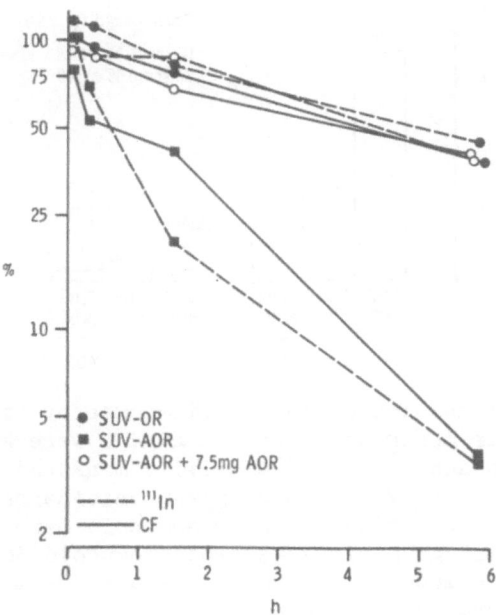

Fig. 8.   Plasma clearance of asialoorosomucoid liposomes. Mice in
          groups of four were injected intravenously with orosomucoid
          liposomes (●), asialoorosomucoid liposomes (■) or 4 mg of
          free asialoorosomucoid followed 12 min later with asialo-
          orosomucoid liposomes (O). Values (at time intervals after
          injection) of latent CF (continuous lines) and [111]In (broken
          lines) are % of injected latent dye or radioactivity per
          total mouse plasma (from Gregoriadis and Senior, 1984).

1983; Spanjer and Scherphof, 1983) or asialoorosomucoid coupled via
SPDP (Gregoriadis and Senior, 1984). The choice of small liposomes
as opposed to large ones was based on previous findings (Ashwell and
Morell, 1974) that the galactose receptor is located on the surface
of the hepatic parenchymal cells, access to which (through the
fenestrae) is possible only for vesicles with a diameter smaller
than about 100nm. In view of the recent discovery (Teradaira et al.,
1983) that the galactose receptor also exists on the surface of the
Kupffer cells and that this receptor binds galactose terminating
molecules in both the particulate and free forms, the rationale in
using small liposomes may now be somewhat redundant. However, small
liposomes may target preferentially to the hepatic parenchymal cells
even though Kupffer cells also bear the receptor.

     Small unilamellar liposomes composed of DSPC, cholesterol and
DPPE modified with SPDP to give PDP-DPPE (1:1:0.15 molar ratios)
contained 0.2 M CF and tracer ^{111}In-labelled bleomycin. These

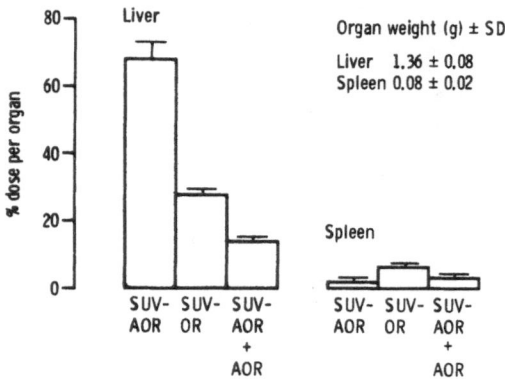

Fig. 9.   Uptake of asialoorosomucoid liposomes by tissues. Rats
from experiments described in Fig. 8 were killed 6 h post
injection and ^{111}In radioactivity measured in the liver
and spleen. SUV-AOR, mice given asialoorosomucoid liposomes; SUV-OR, mice given orosomucoid liposomes; SUV-AOR
+ AOR, mice given free asialoorosomucoid followed by
asialoorosomucoid liposomes. Values are % ± S.D. of the
dose recovered per organ.

liposomes were subsequently interacted with SPDP-modified [125]I-
labelled orosomucoid (Gregoriadis and Senior, 1984). The preparation was then divided into two parts, one part being desialylated
so as to expose the penultimate galactose residues of the protein.
The two preparations, orosomucoid and asialoorosomucoid-bearing
liposomes, were then used for in-vivo work. Mice were injected
intravenously with orosomucoid-bearing liposomes, asialoorosomucoid-
bearing liposomes or with free asialoorosomucoid followed 12 min by
asialoorosomucoid-bearing liposomes.

        According to Fig. 8, clearance rates of both latent CF and
[111]In (orosomucoid liposomes) from the circulation are practically
identical exhibiting a half-life of about 4.5-5 h. This half-life
is less than that expected from cholesterol-rich DSPC liposomes
injected at a dose of 0.8 mg per mouse (see earlier) and is likely
to reflect the chemical modification of their surface involving
coupling with orosomucoid via SPDP. When the desialylated preparation is injected, half-life is reduced to less than 30 min,
presumably because of rapid binding of the vesicles to the galactose
receptor. Confirmation of such selective binding comes from the
patterns of clearance obtained by administering free asialoorosomucoid which competes for binding sites with asialoorosomucoid liposomes given later on: rates of clearance for both markers returned
to those originally seen with orosomucoid liposomes (Fig. 8). The
use of free intact orosomucoid in a similar competition experiment
did not have the same effect (not shown).

Assay of ^{111}In in the tissues 6 h post injection (by which time most of the injected dose of asialoorosomucoid liposomes has been eliminated from the blood; see Fig. 8) revealed that 68% of the radioactivity was recovered in the liver (Fig. 9). Hepatic uptake of ^{111}In from orosomucoid liposomes (corrected for contamination by blood radioactivity) was 29% of the dose at 6 h, with the difference (39%) between the two values probably being the minimum proportion of the dose interacted with the galactose receptor. Unexpectedly, however, in the competition experiment hepatic values did not return to the "normal" level of 29% but, instead, diminished further to 15%. It is therefore conceivable that the difference (53%) between the values of hepatic uptake of asialoorosomucoid liposomes without and with administered free asialoorosomucoid represents more truly the extent of binding of liposomes to the galactose receptor. This is because asialoorosomucoid liposomes may not be intercepted by the RES as readily as orosomucoid liposomes. Furthermore, in view of the possibility that free asialoorosomucoid may not suppress completely binding of liposomes to the receptor, some of the observed hepatic uptake in the competition experiment could be attributed to specific binding by the receptor. Fig. 9 also shows that the shift of non-specific uptake of orosomucoid liposomes by the liver to specific uptake by cells expressing the galactose receptor, induced by desialylation, is reflected in the changes of radioactivity values in the spleen. Thus, although 6.5% of the dose is recovered in the tissue after injection of orosomucoid liposomes, this value is reduced to 2.0% for asialoorosomucoid liposomes. Interestingly, it remains low (3.1%) in the competition experiment, supporting the notion above that asialoorosomucoid liposomes may have a reduced affinity for the RES.

THE USE OF MONOCLONAL ANTI-THY$_1$ IgG$_1$ FOR THE TARGETING OF LIPOSOMES

IN VITRO AND IN VIVO

Antibodies, because of their great target versatility, have been the ligand of choice in most targeting work with liposomes. To this end, the advent of hybridoma technology has added further impetus, especially since targeting of liposomes with antibodies may, in some instances, have certain advantages over the use of antibodies coupled to drugs directly (Gregoriadis, 1983). For instance, liposomes can incorporate large quantities of a wide range of drugs, keep drugs in isolation from the biological milieu (e.g. blood) and deliver them to cells, potentially by a single immunoglobulin molecule per vesicle. In addition, a firmer and more specific binding to cells may be achieved with liposomes bearing antibodies against more than one type of antigenic determinant.

Two aspects must be considered in using antibody-bearing liposomes in-vivo. Firstly, as already discussed, it is logical to

expect that binding of liposomes to cells will occur more effectively
when the former retain both their drug and ligand components in the
presence of blood and circulate in it for periods long enough to
ensure lengthy exposure of the target to the vesicles. As shown
above, vesicle stability and half-life in the circulation can be
considerably augmented by the correct choice of lipids. Secondly,
although recent findings suggest that transcapillary passage of
small liposomes may be possible with some tumours (Proffitt et al.,
1983; Hashimoto et al., 1983; Mayhew and Rustum, 1983), it is more
likely that in-vivo, liposomes will best target to cells within the
vasculature or in tissues containing sinusoidal capillaries (Poste,
1983). Both these aspects were taken into account in work described
below and, in more detail, elsewhere (Wolff et al., 1984).

Preparation of Liposomes Bearing Anti-Thy$_1$ IgG$_1$

Small unilamellar liposomes containing quenched CF alone, or
together with tracer ^{111}In-Ca-diethylene triaminepentaacetic acid
(^{111}In-Ca-DPTA) were composed of DSPC, DPPE and cholesterol (molar
ratio of total phospholipid to cholesterol, 1:1). Heterobifunctional
reagents used to activate DPPE before or after its incorporation into
liposomes were N-hydroxysuccinimide ester of iodoacetic acid (NHIA),
N-succinimidyl-4-(2-bromoacetylamino) benzoate (SBAB) and SPDP.
^{125}I-labelled rat anti-Thy$_1$ IgG$_1$ was activated with SPDP prior to
its use. Of the three types of activated liposomes, those activated
with SPDP gave best values in terms of extent of binding (Table 3).

Interaction of anti-Thy$_1$ IgG$_1$-Bearing Liposomes with AKR-A Cells in

vitro

Interaction of IgG-bearing liposomes with cells in-vitro was
carried out with vesicles made of DSPC, DPPE (modified with SPDP
prior to its incorporation into liposomes) and cholesterol (molar
ratios 0.9:0.1:1 or 0.99:0.01:1) or with vesicles made of DSPC,
DPPE (modified with SBAB after its incorporation into liposomes)
and cholesterol (molar ratios 0.9:0.1:1). Exposure of an increasing
number of AKR-A cells expressing the cross-reacting mouse Thy$_{1.1}$
antigen on their surface to anti-Thy$_1$ IgG$_1$-bearing liposomes revealed
that at least 69.5% of the liposomes used contained enough IgG on
their surface to ensure binding to cells (Table 4). Furthermore,
the role of the anti-Thy$_1$ IgG$_1$ in promoting association of the liposomal moiety with the AKR-A cells specifically was supported by the
low association values observed for liposomes devoid of IgG and from
experiments where anti-Thy$_1$ IgG$_1$-bearing liposomes presented to AKR-A
cells and to another cell line (EL4-Tc) which does not express the
Thy$_{1.1}$ antigen, bound considerably more (3-4 fold) to the former
(Wolff et al., 1984). Interestingly, binding of the IgG$_1$-bearing
liposomes to AKR-A cells was equally effective with both types of
liposomes used, one activated with SBAB, the other with SPDP, in

Table 3. Anti-Thy$_1$ IgG$_1$ on the surface of liposomes

IgG-bearing liposomes	IgG$_1$ bound ($\mu g/\mu$mol phospholipid)	IgG$_1$ molecules per vesicle
IgG (SUV-IA)[a]	4.2	0.6
IgG (SUV-BAB)[b]	6.7	0.9
	9.2	1.3
	17.1	2.3
IgG (PDP-DPPE SUV)[c]	30.0	4.1
	32.0	4.3
	49.3	6.7
	57.4	7.9

Estimation of the number of IgG molecules (assumed mol. wt. 150,000) per vesicle was based on an average vesicle diameter of 60 nm, a surface of the phospholipid/cholesterol unit of 83Å^2 and on the localisation of two-thirds of the phospholipid in the outer bilayer. IgG-bearing liposomes were composed of DSPC, activated DPPE and cholesterol (0.99:0.01:1 or 0.9:0.1:1 molar ratios) (from Wolff and Gregoriadis, 1984).

[a] IgG-bearing liposomes with DPPE activated with NHIA after incorporation into liposomes.

[b] IgG-bearing liposomes with DPPE activated with SBAB after incorporation into liposomes.

[c] IgG-bearing liposomes with DPPE activated with SPDP before incorporation into liposomes.

spite of the reduced amount of IgG (1.3-2.3 molecules per vesicle, Table 4) in the former preparation.

In other experiments, the effect of plasma on the binding of IgG-bearing liposomes to the AKR-A cells was investigated (Table 5). As already noted with other targeted liposomes (Gregoriadis et al., 1981; Heath et al., 1980; Martin and Papahadjopoulos, 1982) the presence of plasma reduced binding of liposomes to the cells only modestly. Although it is conceivable that a variety of plasma

Table 4. Effect of cell number on the binding of free and
liposomal anti-Thy$_1$ IgG$_1$ to AKR-A cells

Preparation	Cell Number	IgG bound (% of added)	CF bound (% of added)
Free IgG	10^6	6.8 ± 0.6	-
	7.5×10^6	35.5 ± 0.3	-
	10^7	56.4 ± 3.9	-
	1.5×10^7	59.6 ± 1.9	-
IgG(PDP-DPPE SUV)[a]	10^6	15.2 ± 2.0	14.3 ± 1.6
	10^7	68.2 ± 7.0	69.5 ± 6.9
IgG(PDP-DPPE SUV)[b]	10^6	14.2 ± 0.5	11.9 ± 0.9
	10^7	68.1 ± 4.5	60.0 ± 3.6
PDP-DPPE SUV[b]	5×10^6		7.7 ± 1.2

AKR-A cells were incubated at $4°C$ for 1 h with 1 μg of anti-
Thy$_1$ ^{125}I-IgG$_1$, free or coupled to CE-containing liposomes
composed of DSPC, PDP-DPPE and cholesterol (0.99:0.01:1 or
0.9:0.1:1 molar ratios) or with similar liposomes devoid of
IgG. All values are means $\pm$ S.D. from three separate experiments. For other details see legend to Table 3 (from Wolff
and Gregoriadis, 1984).

[a]Liposomes contained 10% (of total phospholipid) DPPE

[b]Liposomes contained 1% (of total phospholipid) DPPE

components may adhere onto or interact with uncoupled pyridyldithio-
propionyl groups on the surface of liposomes and, in this way, mask
the antigen recognition sites of the IgG, it is also possible that
there may have been some cleavage of the disulphide bond between
the (SPDP-modified) IgG and liposomes by plasma reducing agents.
The latter possibility is enhanced by data presented in Table 6
from experiments in which control (IgG-free) and IgG-bearing liposomes were chromatographed following their exposure to plasma.
Although loss of IgG after 0.5 h was only 15%, it increased to
nearly 50% on longer (24 h) incubation. Table 7 also shows that
cholesterol-rich DSPC liposomes, known to retain entrapped solutes
fully upon extensive incubation with plasma, behave similarly when

Table 5. Effect of plasma on the binding of free and lipo-
          somal anti-Thy$_1$ ^{125}I-IgG$_1$ to AKR-A cells

Preparation	Incubation buffer		Plasma	
	^{125}I-IgG bound	CF bound	^{125}I-IgG bound	CF bound
Free IgG	56.4±3.9[a]	-	55.4±4.1[a]	-
IgG-SUV[b]	68.1±4.5	60±3.6[a]	51.3±2.2	48.7±0.2[a]

AKR-A cells (10^7) were incubated at $4\,^{\circ}$C for 1 h in incubation buffer or in mouse plasma with free ($1\,\mu$g) or liposomal (0.86-$0.90\,\mu$g) anti-Thy$_1$ ^{125}I-IgG$_1$. Liposomes were
composed of DSPC, PDP-DPPE and cholesterol (0.99:0.01:1).
Values are means ± S.D. from three separate experiments.
For other details see the legend to Fig. 3 (from Wolff and
Gregoriadis, 1984).

[a]Values denote % bound of added.

[b]IgG(PDP-DPPE SUV).

SPDP-modified DPPE replaces some of the DSPC component. On the
other hand, the coupling of IgG onto the liposomal surface does
lead to some loss of vesicle stability (CF and ^{111}In markers) in
the presence of plasma but only upon prolonged (24 h) exposure
(Table 6).

Interaction of Anti-Thy$_1$ IgG$_1$-Bearing Liposomes with AKR-A Cells

in-vivo

     In-vivo studies with anti-Thy$_1$ ^{125}I-IgG$_1$-bearing liposomes were
carried out using liposomes composed of DSPC, a small proportion of
DPPE (in the form of PDP-DPPE) and cholesterol (molar ratios 0.99:
1:0.01) and containing quenched CF and ^{111}In-Ca-DTPA. Since it was
expected that liposomes given by the intravenous route would interact
efficiently only with accessible target cells, i.e. those within the
intravascular space, mice injected intravenously with AKR-A cells
were chosen as a model system for targeting in-vivo. It was reasoned
that after the injection of cells a proportion of these would still
be available in the circulation to interact with anti-Thy$_1$ IgG$_1$-
bearing liposomes injected subsequently by the same route.

     Data published elsewhere (Wolff and Gregoriadis, 1984) have
established the following (a) on the basis of the rates of clearance

Table 6. Stability of liposomes in the presence of plasma

Incubation time (h)	Liposomes		IgG-liposomes		
	Latent CF (%)	Retained ^{111}In (%)	Latent CF (%)	Retained ^{111}In (%)	Retained ^{125}I (%)
0.5	99.8	96.0	100.0	95.5	85.3
5.5	99.3	95.1	99.6	93.3	78.5
24.0	98.2	91.0	72.8	71.9	54.3

IgG-free or anti-Thy$_1$ ^{125}I-IgG$_1$-bearing liposomes composed of DSPC, PDP-DPPE and cholesterol (0.99:0.01:1) contained CF and ^{111}In-Ca-DTPA were incubated with mouse plasma. CF latency is expressed as % of latency in the original preparations (97.9-98.1%). ^{111}In and ^{125}I radioactivity retained by liposomes upon incubation with plasma was estimated after passing samples of the incubation mixtures through CL-4B Sepharose minicolumns and expressed as % of the applied radioactivity recovered with liposomes (from Wolff and Gregoriadis, 1984).

of the two aqueous phase markers (CF and ^{111}In) and ^{125}I-radioactivity representing the IgG, anti-Thy$_1$ IgG$_1$-bearing liposomes bound onto AKR-A cells present in the circulation, within 1-2 h after injection; (b) experiments with mice that had not been treated with cells showed that some of the injected liposomes also bound to the Fc receptors of the RES (via the Fc portions of the IgG) and, probably, to thymocytes. It is not known whether such binding also occurred when AKR-A cells were present in the circulation; (c) on the basis of blood levels of ^{99m}Tc-labelled AKR-A cells and ^{111}In radioactivity, it appeared that association of cells with liposomes enhanced the rapid removal (mainly by the liver) of both interacting moieties.

CONCLUSIONS

There is now overwhelming evidence that liposomal stability can be arranged so that under longterm storage, liposomes remain virtually unchanged with full retention of drug contents and, in vivo, release or retain their drug content according to what is required of them. Other developments in liposome technology

(Gregoriadis, 1984) have led to simple techniques for high yield drug entrapment, appropriate for scaled up preparations. Such developments have paved the way for the widespread acceptance of liposomes as a delivery system for the treatment of diseases involving the RES, as diagnostic agents in-vitro and in-vivo including the imaging of tissues and for other uses in which a carrier system of such qualifications is required. Now it is becoming apparent that liposomes may also serve as targeting devices to direct drugs to tissues other than those of the RES. In the case of liposomes which can target drugs to the hepatic parenchymal cells, a role in the treatment of diseases implicating these cells (e.g. liver stage of malaria, other microbial and viral diseases) is a distinct possibility worthy of further investigation. The main limitation of the targeted liposome system is its apparent inability to escape the vascular space. However, there are numerous situations where drug delivery will be needed intravascularly or, indeed, in restricted locations such as the gut, or the lymphatic system.

ACKNOWLEDGEMENTS

     We thank Dr. P. Thorpe, Imperial Cancer Research Fund Laboratory for providing the anti-Thy1 IgG1 and AKR-A cells and Mrs. M. Moriarty for secretarial assistance. B.W. gratefully acknowledges support from the British Council.

REFERENCES

Allen, T.M., 1981, A study of phospholipid interaction between high
        density lipoprotein and small unilamellar vesicles, Biochim.
        Biophys. Acta, 640:385.
Ashwell, G. and Morell, A.G., 1974, The role of surface carbo-
        hydrates in the hepatic recognition and transport of cir-
        culating glycoproteins, Adv. Enzymol , 41:99.
Gregoriadis, G., 1974, Structural requirements for the specific
        uptake of macromolecules and liposomes by target tissues,
        in: "Enzyme Therapy in Lysosomal Storage Diseases", J.M.
        Tager, G.J.M. Hooghwinkel and W.Th. Daems, eds., North-
        Holland Publishing Co., Amsterdam.
Gregoriadis, G., 1981, Targeting of drugs: Implications in medicine,
        The Lancet, 2:241.
Gregoriadis, G., 1983, Targeting of drugs with molecules, cells and
        liposomes, TIPS, 4:304.
Gregoriadis, G., ed., 1984, "Liposome-Technology", vols. I-III, CRC
        Press Inc., Boca Raton.
Gregoriadis, G. and Davis, C., 1979, Stability of liposomes in-vivo
        and in-vitro is promoted by their cholesterol content and
        the presence of blood cells, Biochem. Biophys. Res. Comm.,
        89:1287.

Gregoriadis, G. and Meehan, A., 1981, Interaction of antibody-bearing small unilamellar liposomes with antigen-coated cells: the effect of antibody and antigen concentration on the liposomal and cell surface respectively, Biochem. J., 200:211.

Gregoriadis, G. and Neerunjun, D., 1975, Homing of liposomes to target cells, Biochem. Biophys. Res. Comm., 65:537.

Gregoriadis, G. and Senior, J., 1980, The phospholipid component of small unilamellar liposomes controls the rate of clearance of entrapped solutes from the circulation, FEBS Lett., 119:43.

Gregoriadis, G. and Senior, J., 1984, Targeting of small unilamellar liposomes to the galactose receptor in-vivo, Biochem. Soc. Trans., 12:337.

Gregoriadis, G., Neerunjun, D.E. and Hunt, R., 1977, Fate of a liposome-associated agent injected into normal and tumour-bearing rodents. Attempts to improve localization in tumour tissues, Life Sci., 21:357.

Gregoriadis, G., Mah, M.M. and Meehan, A., 1981, Interaction of antibody-bearing small unilamellar liposomes with target free antigen in-vitro and in-vivo: Some influencing factors, Biochem. J., 200:203.

Hashimoto, Y., Sugawara, M., Masuko, T. and Hojo, H., 1983, Anti-tumour effect of actinomycin D entrapped in liposomes bearing subunits of tumour-specific monoclonal immunoglobulin M antibody, Cancer Res., 43:5328.

Heath, T.D., Fraley, R.T. and Papahadjopoulos, D., 1980, Antibody targeting of liposomes: Specific interaction of vesicles conjugated to anti-erythrocyte $F(ab')_2$, Science, 210:539.

Hwang, K.J., Luk, K-F.S. and Beaumier, P.L., 1980, Hepatic uptake and degradation of unilamellar sphingomyelin/cholesterol liposomes: A kinetic study, Proc. Natl. Acad. Sci. USA, 77: 4030.

Kirby, C. and Gregoriadis, G., 1980, The effect of the cholesterol content of small unilamellar liposomes on the fate of their lipid components in-vivo, Life Sci., 27:2223.

Kirby, C. and Gregoriadis, G., 1981, Plasma-induced release of solutes from small unilamellar liposomes is associated with pore formation in the bilayers, Biochem. J., 199:251.

Kirby, C. and Gregoriadis, G., 1983, The effect of lipid composition of small unilamellar liposomes containing melphalan and vin-cristine on drug clearance after injection into mice, Biochem. Pharmacol., 32:609.

Kirby, C. and Gregoriadis, G., 1984a, A simple novel method for efficient drug entrapment in liposomes, in: "Liposome Technology", Vol. I, G. Gregoriadis, ed., CRC Press Inc., Boca Raton.

Kirby, C. and Gregoriadis, G., 1984b, Incorporation of Factor VIII into liposomes, in: "Liposome Technology", Vol. II, G. Gregoriadis, ed., CRC Press Inc., Boca Raton.

Kirby, C. and Gregoriadis, G., 1984c, Preparation of liposomes containing Factor VIII for oral treatment of haemophilia, J. Microencap., 1:33.

Kirby, C., Clarke, J. and Gregoriadis, G., 1980, Cholesterol content of small unilamellar liposomes controls phospholipid loss to high density lipoproteins in the presence of serum, FEBS Lett., 111:324.

Kirby, C., Clarke, J. and Gregoriadis, G., 1980, Effect of the cholesterol content of small unilamellar liposomes on their stability in-vivo and in-vitro, Biochem. J., 186:591.

Krupp, L., Chobanian, A.V. and Brecher, I.P., 1976, The in-vivo transformation of phospholipid vesicles to a particle resembling HDL in the rat, Biochem. Biophys. Res. Commun., 72:1251.

Martin, F.J. and Papahadjopoulos, D., 1982, Irreversible coupling of immunoglobulin fragments to preformed vesicles, J. Biol. Chem., 257:286.

Mayhew, E. and Rustum, Y., 1983, Effect of liposome-entrapped chemotherapeutic agents on mouse primary and metastatic tumours, Biol. Cell., 47:81.

Morell, A.G., Gregoriadis, G., Scheinberg, I.H., Hickman, J. and Ashwell, G., 1971, The role of sialic acid in determining the survival of glycoproteins in the circulation, J. Biol. Chem., 246:1461.

van Oss, C.J., Gillman, C.F., Bronson, P.M. and Border, J.R., 1974, Phagocytosis-inhibiting properties of human serum alpha-1 acid glycoprotein, Immunol. Comm., 3:321.

Patel, H.M., Tüzel, N.S. and Ryman, B.E., 1984, Inhibitory effect of cholesterol on the uptake of liposomes by liver and spleen, Biochim. Biophys. Acta, 761:142.

Poste, G., 1983, Liposomes targeting in-vivo: Problems and opportunities, Biol. Cell., 47:19.

Proffitt, R.T., Williams, L.E., Presant, C.A., Tin, G.W., Uliana, J.A., Gamble, R.C. and Baldeschwielar, J.D., 1983, Liposomal blockade of the reticuloendothelial systems: Improved tumour imaging with small unilamellar vesicles, Science, 220:502.

Scherphof, G., Roerdink, F., Waite, M. and Parks, I., 1978, Disintegration of phosphatidylcholine liposomes in plasma as a result of interaction with high density lipoproteins, Biochim. Biophys. Acta, 542:296.

Scherphof, G., Damen, J. and Wilschut, J., 1984, Interactions of liposomes with plasma proteins, in: "Liposome Technology", G. Gregoriadis, ed., CRC Press Inc., Boca Raton.

Senior, J. and Gregoriadis, G., 1982a, Stability of small unilamellar liposomes in serum and clearance from the circulation: the effect of the phospholipid and cholesterol components, Life Sci., 30:2123.

Senior, J. and Gregoriadis, G., 1982b, Is half-life of circulating small unilamellar liposomes determined by changes in their permeability?, FEBS Lett., 145:109.

Senior, J. and Gregoriadis, G., 1984, Methodology in assessing liposomal stability in the presence of blood, clearance from the circulation of injected animals, and uptake by tissues, in: "Liposome Technology", Vol. III, G. Gregoriadis, ed., CRC Press Inc., Boca Raton.

Senior, J., Gregoriadis, G. and Mitropoulos, K., 1983, Stability and clearance of small unilamellar liposomes: Studies with normal and lipoprotein-deficient mice, Biochim. Biophys. Acta, 760:111.

Spanjer, H.H. and Scherphof, G., 1983, Targeting of lactosylceramide-containing liposomes to hepatocytes in-vivo, Biochim. Biophys. Acta, 734:40.

Szoka, F. and Mayhew, E., 1983, Alteration of liposome disposition in-vivo by bilayer situated carbohydrates, Biochem. Biophys. Res. Commun., 110:140.

Teradaira, R., Kolb-Bachofen, V., Schlepper-Schafer, J. and Kolb, H., 1983, Galactose-particle receptor on liver macrophages. Quantitation of particle uptake, Biochim. Biophys. Acta, 759: 306.

Toonen, P.A.H.M. and Crommelin, D.J.A., 1983, Immunoglobulins as targeting agents for liposome-encapsulated drugs, Pharm. Weekblad Scientific Edition, 5:269.

Tümer, A., Kirby, C., Senior, J. and Gregoriadis, G., 1983, Fate of cholesterol-rich unilamellar liposomes containing ^{111}In-labelled bleomycin after subcutaneous injection into rats, Biochim. Biophys. Acta, 760:119.

Tyrrell, D.A. and Ryman, B.E., 1980, Liposomes: Bags of potential, Essays Biochem., 16:49.

Wolff, B. and Gregoriadis, G., 1984, The use of monoclonal anti-Thy_1, IgG_1 for the targeting of liposomes to AKR-A cells in-vitro and in-vivo, Submitted.

INTERACTIONS OF LIPOSOMES WITH LIPOPROTEINS AND LIVER CELLS

IN VIVO AND IN VITRO STUDIES

Gerrit Scherphof, Jan Damen[1], Jan Dijkstra[2],
Frits Roerdink and Halbe Spanjer

Laboratory of Physiological Chemistry
University of Groningen
Bloemsingel 10
9712 KZ Groningen, The Netherlands

[1]The Netherlands Cancer Institute, Division of Cell Biology
 Plesmanlaan 121, 1066 CX Amsterdam, The Netherlands

[2]Department of Pharmacology, School of Pharmacy
 University of California, San Francisco CA 94143, U.S.A.

INTRODUCTION

Detailed knowledge on the in-vivo behaviour of liposomes and
of the factors influencing such behaviour is a necessity for any
serious attempt to apply liposomes as an in-vivo carrier system for
both therapeutics and diagnostics. For several years now our group
has made an effort to study liposomes in-vivo and, in doing so, we
have paid special attention to the interaction with blood components
such as lipoproteins and to the role of the liver and its various
cell populations in the elimination of liposomes from the bloodstream and their subsequent processing.

This paper is meant to present an overview of our main observations and conclusions with an emphasis on some recent observations.

INTRAHEPATIC FATE OF LIPOSOMAL LABELS

One of our earliest observations concerned the existence of a
discrepancy between the intrahepatic fate of two radioactive liposomal labels: ^{125}I-labeled bovine serum albumin as a marker for the
encapsulated aqueous space and (^{14}C-Me-choline)-labeled phosphatidylcholine as a marker of the liposomal membrane. The ^{125}I-label

Table 1. Shift in distribution of ^{14}C-radioactivity
          between parenchymal and non-parenchymal rat
          liver cells.

Multilamellar liposomes consisting of phosphatidylcholine, cholesterol and
dicetylphosphate (6:3:1 molar ratio) were prepared by extrusion of an un-
sonicated aqueous dispersion of this lipid mixture through a series of
polycarbonate membrane filters of 1.0, 0.8, 0.6 and 0.4 μm pore-size,
respectively. An aliquot (1 μmol of lipid per 100 g rat) of these liposomes was injected and at the times indicated, parenchymal as well as total
non-parenchymal cells were isolated from the same liver following collagenase
perfusion. The non-parenchymal cells were not fractionated further. [14]C-
radioactivity was measured in both cell fractions as well as in the non-
fractionated total liver cell suspension. Results were calculated on the
basis of known cell numbers of 100 g rat liver as percent of injected dose.

Minutes after injection	10	20	30	60	120	240
Whole liver cell suspension	32	38	40	39	42	39
Hepatocytes	19	24	21	30	32	40
Non-parenchymal cells	11	18	19	16	9	6

rapidly accumulated mainly in the non-parenchymal cell fraction of
the liver, reaching a maximum after about 15-20 min, after which a
gradual decline of label content occurred due to intracellular
proteolysis and release of low-molecular weight degradation products
from the cells.[1] The ^{14}C-label, on the other hand, was recovered
to a large extent in the parenchymal cell fraction, particularly at
longer times after injection (Table 1). This time-dependent shift
of lipid radioactivity from the non-parenchymal to the parenchymal
cell fraction, which led us to postulate the existence of an intercellular communication mechanism between the two cell populations
in the liver, was also found when, instead of ^{14}C-phosphatidyl-
choline, ^{14}C-sphingomyelin was used as a liposomal lipid marker.[2]
In this case an additional interesting feature of the transfer mechanism became apparent; the chemical nature of the radioactivity found
in the hepatocytes (i.e. the parenchymal cells) was sphingomyelin.
This suggests that the label transfer takes place by way of a transfer of intact phospholipid molecules rather than through water-
soluble degradation products as would be easier to understand. If
the label entered the hepatocytes as, for instance, choline released
as a sphingomyelin degradation product from Kupffer cells, which
initially had internalized the liposomes, we would have expected the
hepatocytes to incorporate this label chiefly into phosphatidylcholine: its rate of biosynthesis far exceeds that of sphingomyelin.

The apparent discrepancy between lipid label and encapsulated label was not exclusive for the use of albumin as an encapsulated marker. With the non-degradable ^{125}I-labeled poly(vinylpyrrolidone) (PVP) as a marker we also found the label mainly in the non-parenchymal cells of the liver.[3] With this marker we further extended our observations in that we pin-pointed the localization of the label in the non-parenchymal cell fraction to the Kupffer cells (Table 2). The other major constituents of the non-parenchymal liver cell fraction, i.e. the endothelial cells, were virtually devoid of radioactivity, indicating that these cells do not take part in the elimination of liposomes from the blood. Also the use of a morphological marker of the encapsulated volume, such as horse radish peroxidase (HRP), led us to the conclusion that both in rats[4] and in mice[5] intravenously injected liposomes are mainly taken up by Kupffer cells.

The shift of ^{14}C-lipid radioactivity from non-parenchymal to parenchymal cells takes place in a period when radioactivity is no longer found in the blood and total liver uptake has reached a maximum value of approximately 50% of the injected dose (Table 1). Theoretically it is possible that, purely by chance, the loss of radioactivity from the non-parenchymal cells matches the increase in the parenchymal cells without the existence of a mechanistic correlation between the two cell populations. The loss of radioactivity from the non-parenchymal cells can obviously be explained by a release of water-soluble degradation products such as choline, which are probably rapidly eliminated from the blood.

The increase of radioactivity in the parenchymal cells could then only be explained if one postulates the existence of a latent compartment containing liposomes which have been rapidly eliminated from the blood but, in a secondary process, can be slowly released from this compartment and subsequently taken up by the hepatocytes. Conceivably, the physical reality of such a compartment could be formed by the surface of the vascular system, either throughout the body, or limited to certain organs, including the liver itself. However, if such a mechanism were operative, one would not expect it to be confined to the lipid label of the liposomes; also liposomal aqueous-space markers such as PVP or HRP would be expected to display the shift kinetics, which however, is not the case. One further possibility remains to explain the kinetics of the lipid label appearance in parenchymal and non-parenchymal cells, besides a real intercellular transfer of intact phospholipid molecules; that is the possibility that lipid label enters the parenchymal cells through the mediation of plasma lipoproteins. By exchanging non-radioactive lipoprotein lipid for radioactive liposomal phospholipid the liposomes, particularly those that are not rapidly internalized by (Kupffer) cells and for some time remain part of a latent compartment as discussed above, might lose lipid label in favor of one or more lipoprotein fractions which then might donate this label to

Table 2. Cellular distribution of 125I-labeled poly(vinyl pyrrolidone)- containing liposomes in rat liver following intravenous injection.

Multilamellar or large unilamellar vesicles composed of sphingomyelin/cholesterol (3:2 molar ratio) (Expt. 1) or phosphatidylcholine/cholesterol/dicetylphosphate (6:3:1 molar ratio) (Expts. 2 and 3) and containing 125I-labeled poly(vinyl pyrrolidone) in the aqueous space were injected intravenously at a dose of 3-7 μmol/rat. 1 h after injection, either parenchymal or non-parenchymal cells of the liver were isolated. The non-parenchymal cell fraction was separated into an endothelial cell fraction and a Kupffer cell fraction by elutriation centrifugation. The separated non-parenchymal cell fractions were stained for peroxidase activity to estimate the extent of contamination in the Kupffer (peroxidase-positive) and endothelial (peroxidase-negative) cell fraction. Cell recoveries were estimated on the basis of 450.10^6 parenchymal cells and 194.10^6 non-parenchymal cells per 100 g rat body weight and a 1:3 ratio of Kupffer and endothelial cells in whole liver. Assuming, on the basis of the raw data, that only Kupffer cells contained radioactive material, the radioactivities per 10^6 cells were corrected for contamination by the other cell type.

	Cell type isolated	Cell recovery (%)	Peroxidase positive cells (%)	dpm per 10^6 cells	dpm per 10^6 cells corrected for contamination	Percentage of injected dose in total cell fraction
Expt. 1	Endothelial	35.1	4.7	141	27	1.1
	Kupffer	28.3	77.1	1858	2410	33.2
	Parenchymal	44.4	-	7.7	-	0.7
Expt. 2	Endothelial	38.2	8.1	121	-34	0
	Kupffer	26.7	85.3	1917	2248	27.7
	Parenchymal	48.0	-	17.8	-	2.0
Expt. 3	Endothelial	24.7	17.0	32	-0.5	0
	Kupffer	28.9	79.0	191	242	21.9
	Parenchymal	45.0	-	6.8	-	3.4

the parenchymal cells. As long as the existence of such a hypothetical pool of non-internalized liposomes somewhere outside the bloodstream has not been established and its physical nature not characterized, it will be difficult to fully exclude such a mechanism of lipid
label uptake by the hepatocytes. As will be seen in the following
section, however, the transfer of liposomal phospholipid label to
plasma lipoproteins is of limited significance for the relatively
large liposomes used in the studies discussed above, at least under
in-vitro conditions.

It remains to be seen if this virtual lack of phospholipid
transfer from large liposomes to lipoproteins also applies under in-
vivo conditions where interactions with large areas of cell surfaces
could play an additional role. Cell-surface induced leakage of
liposomal contents has been reported by a number of investigators.[6,7]

EFFECTS OF PLASMA AND PLASMA LIPOPROTEINS ON LIPOSOME INTEGRITY

It is well established now that the phospholipid constituents
of liposomes are susceptible to exchange with plasma lipoprotein
phospholipids.[8-13] This process, which can lead to massive release
of aqueous liposomal contents, involves either a net mass transfer
of liposomal phospholipid to, mainly, high density lipoprotein (HDL)
or an exchange between the liposomal and lipoprotein phospholipid
pools. In the latter case we will observe only the release of liposomal label with a concomitant decrease in specific radioactivity
of the liposomal phospholipid.

Cholesterol content of the liposomes is one of the variables
that determine the relative contributions of net transfer and exchange to the effects of interaction between liposome and lipoprotein.[11-14] As is seen in Fig. 1, with increasing cholesterol content in small unilamellar vesicles carrying egg-yolk phosphatidylcholine as the phospholipid constituent, both the release of encapsulated solute and the loss of lipid label diminish. While at 35
mol % of cholesterol, solute release turns out to be nearly completely suppressed, there is still substantial transfer of phosphatidylcholine label. By incubating such cholesterol-rich liposomes with HDL, labeled[15] in-vivo with ^{32}P, we were able to demonstrate directly that under these conditions nearly equal amounts of
liposomal^{14}C-phospholipid were exchanged for ^{32}P-labeled lipoprotein
phospholipid (Table 3). It is of interest to note that, when the
liposomal phosphatidylcholine is replaced by sphingomyelin, also
labeled in the choline moiety with ^{14}C, the transfer of label to the
lipoprotein is nearly completely abolished (Fig. 1); apparently this
phospholipid species is virtually not susceptible to exchange.

The exchange of phosphatidylcholine between cholesterol-rich
liposomes and HDL is limited to the outer half of the liposomal membrane as could be demonstrated by making use of the possibility to

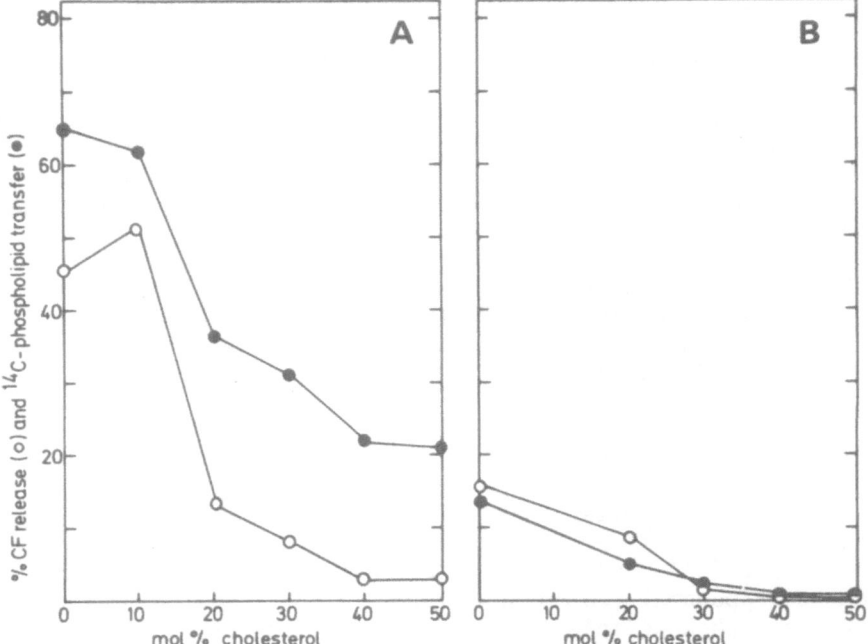

Fig. 1.   Influence of the liposomal cholesterol content
           on the release of liposomal constituents.
           Carboxyfluorescein release (O) and phospholipid
           transfer to HDL (●) from liposomes consisting
           of (^{14}C)-phosphatidylcholine (A) or (^{14}C)
           sphingomyelin (B) with various amounts of chol-
           esterol was determined after 30 min incubation
           with rat plasma (0.5μmol SUV lipid per ml of
           plasma).

hydrolyze specifically the phospholipid in this outer monolayer with
phospholipase A$_2$.[16] As we showed previously,[17] phospholipases A$_2$
can selectively degrade outer monolayer phospholipids in a liposomal
bilayer without loss of the permeability properties of the liposomal
membrane; an encapsulated solute, such as carboxyfluorescein, is
fully retained under conditions where all phospholipid in the outer
half of the membrane is hydrolyzed. Apparently, the reaction products formed, i.e. free fatty acids and lysophospholipid fulfil all
the requirements to constitute at least one half of a membrane
bilayer. Only upon addition of serum albumin, which extracts the
reaction products from the liposomes, the hydrocarbon chains of the
non-degraded inner half of the bilayer became exposed to the aqueous
environment and the liposome collapses instantaneously with concomitant release of all contents.

Table 3. Bidirectional transfer of phospholipids between
          liposomes and HDL.

Liposomes consisting of 0.5 μmol ^{14}C-labeled phospholipid and various amounts
of cholesterol, and with carboxyfluorescein entrapped, were incubated with
^{32}P-labeled HDL and the d > 1.25 g/ml fraction (amounts equivalent to 1.0 ml
plasma) in a final volume of 2.0 ml for 30 min at 37°C.

Liposomal lipid composition	Transfer of ^{14}C-labeled phospholipid to HDL (nmol)	Transfer of ^{32}P-labeled phospholipid to liposomes (nmol)	Carboxyfluorescein release (%)
^{14}C-phosphatidyl-choline	260	22	46
^{14}C-phosphatidyl-choline/cholesterol (3:2)	75	53	3.5
^{14}C-sphingomyelin/cholesterol (3:2)	8	13	0.4

As is schematically outlined in Fig. 2, we subjected cholesterol-
rich small unilamellar vesicles, which had been incubated with HDL to
exchange part of their (labeled) phosphatidylcholine with that of
the lipoprotein, to the action of bee-venom phospholipase A_2. If the
phospholipid in the outer half of the liposomal membrane were available for this exchange, the specific radioactivity of the phospholipid in the inner half would remain unchanged while that of the
phospholipid in the outer half would decrease. The extent of this
decrease can be predicted from the total amount of radioactivity
transferred to the lipoprotein and the fraction of liposomal phospholipid which is in the outer half of the membrane, i.e. the fraction
which can be hydrolyzed by the phospholipase: approx. 2/3 for vesicles of the size used in these experiments. Experimentally, the
specific radioactivities of the phospholipid in the outer and inner
halves of the membrane can be determined after separating the two
phospholipid pools by thin layer chromatography after the phospholipase treatment. The outer-half phospholipid is represented by the
lyso derivative, while the non-degraded phosphatidylcholine represents the inner-half phospholipid.

Table 4 shows that calculated and experimental values coincide
reasonably well. Particularly the values for the non-degraded, i.e.
the inner-half phosphatidylcholine, agree very closely. This by
itself already demonstrates that the phospholipid in the inner monolayer of the liposomal membrane is not accessible to interaction

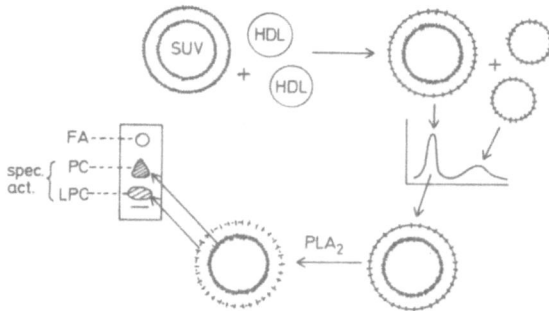

Fig. 2.   Schematic representation of the experimental
approach to demonstrate asymmetric phospholipid exchange.   Phosphatidyl(Me-^{14}C)choline/
cholesterol (1:1) liposomes are incubated with
HDL such that outer-monolayer phospholipid is
diluted with cold HDL-phospholipid; liposomes
are separated from HDL; liposomes are treated
with phospholipase such that all outer-monolayer phospholipid is hydrolyzed; liposomes
are extracted and lipid extract is chromatographed to separate phosphatidylcholine from
its lysoderivative (and fatty acid); specific
radioactivities of both phospholipids are determined.

with the lipoprotein. One explanation for the less accurate conform-
ity between calculated and experimental values of the outer-half
phospholipid may be that some of the liposomal phosphatidylcholine
is exchanged for other phospholipid species from the lipoprotein;
this would obviously lead to a somewhat smaller decrease in specific
activity since the liposomal phosphatidylcholine would become less
diluted with lipoprotein phosphatidylcholine.

As was alluded to in the preceding section, the destructive
action of plasma lipoproteins on liposomes is dependent upon liposome size. The smaller the size of the liposome the larger the extent to which its phospholipid is transferred to the lipoprotein and,
concomitantly, the larger the extent to which encapsulated solute is
released. This is clearly illustrated in Table 5 which shows the
percent of radioactive phosphatidylcholine transfer to plasma HDL
and the percent of carboxyfluorescein release from small unilamellar
egg phosphatidylcholine vesicles of different size. The various
vesicle populations, whose dimensions were calculated from the meas-

Table 4. Phospholipase treatment after incubation with HDL

Vesicles (3 μmol total lipid, including 1200 nmol ^{14}C-phosphatidylcholine, approximately 0.125 μCi) were incubated with HDL (5 mg of protein) for 3-4 h at 37°C in 0.5 ml lipoprotein-free plasma supplemented with Tris-buffered saline to a final volume of 2.0 ml. At the end of the incubation the mixture was cooled and chromatographed on BioGel A 1.5 m. The column fractions were assayed for radioactivity and the percent of label transfer to HDL was estimated. The void-volume fractions containing the liposomes were pooled and concentrated. An aliquot of the concentrate (approximately 1 μmol lipid) was treated with phospholipase A₂ for 30 min to allow maximal degradation of phosphatidylcholine (60-65%). After addition of excess EDTA to inactivate the phospholipase the lipids were extracted. Phosphatidyl-choline and lysophosphatidylcholine were separated by thin layer chromato-graphy. The spots were assayed for phosphorus and radioactivity content.

LIPOSOMAL SPECIFIC RADIOACTIVITIES (dmp/nmol P)

Exp. no.	PC* after incubation with HDL found	PC* after incubation with HDL calculated	% PC hydrolyzed	PC after PLA₂* found	PC after PLA₂* calculated	LPC* after PLA₂ found	LPC* after PLA₂ calculated
1	-	152	64.6	183	187	150	129
2	163	165	61.5	211	203	149	146
3	209	205	63.5	271	281	197	178

PC*, phosphatidylcholine; LPC*, lysphosphatidylcholine; PLA₂*, phospho-lipase A₂.

Table 5. Size-dependent plasma susceptibility of CF-containing ^{14}C-phosphatidylcholine SUV.

SUV contained (Me-^{14}C) egg lecithin and 100 mM carboxyfluorescein. 0.46 μmol-lipid aliquots of the separated fractions constituting the second peak eluting from a Sepharose 2B column (the "SUV peak") were incubated individually with 0.8 ml of rat plasma for 60 min at 37°C in a final volume of 1.2 ml.

Trapped volume (1/mol)	0.16	0.17	0.19	0.23	0.31	0.41
% ^{14}C transfer	90.0	86.3	84.6	71.3	69.6	61.7
% CF release	89.5	90.0	89.7	79.3	68.2	58.2

ured encapsulated volume per quantity of lipid as described by
Wilschut,[18] were obtained as fractions of a Sepharose 2B column,
together constituting the "SUV-peak". This peak does not represent
a homogeneous population of vesicles as is clearly demonstrated in
Fig. 3. Particularly in the first half of the peak there is a considerable shift in trapped volume from about 0.6 1/mol down to 0.2
1/mol. The second half of the peak represents a much more homogeneous population of vesicles: trapped volume decreases only from
0.19 1/mol mid-way the peak to 0.17 1/mol in the last fractions.

In addition to a clear-cut size dependence of lipoprotein-in-
duced phospholipid and solute release, Table 5 also suggests the
existence of a close relationship between the two parameters of vesicle destruction, i.e. phospholipid transfer and caboxyfluorescein
release. We further substantiated this observation by carrying out

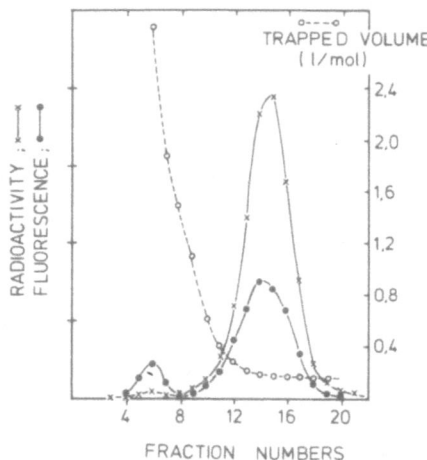

Fig. 3.   Heterogeneity of carboxyfluorescein-containing
          sonicated phosphatidylcholine vesicles on
          Sepharose 2B-CL. (Me-^{14}C)egg lecithin was dis-
          persed in 1.2 ml of 100 mM carboxyfluorescein
          (CF), pH 7.4 and sonicated for 3 h in a bath-
          type sonicator. Non-encapsulated CF was re-
          moved by gel filtration on Sephadex G-100.
          Pooled void-volume fractions were chromato-
          graphed on a Sepharose 2B-CL column, 28 cm x
          1.0 cm. Column fractions were collected and
          assayed for radioactivity, and fluorescence
          after addition of Triton X-100 (1% v/v, final
          concentration) to break up the vesicles and
          relieve selfquenching of the fluorophore.

a large number of incubations of homogeneous vesicles (second half
of a Sepharose-2B peak) in plasma at various plasma : liposomal-
phospholipid ratios. Low ratios would produce a relatively small
effect, high ratios would result in large extents of solute and
phospholipid release. Over a range from 10-90% we found a linear
relationship between the percent of solute release and the percent
of phospholipid transfer. The slope of this line deviated from unity
only by 7%, solute release just being slightly in excess of lipid
transfer. A similar result, although with fewer experiments, was
obtained when ^{125}I-labeled poly(vinylpyrrolidone) was used as an
encapsulated solute. We interpret these results to indicate that
the interaction of the lipoprotein with the liposome leads to an all-
or-nothing effect. When a vesicle is damaged by interaction with the
lipoprotein all of its phospholipid is accomodated by lipoprotein
particles and all of its contents are released. It will probably
require several lipoprotein particles to accomodate all the lipid of
one vesicle and the lipoprotein particles may become saturated with
liposomal phospholipid. (Cf. our previously published calculations
[19]). The more lipid they acquire the less agressive they will become
towards additional vesicles; and once attacked by a lipoprotein particle, a vesicle may become more susceptible to further destruction
by additional lipoprotein molecules. These considerations may help
to explain why complete destruction of all phospholipid vesicles
present requires a very large excess of lipoprotein over vesicles.
The excessive acquirement by HDL of liposomal phospholipid can actually be demonstrated by observing, upon gel filtration, a gradual
shift towards higher molecular weight of the elution volume of radio-
activity-containing HDL following plasma incubations with increasing
amounts of liposomes.[19]

The view expressed here on the way the lipoprotein attacks phospholipid vesicles is different from the one presented by Kirby and
Gregoriadis,[20] who suggested that the lipoprotein forms pores in the
vesicles from which encapsulated solutes can leak at molecular
weight-dependent rates. As indicated above, this can not be recon-
ciled with our observations (however, see chapter by Gregoriadis et
al., this book). Possibly, the results obtained by Kirby and
Gregoriadis have been influenced by the heterogeneity of the vesicle
populations they used; if from a heterogeneous vesicle preparation
the smaller vesicles are destroyed, there is a relative excess of
phospholipid loss over solute release and the remaining vesicles
have a relatively larger trapped volume: lipid mass ratio. When
using large unilamellar liposomes (trapped volume 4.4 l/mol) we
found that the extent of phospholipid release was greatly reduced
as compared with our observations on small vesicles. However, solute
release was still high; in other words, the close relationship between phospholipid and solute release no longer appeared to exist
when the vesicles were beyond a certain size limit. It would seem
that for the larger vesicles the casues of solute release are different from those for small vesicles. Interaction with plasma (lipo)

proteins probably remains limited to relatively subtle perturbations
of the membrane structure causing an increase in permeability. Possibly, it is the limited bilayer curvature in these vesicles that
prevents a more invasive interaction of the lipoproteins with the
liposomal membrane; a similar difference between small and large vesicles was reported by us previously in a study on phospholipase A_2-
susceptibility of phospholipid vesicles.[21] With large multilamellar
vesicles phospholipid transfer to lipprotein is also low as compared
with experiments with small vesicles. In addition to the effect of
the curvature of the liposomal membranes we may expect the presence
of multiple bilayers to influence the interaction; the inner bilayers
are obviously not in immediate contact with the plasma. This limited
interaction does also come to expression in terms of solute release.
Also in this respect multilamellar vesicles are more resistant to
plasma interaction than the unilamellar vesicles, either large or
small. It is probably only the outermost aqueous compartment of the
multilamellar liposomes that is subject to release of contents.

To conclude this section, it should be noted that the interaction of HDL with liposomes is profoundly influenced by the presence
of non-lipoprotein constituents in plasma.[14] We partially purified
a protein factor from the lipoprotein-free fraction of plasma that
strongly enhances both the net transfer of phospholipid from pure-
phosphatidylcholine liposomes to HDL as well as the exchange of this
phospholipid between cholesterol-rich liposomes and isolated HDL.[22]
This factor is probably the same as the phospholipid exchange protein
partially purified recently from plasma by others.[23]

INTERACTION OF LIPOSOMES WITH KUPFFER CELLS IN MONOLAYERS

The results of our in-vivo studies on hepatic uptake and processing of liposomes led us to initiate an investigation of the
interaction of liposomes with monolayers of isolated Kupffer cells
in maintenance culture.[24]

After two days in culture the cells have optimally recovered
from the damage they experienced during the isolation procedure,
which involves excessive treatment with proteolytic enzymes. Liposome uptake values, expressed as nmol of liposomal lipid associated
with 1 mg of cell protein, were calculated from the measurement of
^{3}H and ^{14}C radioactivity of cells incubated with large unilamellar
liposomes (mean diameter 260 nm) labeled in the lipid moiety with
egg phosphatidyl(Me-^{14}C)choline and in the aqueous volume with (^{3}H)-
inulin. After 1 day in culture uptake is approx. one fifth that
after 2 days; upon further extending culture time no further increase
in liposome uptake rate is found (Fig. 4). The uptake of both labels
displays saturation kinetics[2] with maximal uptake rates amounting to
approx. 2 nmol of liposomal phospholipid per hour per 10^6 cells.
This is equivalent to about 1500 vesicles per cell. Upon incubation
beyond one hour the association of ^{14}C-label gradually lags behind

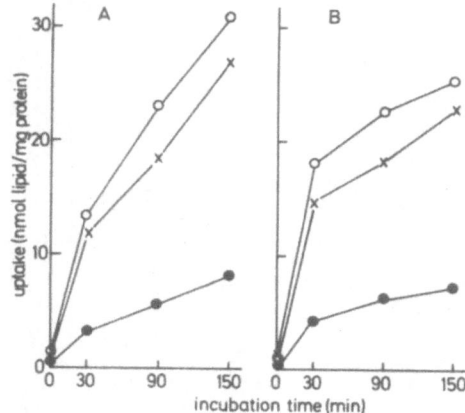

Fig. 4.   Dependence of liposome uptake on time by
          Kupffer cells in culture. Large unilamellar
          liposomes (LUV; 64 nmol of total lipid)
          labeled with (^{14}C)lecithin and (^{3}H)inulin
          were incubated at 37°C for the indicated times
          with Kupffer cells which had been in culture
          for 24 h (●—●), 46 h (X—X) or 70 h (O—O) foll-
          owing their isolation from rat livers. Uptake,
          either calculated from ^{3}H-label (A) or from
          ^{14}C-label (B) was expressed as nmol total lipo-
          somal lipid per mg cell protein.

that of ^{3}H-label. This becomes particularly apparent when observing
the ^{3}H/^{14}C ratio of the cells during a 4 h incubation, as is shown
in Fig. 5. Starting at 30 min after the beginning of the incubation,
the isotopic ratio gradually increases from 5 to 8. When the incubation is performed at 4°C no change in isotopic ratio is observed
while in the presence of metabolic inhibitors the ratio slowly decreases from 5 to slightly less than 4.5. Our interpretation of
these results is that at 37°C at least a fraction of the cell-associated liposomes become internalized upon which the phospholipid is
degraded and water-soluble degradation products leave the cells.
At 4°C, when the total amount of liposomal label associating with
the cells is less than 25% of that found associated at 37°C, there
is no lipid degradation and release of products, presumably because
the liposomes are not internalized under those conditions. When
antimycin A and NaF are present, label association is also reduced
to less than 20% of control levels and the lack of increase in isotopic ratio indicates that, also under these conditions, no lipid
is degraded. These results indicate that the uptake mechanism is

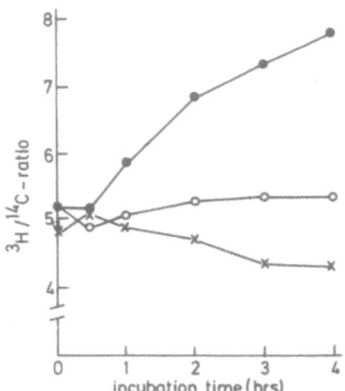

Fig. 5. Effects of temperature and metabolic inhibitors
on liposome uptake by Kupffer cells. Liposomes
(LUV; 60 nmol lipid) labeled with (^{14}C)PC and
(^{3}H)inulin were incubated with 1.6 x 10^6 cells
for the indicated times at 37°C (●—●), at 4°C
(O—O) or at 37°C in the presence of antimycin A
(1μg/ml) and NaF (10 mM) (X—X). ^{3}H/^{14}C ratios
were calculated from the cell-associated radioactivities.

mainly an energy-dependent form of adsorptive endocytosis.

This impression was confirmed by morphological examination of
the cells with fluorescence and electron microscopy following incubation of the cells with liposomes containing FITC-dextran and horse
radish peroxidase, respectively. Fluorescent material was observed
in the cells within large vacuoles, leaving the nucleus free, and in
the electron microscope peroxidase reaction, product was also seen
to accumulate in numerous electron-dense vacuoles, until within 90
min most cells were filled with electron-dense material.[25] The lysosomal nature of the intracellular sites where the liposomes end up
was confirmed by the effect of lysosomotropic agents on uptake of
bio-degradable liposomal markers and subsequent release of breakdown
products. Ammonium chloride and chloroquine are known to accumulate
in the acid environment of the lysosomal system; the consequent rise
in lysosomal pH inhibits the action of lysosomal enzymes.[26-28] The
effects of ammonium chloride on uptake and release of ^{14}C-lipid and
^{125}I-albumin label and on their acid- and water-solubilities, respectively, are demonstrated in Tables 6 and 7. In Table 6 it is
seen that the presence of ammonia during uptake greatly enhances the
cell-associated amount of protein label, particularly at prolonged

Table 6. Effect of NH_4Cl on the water- and acid-solubilities of cell-associated lipid (^{14}C) and protein (^{125}I) label.

Large unilamellar liposomes (average diameter 260 nm; 70 nmol of total lipid) consisting of ^{14}C-egg lecithin (labeled in the choline moiety), cholesterol and phosphatidylserine in a molar ratio of 4:5:1 and containing ^{125}I-labeled albumin were incubated in serum-free medium with approximately 1.6×10^6 Kupffer cells in mono-layer maintenance culture, in presence or absence of 10 mM NH_4Cl. At the times indicated, the total amounts of ^{14}C and ^{125}I label associated with the cells were measured and the radioactivities were converted into liposome equivalents expressed as nmol of total liposomal lipid. Trichloroacetic acid (10%) solubility and water solubility were determined for ^{125}I and ^{14}C label as a measure of protein and lipid degradation, respectively.

Incubation time (h)	Ammonia during uptake	Protein label uptake[a]	Acid solubility protein label (%)	Lipid label uptake[a]	Water solubility lipid label (%)
0.5	absent	4.7	11	7.6	10
0.5	present	5.3	3	7.1	2
2.5	absent	3.7	16	12.0	20
2.5	present	10.2	5	14.1	7

[a]Expressed as nmol of total lipid per mg of cell protein

Table 7. Effect of the presence of NH_4Cl on the release of liposomal labels from Kupffer cells after medium change.

Liposomes containing ^{14}C-label in the phosphatidylcholine and ^{125}I-label in the encapsulated albumin were incubated under the conditions as described in Table 6 in presence of 10 mM NH_4Cl. After 1 h the medium containing the liposomes was removed and replaced by a liposome-free medium either with or without 10 mM NH_4Cl. Liposome equivalents of cell associated radioactivities, all expressed as nmol of total liposomal lipid were calculated from both labels and differentiated according to solubilities in trichloracetic acid, chloroform and water to determine the extents of protein and lipid degradation.

Time after medium change (h)	Ammonia after medium change	Protein label release[a]			Lipid label release[a]		
		total	acid soluble	acid insoluble	total	water soluble	chloroform soluble
0.5	absent	5.8	4.9	0.9	2.0	1.8	0.2
0.5	present	0.7	0.3	0.4	0.7	n.d.	n.d.
2.0	absent	9.1	7.8	1.3	4.9	4.4	0.5
2.0	present	1.4	0.6	0.8	1.2	n.d.	n.d.

[a]Expressed as nmol of total liposomal lipid per mg of cell protein; n.d. not determined.

incubation; apparently the ammonia efficiently inhibits intracellular protein degradation and release of breakdown products. Concomitantly, the fraction of the cell-associated activity that is acid-soluble, is also reduced in presence of ammonia. The effect of ammonia on lipid label uptake, although observable at 150 min, is much less pronounced; the reason for this will become clear from other experiments described below. The amount of water-soluble lipid label, representative for lipid degradation products, is more clearly affected by ammonia. Since only a small fraction of the added liposomes becomes cell-associated, the release of labeled liposomal degradation products can only be measured after the incubation medium containing the liposomes has been removed. When, after an uptake period, the medium is replaced by a liposome-free medium, the release of labeled degradation products in the "clean" new medium can be monitored upon continued incubation.

Table 7 shows the results of such an experiment in which, again, liposomes were used with (Me-^{14}C)-labeled phosphatidylcholine as a lipid marker and radio-iodinated albumin as a degradable aqueous space marker. After an uptake period of one hour in the presence of ammonia, during which approx. 10 nmol of lipid was taken up per mg of cellular protein, the incubation was continued for another 2 h in fresh medium without liposomes either in the presence or absence of ammonia. Without ammonia, about 90% of the protein label was released from the cells in 2 h, 85% of which was acid-soluble. The presence of ammonia reduced the amount released to 1.4 nmol i.e. about 15%, not even half of which was acid-soluble. The acid-insoluble radioactivity recovered in the medium is assumed to represent liposomes released from the cell-surface and/or whole cells detached from the culture dishes during the prolonged incubation. For the lipid label the results are not as pronounced; only about 40% of the label is released from the cells, 90% of it in water-soluble, i.e. degraded, form. This can not be ascribed to intracellular accumulation of degradation products because of an impermeability of the plasma membrane since only 10-20% of the cell-associated ^{14}C-label was water-soluble (Table 6). Chloroquine was found to have similar effects as ammonia, the main difference being that the ammonia effects are more readily reversible.

The results described thus far seem to indicate that, after internalization of the liposomes, entrapped protein is more extensively degraded in the lysosomes than the phospholipid constituents. Upon closer inspection, however, we found that much more lipid degradation is taking place than our initial experiments suggested. It turned out that a considerable fraction of the water-soluble labeled degradation products is reutilized for synthesis of cellular phosphatidylcholine, thus masking the true extent of liposomal lipid degradation. This became apparent when the labeled phosphatidyl-choline in the liposomes was replaced by sphingomyelin, also labeled in the choline moiety with ^{14}C. Whereas it was not possible to

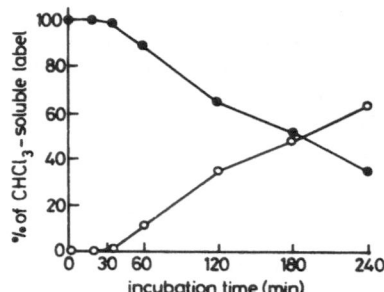

Fig. 6. Intracellular conversion of liposomal choline-
         labeled sphingomyelin (SM) into phosphatidyl-
         choline (PC) at 37°C after binding of lipo-
         somes at 4°C. 11 x 10⁶ Kupffer cells were pre-
         incubated at 4°C with liposomes (490 nmol total
         lipid) consisting of (^{14}C)SM, cholesterol and
         PS (molar ratio 4:5:1). After 90 min (indicated
         as zero time) the medium was replaced by a lipo-
         some-free medium and the incubation was continued
         at 37°C. At the times indicated the relative
         amounts of label in the chloroform-soluble cell
         extracts associated with SM (●) and PC (O) were
         determined via thin-layer chromatography.

separate the liposomal egg phosphatidylcholine from the cellular
phosphatidylcholine in the solvent system used, it was easy to separate the liposomal sphingomyelin from the cellular phosphatidylcholine. Thus we found that during the course of the incubation
the proportion of cell-associated radioactive (liposomal) sphingomyelin gradually decreases in favor of the (cellular) phosphatidylcholine fraction (Fig. 6). This process starts after a lag period
of about 30 min, the time required for the internalized liposomes
to reach the lysosomal system, for the lipid to be hydrolyzed and
for the degradation products to be incorporated into the cellular
phosphatidylcholine pool.

       Since the degradation of sphingomyelin requires other enzymes
than that of phosphatidylcholine one could argue that our observations on the conversion of sphingomyelin label to phosphatidlycholine
label are not necessarily applicable when ^{14}C-phosphatidylcholine is
the liposomal phospholipid. Therefore, we repeated the experiment
with liposomes containing dimyristoyl phosphatidyl (Me-^{14}C)choline
(^{14}C-DMPC) as the main phospholipid constituent. By virtue of its
relatively short hydrocarbon chains DMPC can be separated from cellular phosphatidylcholine with the solvent system we used. Thus, a

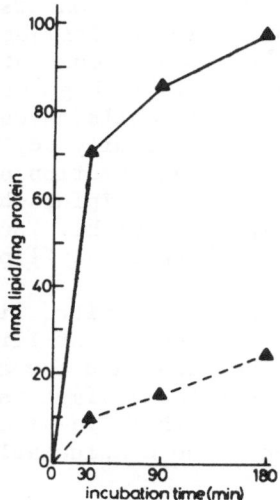

Fig. 7.   Preferential transfer of choline-labeled
          lysophosphatidylcholine from liposomes to
          Kupffer cells. Liposomes (900 nmol lipid)
          composed of egg PC, cholesterol, PS and (^{14}C)
          choline labeled 1-acyl LPC (molar ratio 4:5:
          1:0.15) and containing encapsulated (^{3}H)inulin
          were incubated at 37°C with 1.8 x 10^{6} Kupffer
          cells. At the times indicated the total
          amounts of the cell-associated labels were
          determined and expressed as nmol of total
          liposomal lipid per mg protein. Uptake was
          calculated from ^{14}C-label (——) and from ^{3}H-
          label (---).

similar shift of cell-associated radioactivity from the liposomal
phospholipid to the cellular phosphatidylcholine could be observed,
indicating that also for phosphatidylcholine the extent of hydrolysis
is higher than was suggested by the initial experiments. Both con-
versions, that of sphingomyelin as well as that of DMPC, were inhibited by ammonium chloride, confirming that the lysosomal compartment
is involved in the overall process. In an attempt to further characterize the mechanisms by which Kupffer cells are able to process
liposomal lipids we investigated the fate of labeled lysophosphatidylcholine (^{14}C-LPC) as a constituent of the liposomal membrane.
As an internal marker, the liposomes also contained ^{3}H-inulin. We
found that the uptake of ^{14}C-LPC exceeded ^{3}H-inulin uptake almost
five-fold (Fig. 7). Analysis of the nature of the cell-associated
^{14}C-label revealed that nearly all of it was converted into phos-

phatidylcholine. Neither uptake nor conversion were inhibited by
ammonium chloride or chloroquine, indicating that the lysosomal
system is not involved. Presumably, most of the lysophosphatidylcholine, by virtue of its relatively high water-solubility, is transferred from the liposomes to the cells through the aqueous phase.
After arrival in the cell it is readily acylated to form phosphatidylcholine. These experiments should caution us for the presence of
small contaminations of lysophospholipid, when labeled phospholipid
is used as a marker for liposome uptake. Such contaminations may
lead to disproportionally high rates of "liposome uptake".

    To conclude this section on in-vitro liposome uptake by Kupffer
cells, we compared the rate of uptake of large and small unilamellar
liposomes of identical composition and found that the larger ones
were taken up approximately twice as fast as the small ones. This
difference may help to explain that, in-vivo, large liposomes are
generally cleared from the blood considerably faster than small
liposomes.[29]

IN VIVO FATE OF SMALL UNILAMELLAR VESICLES

    Although small unilamellar lipid vesicles (SUV, diameter 25-80
nm) have the disadvantage of a relatively small encapsulated volume
per quantity of lipid, they probably stand a better chance of escaping from the vascular bed after intravenous administration, than
larger vesicles such as LUV or MLV.[30] Particularly in the liver
where the endothelial lining of the sinusoids is perforated with
fenestrations,[31] small liposomes may gain access to the parenchymal
tissue. Since these parenchymal cells, the hepatocytes, are the
cells responsible for most of the specific liver functions and constitute about 80% of the total liver mass, it would be of interest
to see if small liposomes, in contrast to the larger ones (see above)
do indeed reach the hepatocytes and whether they are taken up by
them. In order to avoid potential difficulties in interpretation
due to the use of phospholipids as liposomal markers (see above) we
labeled the liposomes in their aqueous space with ^{3}H-inulin. Free
inulin is rapidly cleared from the blood through the kidneys, it
is non-degradable once it gets inside a cell and it has a long
intracellular half life.

    As noted above, small liposomes have been found to display a
relatively long half-life in blood, as compared to larger vesicles.
[29,30] Within the class of small unilamellar vesicles, however,
considerable variations in half-life exist, depending on the lipid
composition of the vesicles. Fig. 8 shows that the half-life of
SUV, at a dose of 10μ mol of total lipid per 100 g body weight,
varies from as long as 12 h for sphingomyelin/cholesterol (1:1),
to 3.5 h for the same composition with an additional 10 mol % of
phosphatidylserine (PS), to 1 h when, in addition to the PS, 8 mol
% lactosylceramide is present. These results clearly show that the

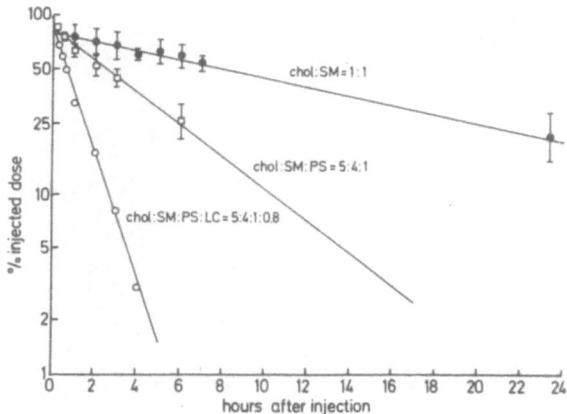

Fig. 8. Blood elimination of small unilamellar
sphingomyelin/cholesterol vesicles; influence
of negative charge (10% phosphatidylserine)
and a glycolipid ligand (8% lactosylceramide).
Rats were injected with a 0.5 ml dose of vesicles (10 mol of total lipid per 100 g body
weight) containing (^{3}H)inulin as an aqueous
solute.

incorporation of negative charge or of a specific ligand for a cell-
surface receptor[32] can drastically influence in-vivo behaviour of
liposomes. The effect of negative charge has been reported for
large liposomes several years ago by pioneer investigators of the
in-vivo fate of liposomes.[33,34] Also the effect of incorporation
of an asialoglycoprotein (galactose terminal residues) or a galacto-
lipid has been described by three groups of investigators.[35-38]
Two of those groups, using multilamellar liposomes, reported, in
addition to an increased rate of blood elimination, an increase in
hepatic uptake and ascribed this to enhanced uptake by the hepatocytes;[35,37] for many years hepatocytes have been known to possess
a cell-surface receptor specific for terminal non-reducing galactose
residues.[32]

    We also investigated hepatic uptake and intrahepatic distribu-
tion of lactosylceramide-containing small unilamellar liposomes following intraveneous injection into rats. For budgetary reasons (the
glycolipid is very expensive) we lowered the injected dose to 2 μmol
total lipid per 100 g body weight. Obviously, at this dose, the
elimination rate from blood is considerably faster: $T_{1/2}$ = 32 min
for control vesicles and 12 min for lactosyleceramide-containing
vesicles. Fig. 9 shows details of the clearance curves. Because

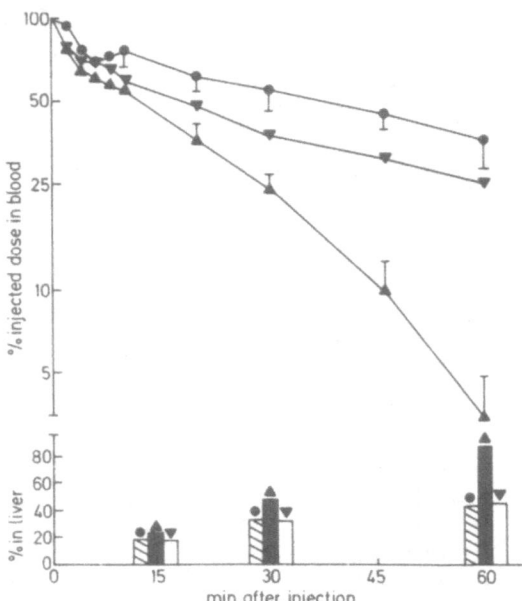

Fig. 9. Elimination of lactosylceramide vesicles from
         blood and uptake by liver: effect of asialo-
         fetuin. Rats were injected intracardially with
         ^{3}H-inulin-containing liposomes (2μmol total
         lipid per 100 g body weight). Radioactivities
         were measured in blood (top; curves) or in liver
         homogenates (bottom; bars). Points represent
         means for at least two determinations and bars,
         where presented, indicate the standard devia-
         tions. ●—● , control vesicles; ▲-▲ lactosyl-
         ceramide vesicles; ▼—▼, lactosylceramide ves-
         icles injected into animals which had received
         10 mg of asialofetuin per 100 g body weight 1
         min prior to administration of the vesicles.

in these experiments we made use of an animal model which allowed us
to draw frequent blood samples without disturbing the animal,[39] we
were able to see a phenomenon which may be the expression of the
existence of the "latent compartment" discussed earlier. During the
first 15 min after injection there is a "dip" in radioactivity content of the blood, suggesting that a fraction of the liposomes,
after a rapid initial removal from the blood comes back in circulation a few minutes later, following which they are truly eliminated.
Fig. 9 also shows that desialylated fetuin, a glycoprotein exposing

Table 8. Intrahepatic cellular distribution of control
and lactosylceramide vesicles containing ^{3}H-
inulin after I.V. injection.

One h after injection of control or lactosylceramide (LC) vesicles (2 μmol
of lipid per 100 g body weight; composition phosphatidylcholine:cholesterol:
phosphatidylserine (4:5:1) with or without 0.8 mol % LC), whole liver, parenchymal cells or non-parenchymal cells were analyzed for radioactivity content.
Uptake is expressed as percentage of injected dose recovered in total liver
and total cell fractions, assuming 450 x 10^6 parenchymal cells and 194 x 10^6
non-parenchymal cells per 100 g of rat.

Liposomes	Liposome-encapsulated ^{3}H-inulin uptake in		
	Total liver	Parenchymal cells	Non-parenchymal cells
Control-SUV	41.2 ± 6.3 (n=4)	20.3 ± 3.0 (n=3)	19.5 ± 3.9 (n=3)
LC-SUV	84.7 ± 3.7 (n=6)	48.0 ± 2.7 (n=4)	27.0 ± 5.8 (n=3)

non-reducing terminal galactose residues, effectively retards the
blood elimination rate of the lactosylceramide vesicles to the level
of the control vesicles. This supports the assumption that the rapid
elimination of the lactosylceramide vesicles involves the galactose
receptor. The bottom part of Fig. 9 represents the total liver
uptake values of the two types of liposomes at three times after
injection. The specific involvement of the liver in the elimination
of the glycolipid-containing vesicles becomes more pronounced with
time. At 1 hour after injection the incorporation of the glycolipid
has resulted in a total liver uptake of over 80% of the injected dose.
The increment in hepatic uptake due to the lactosylceramide is in all
cases virtually cancelled by the presence of the asialofetuin.

Upon isolation of parenchymal or non-parenchymal cell fractions
from livers of rats injected with control or lactosylceramide liposomes, we found that the increase in hepatic uptake as a result of
the presence of the glycolipid was almost fully accounted for by an
enhanced uptake by the hepatocytes (Table 8). Whereas the average
uptake in the hepatocyte fraction increased very significantly from
20 to almost 50% of the injected dose, the increase in the non-
parenchymal cell fraction amounted only to an insignificant 7%. In
two separate experiments we demonstrated that, also for the lactosylceramide vesicles, the radioactivity in the non-parenchymal cell
fraction was fully accounted for by the uptake in the Kupffer cells;
no significant amounts of radioactivity were recovered in the endothelial cells, in line with out earlier observations on large liposomes.[3]

The increase in average Kupffer cell uptake as a result of
lactosylceramide incorporation (Table 8), although not statistically
significant, raises the question to what extent Kupffer cell uptake
of such vesicles is accomplished by way of the galactose-specific
receptor that recently has been discovered on the surface of these
cells.[40],[41] In order to obtain more decisive evidence of the possible participation of such a receptor it would be desirable to have
available a type of liposome which, without any specific ligand,
would display a relatively low affinity for Kupffer cells. Small
increases in Kupffer cell uptake, due to ligand incorporation, would
thus be more reliably detectable. Despite the ease with which small
liposomes supposedly have access, through the 100-200 nm fenestrations,[42] to the hepatocytes and the observed lower uptake rates by
Kupffer cells in-vitro (see earlier), our control SUV still ended up

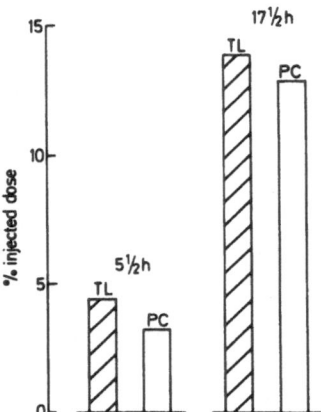

Fig. 10. Preferential uptake of sphingomyelin/
          cholesterol vesicles by liver parenchymal
          cells. Two rats were injected with SM/Chol
          (1:1) SUV containing (^{3}H)-inulin at a dose
          $10\,\mu$mol of total lipid per 100 g body weight.
          5.5 and 17.5 hours after injection the livers
          were perfused with collagenase. Radioactivity
          was determined in the total cell suspension,
          representing whole liver, and in the hepato-
          cyte fractions isolated from it. Assuming 450
          x 10^6 hepatocytes to be present per 100 g rat,
          the amount of radioactivity in the total hepa-
          tocyte population was calculated. Results are
          expressed as percent of injected dose found
          in whole liver and total hepatocyte population.
          Hatched bars, total liver (TL); open bars,
          parenchymal cells (PC).

in Kupffer cells to about equal extents as in hepatocytes (see Table 8). This means that, on a per cell basis, the Kupffer cells are still far more active in internalizing these small vesicles than the parenchymal cells. It occurred to us that the very long half-life of the cholesterol/sphingomyelin liposomes as presented in Fig. 8 and as also observed by other groups[43,44] may be the result of a relative lack of affinity of the Kupffer cells for such vesicles. Indeed, we found 80-90% of the total hepatic liposomal radioactivity associated with the parenchymal cell fraction (Fig. 10). By incorporating lactosylceramide into those vesicles we observed, in addition to a decreased half-life in blood, a considerable increase in the relative amount of radioactivity in the non-parenchymal cell fraction. These results are compatible with the existence on the Kupffer cells of a galactose-specific receptor which is able to recognize the sugar moiety of the liposomal lactosylceramide.

To conclude this final section we want to put forward that, thus far, we have presented little, if any, evidence that the liposomal material which we find associated with the parenchymal cells is actually internalized by these cells. Because of the small encapsulated volume of SUV and because any liposomal material taken up by the parenchymal cells is vastly diluted in the large mass of liver cells it is virtually impossible to obtain solid morphological evidence. Although from in-vitro work [45,46] evidence of intracellular uptake of liposomes by hepatocytes is available, particularly for the contents of large multilamellar vesicles, no such information is at hand for in-vivo work with small or large vesicles.

Table 9. Effect of in-vivo chloroquine treatment of
rats on intracellular conversion of phospholipids in hepatocytes after I.V. injection
of sphingomyelin-labeled liposomes.

Rats were injected with SUV (5 μmol of total lipid per 100 g body weight), composed of (^{14}C)choline-labeled sphingomyelin/cholesterol/phosphatidylserine (4:5:1). Pretreatment of animals with chloroquine consisted of three consecutive i.p. injections of 5, 2.5 and 2.5 mg/100 g body weight at 2 h and 1 h before and at 1 h after injection of the liposomes. Control animals received 0.9% NaCl. 2 h after liposome injection, liver parenchymal cells were isolated and analyzed for radioactivity. Aliquots of the cell suspensions were extracted with chloroform/methanol and the lipid extract was fractionated by thin layer chromatography. The sphingomyelin and phosphatidylcholine spots were analyzed for radioactivity.

Treatment	Total lipid uptake (nmol/10^6 cells)	% of total lipid label in sphingomyelin	phosphatidylcholine
None	1.03	14.2	82.7
Chloroquine	0.94	47.2	47.5

Recently, we observed that the conversion of choline-labeled liposomal sphingomyelin into labeled phosphatidylcholine, which takes place in the hepatocytes following intravenous injection of the liposomes, is effectively blocked when the animals are treated with the lysosomotropic agent chloroquine prior to liposome injection (Table 9). This could be taken as evidence that the liposomal sphingomyelin label passes through the lysosomal compartment in the hepatocytes. This can best be understood if one assumes that the liposome is taken up as an entity by way of an endocytic process; if, on the other hand, the sphingomyelin would gain access to the interior of the hepatocyte by way of a lipid exchange mechanism at the cell surface, lysosomal involvement in the conversion to phosphatidylcholine would be much less obvious.

ACKNOWLEDGEMENTS

The work described in this paper could not have been carried out without the enthusiastic and skilful assistence of Bert Dontje, Mieke van Galen, Henriette Morselt, Joke Regts and Jan Wijbenga. The cooperation with Dr. Eddie Wisse and Dr. Caesar Hulstaert and the unidentified contributions to this work by Dr. Harma Ellens, Dr. Jan Wilschut, Dr. Dick Hoekstra, Ger Hartman and Ben Bolscher are gratefully acknowledged. The research on which this paper is based was financially supported by the Netherlands Organization for the Advancement of Pure Research (ZWO) under the auspices of the Foundation of Fundamental Medical Research (FUNGO) and by the Dutch Cancer Foundation, Koningin Wilhelmina Fonds (KWF). We thank Rinske Kuperus for her expeditious help in preparing the manuscript.

REFERENCES

1.   G. Scherphof, F. Roerdink, D. Hoekstra, J. Zborowski and E. Wisse, Stability of liposomes in presence of blood constituents: consequences for uptake of liposomal lipid and entrapped compounds by rat-liver cells, in "Liposomes in Biological Systems" G. Gregoriadis and A.C. Allison, eds., John Wiley, Chichester. (1980).
2.   G. Scherphof, F. Roerdink, J. Dijkstra, H. Ellens, R. de Zanger and E. Wisse, Uptake of liposomes by rat and mouse hepatocytes and Kupffer cells, Biol. Cell. 47:47 (1983).
3.   F. Roerdink, J. Dijkstra, G. Hartman, B. Bolscher and G. Scherphof, The involvement of parenchymal, Kupffer and endothelial liver cells in hepatic uptake of intravenously injected liposomes. Effects of lanthanum and gadolinium salts, Biochim. Biophys. Acta. 677:79 (1981).
4.   F.H. Roerdink, E. Wisse, H.W.M. Morselt, J. van Meulen and G.L. Scherphof, Cellular distribution of intravenously injected protein-containing liposomes in the rat liver, in "Kupffer cells and other liver sinusoidal cells" E. Wisse and D. Knook, eds.,

Elsevier/North-Holland, Amsterdam. (1977).

5. H. Ellens, H.W.M. Morselt, B.H.J. Donje, D. Kalicharan, C.E.
    Hulstaert and G.L. Scherphof, Effects of liposome dose and the
    presence of lymphosarcoma cells on blood clearance and tissue
    distribution of large unilamellar liposomes in mice, Cancer
    Res., 43:2927 (1983).

6. J. van Renswoude and D. Hoekstra, Cell-induced leakage of lipo-
    some contents, Biochemistry. 20:540. (1981).

7. R. Fraley, R.M. Straubinger, G. Rule, E.L. Springer and D.
    Papahadjopoulos, Liposome-mediated delivery of deoxyribonucleic-
    acid to cells: Enhanced efficiency of delivery related to lipid
    composition and incubation conditions, Biochemistry.20:6978
    (1981).

8. L. Krupp, A.V. Chobanian and P.I. Brecher, The in-vivo trans-
    formation of phospholipid vesicles to a particle resembling
    HDL in the rat, Biochem. Biophys. Res. Commun., 72.1251 (1976).

9. G. Scherphof, F. Roerdink, M. Waite, and J. Parks, Disintegra-
    tion of phosphatidylcholine liposomes as a result of inter-
    action with high-density lipoproteins, Biochim. Biophys. Acta,
    542:296 (1978).

10. A.R. Tall and D.M. Small, Solubilisation of phospholipid mem-
    branes by human plasma high-density lipoproteins, Nature.
    265:163 (1977).

11. M.C. Finkelstein and G. Weissmann, Enzyme replacement via lipo-
    somes; variations in lipid composition determine liposomal
    integrity in biological fluids, Biochim. Biophys. Acta 587:202
    (1979).

12. C. Kirby, J. Clarke and G. Gregoriadis, Effect of the chol-
    esterol content of small unilamellar liposomes on their stab-
    ility in vivo and in vitro, Biochem. J. 186:591 (1980).

13. T.M. Allen and L.G. Cleland, Serum-induced leakage of liposome
    contents, Biochim. Biophys. Acta. 597:418 (1980).

14. J. Damen, J. Dijkstra, J. Regts and G. Scherphof, Effect of
    lipoprotein-free plasma on the interaction of human plasma
    high density lipoprotein with egg yolk phosphatidylcholine
    liposomes, Biochim. Biophys. Acta.620:90 (1980).

15. J. Damen, J. Regts and G. Scherphof, Transfer and exchange of
    phospholipid between small unilamellar liposomes and rat plasma
    high density lipoproteins: dependence on cholesterol and phos-
    pholipid composition, Biochim. Biophys. Acta. 665:538 (1981).

16. G. Scherphof, B. van Leeuwen, J. Wilschut and J. Damen,
    Exchange of phosphatidylcholine between small unilamellar lipo-
    somes and human high density lipoprotein exclusively involves
    the phospholipid in the outer monolayer of the liposomal mem-
    brane, Biochim. Biophys. Acta. 732:595 (1983).

17. J.C. Wilschut, J. Regts and G. Scherphof, Action of phosolipase
    A$_2$ on phospholipid vesicles: Preservation of the membrane perm-
    eability barrier during asymmetric bilayer degradation, FEBS
    Lett. 98:181 (1979).

18. J.C. Wilschut, Preparation and properties of phospholipid ves-

icles, in "Liposome Methodology", D. Leserman and J. Barbet, eds., Inserm, Paris, (1982).

19. G.L. Scherphof, J. Damen and J. Wilschut, Interactions of liposomes with plasma proteins in "Liposome Technology", G. Gregoriadis, ed., CRC Press, Boca Raton (1984).

20. C. Kirby and G. Gregoriadis, Plasma-induced release of solutes from small unilamellar liposomes is associated with pore formation in the bilayers, Biochem J. 199:251 (1981).

21. J.C. Wilschut, J. Regts, H. Westenberg and G. Scherphof, Action of phospholipases A_2 on phosphatidylcholine bilayers. Effect of the phase transition, bilayer curvature and structural defects, Biochim. Biophys. Acta.508:185 (1978).

22. J. Damen, J. Regts and G. Scherphof, Transfer of ^{14}C phosphatidylcholine between liposomes and human plasma high density lipoprotein: partial purification of a transfer stimulating plasma factor using a rapid transfer assay, Biochim. Biophys. Acta. 712:444 (1982).

23. A.R. Tall, L.R. Forester and G.L. Bongiovanni, Facilitation of phosphatidylcholine transfer into high density lipoproteins by an apolipoprotein in the density 1.20-1.26 g/ml fraction of plasma J. Lipid Res. 24:277 (1983).

24. J. Dijkstra, W.J.M. van Galen, F.H. Roerdink, D. Regts and G.L. Scherphof, Uptake of liposomes by Kupffer cells in vitro, in "Sinusoidal Liver Cells", D.L. Knook and E. Wisse, eds., Elsevier Biomedical Press, Amsterdam, (1982).

25. J. Dijkstra, W.J.M. van Galen, C.E. Hulstaert, D. Kalicharan, F.H. Roerdink and G.L. Scherphof, Interaction of liposomes with Kupffer cells in vitro, Exp. Cell Res. 150:161 (1984).

26. D.J. Reijngoud, P.S. Oud, J. Kas and J.M. Tager, Relationship between medium pH and that of the lysosomal matrix as studied by two independent methods, Biochim Biophys. Acta. 448:290 (1976).

27. S. Ohkuma and B. Poole, Fluorescence probe measurement of the intralysosomal pH in living cells and the perturbation of pH by various agents, Proc. Natl. Acad. Sci. USA, 75:3327 (1978).

28. P.O. Seglen, B. Grinde and A.E. Solheim, Inhibition of the lysosomal pathway of protein degradation in isolated rat hepatocytes by ammonia, methylamine, chloroquine and leupeptin, Eur. J. Biochem. 95:215 (1979).

29. R.L. Juliano and D. Stamp, The effect of particle size and charge on the clearance rates of liposomes and liposome encapsulated drugs, Biochem Biophys. Res. Commun. 63:651 (1975).

30. G.H. Hinkle, G.S. Born, W.V. Kessler and S.M. Shaw, Preferential localization of radiolabeled liposomes in liver, J. Pharm. Sci. 67:795 (1978).

31. E. Wisse, An electron microscopic study of the fenestrated endothelial lining of rat liver sinusoids, J. Ultrastruct. 31:125 (1970).

32. G. Ashwell and A.G. Morell, The role of surface carbohydrates in the hepatic recognition and transport of circulating glyco-

proteins, Adv. Enzymol. 41:99 (1974).

33. G. Gregoriadis and D.E. Neerunjun, Control of the fate of hepatic uptake and catabolism of liposome-entrapped proteins injected into rats. Possible therapeutic applications, Eur. J. Biochem. 47:179 (1974).

34. L.D. Steger and R.J. Desnick, Enzyme therapy VI. Comparative in-vivo fates and effects on lysosomal integrity of enzyme entrapped in negatively and positively charged liposomes, Biochim. Biophys. Acta. 464:530 (1977).

35. G. Gregoriadis and D.E. Neerunjun, Homing of liposomes to target cells, Biochem. Biophys. Res. Commun. 65:537 (1975).

36. P. Ghosh and B.K. Bachhawat, Grafting of different glycosides on the surface of liposomes and its effect on the tissue distribution of ^{125}I-labeled γ-globulin encapsulated in liposomes, Biochim. Biophys. Acta. 632:562 (1980).

37. P. Ghosh, P.K. Das and B.K. Bachhawat, Targeting of liposomes towards different cell types of rat liver through the involvement of liposomal surface glycosides, Arch. Biochem. Biophys. 213:266 (1982).

38. F.C. Szoka and E. Mayhew, Alteration of liposome disposition in-vivo by bilayer situated carbohydrates, Biochem. Biophys. Res. Commun. 110:140 (1983).

39. H. Spanjer and G. Scherphof, Targeting of lactosylceramide containing liposomes to hepatocytes in-vivo, Biochim. Biophys. Acta. 734:40 (1983).

40. V. Kolb-Bachofen, J. Schlepper-Schäffer, W. Vogell and H. Kolb, Electron microscopic evidence for an asialoglycoprotein receptor on Kupffer cells: localization of lectin-mediated endocytosis, Cell. 29:859 (1982).

41. E. Müller, M.W. Franco and R. Schauer, Involvement of membrane galactose in the in-vivo and in-vitro sequestration of desialylated erythrocytes, Hoppe Seyler's Z. Physiol. Chem.362:1615. (1981).

42. E. Wisse, R. de Zanger and R. Jacobs, Lobular gradients in endothelial fenestrations and sinusoidal diameter favor centrolubular exchnage processes: a scanning EM study in "Sinusoidal Liver Cells" D.L. Knook and E. Wisse, eds., Elsevier Biomedical Press, Amsterdam, (1982).

43. M.R. Mauk, R.C. Gamble and J.D. Baldeschwieler, Vesicle targeting:timed release and specificity for leukocytes in mice by subcutaneous injection, Science 207:309 (1980).

44. G. Gregoriadis, C. Kirby, P. Large, A. Meehan and J. Senior, Targeting of liposomes: study of influencing factors, in "Targeting of Drugs", G. Gregoriadis, J. Senior and A. Trouet, eds., Plenum, New York, (1982).

45. D. Hoekstra, R. Tomasini and G. Scherphof, Interaction of phospholipid vesicles with rat hepatocytes in primary monolayer culture, Biochim. Biophys. Acta. 542:456 (1978).

46. D. Hoekstra, and G. Scherphof, Effect of foetal calf serum and serum protein fractions on the uptake of liposomal phosphatidylcholine by rat hepatocytes in primary monolayer culture, Biochim. Biophys. Acta. 551:109 (1979).

ENDOCYTOSIS OF LIPOSOMES AND INTRACELLULAR FATE OF ENCAPSULATED

MOLECULES: STRATEGIES FOR ENHANCED CYTOPLASMIC DELIVERY

Robert M. Straubinger, Keelung Hong, Daniel S. Friend,[+]
Nejat Duzgunes and Demetrios Papahadjopoulos

Departments of Pharmacology and [+]Pathology
Cancer Research Institute, M-1282
University of California San Francisco
San Francisco, California, 94143

INTRODUCTION

Liposomes have evoked considerable interest as carriers of macromolecules, based on their ability to encapsulate a wide variety of biologically active materials and deliver them, functionally intact, to particular in-vivo or in-vitro compartments. Liposomes are adaptable to the requirements of a wide range of experimental conditions, since it is possible to alter such parameters as liposome size, surface charge, bilayer fluidity, and stability. Recently, several groups have shown that it is possible to couple covalently to the liposome surface a variety of ligands, such as immunoglobulins, to promote specificity of liposome-cell interaction and to increase the number of liposomes bound to or internalized by cells (Heath et al., 1980; Leserman et al., 1980; Heath et al., 1983).

In order to exploit the carrier potential of liposomes, it has been necessary to develop a clearer understanding of the mechanism of liposome-cell interaction and the intracellular fate of liposome-entrapped molecules. Early studies implicated liposome-plasma membrane fusion (Weinstein et al., 1977; Poste and Papahadjopoulos, 1978; Pagano et al., 1978; Kimelberg and Mayhew, 1978; Pagano and Weinstein, 1978; Poste, 1980), but recent evidence (Szoka et al., 1979, 1980a; Struck et al., 1981) suggests that liposomes do not fuse with the plasma membrane without perturbations such as polyethylene glycol treatment (Szoka et al., 1981) or the inclusion of viral proteins thought to promote membrane fusion (Uchida et al., 1979; Volsky et al., 1979; Huang et al.,

1980; Vainstein et al., 1981). Endocytosis also has been advocated as the dominant mechanism of delivery (Kimelberg and Mayhew, 1978; Finkelstein and Weissman, 1978) but has not been regarded as a useful route by which liposomes could enter non-phagocytic cells or by which large or labile molecules could gain access to the cytoplasm.

In our earlier studies (Wilson et al., 1979; Fraley et al., 1980, 1981) we quantified liposome-cell association, cell-mediated leakage, and other parameters which alter the efficiency of delivery of biologically functional macromolecules. These studies indicated that the negative charge promotes avid binding to cells, and a high mole percent of cholesterol reduces cell-mediated leakage of lipo-some contents (Fraley et al., 1981). For the present study, negatively-charged liposomes were produced by the reverse-phase evaporation (REV) method (Szoka and Papahadjopoulos, 1978) and subsequently extruded through controlled-pore polycarbonate mem-branes (0.2 μm diameter) to assure a more homogeneous size distrib-ution (Szoka et al., 1980b). Such vesicles have been found to be superior to positively-charged or neutral liposomes in functional delivery of cytotoxic drugs (T. Heath, in preparation) and macro-molecules such as DNA and RNA to a variety of animal (Fraley et al., 1981; Schaefer-Ridder et al., 1982) and plant cells (Fraley et al., 1982).

In our system (Fraley et al., 1982), it was found that under certain conditions, chloroquine (CLQ) or NH_4Cl increased the functional intracellular delivery of liposome-encapsulated Simian Virus 40 (SV 40) DNA to African Green Monkey kidney (CV-1) cells. These weakly basic compounds raise the pH of acidic intracellular compartments (de Duve et al., 1974; Ohkuma and Poole, 1978), and our results suggested that such lysosomotropic agents enhanced SV 40 expression by inhibiting degradation of liposome-delivered DNA by lysosomal nucleases (de Duve et al., 1962; Arsenis et al., 1970). In addition, it was found that the pathway for liposome-mediated DNA delivery involved an energy-dependent step, since the metabolic inhibitors sodium azide and 2-deoxyglucose reduced DNA expression under the same conditions in which lysosomotropic agents enhanced expression (Fraley et al., 1982).

In order to examine further the role of the endocytic path-way in determining the intracellular disposition of liposome-encapsulated molecules, we have recently examined the fate of fluorescent and electron-dense probes encapsulated within the liposome aqueous space (Hong et al., 1983; Straubinger et al., 1983). Two of the fluorescent compounds are related structurally. The first, 5(6)-carboxyfluorescein (CF), is a trivalent anion at neutral pH which becomes electrically neutral at acidic pH (pK at 6.7, 4.4 (Heiple and Taylor, 1982), and approximately 3.5). The second compound, calcein (CAL; bis-(N,N'-di-(carboxymethyl) aminomethyl)fluorescein; Wallach and Steck, 1963) is more strongly charged than CF as a result of two methylimino diacetic acid residues (carboxyl pK < 4; methylimino pK 10-12; Wallach et al., 1959). Hence CAL contains multiple charged groups in the pH range encountered intracellularly (pH > 4.6; de Duve et al., 1974; Ohkuma and Poole, 1978). CF has been used in several studies as a marker for liposome delivery to cells (Blumenthal et al., 1977; Weinstein et al., 1977; Hagins and Yoshikami, 1978; Szoka et al., 1979), and CAL has been used previously to quantify serum-mediated liposome leakage (Allen and Cleland, 1980). A third compound, fluorescein isothiocyanate dextran (FTC-D) has an average molecular weight of 20K dalton and was selected because it is a large, membrane-impermeant macromolecule which cells do not degrade readily. For ultrastructural studies, we devised a method to load liposomes with colloidal gold (Hong et al., 1983) to overcome the problem of resolving intracellular liposomes from other vesicular structures.

EXTRA- AND INTRACELLULAR DISTRIBUTION OF FLUORESCENT LIPOSOME

MARKERS

When CV-1 cells were exposed for 30 min (37°C) to phosphatidyl-serine;cholesterol (PS:Chol; 2:1 mole ratio) liposomes containing CF and then washed free of unbound liposomes, we observed both vesicular (punctate) and diffuse cell-associated fluorescence (Fig. 1a). Diffuse fluorescence represents CF free in the cyto-plasm (Weinstein et al., 1977), while vesicular fluorescence represents dye in surface-bound and intracellular liposomes.

Cells treated with liposomes containing CAL (Fig. 1c) or FTC-D (not shown) displayed only vesicular fluorescence, suggesting that these molecules gain access to the cytoplasm less readily than CF. It is unlikely that the difference between CF and CAL or FTC-D distribution can be explained by differential liposome-cell binding and internalization, since liposome-encapsulated molecules are not expected to influence the process of liposome-cell interaction (Fraley et al., 1981).

PROPERTIES OF THE ENCAPSULATED FLUORESCENT PROBES

In order to investigate the mechanism behind the differential CF and CAL localization, we compared the pH dependence of efflux of the two dyes from liposomes by dialyzing parallel preparations against isotonic buffers over a pH range of 4 to 7.5. Fig. 2 shows comparable CF and CAL leakage near neutral pH, where both molecules bear multiple negative charges. Below pH 5.5, the pH-dependence of CF leakage increased drastically, and at pH 4.5, the cumulative leakage was 17 times greater than that at neutrality. Over the

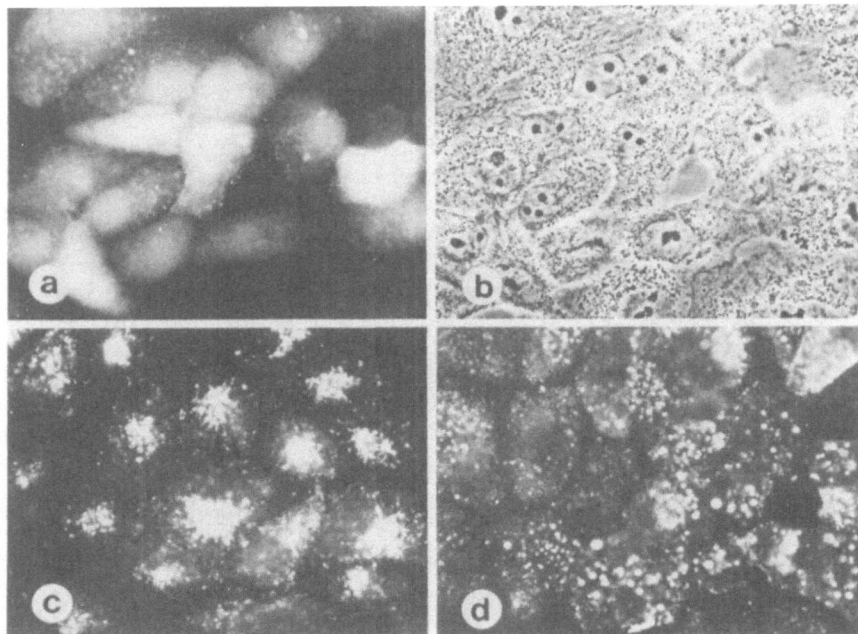

Fig. 1.   Intracellular Localization of Fluorescent Liposome Markers.
CV-1 cells were incubated for 30 min with CF- or CAL-
containing liposomes, washed in buffer to remove free
liposomes, and examined in buffer using a water-immersion
objective (25x). (a) Diffuse (cytoplasmic) and vesicular
fluorescence from CF-containing liposomes. (b) Phase-
contrast image of the field shown in (a). (c) Vesicular
fluorescence from CAL-containing liposomes. (d) Cells
treated with CF-containing liposomes as in (a) but pretreated for 30 min with 10C µM chloroquine. (Reprinted
with permission from Straubinger et al., 1983.)

same range, CAL leakage changed less than a factor of two. Significantly, the difference between CF and CAL leakage was approximately
tenfold at pH 4.6, the estimated lysosomal pH (Ohkuma and Poole,
1978).

From the molecular structures of CF and CAL (inset, Fig. 2),
one can rationalize the markedly different membrane permeability
properties of the two compounds. CF is a small (376 dalton) acidic
molecule that is partially protonated at pH 4 to 5. The fully protonated species of CF would be electrically neutral and uncharged.
CAL bears a resemblance to CF, but would be expected to be a more
acidic molecule than CF, and even the electrically neutral species
would bear several charged groups. Therefore, at the pH expected

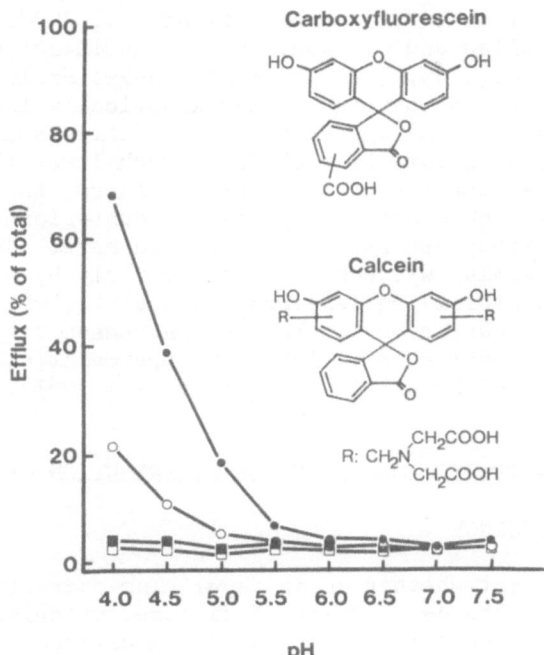

Fig. 2.   pH Dependence of CF and CAL Efflux from Liposomes.
           PS:Chol liposomes (2:1) containing either CF or CAL were
           dialyzed against isotonic citrate-phosphate buffer at the
           indicated pH. The data are presented as the cumulative
           efflux over 8 h (CF: open circles; CAL: open squares) and
           22 h (CF: filled circles; CAL: filled squares). Inset:
           Molecular structures of CF and CAL. (Reprinted with per-
           mission from Straubinger et al., 1983.)

for the endocytic pathway (4.6-5.5: de Duve et al., 1974; Ohkuma
and Poole, 1978), one would anticipate a greater rate of efflux
for CF, the uncharged, neutral form of which would be expected to
have a comparatively higher lipid solubility than the zwitterionic
(or negatively charged) form of CAL.

    Although dye efflux in this simple model system (Fig. 2) was
relatively slow (over 8 to 22 hours), and diffuse cytoplasmic CF
fluorescence (Fig. 1a) was visible within an hour, it is probable
that the model does represent the situation responsible for differential intracellular CF and CAL distribution. First, slow efflux
from a small number of endocytosed liposomes could result in
observable cytoplasmic fluorescence, given the relatively large
aqueous volume of the liposomes used in the present studies.
Second, temperature (Szoka et al., 1979), a variety of proteins

(Papahadjopoulos et al., 1973; Scherphof et al., 1978; Gregoriadis and Davis, 1979; Allen and Cleland, 1980), and intact cell membranes (Wilson et al., 1979; Szoka et al., 1979; Fraley et al., 1981; Van Renswoude and Hoekstra, 1981) drastically increase the rate of efflux of liposome contents. Thus, dialysis in a simple buffer at 19°C gives a rate of dye efflux much lower than would be expected at 37°C in the presence of cells. Third, the endocytic pathway may contain more membrane-permeant counterions than supplied by the citrate-phosphate buffer used here. Indeed, we observed that CF efflux was enhanced considerably by modification of the incubation medium; buffers such as acetate, which supply a membrane-permeant counterion to protons, increased the rate of CF efflux if substituted for the relatively impermeant citrate-phosphate buffer used here (Straubinger et al., 1983).

EFFECT OF LYSOSOMOTROPIC AGENTS ON THE DISTRIBUTION OF CELL-

ASSOCIATED FLUORESCENCE

In order to test whether an intracellular compartment of low pH is required for the development of diffuse, cytoplasmic CF fluorescence, cells were treated with 100 μM chloroquine or 10 mM NH4Cl before addition of CF-containing liposomes. Either compound inhibited the emergence of diffuse CF fluorescence, and cells showed only vesicular fluorescence (Fig. 1d; compare with Fig. 1a), confirming that CF transfer to the cytoplasm was dependent on low pH in the endosomal compartment.

EFFECT OF GLYCEROL TREATMENT

As we reported previously (Fraley et al., 1981), CV-1 cells may be treated briefly with glycerol after liposome-cell interaction, resulting in a substantial increase both in the infectivity of encapsulated SV 40 DNA and in diffuse, cytoplasmic CF fluorescence. More recently (Straubinger et al., 1983), we found that treatment of the cells with chloroquine or NH4Cl inhibited the emergence of diffuse CF fluorescence in glycerol-treated cells. As in the case of chloroquine treatment without glycerol, only vesicular fluorescence was apparent. Since one of the steps in the glycerol-dependent pathway involves an acidic, chloroquine-sensitive compartment, these results suggest that glycerol, which is known to cause non-specific internalization of plasma membrane (Norberg, 1970), enhances intracellular delivery of encapsulated molecules by promoting increased endocytosis of liposomes, rather than by permeabilization of cell membranes.

Neither encapsulated CAL nor FTC-D gave rise to diffuse fluorescence following glycerol treatment (not shown). Since these probes probably would cross membranes damaged by osmotic shock (Okada and Rechsteiner, 1982), this result is further evidence that major disruption of cellular membranes is not the mechanism by which glycerol enhances intracellular delivery of liposome contents.

EFFECT OF METABOLIC INHIBITORS

The development of diffuse, cytoplasmic CF fluorescence was inhibited in the presence of metabolic poisons. Five mM sodium azide and 50 mM 2-deoxyglucose prevented diffuse CF fluorescence, either with or without glycerol treatment. CF-, CAL-, or FTC-D-containing liposomes appeared as dimly fluorescent vesicles which covered cells uniformly, although some cells also displayed large fluorescent aggregates, especially at the cell periphery. These aggregates appeared to be in the process of being shed from the cell surface (not shown).

INTRACELLULAR FATE OF LIPOSOME CONTENTS

Twelve to 24 hours after cells were treated with CF-containing liposomes, most vesicular and diffuse fluorescence had disappeared. However, large fluorescent vesicles persisted in the perinuclear region of cells which were continuously exposed to chloroquine or NH_4Cl, suggesting that preventing restoration of the acidic endosomal environment reduced the rate at which CF diffused to the cytoplasm and ultimately to the extracellular medium.

In contrast, intracellular (vesicular) CAL fluorescence persisted in the absence of lysosomotropic agents. Likewise, FTC-D accumulated in the perinuclear region and dextran fluorescence could be observed over three days of continued culture. Thus water soluble molecules which do not become significantly more membrane-permeant at low pH (either because of charged groups or large size) do not escape the endocytic pathway at an appreciable rate.

ULTRASTRUCTURAL STUDIES: PREPARATION OF COLLOIDAL GOLD-CONTAINING LIPOSOMES

At the ultrastructural level, it has been difficult to trace unambiguously the path of liposome internalization for lack of a suitable marker. Ferritin and horseradish peroxidase have been used to follow liposome fate in-vitro (Magee et al., 1974; Weissmann et al., 1977; Wu et al., 1981). However, the endogenous peroxidase activity of many tissues, as well as the natural occurrence of intracellular ferritin complicates the use of such markers. Colloidal gold particles are an attractive histological marker,

owing to their high electron density, uniform size and shape, and
versatility (Faulk and Taylor, 1971; Geoghegan and Ackerman, 1977;
Handley et al., 1981). However, colloidal gold is unsuitable for
encapsulation in liposomes, as it precipitates at the high concentrations necessary to load a large fraction of a liposome population.
Recently we developed a method for the preparation of liposomes containing gold granules (Hong et al., 1983) and demonstrated the
utility of gold-containing liposomes as histochemical markers of
liposome uptake in cells (Straubinger et al., 1983).

For each preparation of gold-liposomes, gold chloride-citrate
solution was freshly prepared by adding chloroauric acid (12.72 mM
$HAuCl_4$ in distilled water) to a basic citrate solution (13.6 mM
trisodium citrate, 3.33 mM K_2CO_3) immediately before preparation
of the liposomes. The final gold chloride-citrate solution consisted
of 3.18 mM HAuCl , 2.5 mM K_2CO_3, 10.2 mM trisodium citrate and the
pH was maintained at 6.0-6.2 to avoid premature nucleation of gold
sols during liposome preparation.

The gold chloride-citrate solution was encapsulated in vesicles
prepared by the reverse-phase evaporation method (Szoka and Papahadjopoulos, 1978) with minor modifications. Specifically, 10 μmol
phospholipid was dissolved in 1 ml of freshly water-washed diethyl
ether. Gold chloride-citrate solution (0.5 ml) was added, and the
mixture emulsified by sonication for 3 min at 25°C in a bath-type
sonicator. Ether was removed under reduced pressure at 30°C, and
the resulting vesicles were extruded through polycarbonate membranes
of 0.2 μm pore diameter (Olson et al., 1979) to ensure a narrow size
distribution. Gold sols were formed by incubating the liposome
suspension at 37°C for 0.5-1 h.

Unentrapped gold granules were removed by passing the liposome
suspension through a Sephacryl S-1000 column (1 cm x 10 cm) which
was pre-equilibrated with buffer (5 mM N-2-hydroxyethylpiperazine-
N'-2-ethanesulfonic acid, 100 mM NaCl, pH 7.0). To avoid loss of
liposomes on the Sephacryl gel, the column was presaturated with
liposomes prepared with buffer instead of gold chloride. A short
DEAE cellulose column (1 x 4 cm) may be used as an alternative
method to separate the gold-liposomes from free or adherent gold
sols; since gold sols carry a net negative charge in water, they
bind strongly to DEAE cellulose. The DEAE cellulose column was
also pretreated with buffer-loaded liposomes to avoid loss of
anionic phospholipid on the DEAE cellulose. The recovery of liposomes from either column was high with more than 70% of the phospholipid recovered. Gold sols can be quantified by measuring absorbance at 525 nm after the liposomes are dissolved in 1% Triton X-100.

By negative-stain electron microscopy, gold sols were of
uniform size (approximately 15 nm diameter) and shape, and entrapped
inside 80-90% of the liposomes. By comparison, less than 10% of

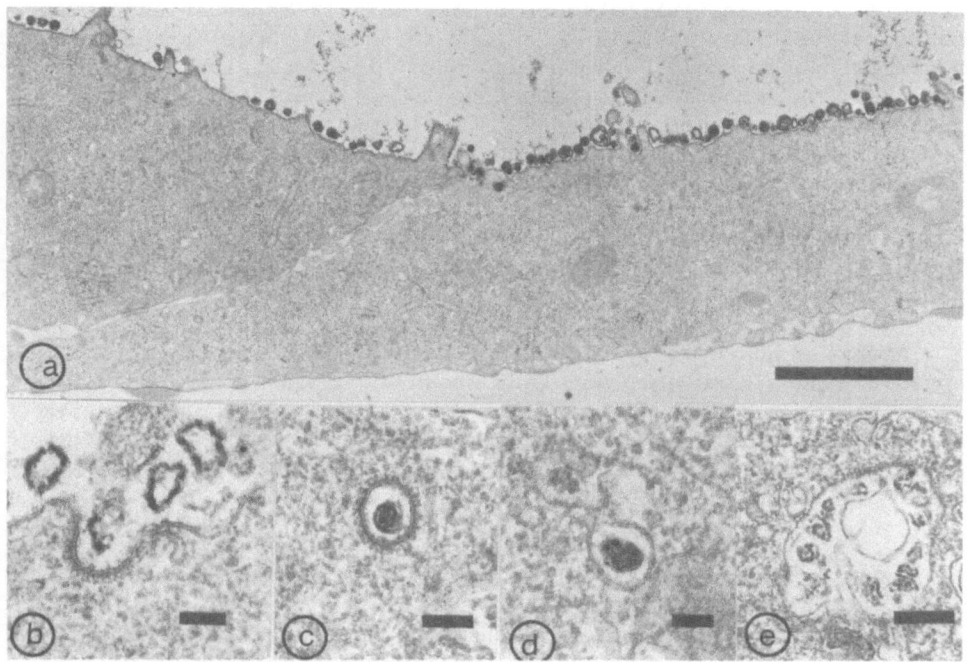

Fig. 3. Ultrastructural Localization of Gold-containing Liposomes
in vitro. Cells were treated with gold-containing liposomes for 30 min at 37°C and fixed. (a) Liposomes bound
to the apical surface of CV-1 cells (bar: 1.0 μM).
(b) Liposomes in coated pits (bar: 0.1 μ m). (c) Liposome
in coated vesicle (bar: 0.2 μm). (d) Liposome in smooth
(coatless) vesicle; gold is visible in liposome (bar: 0.1
μm). (e) Colloidal gold and liposomes accumulated in a
large endocytic vacuole (endosome) (bar: 0.2 μm).

the liposomes contained gold when prepared in the presence of preformed, monodispersed gold colloid (Frens, 1973).

ULTRASTRUCTURAL EVIDENCE FOR LIPOSOME ENDOCYTOSIS IN VITRO

To resolve the pathway of endocytosis, CV-1 cells were fixed
after incubation with colloidal gold-containing liposomes at 4°C
or at 37°C (Straubinger et al., 1983). Under conditions in which
endocytosis is inhibited, liposomes were observed to be bound to
the plasma membrane but were not internalized (not shown). The
diameter of surface-bound liposomes ranged from 0.08 to 0.25 μ m.
At 37°C liposomes were observed to be bound to the plasma membrane
(Fig. 3a) and individual liposomes were visible in coated pits on
the cell surface (Fig. 3b) as well as in coated vesicles (Fig. 3c).
Liposomes or gold particles were not found in the coatless micro-

endocytic vesicles implicated in pinocytosis or in other endocytic transport vesicles. Single gold-containing liposomes appeared to progress through small, smooth (uncoated) intracellular vesicles (Fig. 3d), and groups of liposomes could be identified in larger endosomes or lysosomes (Fig. 3e). Significantly, intracellular liposomes were generally 0.1 μm or less in diameter, a smaller mean diameter than those bound to the cell surface (Straubinger et al., 1983). This observation suggested that optimal liposome-mediated intracellular delivery may depend on producing liposomes of a size less than the maximum diameter which the coated vesicle system can accommodate.

After 14 hours at 37°C, surface-bound liposomes were not observed on the plasma membrane (not shown). Liposomes and gold were identified in vesicles in the Golgi region, and many gold granules without distinguishable liposome membranes could be found in dense bodies (secondary lysosomes). It is probable that extensive metabolism of the endocytosed liposomes had occurred over 14 hours.

ULTRASTRUCTURAL EVIDENCE FOR LIPOSOME ENDOCYTOSIS IN VIVO

Preliminary results suggest that the coated vesicle pathway is a mechanism by which cells take up liposomes in vivo as well as in-vitro. We examined tissues from mice two minutes after an intravenous injection of colloidal gold-containing liposomes. In the liver, which is the major organ responsible for clearance of liposomes of the type used in these studies (Gregoriadis and Ryman, 1972; reviewed in Mayhew and Papahadjopoulos, 1983), we observed gold-containing liposomes on the plasma membrane and within sinusoidal lining cells. Fig. 4a shows liposomes bound to the cell surface and in small endocytic vesicles within the cell. Fig. 4b shows gold-containing liposomes within a large Kupffer-cell endosome. Apparently a large fraction of the injected liposomes were stable and retained gold particles during two minutes of circulation. We are currently expanding these studies to examine liposome uptake in several other tissues, particularly in the lung and spleen.

STRATEGIES FOR IMPROVING CYTOPLASMIC DELIVERY

From these observations, it appears that liposomes are endocytosed by coated pits, similarly to many macromolecules with specific surface-bound receptors, such as low density lipoprotein (LDL), transferrin, α-2 macroglobulin, and others (Goldstein et al., 1979; Pastan and Willingham, 1981). Furthermore, the intracellular fate of endocytosed liposomes is also similar to that of some macromolecules which enter cells by the same pathway: they progress from coated pits to coated vesicles, to uncoated vesicles, and to large vesicles that are endosomes or lysosomes. Eventually, most liposomes and their contents can be found in the trans-Golgi region, either in dense bodies (secondary lysosomes), or in other vesicles.

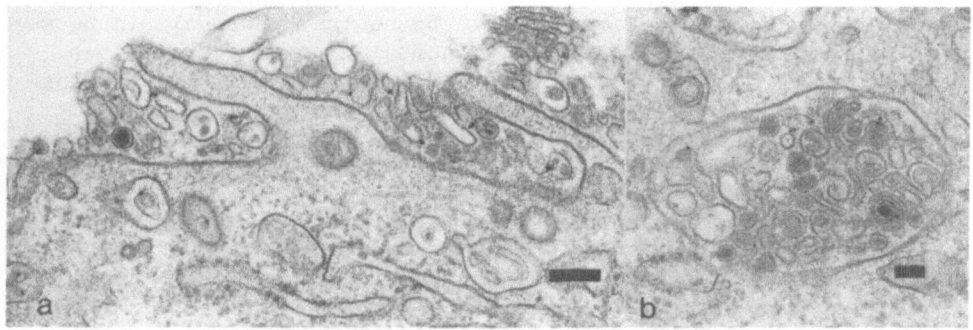

Fig. 4.   Ultrastructural Localization of Gold-containing Liposomes
in vivo. Liposomes were injected into the tail vein of
mice; animals were sacrificed after 2 min and fragments
of the liver were fixed. (a) Gold-containing liposomes
bound to the surface of a Kupffer cell and within cytoplasmic vesicles (bar: 0.2 μm); (b) Liposomes and gold
within an endosomal or lysosomal vacuole (bar: 0.1 μm).

It is clear now that liposome-plasma membrane fusion is an infrequent event, and that only certain types of liposome-encapsulated materials can escape with high frequency from intracellular vacuoles. Therefore, the next logical step in improving liposome-mediated delivery to the cytoplasm is to enhance the escape of encapsulated molecules from the endocytic pathway. One strategy is to design liposomes which respond to acidification by undergoing a structural change such as bilayer destabilization or membrane fusion. Several lipid enveloped viruses, including Semliki Forest, Vesicular Stomatitis, and Influenza viruses, carry glycoproteins which appear to allow endocytosed virions to fuse with cellular membranes at low pH (Helenius et al., 1980; Miller and Lenard, 1980; White et al., 1981). Since it is somewhat cumbersome to isolate and reconstitute the specific viral proteins in liposomes, a more attractive approach was to examine pure lipid systems. Presently we are studying the properties of liposomes composed of the neutral phospholipid phosphatidylethanolamine (PE) as the principal liposome constituent. Protonation of its primary amino group (pK > 8.0; Papahadjopoulos, 1968) makes pure PE unable to form stable bilayer vesicles at neutral pH, and the addition of PE to liposomes enhances membrane fusion in model systems (Duzgunes et al., 1981). To enable formation of stable PE-rich liposomes near neutral pH, we chose a negatively charged amphiphile, oleic acid (OA; cis 18:1). Since the pK of OA in phospholipid membranes is close to neutral pH (von Tscharner and Radda, 1981), protonation of OA at a pH within the range expected for the endocytic pathway (pH > 4.6; de Duve, 1974; Ohkuma and Poole, 1978) should result in liposome destabilization, collapse, and/or membrane-membrane fusion. The probes for intracellular liposome-mediated delivery are CAL and FTC-D, molecules which are not delivered to the cytoplasm by liposome compositions studied previously.

CHARACTERIZATION OF pH-SENSITIVE LIPOSOMES

We evaluated the stability of OA:PE liposomes (3:7 mol ratio) as a function of pH. Control liposomes were composed of PS:PE and OA:phosphatidylcholine (PC). Calcein was captured in liposomes at 60 mM, a concentration at which fluorescence is self-quenched;

release of the dye and dilution into the external medium relieves
quenching and results in an increase in dye fluorescence. Aliquots
of liposomes containing calcein were injected into isotonic (295 ±
5 mOsm/kg) saline buffered with 10 mM citrate and 10 mM phosphate
over the pH range of 4.0 to 8.0. When CAL release from PS:PE was
compared with that from OA:PE liposomes, there were striking differences in liposome stability as a function of pH. PS:PE liposomes
showed no pH-dependent leakage; the baseline fluorescence detected
resulted from incomplete quenching of the liposome-entrapped dye.
In contrast, pH-dependent release from OA:PE liposomes was observed.
At pH 7.0 or above, CAL leakage was negligible. Below pH 7, the
rate of leakage increased with decreasing pH; at pH 5 nearly 90%
had leaked within 4 minutes. It is also clear from additional experiments that OA and PE together were required for the pH dependent
leakage, since PS or PC liposomes containing OA were not pH-sensitive.

INTRACELLULAR DELIVERY OF LIPOSOME-ENTRAPPED MOLECULES

In preliminary experiments, liposomes composed of OA:PE (3:7
mol ratio) gave rise to both vesicular and diffuse intracellular
CAL fluorescence (not shown). The intensity of diffuse fluorescence,
as well as the proportion of cells displaying such fluorescence, was
observed to vary with the confluence of CV-1 monolayers. In confluent cell monolayers, a high proportion of cells (35-70%) showed
unambiguous diffuse CAL fluorescence. Glycerol treatment enhanced
considerably the intensity of diffuse cytoplasmic CAL fluorescence
in cells treated with OA:PE liposomes. The emergence of diffuse
fluorescence was energy dependent, both in the presence or absence
of glycerol treatment; treatment of cells with 5 mM azide and 50 mM
2-deoxyglucose inhibited cytoplasmic CAL fluorescence and only
vesicular fluorescence was observed (not shown).

Fluoresceinated dextran (FTC-D; avg. 18K daltons) also was
encapsulated in liposomes composed of OA and PE, in order to determine if such liposomes would deliver large water-soluble macromolecules to cells. In general, cells showed only vesicular
fluorescence. However, brief glycerol treatment resulted in the
appearance of diffuse, cytoplasmic dextran fluorescence in many

cells. This observation suggests that OA:PE liposomes will be useful for intracellular delivery of large, water-soluble molecules.

SUMMARY

We have presented evidence that endocytosis by the coated vesicle system can be a major route by which liposomes and their contents enter cells in-vitro. Preliminary results suggest that this route is not unique to cells in culture, since coated pit endocytosis is apparently the mechanism of hepatic uptake of liposomes injected intravenously into mice. A number of interesting questions remain unresolved. One is the nature of the liposome binding site and the mechanism by which liposomes become localized in coated pits. It is unclear whether negatively-charged liposomes bind fortuitously to a specific plasma membrane receptor, whether liposomes bind nonspecifically to coated pits, or whether liposome binding constitutes a signal for coated pit formation and internalization.

We are currently attempting to exploit characteristics of the endocytic pathway to enhance cytoplasmic delivery of liposome contents which are otherwise unable to escape from intracellular vesicles. We have used CAL and FTC-D as model compounds for highly charged and large molecules, and have designed liposomes which promote delivery of these compounds to the cytoplasm. It is anticipated that this approach may be combined eventually with developing methods for ligand-mediated targeting to specific cell types, thus extending the range of biologically active agents which may be delivered cytoplasmically.

ACKNOWLEDGEMENTS

We gratefully acknowledge helpful discussions with Dr. Timothy Heath, and thank Dr. Frank Szoka for his generosity in making available the fluorescence microscope. We also thank Ms. Ivy Hsieh and Ms. Ninfa Lopez for excellent technical assistance. This work was funded by grants GM 26369, CA 25526 (to D.P.) and HD 10445 (to D.S.F.) from the National Institutes of Health.

REFERENCES

Allen, T.M. and Cleland, L.G., 1980, Serum-induced leakage of liposome contents, Biochim. Biophys. Acta, 597:418.
Arsenis, C., Gordon, J. and Touster, O., 1970, Degradation of nucleic acids by lysosomal extracts of rat liver and Ehrlich ascites tumor cells, J. Biol. Chem., 245:205.
Blumenthal, R., Weinstein, J.N., Sharrow, S.O. and Henkart, P.,

1977, Liposome-lymphocyte interaction: saturable sites for transfer and intracellular release of liposome contents, Proc. Nat. Acad. Sci. USA, 75:5603.

de Duve, C., Wattiaux, R. and Baudhuin, P., 1962, Distribution of enzymes between subcellular fractions in animal tissues, Adv. Enzymol., 24:291.

de Duve, C., de Barsy, T., Poole, B., Trouet, A., Tulkens, P. and van Hoof, F., 1974, Lysosomotropic agents, Biochem. Pharmacol., 23:2495.

Duzgunes, N., Wilschut, J., Fraley, R.T. and Papahadjopoulos, D., 1981, Studies on the mechanisms of membrane fusion: Role of headgroup composition on calcium- and magnesium-induced fusion of mixed phospholipid vesicles, Biochim. Biophys. Acta, 642:182.

Faulk, W.P. and Taylor, G.M., 1971, An immunocolloid method for the electron microscope, Immunochemistry, 8:1081.

Finkelstein, M. and Weissman, G., 1978, The introduction of enzymes into cells by means of liposomes, J. Lipid Res., 18:289.

Fraley, R., Delaporta, S. and Papahadjopoulos, D., 1982, Liposome-mediated delivery of TMV RNA into tobacco protoplasts: a sensitive assay for monitoring liposome-protoplast interactions, Proc. Nat. Acad. Sci. USA, 79:1859.

Fraley, R., Straubinger, R., Rule, G., Springer, L. and Papahadjopoulos, D., 1981, Liposome-mediated delivery of DNA to cells: enhanced efficiency of delivery by changes in lipid composition and incubation conditions, Biochemistry, 20:6978.

Fraley, R., Subramani, S., Berg, P. and Papahadjopoulos, D., 1980, Introduction of liposome-encapsulated SV40 DNA into cells, J. Biol. Chem., 255:10431.

Frens, G., 1973, Controlled nucleation for the regulation of the particle size in monodisperse gold suspensions, Nature, 241:20.

Geoghegan, W.D. and Ackerman, G.A., 1977, Adsorption of horseradish peroxidase, ovomucoid and anti-immunoglobulin to colloidal gold for the indirect detection of concanavalin A, wheat germ agglutinin and goat anti-human immunoglobulin G on cell surfaces at the electron microscopic level: a new method, theory and application, J. Histochem. Cytochem., 25:1187.

Goldstein, J., Anderson, R.G.W. and Brown, S., 1979, Coated pits, coated vesicles and receptor-mediated endocytosis, Nature, 279:679.

Gregoriadis, G. and Davis, C., 1979, Stability of liposomes in-vivo and in-vitro is promoted by their cholesterol content and the presence of blood cells, Biochem. Biophys. Res. Commun., 89:1287.

Gregoriadis, G. and Ryman, B.E., 1972, Fate of protein-containing liposomes injected into rats, Eur. J. Biochem., 24:485.

Hagins, W.A. and Yoshikami, S., 1978, Intracellular transmission of visual excitation in vertebrate retinal photoreceptors:

electrical effects of chelating agents introduced into rods
by vesicle fusion, in: "Vertebrate Photoreceptors", Fatt, P.
and Barlow, H.B., eds., Academic Press, New York.

Handley, D.A., Arbeeny, C.M., Witte, L.D. and Chien, S., 1981,
Colloidal gold-low density lipoprotein conjugates as membrane
receptor probes, Proc. Natl. Acad. Sci. USA, 78:368.

Heath, T., Fraley, R. and Papahadjopoulos, D., 1980, Antibody targeting of liposomes: cell specificity obtained by conjugation
of F(ab')$_2$ to the vesicle surface, Science, 210:539.

Heath, T., Montgomery, J.A., Piper, J.R. and Papahadjopoulos, D.,
1983, Antibody-targeted liposomes: increase in specific
cytotoxicity by a liposome dependent drug, Proc. Natl. Acad.
USA, 80:1377.

Heiple, J.M. and Taylor, D.L., 1982, An optical technique for
measurement of intracellular pH in single living cells,
in: "Intracellular pH: Its Measurement, Regulation, and
Utilization in Cellular Functions", Nuccitelli, R. and
Deamer, D.W., eds., A.R. Liss, New York.

Helenius, A., Kartenbeck, J., Simons, K. and Fries, E., 1980, On
the entry of Semliki Forest virus into BHK-21 cells, J. Cell
Biol., 84:404.

Hong, K., Friend, D.S., Glabe, C.G. and Papahadjopoulos, D.P., 1983,
Liposomes containing colloidal gold are a useful probe of
liposome-cell interactions, Biochim. Biophys. Acta, 732:320.

Huang, R.T.C., Wahn, K., Klenk, H.-D. and Rott, R., 1980, Fusion
between cell membranes and liposomes containing the glycoproteins of influenza virus, Virology, 104:294.

Kimelberg, H.K. and Mayhew, E.G., 1978, Properties and biological
effects of liposomes and their uses in pharmacology and
toxicology, CRC Crit. Rev. Toxicol., 6:25.

Leserman, L., Barbet, J., Kourilsky, R. and Weinstein, J., 1980,
Targeting to cells of fluorescent liposomes covalently
coupled with monoclonal antibody or protein A, Nature, 288:
602.

Magee, W.E., Goff, C.W., Schoknecht, J., Smith, M.D. and Cherian,
K., 1974, The interaction of cationic liposomes containing
entrapped HRP with cells in culture, J. Cell Biol., 63:492.

Mayhew, E. and Papahadjopoulos, D., 1983, Use of liposomes as a
drug carrier system in-vivo, in: "Liposomes", Ostro, M.J.,
ed., Marcel Dekker, New York.

Miller, D. and Lenard, J., 1980, Inhibition of vesicular stomatitis
virus infection by spike glycoprotein, J. Cell Biol., 84:430.

Norberg, B., 1970, Amoeboid movements and cytoplasmic fragmentation
of glycerinated leucocytes induced by ATP, Exp. Cell Res.,
59:11.

Ohkuma, S. and Poole, B., 1978, Fluorescence probe measurement of
the intralysosomal pH in living cells and the perturbation
by various agents, Proc. Natl. Acad. Sci. USA, 75:3327.

Okada, C.Y. and Rechsteiner, M., 1982, Introduction of macromolecules into cultured mammalian cells by osmotic lysis of

pinocytic vesicles, Cell, 29:33.

Olson, F., Hunt, C., Szoka, F., Vail, W. and Papahadjopoulos, D., 1979, Preparation of liposomes of defined size distribution by extrusion through polycarbonate membranes, Biochim. Biophys. Acta, 557:9.

Pagano, R.E., Sandra, A. and Takeichi, M., 1978, Interactions of phospholipid vesicles with mammalian cells, Ann. N.Y. Acad. Sci., 308:185.

Pagano, R.E. and Weinstein, J.N., 1978, Interaction of liposomes with mammalian cells, Ann. Rev. Biophys. Bioeng., 7:435.

Papahadjopoulos, D., 1968, Surface properties of acidic phospholipids: interaction of monolayers and hydrated liquid-crystals with uni- and bi-valent metal ions, Biochim. Biophys. Acta, 163:240.

Papahadjopoulos, D., Cowden, M. and Kimelberg, H., 1973, Role of cholesterol in membranes: effects on phospholipid-protein interactions, membrane permeability and enzymatic activity, Biochim. Biophys. Acta, 330:8.

Pastan, I.H. and Willingham, M.C., 1981, Receptor-mediated endocytosis of hormones in cultured cells, Ann. Rev. Physiol., 43:239.

Poste, G., 1980, The interaction of lipid vesicles (liposomes) with cultured cells and their use as carriers for drugs and macromolecules, in: "Liposomes in Biological Systems", Gregoriadis, G. and Allison, A.C., eds., Wiley & Sons, Chichester.

Poste, G. and Papahadjopoulos, D., 1978, The influence of vesicle membrane properties on the interaction of lipid vesicles with cultured cells, Ann. N.Y. Acad. Sci., 308:164.

Ralston, E., Hjelmeland, L.M., Klausner, R.D., Weinstein, J.N. and Blumenthal, R., 1981, Carboxyfluorescein as a probe for liposome-cell interactions: Effect of impurities and purification of the dye, Biochim. Biophys. Acta, 649:133.

Schaefer-Ridder, M., Yang, Y. and Hoffschneider, P.H., 1982, Liposomes as gene carriers: efficient transformation of mouse L cells by thymidine kinase gene, Science, 215:166.

Scherphof, G., Roerdink, F., Waite, M. and Parks, J., 1978, Disintegration of phosphatidylcholine liposomes in plasma as the result of interaction with high-density lipoproteins, Biochim. Biophys. Acta, 296:296.

Straubinger, R.M., Hong, K., Friend, D.S. and Papahadjopoulos, D., 1983, Endocytosis of liposomes and intracellular fate of encapsulated molecules: Encounter with a low pH compartment after internalization in coated vesicles, Cell, 32:1069.

Struck, D.K., Hoekstra, D. and Pagano, R.E., 1981, Use of resonance energy transfer to monitor membrane fusion, Biochemistry, 20:4093.

Szoka, F.C., Jacobson, K. and Papahadjopoulos, D., 1979, The use of aqueous space markers to determine the mechanism of interaction between phospholipid vesicles and cells, Biochim. Biophys. Acta, 551:295.

Szoka, F.C., Jacobson, K., Derzko, Z. and Papahadjopoulos, D., 1980a,
    Fluorescence studies on the mechanism of liposome-cell inter-
    actions in-vitro, Biochim. Biophys. Acta, 600:1.
Szoka, F., Olson, F., Heath, T., Vail, W., Mayhew, E. and Papahad-
    jopoulos, D., 1980b, Preparation of unilamellar liposomes of
    intermediate size by a combination of reverse phase evapor-
    ation and extrusion through polycarbonate membranes, Biochim.
    Biophys. Acta, 601:559.
Szoka, F., Magnusson, K.-E., Wojcieszyn, J., Hou, Y., Derzko, Z.
    and Jacobson, K., 1981, Use of lectins and polyethylene
    glycol for fusion of glycolipid-containing liposomes with
    eukaryotic cells, Proc. Natl. Acad. Sci. USA, 78:1685.
Szoka, F. and Papahadjopoulos, D., 1978, Procedure for preparing
    liposomes with large internal aqueous space and high capture
    by reverse-phase evaporation, Proc. Natl. Acad. Sci. USA, 75:
    4194.
Uchida, T., Kim, J., Yamaizumi, M., Miyake, Y. and Okada, Y., 1979,
    Reconstitution of lipid vesicles associated with HJV (Sendai
    virus) spikes, J. Cell Biol., 80:10.
Vainstein, A., Atidia, J. and Loyter, A., 1981, Reconstituted
    Sendai virus envelopes as a biological carrier for micro-
    injection of proteins and DNA molecules into animal cells,
    in: "Liposomes, Drugs and Immunocompetent Cell Functions",
    Nicolau, C. and Paraf, A., eds., Academic Press, New York.
Van Renswoude, J. and Hoekstra, D., 1981, Cell-induced leakage of
    liposome contents, Biochemistry, 20:540.
Volsky, D.J., Cabantchik, M., Beigel, M. and Loyter, A., 1979,
    Implantation of the isolated human erythrocyte anion channel
    into plasma membranes of Friend erythroleukemic cells by use
    of Sendai virus envelopes, Proc. Natl. Acad. Sci. USA, 76:5440.
von Tscharner, V. and Radda, G.K., 1981, The effect of fatty acids
    on the surface potential of phospholipid vesicles measured
    by condensed phase radioluminescence, Biochim. Biophys. Acta,
    643:435.
Wallach, D.F.H. and Steck, T.L., 1963, Fluorescence techniques in
    the microdetermination of metals in biological materials,
    Anal. Chem., 35:1035.
Wallach, D.F.H., Surgenor, D.M., Soderberg, J. and Delano, E., 1959,
    Preparation and properties of 3,6-dihydroxy-2,4-bis-(N,N'-
    di-(carboxymethyl)-aminomethyl) fluoran: Utilization for the
    ultramicrodetermination of calcium, Anal. Chem., 31:456.
Weinstein, J.N., Yoshikami, S., Henkart, P., Blumenthal, R. and
    Hagins, W.A., 1977, Liposome-cell interaction: transfer and
    intracellular release of a trapped fluorescent marker,
    Science, 195:489.
Weissmann, G., Cohen, C. and Hoffstein, S., 1977, Introduction of
    enzyme by means of liposomes into non-phagocytic human cells
    in-vitro, Biochim. Biophys. Acta, 498:375.
White, J., Matlin, K. and Helenius, A., 1981, Cell fusion by
    Semliki Forest, influenza, and vesicular stomatitis viruses,
    J. Cell Biol., 89:674.

Wilson, T., Papahadjopoulos, D. and Taber, R., 1979, The introduction
    of poliovirus RNA into cells via lipid vesicles (liposomes)
    Cell, 17:77.
Wu, P., Tin, G.W. and Baldeschwieler, J.D., 1981, Phagocytosis of
    carbohydrate-modified phospholipid vesicles by macrophage,
    Proc. Natl. Acad. Sci. USA, 78:2033.

LIPOSOMES IN LEISHMANIASIS: THE LYSOSOME CONNECTION

Carl R. Alving[1], John S. Weldon[2], John F. Munnell[2]
and William L. Hanson[3]

[1]Department of Membrane Biochemistry
Walter Reed Army Institute of Research
Washington, DC 20307; [2]Department of
Anatomy and Radiology, and [3]Department
of Parasitology, College of Veterinary
Medicine, University of Georgia, Athens
GA 30602

INTRODUCTION

Leishmaniasis is a disease caused by protozoan parasites that live almost exclusively in macrophages in the reticuloendothelial (RE) system. Depending on the species of parasite, the area of the world, and the site of infection, three different clinical presentations.are generally recognized: cutaneous ("oriental sore", "chiclero ulcer"), mucocutaneous ("espundia"), and visceral leishmaniasis ("kala azar"). The visceral disease, which affects RE cells in the liver, spleen, and bone marrow, is the most severe. In its untreated state, over a short or long period of time, visceral leishmaniasis is thought to be virtually 100% fatal. (See Faust et al., 1968 and Lainson and Shaw, 1978 for complete descriptions of the various forms of leishmaniasis.) The disease in all of its forms afflicts as many as 100 million people in tropical and subtropical areas throughout the world, and research on leishmaniasis is included as part of the "Six Diseases Program" of the World Health Organization (along with malaria, schistosomiasis, filariasis, trypanosomiasis, and leprosy).

At the present time there is no widely available vaccine, and

This work was supported in part by United States Army Medical Research and Development Command, Contract DH 17-D-17-C 5011.

the major medical means for controlling the disease is chemotherapy
(Steck, 1974). However treatment of leishmaniasis presents special
therapeutic problems because the drugs that are used have great
potential for toxicity. The toxicity problem was more severe in
the past when trivalent antimonials were used. Trivalent antimonials have considerable cardiotoxicity and have caused many deaths
(Mainzer and Krause, 1940). The pentavalent antimonials, which are
still considered to be experimental drugs in the United States, are
less toxic than trivalents in the dosages used, but are still toxic
(de Lacerda, Jr. et al., 1965). Part of the toxicity of pentavalent
antimonials may derive from conversion to the trivalent form in the
body (Steck, 1974). In our experience with experimental hamster
infections, if rapid death after intracardial injection is used as
a measure of toxicity, and subsequent reduction of hepatic parasite
counts and deaths due to leishmaniasis is used as a measure of
efficacy, then the efficacious dose of pentavalent antimonial drug
approximately equals the toxic dose (Alving et al., 1978a). By
these criteria the therapeutic index, defined as the ratio of toxic
dose/efficacious dose, was therefore only approximately 1.

Because of potential toxicity of pentavalent drugs, doses
given to patients are limited, and treatment failures are common
(Kinnamon et al., 1979). Limited doses of administered drugs might
also be expected to promote emergence of drug-resistant organisms.
Although antimonial drug resistance is usually estimated to be
approximately 5% to 10% of cases, 46% resistance was reported in
one hospital in India in 1975 and 1976 (Sanyal and Arora, 1979).

Several years ago three laboratories independently reported
that liposome-encapsulated antimonial drugs were hundreds of times
(as much as 865 times) better than unencapsulated drugs in treatment
of experimental visceral leishmaniasis (Alving et al., 1978a, b;
Black et al., 1977; New et al., 1978). By the above mentioned
criteria for efficacy and toxicity, based on calculations from
published data (Alving et al., 1978a), under the conditions used
the therapeutic index is estimated to have been approximately 30
to 40. Subsequent investigation has extended knowledge of enhanced
efficacy to other drugs (Alving et al., 1980; New et al., 1981a),
to treatment of chronic infections (Alving, 1982) and extremely
virulent infections (Alving et al., 1984), to treatment of dogs
(Chapman et al., 1984), and to treatment of experimental cutaneous
leishmaniasis (New and Chance, 1980; Chance and New, 1980; New et
al., 1981b, c). Reviews on the various therapeutic developments
have been published by Alving and Steck (1979), and Alving (1984).
Detailed methods for preparation of liposomes for use in leishmaniasis have been described by Alving and Swartz (1984).

The original reason that liposomes were tested for treatment
of leishmaniasis was because of previous reports that indicated
that parenterally-administered liposomes were cleared at least

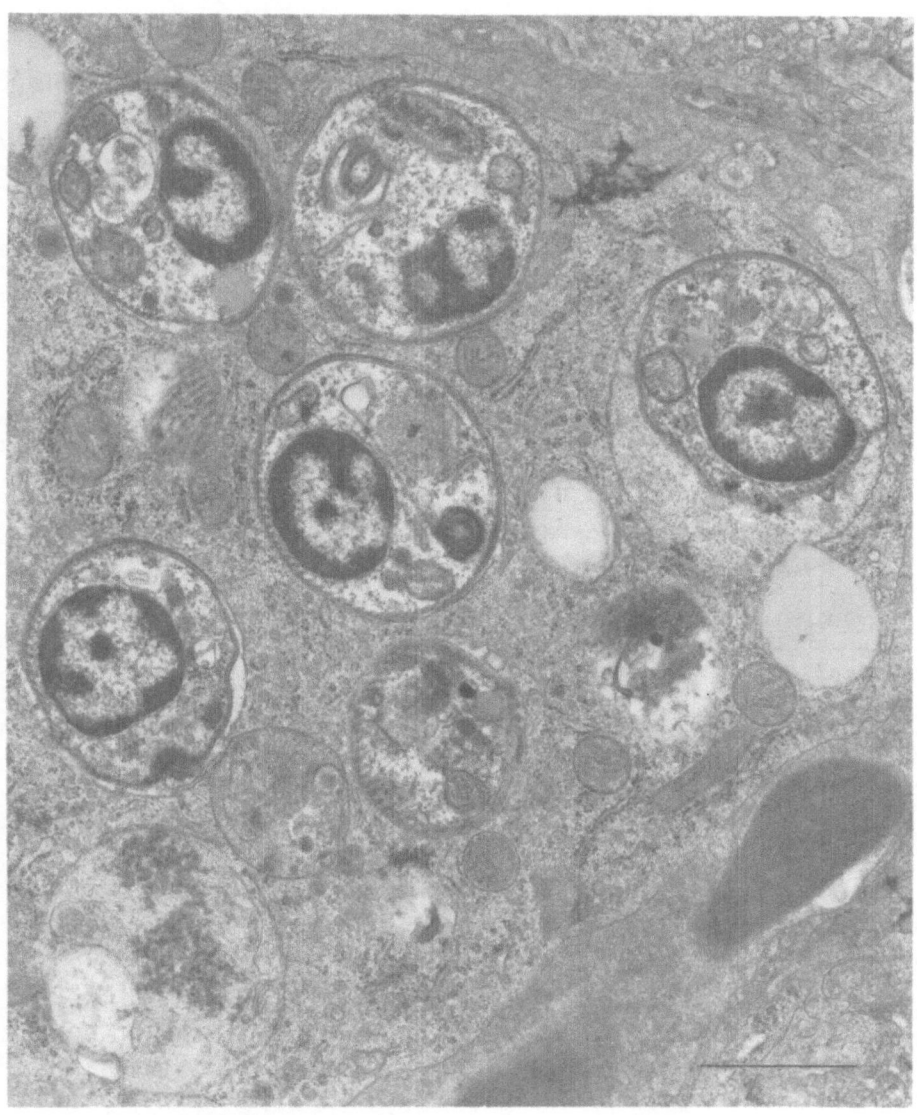

Fig. 1. Section of a hamster Kupffer cell containing at least
         six Leishmanial parasites. The hamster had been in-
         fected 17 days previously with Leishmania donovani and
         had not been injected with liposomes. The membranes
         of the parasitophorous vacuoles can be sometimes readily
         seen. The membranes often are greatly distended with
         lysosomal contents following fusion with a lysosome
         (eg., parasite at far right; also, see Figs. 4 and 5).

partially by cells in the RE system (reviewed by Tyrrell et al., 1976). Based on this, although it was not surprising to us that liposome-encapsulated drugs were more efficacious than unencapsulated drugs, the huge magnitude of the enhanced therapeutic response obtained was unexpected. Recently we have proposed a hypothesis, based on electron microscopic evidence, that part of the increased efficacy of liposomes may be due to delivery of liposomes to the vicinity of the intracellular parasite, and that this is followed by delivery of liposomes to the lysosomes of the parasites themselves (Weldon et al., 1983). This hypothesis is reviewed in this paper and further evidence for it is presented.

STATUS OF INTRACELLULAR LEISHMANIA

Forty nine days after parenteral injection of 10 million Leishmania donovani amastigotes, as many as 5 billion to 300 billion parasites are found in the hamster liver. The parasites appear exclusively in macrophages, and, as shown in Fig. 1, some individual cells may even have numerous organisms. Each of the parasites is enclosed in a so-called parasitophorous vacuole (Alexander and Vickerman, 1975; Chang and Dwyer, 1976) (Fig. 1). The vacuolar membrane has the capacity to fuse with lysosomal membranes, and after doing so the parasite, now bathed in lysosomal contents, remains unharmed (Fig. 1). Indeed, it has even been suggested that lysosomal contents may have a nourishing effect on Leishmania (Alexander and Vickerman, 1975). The ability of leishmanial parasitophorous vacuoles to fuse with lysosomes formed the basis of the original concept that the parasite might be vulnerable to drugs delivered to lysosomes and then to the parasite by liposomes.

STATUS OF INTRACELLULAR LIPOSOMES

Several hours after injection of liposomes, large numbers of vesicles are found in macrophages. The liposomes are found either in phagosomes or in lysosomes. In many instances the liposomes have not suffered any obvious damage or disruption and still retain the slightly electron-dense antimonial drug (Fig. 2). The lysosomes containing liposomes commonly come into close apposition to leishmanial parasites (Alving et al., 1978) (Fig. 3). Instances of apparent fusion of the lysosomal membranes with parasitophorous vacuolar membranes are often observed (Weldon et al., 1983) (Fig. 4). The above interactions result in a situation in which the liposome is actually enclosed, along with other lysosomal material, within the parasitophorous vacuole (Fig. 5). After this point the only barriers between the antimonial drug and the parasite cytoplasm are the gradually decaying liposomal membranes and the parasite plasma membrane (Fig. 6).

It is evident that the liposomes are "lysosomotropic" drug-carrying particles, as defined by de Duve et al. (1974). However,

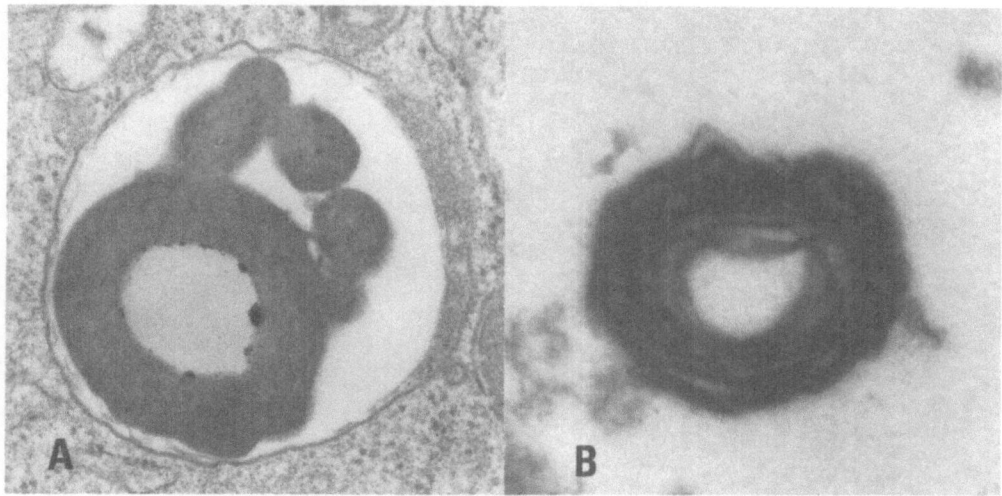

Fig. 2.   Comparison of liposomes containing an antimonial drug
          (A) and empty liposomes containing saline (B) approx-
          imately one hour after intracardial injection into a
          hamster. The drug-loaded liposome (A) had been phago-
          cytosed but showed few signs of degradation. The
          amorphous electron dense material in the liposome was
          frequently seen, but was only observed with liposomes
          containing antimonial drug, and it presumably represented
          the drug itself. The empty liposome (B) was found free
          in a liver sinusoid, and it lacked electron dense
          material. From Weldon et al., 1983.

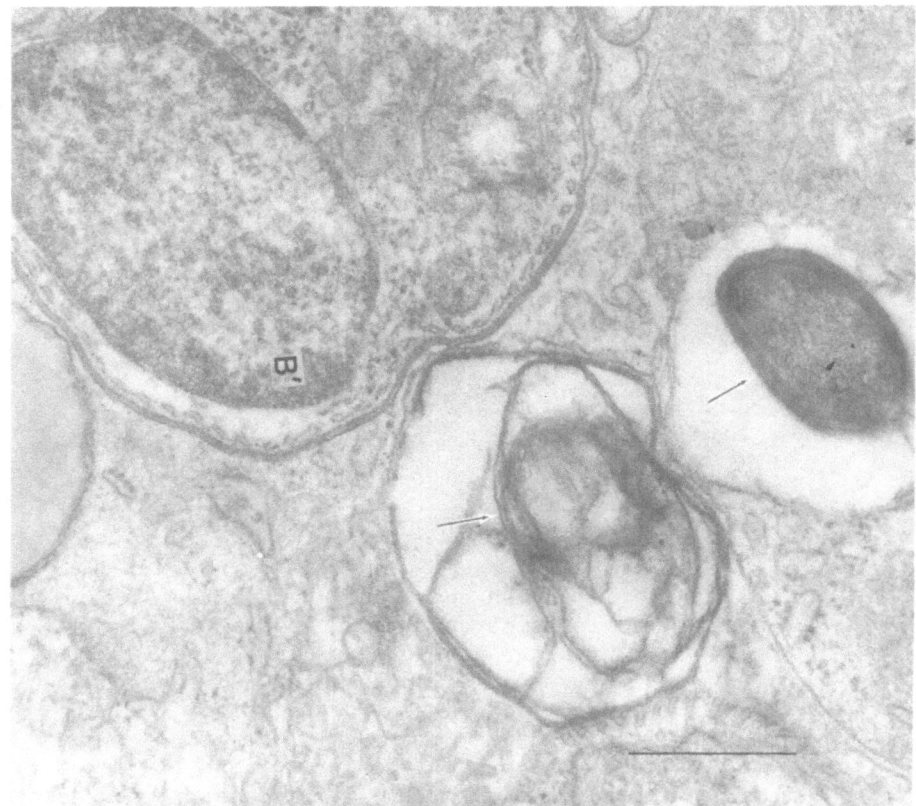

Fig. 3.   Approach of two intracellular vacuoles containing drug-
laden liposomes (arrows) to the immediate vicinity of
a parasite. The lower vacuole is beginning the earliest
stage of interaction with the parasitophorous vacuole.
Bar = 0.5 μm. From Weldon et al., 1983.

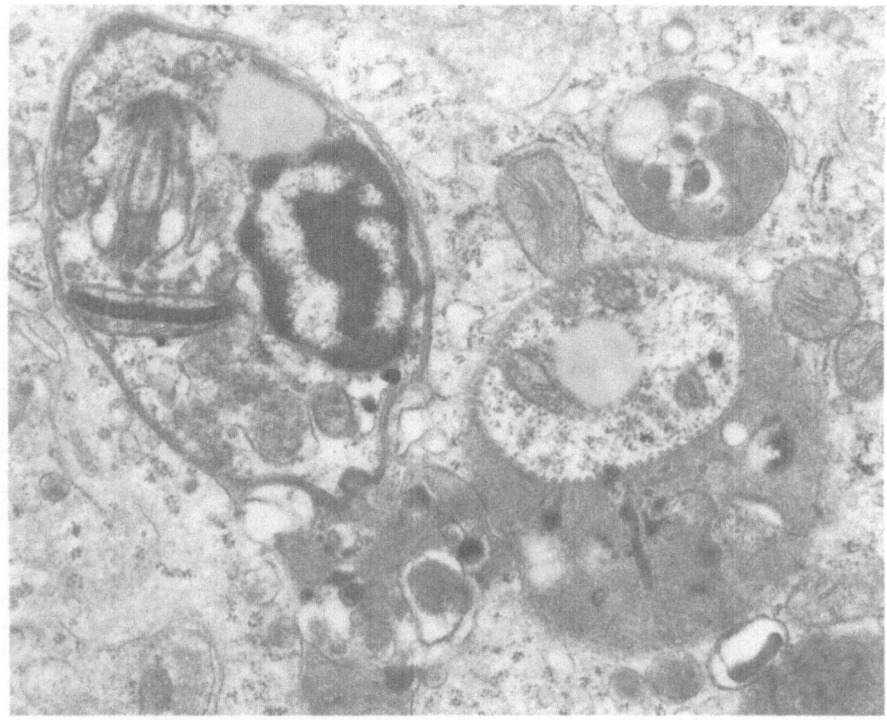

Fig. 4. Apparent fusion of secondary lysosomes with parasito-
         phorous vacuoles. The secondary lysosomes contain a
         variety of lysosomal debris, but in these cases lipo-
         somes were not present.

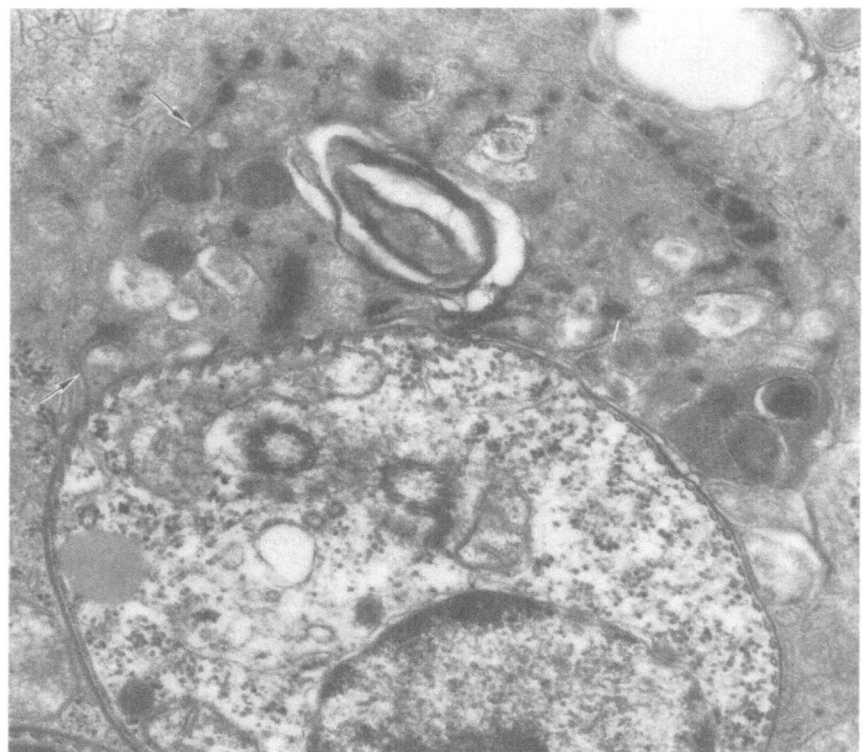

Fig. 5. Drug-laden liposome enclosed within a swollen parasito-
        phorous vacuole. A single vacuolar membrane (arrows)
        surrounds the parasite and lysosomal debris that in-
        cludes a liposome. From Weldon et al., 1983, and
        Alving, 1983. Bar: 0.5μm.

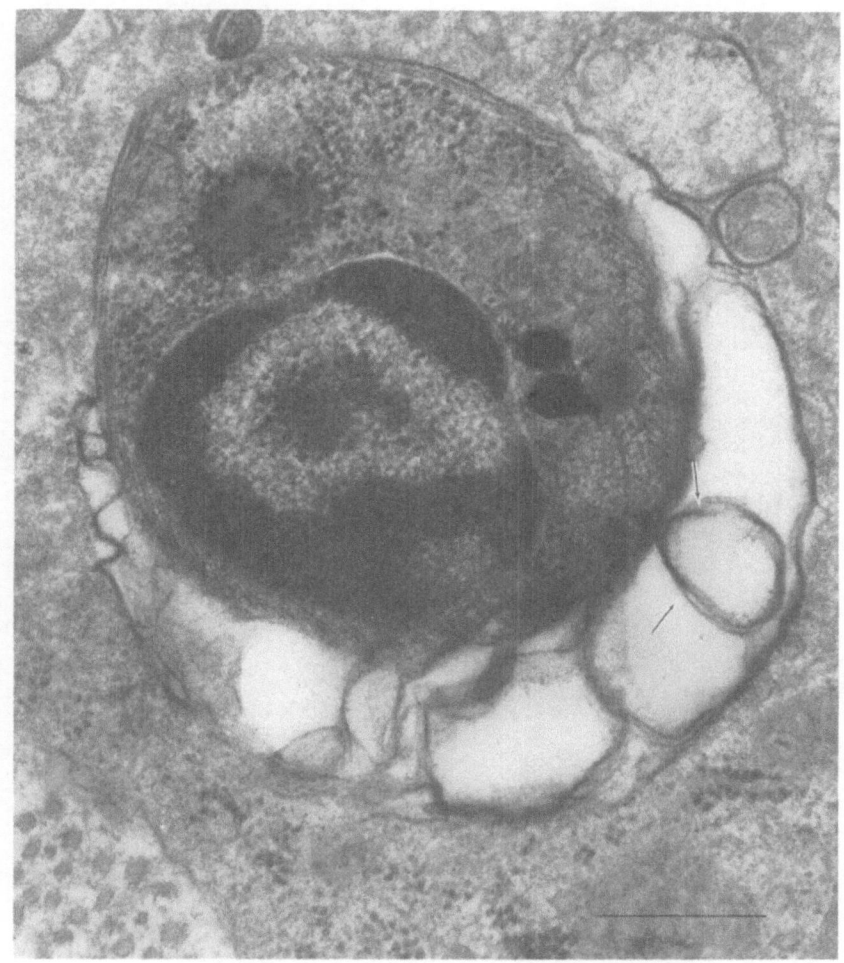

Fig. 6. Close apposition of liposomes with a parasite within
         a parasitophorous vacuole. At least one intact lipo-
         some (arrows), and possibly other liposomes, are
         present within the vacuole. Marker = 0.5 μm.

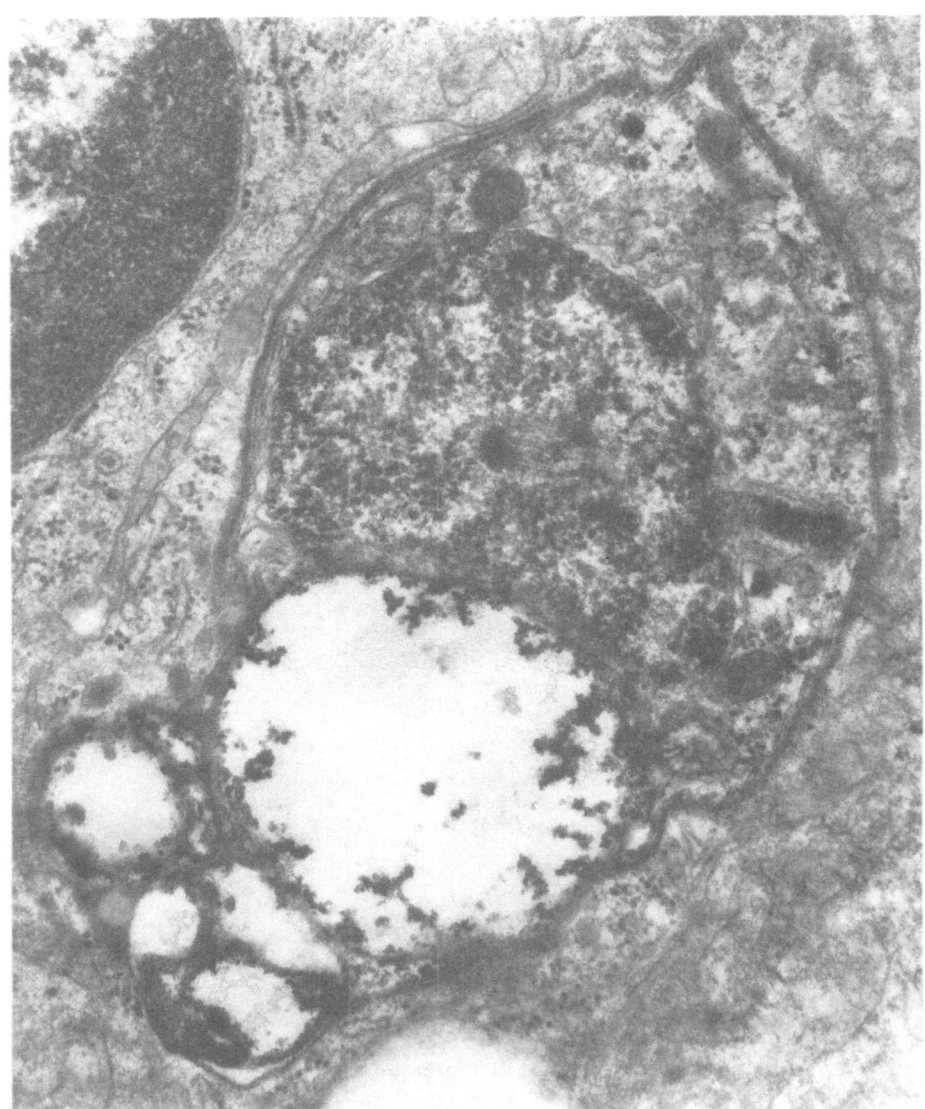

Fig. 7. Dramatic breakup of a Leishmanial parasite in a hamster
Kupffer cell after treatment of the hamster with liposome-encapsulated antimonial drug. Remnants of liposomes are present in the lower right corner.

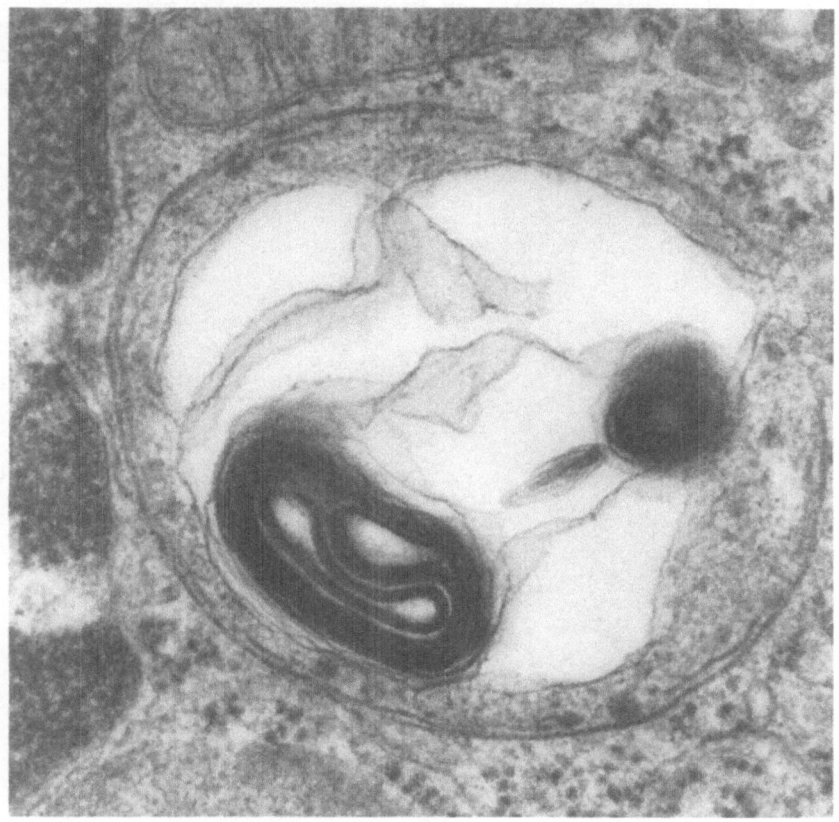

Fig. 8. Two well-defined liposomes situated inside of what
         appears to be an end product of degradation of the
         parasite. The parasite is identified by the remain-
         ing microtubules around its periphery. From Alving,
         1983.

in leishmaniasis the lysosomes themselves also serve as active
carriers for delivering the liposomes to the parasites. Therefore
this process might be viewed as a "lysosomotropic-parasitotropic"
mechanism. Release of antimonial drug in the vicinity of the
parasite results in disintegration of the organism (Fig. 7).

The leishmanial parasite is also known to contain lysosomes of
its own (McAlpine, 1970). Although the penetration of liposomes
into parasites is not as readily observed as ingestion of liposomes
by macrophages, such a process evidently does occur. This final
stage in the intracellular odyssey of liposomes is illustrated in
Fig. 8, and possibly also in Fig. 7.

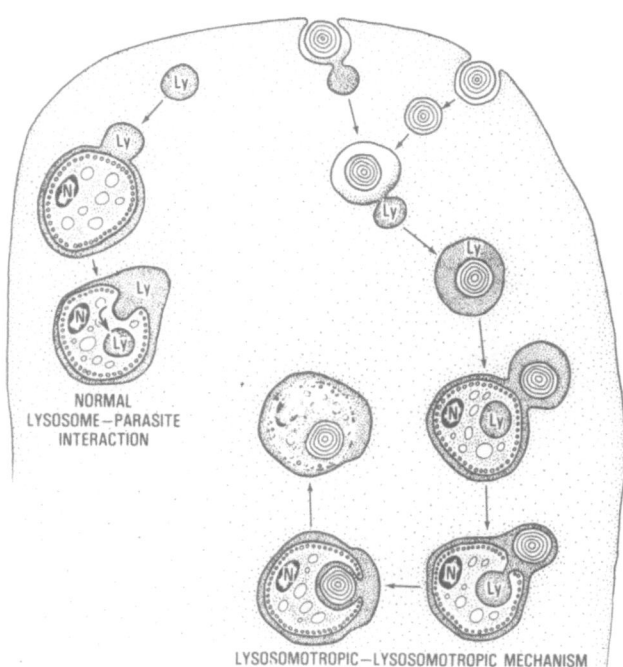

Fig. 9. Proposed lysosomotropic-parasitotropic mechanism for
         delivery of liposome-encapsulated drugs to Leishmania
         within macrophages. The process involves uptake of
         the phagocytosed liposome by a lysosome (Ly), fusion
         of the lysosome with the parasitophorous vacuole, and
         delivery of the liposome to a lysosome within the para-
         site, resulting in disintegration of the parasite.
         Release of liposome-entrapped drug might result in
         parasite death at any step along the way after initial
         ingestion of the liposome by the macrophage. N =
         nucleus.

Because of the close and continuing communications and fusions between parasitophorous vacuoles and lysosomes, liposomes are brought into a steady flow into the immediate vicinity of the parasites, and even within the parasitophorous vacuoles enclosing the parasites. In the most intimate situation the liposomes apparently are even ingested by the parasites and probably enter into the lysosomes within the parasites themselves. By analogy with previously coined terminology we refer to this last step (schematically illustrated in Fig. 9) as a "lysosomotropic-lysosomotropic" mechanism, and we refer to the overall process as a "lysosomotropic-parasitotropic" mechanism of drug delivery.

The leishmanial system is an excellent example in which lysosomes play a vital role in the pathogenesis of the disease, apparently serving as a major carrier of nutrition for the parasite. Utilization of the "lysosomotropic-parasitotropic" carrier pathway as a unique pharmacological tool for "endogenous targeting" of liposomes containing deadly drugs results in "poisoned parasites". The high efficiency and precision of this process, involving liposomes as drug carriers and lysosomes as liposome carriers, and possibly reaching even into the lysosomes of the parasites themselves, probably accounts for the remarkable efficacy of liposome-encapsulated drugs in chemotherapy of leishmaniasis.

REFERENCES

Alexander, J. and Vickerman, K. 1975, Fusion of host cell secondary lysosomes with the parasitophorous vacuoles of Leishmania mexicana-infected macrophages, J. Protozool., 22:502.

Alving, C.R. 1982, Therapeutic potential of liposomes as drug carriers in leishmaniasis, malaria and vaccines. In: "Targeting of Drugs", Gregoriadis, G., Senior, J. and Trouet, A., eds., Plenum, New York.

Alving, C.R. 1984, Delivery of liposome-encapsulated drugs to macrophages, Pharmacol. Therapeutics (in press).

Alving, C.R. and Steck, E.A. 1979, The use of liposome-encapsulated drugs in leishmaniasis, Trends Biochem. Sci., 4:N175.

Alving, C.R. and Swartz, Jr., G.M. 1984, Preparation of liposomes for use as drug carriers in treatment of leishmaniasis, in: "Liposome Technology", Gregoriadis, G., CRC Press, Boca Raton.

Alving, C.R., Steck, E.A., Chapman, Jr., W.L., Waits, V.B., Hendricks, L.D., Swartz, Jr., G.M. and Hanson, W.L., 1978a, Therapy of leishmaniasis: superior efficacies of liposome-encapsulated drugs, Proc. Natl. Acad. Sci. USA, 75:2959.

Alving, C.R., Steck, E.A., Hanson, W.L., Loizeaux, P.S., Chapman, Jr., W.L. and Waits, V.B. 1978b, Improved therapy of experimental leishmaniasis by use of a liposome-encapsulated antimonial drug, Life Sci., 22:1021.

Alving, C.R., Steck, E.A., Chapman, Jr., W.L., Waits, V.B.,
    Hendricks, L.D., Swartz, Jr., G.M. and Hanson, W.L. 1980,
    Liposomes in leishmaniasis: therapeutic effects of antimonial
    drugs, 8-aminoquinolines, and tetracycline, Life Sci., 26:2231.
Alving, C.R., Swartz, Jr., G.M., Chapman, Jr., W.L., Waits, V.B.,
    Hendricks, L.D. and Hanson, W.L. 1984, Liposomes in Leishmani-
    asis: effects of parasite virulence on treatment of experi-
    mental leishmaniasis in hamsters, Ann. Trop. Med. Parasit.
    (in press).
Black, C.D.V., Watson, G.J. and Ward, R.J. 1977, The use of
    Pentostam liposomes in the chemotherapy of experimental
    leishmaniasis, Trans. R. Soc. Trop. Med. Hyg., 71:550.
Chance, M.L. and New, R.R.C. 1980, The use of liposome-entrapped
    drugs in the treatment of experimental cutaneous and visceral
    leishmaniasis, in: "The Host Invader Interplay", H. Van den
    Bossche, ed., Elsevier/North-Holland, Amsterdam.
Chang, K. and Dwyer, D.M. 1976, Multiplication of a human parasite
    (Leishmania donovani) in phagolysosomes of hamster macro-
    phages in-vitro, Science, 193:678.
Chapman, Jr., W.L., Hanson, W.L., Alving, C.R. and Hendricks, L.D.
    1984, Antileishmanial activity of liposome-encapsulated
    meglumine antimoniate in the dog, Am. J. Vet. Res. (in press).
de Duve, C., de Barsy, T., Poole, B., Trouet, A., Tulkens, P. and
    van Hoof, F. 1974, Lysosomotropic agents, Biochem. Pharmacol.,
    23:2495.
de Lacerda, Jr., F.S., Germiniani, H., da Mota, C.C.S. and Baranski,
    M.C. 1965, The electrocardiographic changes produced by treat-
    ment with tri- and pentavalent antimony compounds, Rev. Inst.
    Med. Trop. Sao Paulo, 7:210.
Faust, E.C., Beaver, P.C. and Jung, R.C. 1968, The Leishmania para-
    sites of man, Animal Agents and Vectors of Human Disease
    (3rd edn.) p. 34-64.
Kinnamon, K.E., Loizeaux, P.S., Waits, V.B., Steck, E.A., Hendricks,
    L.D., Chapman, Jr., W.L. and Hanson, W.L. 1979, Leishmaniasis:
    military significance and new hope for treatment, Milit. Med.,
    144:660.
Lainson, R. and Shaw, J.J. 1978, Epidemiology and ecology of leish-
    maniasis in Latin America, Nature, 273:595.
Mainzer, F. and Krause, M. 1940, Changes of the electrocardiogram
    appearing during antimony treatment, Trans. R. Soc. Trop. Med.
    Hyg., 33:405.
McAlpine, J.C. 1970, Electronic cytochemical demonstration of a
    lysosome in Leishmania donovani, Trans. R. Soc. Trop. Med.
    Hyg., 64:822.
New, R.R.C. and Chance, M.L. 1980, Treatment of experimental
    cutaneous leishmaniasis by liposome-entrapped Pentostam,
    Acta Tropica, 37:253.
New, R.R.C., Chance, M.L., Thomas, S.C. and Peters, W. 1978, Anti-
    leishmanial activity of antimonials entrapped in liposomes,
    Nature, 272:55.

New, R.R.C., Chance, M.L. and Heath, S. 1981a, Antileishmanial
    activity of amphotericin and other antifungal agents en-
    trapped in liposomes, J. Antimicrobial. Chemother., 8:371.
New, R.R.C., Chance, M.L. and Heath, S. 1981b, The treatment of
    experimental cutaneous leishmaniasis with liposome-entrapped
    pentostam, Parasitology, 83:519.
New, R.R.C., Chance, M.L. and Critchley, M. 1981c, The distribution
    of radiolabelled drug in animals infected with cutaneous
    leishmaniasis: comparison of free and liposome-bound sodium
    stibogluconate, in: "Radionuclide Imaging in Drug Research",
    Wilson, C.G., Hardy, J.G., Davis, S.S. and Frier, M., eds.,
    Croom & Helm, Ltd.
Sanyal, R.K. and Arora, R.R. 1979, Assessment of drug therapy of
    kala-azar in current epidemic in Bihar, J. Com. Dis., 11:198.
Steck, E.A. 1974, The leishmaniases, Prog. Drug Res., 18:289.
Tyrrell, D.A., Heath, T.D., Colley, C.M. and Ryman, B.E. 1976, New
    aspects of liposomes, Biochim. Biophys. Acta, 457:259.
Ward, R.J., Black, C.D.V. and Watson, G.J. 1980, The determination
    of free and liposome-entrapped antimony in biological samples
    and its application to the chemotherapy of experimental leish-
    maniasis, in: "Drug Measurement and Drug Effects in Laboratory
    Health Science", Siest, G. and Young, D.S., eds., Karger, Basel.
Weldon, J.S., Munnell, J.F., Hanson, W.L. and Alving, C.R. 1983,
    Liposomal chemotherapy in visceral leishmaniasis: an ultra-
    structural study of an intracellular pathway, Z. Parasit.,
    69:415.

THROMBOGENIC POTENTIAL OF PHOSPHATIDYLSERINE-CONTAINING LIPOSOMES

G.B. Humphrey, L.V. Allen, R. Blackstock, P.C. Comp,
L.E. De Bault, C.T. Esmon, A. Harriman, F.A. Holloway,
H.F. Krous, M. Mojarad, H.L. Stone and R. Wierimaa

University of Oklahoma Health Sciences Center
and Oklahoma State University

INTRODUCTION

By the spring of 1981, Fidler and Poste and their co-workers
had demonstrated that monocytes could be activated to become tumoricidal after incubation with liposomes containing monocyte/macrophage
activating agents (Fidler, 1975; Fidler et al., 1976). These same
phosphatidylcholine (PC), phosphatidylserine (PS) liposomes containing muramyl dipeptide (MDP) or one of its derivatives could eradicate
spontaneous metastases in mice (Fidler et al., 1981). Furthermore,
no toxicity is observed when liposomes are administered intravenously
to dogs and mice (Hart et al., 1981).

We were of the opinion that the existing literature (through
1981) had not considered the thrombogenic potential of phosphatidylserine (PS)-containing liposomes, nor whether these liposomes might
cause subtle insults (i.e., microthrombi) to the microcirculation.
After we initially evaluated and demonstrated that PS-containing
multilamellar vesicles (MLVs) could enhance in-vitro coagulation,
we then initiated experiments to indirectly evaluate the effect of
liposome administration upon the microcirculation of the ventral
nervous system, lungs and myocardium.

In the evaluation of the central nervous system using animal
psychology models, sufficient animals were studied to allow statistical evaluation. However, only a single dose of MLVs was
evaluated. In the canine pulmonary lymphatic model and the canine
coronary circulation model which were used to evaluate the heart
and lungs, only four dogs were evaluated, which does not allow for
statistical analysis. The naturally occurring phospholipids used

333

in these experiments were isolated from egg (phosphatidylcholine) and beef brain (phosphatidylserine). These experiments were discontinued because we realized that it would be impossible to prove that slow viruses did not exist within beef brain PS, and we had demonstrated that the commercially available PS was in fact not pure. Therefore, a decision was made that beef brain phosphatidylserine should not be used in human trials, and we terminated our experiments until a suitable synthetic phosphatidylserine was identified.

MATERIALS AND ANIMALS

Liposomes were made with PC and PS in a 7:3 molar ratio. These liposomes are multilamellar vesicles (MLV) and were prepared by the same method used by Fidler et al. (1981). The phospholipids were obtained from Avanti Polar Lipids. No macrophage-activating agents were used in these experiments. In all other studies, the aqueous component of the MLVs contained phosphate buffered saline.

Quality control included the evaluation of the purity of the phospholipids, using high pressure liquid chromatography. A 20 cm column was packed with Micro bondapack C_{18}. The mobile phase was 5% methanol in phosphate buffer (pH 3.5) in a 95:5 v/v ratio. The flow rate (at 25°C) for PS was 0.5 ml/min and for PC, 2.0 ml/min. Detection was by ultraviolet light at 254 nm which demonstrated the presence of a more polar contaminant in both PC and PS. The size and structure of our liposomes were evaluated by scanning and transmission electron microscopy. Sizing was also done by laser nephelometry.

Bovine purified activated factor X or X_a was prepared and used in the X_a one stage as previously described (Walker et al., 1979). All animals were maintained as prescribed by the rules and regulations at the University of Oklahoma and Oklahoma State University. The dogs were mongrel, weighing 20 to 25 kg. All mice were of the C6 C3 Fl strain, kindly provided by Dr. Fidler from the Frederick Cancer Center. The rats were of the Sprague Dawley strain, weighing 350 to 375 g, and were maintained on 12 hour light/dark cycles and at 75% of their free-fed weight. The Mongolian gerbils weighed 80 g each and were acclimatized for one month prior to experimentation.

For pathological evaluation, all tissues were fixed in 10% buffered formalin, embedded in paraffin, cut at 5 microns, and stained with H and E or PTAH. Sections were examined without knowledge of experimental manipulation.

Details of the experimental models can be found in the following references: behavioral assessment of rats and gerbils (Holloway and Vardiman, 1971; McCleary, 1966; Silverman, 1978; Irwin, 1962;

van Putten et al., 1979), canine lung model (Mojarad et al., 1983;
Vereim and Ohkuda, 1977), and the canine coronary model (Schwartz
and Stone, 1981).

METHODS AND RESULTS

In-Vitro Coagulation Studies

We used the X_a one stage in-vitro coagulation test to evaluate
whether PC or PS entrapped in MLV could enhance coagulation. This
assay is generally used to evaluate a test citrated plasma, and the
coagulation time is compared to that of a normal citrated plasma
(Walker et al., 1979). In the standard test done at $37^{\circ}C$, plasma
is mixed with a calcium chloride solution, a suspension of phospholipid (equal concentration of PC and PS), and activated bovine
factor X (or X_a) and the clotting time recorded. Concentrations
of these components are given in the background reference. In
these in-vitro experiments, a standard normal canine plasma was
used in all tests. The components of our MLV, PC and PS, as well
as our PC/PS MLV, were substituted for the standard phospholipid

Table 1. Influence of Phospholipids and liposomes on the
          X_a one stage coagulation time

Constituents of X_a one stage		Results	Comments
Constants	Variable	Clotting Time (CT) Seconds	
X_a, Ca^{++} and Normal plasma	Phospholipid (PL)		
X	Standard	8	PL enhanced CT
X	O	18	PL unenhanced CT
X	PC	18	No enhancement
X	PS	8	Enhancement
X	PS, 1:4,000	18	Limiting dilution
X	PC/PS MLV	8	Enhancement
X	PC/PS MLV 1:80	18	Limiting dilution

Table 2. Distribution of pulmonary thrombi in mice

Intravenous challenge	Behaviour post injection (10 minutes' observation)	Necropsy findings			
		Pulmonary thrombi			Pulmonary Hemorrhage
		Arteriolar	Artery	Vein	
Liposomes (20 μg)	3/3 asymptomatic	0/2	0/2	0/2	0/2
X_a, 0.1 μg	3/3 asymptomatic	(1)/3	0/3	0/3	0/3
Liposomes and X_a, 0.05 μg	5/5 asymptomatic	0/5	0/5	0/5	0/5
Liposomes and X_a, 0.1 μg	6/7 lethargic 1/7 asymptomatic	7/7	5/7	0/7	3/7

(1) indicates rare event.

suspension. The results are given in Table 1 and demonstrate that the orientation of phosphatidylserine within the MLV does allow these MLV to enhance the coagulation time. The enhancement was due to PS and not PC.

In-Vivo Studies

We next determined if the interaction of PC/PS liposomes with activated coagulation factors would lead to altered pathological consequences when they were injected intravenously. As demonstrated above and in other reports, bovine factor X_a will coagulate plasma from other species. A pilot study was done in two Pentothal anesthetized dogs who received liposomes intravenously. One received liposomes alone, after which there was no change in vital signs. A second animal received liposomes pre-incubated with X_a and calcium chloride. Tachycardia and tachypnea were noted within a few seconds after injection and the animal died one and one-half minutes post injection. Vascular thrombi were present in the lungs of the latter, but not the former dog. We elected to repeat these experiments in mice where more animals could be evaluated and varying doses used. Pre-incubation of liposomes with X_a caused pulmonary thrombi (Table 2). Thrombi were not demonstrated in any other organ examined, i.e., the brain, heart and kidney. Higher concentrations of X_a (0.2μg of protein) caused death in recipient mice (data not shown).

Behavioural Assessment of Rats

We conducted behavioural assessment procedures in rats because we thought the central nervous system might be a sensitive indicator of insults to the microcirculation such as microthrombi. In the first experiments, 6 rats were trained on a wooden beam, 1.5 m above the floor, 2.1 m long by 4 cm wide. There was a starting platform on one end and a feeding platform on the other end, with a pressure sensitive feeding switch connected to a Hunter clock. Rats were placed on the starting platform and timed for the arrival at a pressure sensitive feeding switch. This simple task is thought to be sensitive to both alterations and motivation and motor coordination. After training, baseline measurements of 10 trials for each rat were recorded. On day 1, all rats were injected with freshly prepared liposomes (90μg lipid per animal) and the rats evaluated 20 min later. Saline control injections were evaluated on day 2 and repeat baseline recorded on day 7. The mean and standard error for running times (in seconds) for baseline, liposome injection, saline injection, and post injection conditions were respectively: 4.8 ± 1.1, 4.6 ± 0.1, 4.7 ± 1.0, and 4.2 ± 0.8 ($p > 0.5$). Thus, there was no change in motivation or coordination as assessed by this test after a single dose of liposomes.

Table 3. Operant performance under several injection
          conditions

Condition	Number Responses in 20 min (Mean/ Standard Error)	% Correct Responses (0-10 seconds)
Phase 1:	(no trauma control - IV liposomes 1st experience for these rats with needle trauma)	
Baseline 1	146.9/4.6*	52.4/4.8*
Liposome	176.5/5.3*	40.8/7.0*
Saline 1	187.9/31.9*	40.0/11.3*
Baseline 2	153.2/5.4*	46.6/4.2*
Saline 2	148.0/13.6*	56.3/3.4*
Caffeine	179.8/11.3, p=0.5	41.8/6.3*
Ethanol	102.7/32.8, p <32.8	29.0/9.3, p <0.06
Phase 2:	(controlled for trauma; needle trauma to tail 5x/week x 3 prior to baseline)	
Baseline 1	144.2/4.2*	68.8/2.8*
Liposome	143.3/14.0*	63.3/6.2*
Saline	150.7/15.7*	61.5/6.4*
Baseline 2	132.0/4.7*	67.8/8.9*

*$p = > 0.05$

A second series of experiments was conducted on 4 rats trained
on the Four Lafayette operant chamber apparatus with food pellet
dispensers. A Rockwell Aim, 65 microprocessor, was used for pro-
gramming. This task is technically referred to as an operant,
schedule-controlled task or differential reinforcement of low rates
of response, the DRL. After a delay of 10 seconds, the animal was
rewarded with food if the lever was pressed between the 10th and

the 20th seconds, but not if the lever was pressed in less than 10
seconds or after 20 seconds (a temporal window). Each animal was
tested for 20 min. This task is sensitive to both rate increasing
and rate decreasing drugs and to lesions in various parts of the
central nervous system (Holloway and Vardiman, 1971; McCleary, 1966).
The observations on all rats for Phase 1 were as follows: day 1,
baseline 1; day 2, 20 min after intravenous liposomes; day 3, 20 min
after intravenous saline; and, day 7, repeat baseline 2 (see Table
3). In this phase, there was no control for I.V. tail vein trauma.
There was no significant difference between the four observations
($p > 0.05$), although a trend for increased responding after liposome injections relative to initial baseline was noted ($p < 0.10$).
Weeks later, the sensitivity of these animals was evaluated, using
a known CNS stimulant (caffeine) and a known CNS depressant (ethanol).
The drug effects noted were consistent with the expected outcome
(Holloway and Vardiman, 1971). In Phase 2, conducted 6 months
later, all animals had a 23 gauge needle inserted into the area of
the tail vein daily for 3 weeks. As before, animals were observed
for baseline 1, injected with freshly prepared liposomes, injected
with saline, and baseline 2. There was no difference in the observations (see lower portion of Table 3, Phase 2, $p > 0.05$).
Thus, these sets of experiments indicate that no significant changes
in motivated operant behaviour were manifest after an initial or
after a subsequent acute injection of liposomes.

Behavioural Assessment of Gerbils

The first experiments were conducted to determine if intravenously administered liposomes influence ventral scent marking or
exploratory behaviour of male gerbils. Both are innate behaviours
and may be tested in the open field condition. Presumably, if the
liposomes had effects upon the motor or the central nervous system
of these subjects, the disruption of these two behavioural patterns
would be expected. The circular open field used in this work had
a diameter of 508 cm, with 21.6 cm walls. The field was marked
into quadrants, and there were two marking posts (1.0 cm x 0.7 cm
high) in each quadrant. In each of these test sessions, the animal
was placed initially in a randomly selected quadrant in the field,
and the stop watch started. The number of times the quadrant line
was crossed during the first 60 seconds of the trial was recorded.
Also, the number of times an animal ventral scent marked any of the
posts during the first 240 seconds was recorded. The observer (AH)
was unaware of the experimental state of the animals under test.
The open field was cleaned with ethanol between trials. Animals
were tested 2, 48 and 96 h post injection. Sixteen animals were
injected with saline (Group A) and 16 with liposomes ($40 \mu g$ of
lipid per gerbil) (Group B). The data for general activity or line
crossing is given in Table 4 and for ventral scent marking in Table
5. There was no indication that the groups differed significantly
with respect to either the rates of line crossing or the rates of

Table 4. General activity rates, as measured by line
          crossings during first min in an open field,
          among groups of gerbils placed individually
          in an open field at different times following
          pretreatment

Groups of gerbils tested individually in an open field

	Group A			Group B	
2 h	48 h	96 h	2 h	48 h	96 h
17.5 ± 4.2	12.1 ± 3.7	9.3 ± 2.8	16.6 ± 4.9	11.5 ± 3.5	9.1 ± 3.6

Values denote line crossings during first 60 sec of
open field tests (mean ± standard deviation).

Table 5. Ventral scent marking by two groups of gerbils
          in which the members were exposed individually
          to an open field for 240 sec periods at differ-
          ent times following pretreatment

Groups of gerbils tested individually in an open field

	Group A			Group B	
2 h	48 h	96 h	2 h	48 h	96 h
1.5 ± 2.4	2.3 ± 3.3	4.1 ± 4.0	2.4 ± 2.2	2.0 ± 2.2	3.3 ± 2.0

Values denote ventral scent markings during 240 sec
open field tests (mean ± standard deviation).

ventral scent marking.

    The purpose of the second experiment was to assess the effects
of liposomes upon cerebral functions in learning. In learning
mazes, rodents show cognitive abilities which may be disrupted upon
non-specific cerebral injury. The Lashley maze (Type III) measures

91.5 cm by 37.0 cm by 7.5 cm high. The four alleys enclosed in the maze lay parallel to one another, and each was 7.5 cm wide. The attached starting box and goal box measure 27 cm by 17 cm by 7.5 cm high. Two days prior to maze learning, 8 subjects were randomly selected for each of the two groups used in the preceeding study. Group A animals received saline prior to the experiment with the open field and Group B, liposomes. At the end of a 48 h period of water deprivation, maze learning tests were begun. Each trial ended when the animal made physical contact with the water dish in the goal box. Three observations were recorded: the time (in seconds) to traverse the maze (Table 6), the number of errors (Table 7) and the number of reversals (Table 8). The animals were tested on 3 consecutive days. The groups did not differ significantly from one another with respect to the speed of maze running ($p > 0.20$). There was a significant decrease in the running time between day 1 and day 3 for both groups ($p < 0.01$), but this may be interpreted as indicating similar rates of learning by the two groups. The same was true for both the number of errors (Table 7) and the number of reversals (Table 8). There were no observed differences in learning between the two groups, but both groups improved after additional trials.

Canine Lung and Coronary Models

The pulmonary microcirculation is the first capillary bed to be traversed by intravenously administered liposomes. We elected to use a canine lung model in which one can monitor not only pulmonary lymph flow, but also pulmonary, arterial and venous pressure. This model is very sensitive to minor changes in blood pressure or any inflammatory reaction within the pulmonary microcirculation. The canine heart model also allows monitoring of central and coronary vascular pressures and, of course, the electrocardiogram. One can also evaluate changes in regional blood flow within the heart, using radiolabeled microspheres. Only four dogs were evaluated before the decision to abandon beef brain PS. Two dogs received 300 mg of phospholipids as MLV, and two dogs received 900 mg. There were no abnormalities observed in any of the functions being monitored in these dogs. For statistical reasons, a minimum of six dogs are generally evaluated at a given dose of the agent under study. Therefore, our results only suggest that these are non-toxic doses of liposomes.

DISCUSSION

Our in-vitro experiments documented that PS in MLV can enhance coagulation and our in-vitro experiments demonstrated that MLV preincubated with X_a can cause thrombosis. Our hypothesis that PS liposomes might be thrombogenic was based on phospholipid protein interactions that are known to be important in enhancing blood

Table 6. Mean times (sec) required by two groups of
          gerbils to traverse a Lashley maze during
          three successive 3 day blocks of trials

| | 3 day Blocks | | | | | |
| | Day 1 | | Day 2 | | Day 3 | |
Group	M	SD	M	SD	M	SD
A	242.0	126.9	129.2	58.4	81.2	61.1
B	245.1	132.4	137.2	53.2	76.5	50.2

Table 7. Mean numbers of errors (entrances into shorter
          cul de sac at choice points) by two groups of
          gerbils tested in a Lashley maze during three
          successive 3 day blocks of trials

| | 3 day Blocks | | | | | |
| | Day 1 | | Day 2 | | Day 3 | |
Group	M	SD	M	SD	M	SD
A	4.4	2.4	1.4	0.9	0.5	0.4
B	1.5	1.1	1.2	0.6	0.8	0.6

Table 8. Mean numbers of reversals (alleys retraced)
          during three successive 3 day blocks of trials
          by gerbils tested in a Lashley maze

| | 3 day Blocks | | | | | |
| | Day 1 | | Day 2 | | Day 3 | |
Group	M	SD	M	SD	M	SD
A	9.6	4.8	6.9	4.5	1.0	1.0
B	11.9	8.7	8.4	4.9	4.0	2.6

coagulation (van Dieijen et al., 1981). For example, Factor X_a
has high affinity for PS-containing membranes. Furthermore, cancer
patients are known to have a wide variety of abnormalities in their
coagulation at the time of diagnosis (Rickles and Edwards, 1983).
Of particular interest and concern is the activation of coagulation
factors. For example, a cancer pro-coagulant has been described
that activates Factor X (Gordon et al., 1975). It can be extracted
from human colon, kidney, breast, and vaginal carcinoma tissue.
Thus, it is possible that liposomes, given intravenously to cancer
patients, could interact with activated clotting factors. We would
like to stress, however, that the preincubation of our MLV with X_a
is obviously an artificial model. Others have argued that negatively
charged liposomes are unlikely to cause thromboembolism because they
lower the concentration of free clotting factors in the plasma, due
to their absorption onto the surface of liposomes (Juliano and Lin,
1980). Our interpretation of these observations is that this absorption of coagulation factors onto the surface of the liposome
would enhance its potential for causing thromboembolism.

Liposomes are obviously a promising vehicle for targeting drug
delivery (Gregoriadis, 1978, 1980), but their introduction to
clinical medicine will necessitate particular attention to possible
unique toxicities that are generally not encountered with a water
soluble drug. Our in-vitro coagulation studies have suggested a
possible unique toxicity, i.e., thromboembolization. While large
thrombi had not been observed in the toxicity studies conducted by
Hart et al. (1981), the presence of microthrombi in the microcirculation might not be evident in a routine necropsy. Our hypothesis was that the central nervous system might be a sensitive
indicator of insults to the microcirculation. In the experiments
described in this report, a single dose of liposomes failed to alter
a number of central nervous system functions. Whether multiple
doses would result in toxicity remains to be determined. Other
investigators have sought and demonstrated unique toxicity such as
the depression of the reticuloendothelial system (Ellens et al.,
1982). Identification of these and other toxicities in animals
will be helpful in monitoring Phase I trials in humans.

CONCLUSIONS

Negatively charged phospholipids are common constituents of
liposomes. We have demonstrated that PC and PS incorporated into
MLV can enhance in-vitro coagulation, thus suggesting a thrombogenic
potential for these agents. It is, therefore, our opinion that the
evaluation of lipid particles such as liposomes will necessitate
preclinical evaluation not normally required for hydrophilic agents.
Using animal behaviour models, we have not observed any toxicity
after a single dose of intravenous PC/PS liposomes. However, preincubation of PC/PS liposomes with X_a results in pulmonary thrombi

and/or death in recipient animals. We are, therefore, of the
opinion that cancer patients with in-vivo evidence of activation
of the coagulation system should be excluded from the initial
Phase I human trials.

REFERENCES

Ellens, H., Mayhew, E. and Rustum, Y.M., 1982, Reversible depression
    of the reticuloendothelial system by liposomes, Biochim.
    Biophys. Acta, 714:479.
Fidler, I.J., 1975, Activation in-vitro of mouse macrophages by
    syngeneic, allogeneic or xenogeneic lymphocyte supernatants,
    J.N.C.I., 55:1159.
Fidler, I.J., Darnell, J.H. and Budmen, M.B., 1976, Tumoricidal
    properties of mouse macrophages activated with mediators
    from rat lymphocytes stimulated with concanavalin A, Cancer
    Res., 36:3608.
Fidler, I.J., Sone, S., Fogler, W.E. and Barnes, Z.L., 1981,
    Eradication of spontaneous metastasis and activation of al-
    veolar macrophages by intravenous injection of liposomes
    containing muramyl dipeptide, Proc. Natl. Acad. Sci. USA,
    78:1680.
Gordon, S.G., Franks, J.J. and Lewis, B., 1975, Cancer procoagulant
    A: a factor X activating procoagulant from malignant tissue,
    Thrombosis Res., 6:127.
Gregoriadis, G., 1980, The liposome drug-carrier concept: its
    development and future, in: "Liposomes in Biological Systems",
    G. Gregoriadis and A.C. Allison, eds., John Wiley & Son Ltd.,
    Chichester.
Gregoriadis, G., 1978, Liposomes in therapeutic and preventive med-
    icine: the development of the drug-carrier concept, in:
    "Liposomes and Their Uses in Biology and Medicine", D. Papa-
    hadjopoulos, ed., New York Academy of Sciences, New York.
Hart, I.R., Fogler, W.E., Poste, G. and Fidler, I.J., 1981, Toxicity
    studies of liposome-encapsulated immunomodulators administered
    intravenously to dogs and mice, Cancer Immunol. Immunother.,
    10:157.
Holloway, F.A. and Vardiman, D.R., 1971, Dose-response effects of
    ethanol on appetitive behaviors, Psychonom. Sci., 24:218.
Irwin, S., 1962, Drug screening and evaluative procedures, Science,
    136:123.
Juliano, R.L. and Lin, G., 1980, The interaction of plasma proteins
    with liposomes: protein binding and effects on the clotting
    and complement systems, in: "Liposomes and Immunobiology",
    B.H. Tom and H.R. Six, eds., Elsevier, North Holland.
McCleary, R.A., 1966, Response-modulating functions of the limbic
    system: initiation and suppression, in: "Progress in Phys-
    iological Psychology", Vol. 1, E. Stellar and J.M. Sprague,
    eds., Academic Press, New York.

Mojarad, M., Hamasaki, Y. and Said, S.I., 1983, Platelet-activating factor increases pulmonary microvascular permeability and induces pulmonary edema. A preliminary report, Clin. Resp. Physiol., 19:253.

Rickles, F.R. and Edwards, R.L., 1983, Activation of blood coagulation in cancer: Trousseau's Syndrome revisited, Blood, 62:14.

Schwartz, P.J. and Stone, H.L., 1981, Left stelectomy and denervation super sensitivity in conscious dog, Am. J. Cardiol., 49:1185.

Silverman, P., 1978, Animal behaviour in the laboratory, Pica Press, New York.

van Dieijen, G., Tans, G., van Rijn, J., Zwaal, R.F.A. and Rosing, J., 1981, Simple and rapid method to determine the binding of blood clotting factor X to phospholipid vesicles, Biochemistry, 20:7096.

van Putten, P.R., May, A. and Jarvik, M.E., 1979, Methods in psycho-pharmacology, in: "Methods of Biobehavioral Research", E.A. Serafetinides, ed., Grune & Stratton, New York.

Vereim, C.E. and Ohkuda, K., 1977, Improved method for cannulation of the right lymph duct in dogs, J. Appl. Physiol., 43:899.

Walker, F.J., Sexton, P.W. and Esmon, C.T., 1979, The inhibition of blood coagulation by activated protein C through the selective inactivation of activated factor V, Biochim. Biophys. Acta, 571:333.

THE USE OF LIPOSOMES IN DIAGNOSTIC IMAGING

Vicente J. Caride

Nuclear Medicine Department
Hospital of St. Raphael
Department of Diagnostic Imaging
Yale University Medical School
New Haven, CT, U.S.A.

INTRODUCTION

The delivery of drugs to organs is determined by blood flow, lipid-water partition coefficient of the drug, ionization, drug stability and chemical structure. Using drug carriers the pharmacokinetics can be changed to make possible the selective drug delivery to normal and pathological tissues. Besides protecting the drug, the carrier facilitates the movement of the drug across anatomic and biochemical barriers to selected cells, tissues or organs (Gregoriadis, 1978). The three elements that influence drug distribution via carriers and are susceptible to manipulation are the drug carrier itself, the route (including the administration route and the path between the site of administration and the final destination) and the target tissue. Drug carriers, route and target have specific chemical, physical and anatomical features that deserve special consideration.

The administration of the drug associated with a carrier should minimize drug accumulation in non-target areas and the site of maximum drug deposition should correspond to the desired target sites. The organ of maximal accumulation for a given carrier and administration route is the natural target for such carrier. Certain targets are not readily accessible and to reach them modification of the carrier or administration route is required. This interplay between drug carrier, administration route and target dictates the various strategies for drug delivery.

In our laboratory we are interested in the use of drug carriers

347

for diagnostic imaging. There are important differences between
drug delivery for therapy and drug delivery for imaging. Selective
chemotherapy aims at the induction of a specific, permanent or
transitory therapeutic response by delivering a drug to a selective
site with minimal (if any) drug deposition in non-target tissues.
In such a case it is desirable that the drug delivery be absolute
and permanent.

In diagnostic imaging, detectability of a tissue is achieved
by modifying the target (i.e. lesion) or the non-target tissue
(i.e. perilesional area), the final effect being signal to noise
enhancement. To accomplish this the imaging agent should be present
in the region of interest long enough to allow the collection of
information needed. Therefore, the induction of permanent or transitory physicochemical changes of the target tissue or of tissues
contiguous to the target to improve detectability of the target is
a viable strategic approach for site directed imaging.

POTENTIAL OF LIPOSOMES AS IMAGING AGENTS

Liposomes can carry imaging agents in their aqueous space or
lipid bilayer. Since the natural target of liposomes is the reticuloendothelial system (RES), liposomes constitute ideal agents to image
the liver and spleen and eventually bone marrow. Modification of
liposomal composition, surface charge, size and the use of different
administration routes and doses, allow further variations in organ
distribution (Abra and Hunt, 1981; Juliano and Stamp, 1975; McDougal
et al., 1975; Richardson et al., 1978). The attachment of surface
active materials, specific tissue or tumour antibodies, will in the
future prove invaluable for selective organ or lesion imaging.
Before we get concerned with liposome targeting, however, liposomes
have to be transformed to contrast agents. To respond to the various
demands of diagnostic imaging, liposomes have to be made radiopaque
(for conventional X-ray, computed tomography, digital radiography),
radioactive (for radioscintigraphy), paramagnetic (for nuclear
magnetic resonance studies) or sonoreflective (for ultrasound
imaging).

RADIOACTIVE LIPOSOMES

Radiolabels can be introduced in the lipid or the aqueous
phase of liposomes. The idea is to achieve a permanent, stable
association between the radiotracer and the liposome. This is a
difficult task, particularly in-vivo, since small molecules entrapped in the aqueous phase tend to leak out as the lipid vesicles
interact with serum proteins and cell membranes (Borowski et al.,
1977; Pagano and Weinstein, 1978). Radiotracers attached to the
lipid phase tend to be more stable but they are also subject to

dissociation and molecular exchange with cell membranes and other macromolecules.

The magnitude of radiotracer-liposome dissociation is critical when using the label to quantitate the organ distribution of liposomes. However, for imaging the requirements are less rigid especially when imaging is performed a short time after the administration of the liposome mixture. Moreover, the rate of leakage of a tracer may be itself an important diagnostic parameter.

The instruments used to detect radioactivity and compose an image offer a general rendition of the distribution of radioactivity in the body. Within limits, the presence of free radiotracer will induce minor deterioration in the image, probably amounting to visualization of organs that are otherwise invisible. Since leakage is almost unavoidable the researcher should select the radiotracer with care to fit specific needs and to facilitate the interpretation of the study. The following list, by no means exhaustive, highlights some of the considerations in radiotracer selection:

1) the tracer should be easy to incorporate into liposome, with high efficiency and reproducibility;
2) the tracer-liposome dissociation in-vitro and in-vivo, and the pharmacokinetics of the free radiotracer should be well characterized;
3) the free tracer and liposomes should have different organ distribution;
4) the radioelement should have gamma emission in an energy range suited for the current imaging devices. The physical, biological and effective half-life of the radiocomplex should be short to minimize radiation exposure;
5) the tracer, once dissociated from liposomes, should be rapidly eliminated from the body;
6) for certain applications, a tracer that reacts at specific cellular sites may be needed. The tracer may be destroyed in certain cells or transformed into a new product whose biologic behaviour differs from the compound. In such cases, it may be possible to follow the new product to study a given function;
7) another alternative is the use of tracers that once released remain permanently associated to local structures.

We investigated the use of liposomes labeled with aqueous and lipid tracers (Espinola et al., 1975). The aqueous tracer Tc 99m Sn-diethylenetriamine pentaacetic acid (Tc 99m DTPA) is removed from the circulation by glomerular filtration and is used to evaluate renal function. Images obtained following administration of Tc 99m DTPA show renal and bladder activity with minimal radioactivity in extrarenal tissues. In contrast, the liposome-Tc 99m DTPA complex accumulates in liver and spleen instead of kidneys and bladder. The Tc 99m DTPA released from liposomes is rapidly eliminated into the urine.

Table 1. Organ distribution in mice of positive and negative MLV labelled with 99m Tc DTPA or III In oxine

Time (h)	Organ	+L (Tc DTPA) n=8	+L (In III Oxine) n=8	-L (Tc DTPA) n=8	-L (In III Oxine) n=10
2	Liver	37 ± 1.5*	33 ± 5	33 ± 3.5	33 ± 0.8
	Spleen	77 ± 14	40 ± 6	62 ± 14	32 ± 2.0
	Bone	4 ± 0.5	1.4 ± 0.15	1.7 ± 0.18	1.8 ± 0.05
	Blood	1.85 ± 0.04	2.4 ± 0.2	1.4 ± 0.3	1.5 ± 0.4
48	Liver	7 ± 0.5	31 ± 3	5 ± 0.5	28 ± 2.5
	Spleen	17 ± 3	28 ± 3.5	16 ± 2	18 ± 1.4
	Bone	2 ± 0.4	1.6 ± 0.3	1.1 ± 0.3	2.3 ± 0.08
	Blood	0.2 ± 0.05	0.16 ± 0.01	0.13 ± 0.01	0.06 ± 0.01

*Data expressed as percent of injected dose per g of tissue. Each value is the mean ± SEM. (From Espinola et al, 1975.)

As a label for lipid bilayer we selected In 111-8-hydroxyquino-
line (In 111-oxine). The lipophilic In 111 oxine molecule enters
the liposome membrane by diffusion. Once inserted in the liposome
membrane the In 111 is no longer accessible to blood transferin,
for which it has a higher affinity. The organ distribution of
liposomes labeled with In 111 oxine is similar to liposomes-Tc 99m
DTPA. However, after endocytosis of the liposome-In 111 oxine
complex, the radioelement will bind molecules with affinity for
iron (i.e. ferritin), and will remain associated to intracellular
sites. The hepatosplenic uptake of the liposome-In 111 oxine
complex remains almost unchanged for several days. Table 1 presents
the percent uptake per g of tissue for both liposome preparations
at various times following injection. As expected, there is a
constant decline with time of Tc 99m DTPA activity, indicative of
continuous leakage of the tracer.

The organ distribution of liposomes can be altered through
modification of liposome size, charge, lipid composition and also
increasing the dose of liposomes administered. Large doses of
liposomes induce a blockade of the reticuloendothelial system,
forcing the lipid vesicles into prolonged circulation times and
interacting with extrahepatic tissues. We tested this approach
using radiolabeled liposomes (Caride et al., 1976). Mice (30 g
body weight) were pretreated with 9.94 micromoles of liposomes
(phosphatydilcholine (PC):cholesterol (CH); molar ratio 7:1) given
intravenously. A dose of Tc 99m DTPA liposomes (0.32 micromoles)
was administered minutes later. The hepatic uptake decreased from
a control value of 44.42% to 31.70% (28% change) and the splenic
uptake increased from 4.19% to 8.35% (a 99% change). A 120% increase
in pulmonary and 72% increase in bone marrow deposition of liposomes
were also observed. This shift of the radioactive liposomes to
extrahepatic tissues is operated through saturation of the reticuloendothelial cells in the liver. Increased liposome uptake by
tumours has been recently reported using a similar strategy
(Proffitt et al., 1983). Whether this approach will have application in humans is not known since administration of large doses
of liposomes may be harmful.

The distribution of liposomes can be altered by administering
the lipid vesicles through different route (McDougall et al., 1975).
Macromolecules injected in subcutaneous tissue diffuse slowly and
do not enter the vascular compartment through capillaries but
through the lymphatic system. Colloidal particles for example are
removed from subcutaneous tissues by lymphatic drainage and retained
in regional lymph nodes. The entry into the lymphatic channels
depends on size and charge of the particles and the removal depends
on lymph flow. To a certain extent, liposomes can be compared with
colloids. Table 2 shows the percent of subcutaneously injected

Table 2. Distribution of colloidal particles in regional
          lymph nodes after subcutaneous injection

Treatment	nm	Percent in lymph nodes (2 h)
*Colloidal gold	5	8.4
*Antimony sulfide colloid	5 - 15	5.3
*Sulfur colloid	400 - 1000	0.9
- MLV	500 - 3000	0.17 ± 0.03

*From Stand and Persson, 1979.

colloidal particles present in regional lymph nodes at 2 h compared
to similar data from our laboratory for negatively charged MLV (PC:
CH:dicetyl phosphate; molar ratio 7:1:2). Scintigraphic images
were obtained with liposomes containing Tc 99m DTPA in the aqueous
compartment, to demonstrate that liposomes can effectively carry a
drug along the lymphatic channels. Control animals, injected in
the footpad with free Tc 99m DTPA, showed rapid excretion of the
radioactivity through the kidneys, indicating that the small molecule enters directly the blood capillaries from the subcutaneous
tissues. In contrast, the footpad administration of liposomes
containing Tc 99m DTPA provides an excellent visualization of
regional lymph nodes, from the injection site to the iliac, peri-
aortic and perirenal lymph nodes (Caride, 1981). The variable
amount of radioactivity observed in the bladder indicates that the
tracer is set free upon arrival of liposomes in the bloodstream or
the lymph nodes, at the site of subcutaneous injection or after
RES metabolization. The demonstration of hepatic radioactivity
indicates that intact liposomes reach the systemic circulation as
well. This route of administration has been studied by many for
therapeutic and diagnostic purposes.

     An intriguing diagnostic application of liposomes is in the
visualization of the gastrointestinal tract. Presently, radiographic studies of oesophagus, stomach, small and large bowel use
radiodense contrast materials that physically fill these hollow
organs. Scintigraphic studies of the gastrointestinal tract use
non-absorbable radiotracers to measure rate of emptying and motility of these organs. Few attempts have been made to offer a more
specific diagnostic procedure.

     We are studying the possibility of developing contrast agents
that will interact specifically with abnormal areas in the gastrointestinal tract. Initial studies are directed to the gastric

mucosa, hoping for a higher affinity of liposomes for damaged than normal mucosa. For the study, necrotic lesions in the stomach are induced by instilling hydrochloric acid in pylorus ligated stomachs of rats. Thirty minutes after this treatment the stomach is washed and 1 ml of liposomal (PC:CH:stearylamine; molar ratio 8:1:1) In 111 oxine, instilled in the organ. Forty five minutes later the stomach is removed, washed several times with distilled water, and counted for radioactivity. In this pilot study scintigraphic images of the stomach showed preferential accumulation of radioactivity in areas of damaged mucosa. Although the initial results are encouraging, additional studies using various liposome formulation, radiotracers, variable exposure times and graded mucosal damage are needed before a final conclusion can be reached. In effect, it may be possible that retention of liposomes in the area occurs through non-specific interactions with denatured proteins, extravasated blood as well as necrotic tissues.

RADIOPAQUE LIPOSOMES

Radiographic contrast media are substances which contain elements whose electron density is markedly higher than the tissue under examination. The administration of contrast agents alters the spatial electron density profile of the organ rendering its contour more apparent. The lumen of the gastrointestinal tract may be filled with aqueous barium sulfate suspension to distend it and make it denser than the normal tissues. Similarly, iodinated benzoic acids can be introduced into the lumen of an arterial or venous vascular system, increasing its density enough to make it apparent by differential X-ray absorption. Iodinated benzoate compounds, once in the vascular system, distribute in the plasma and extracellular spaces. During this phase of distribution, there is a change in radiographic density in the body tissues, the degree varying in individual tissues according to their vascularity and capillary permeability. The contrast material is filtered from the blood by the renal glomeruli, concentrated in the renal tubules and excreted in the urine. During this process, there is selective radiopacity of the kidneys to a degree sufficient to allow detailed radiologic study.

Contrast agents can be administered orally (i.e. barium sulfate) intravascularly, intrathecally, by means of catheters into the lymphatic channels and by a variety of other procedures according to the area of the organism that needs to be studied. Intravascular contrast agents are widely used since they allow the rapid acquisition of images delineating vascular channels, arteries and veins and the transitory visualization of the parenchyma of various organs. However, organs like liver and spleen cannot optimally be studied with the present techniques. The reasons are that there is no contrast agent that is removed from the circulation exclusively by

these organs and most contrast agents allow only a transitory
(seconds to minutes) visualization of the hepatic parenchyma. The
latter factor reduces the time available to obtain images. In fact,
if multiple projections of the organ are needed the patient has to
be re-injected with the contrast agent. Other anatomic structures
are also difficult to evaluate. For example, for the visualization
of lymph nodes, the performance of minor surgery is required to
identify and canulate small lymphatic channels for the injection of
contrast agents. The procedure is time consuming and unpleasant
for the patient. In sum, many current radiological techniques
resort to various modalities of contrast enhancement for the
acquisition of images. Available techniques of contrast enhancement are less than ideal for the study of liver, spleen and lymphatics.

To study liver and spleen, colloidal suspensions, oily emulsions
and liposomes provide prolonged contrast enhancement (Alfidi and
Laval-Jeantet, 1976; Caride et al., 1982; Dean et al., 1980; Fischer,
1977; Havron et al., 1981; Lamarque et al., 1979; Seltzer et al.,
1981; Vermess et al., 1976). Since liposomes remain in these organs
for hours, contrast enhancement can be achieved with a single intravenous injection and the radiological evaluation performed at a
later time. Radiopaque liposomes (ROL) can be made by entrapping
water-soluble radiographic contrast agents or by using radiopaque
lipids in their preparation. Both methods have been proven successful in increasing the radiodensity of tissues. In the first case,
conventional iodinated contrast material has been entrapped inside
liposomes. Computed tomographic studies showed opacification of
blood pool, liver, spleen and kidneys. The visualization of contrast material in the kidneys reflects the leakage of the contrast
agent with consequent contrast redistribution. Another potential
problem is the variability in entrapment of the agent into liposomes (Havron et al., 1981).

We have tested ROL in dogs for contrast enhancement of liver
and spleen (Caride et al., 1982). These studies were performed with
multilamellar liposomes, and in the preliminary evaluation no
attempts were made to maximize liposome deposition in liver and
spleen. Liposomes were prepared with brominated PC with or without
the addition of cholesterol and evaluated in dogs using computed
tomography. A total dose of 9.75 g of PC (or 12.2 mMol) containing
2.63 g of bromine (32.9 mMol) was administered by slow intravenous
injection. The dose of injected lipid per g of body weight was
0.39 mg or 0.487 micromol. After Br ROL administration the animals
were studied at 1, 2, 3 and at 24 h with a Pfizer PZ 6 CT scanner
(scanning time 2.5 seconds, 120 KVP and 100 mA). The linear attenuation coefficient for liver, spleen, gall bladder and muscle was
calculated by selecting regions of interest in those organs when
replaying the study.

Hepatic enhancement at 3 h, expressed as the difference in attenuation in Hounsfield units (ΔHU) before and after Br ROL injection, varies from 4.3 to 13.50 ΔHU. Splenic enhancement was more variable approaching an increase in attenuation of 20 ΔHU in some cases. Of interest is the demonstration of hepatosplenic contrast enhancement 24 h later with values close to those measured at 3 h. In some cases the gall bladder showed increased attenuation suggestive of elimination of the bromine into the bile. Whether the bromine was still associated to the phospholipids is not known.

A separate study was performed to compare multilamellar (MLV) to small unilamellar vesicles (SUV). A dose of 5.6 mg/gram (7 micromol) of lipids containing 0.123 mg of Br was administered intravenously to 30 g mice. Hepatic opacification was higher with SUV ROL (ΔHU:11 at 1 h and 9.1 at 24 h) than with MLV ROL (ΔHU:5 at 24 h). No signs of toxicity were observed in these animals in spite of the large dose of lipids administered.

This already significant change in radiodensity can be further improved using liposomes of different composition to favour liver and spleen uptake. Increasing the specific radiodensity of the preparation will permit the administration of smaller volumes of lipids. Among other advantages the reduction in the dose will favour a more physiologic distribution of liposomes without saturation of the RES or opsonin depletion. Other variations in liposome preparation and administration can prolong circulation time and enhance the radiodensity of organs lacking reticuloendothelial cells (Senior and Gregoriadis, 1982).

In summary, the advantages offered by liposomes in the administration of contrast agents include: specific contrast enhancement of liver and spleen, adjustable clearance from the vascular compartment, adjustable clearance from the tissues, biodegradability (liposomes are prepared from naturally occurring phospholipids), variable administration routes. Further, liposomes are potential carriers for antibodies and molecules with affinity for specific receptors to image organs and lesions.

REFERENCES

Abra, R.M. and Hunt, A., 1981, Liposome disposition in-vivo: III. Dose and vesicle size effects, <u>Biochim. Biophys. Acta</u>, 666: 493.
Alfidi, R.J. and Laval-Jeantet, M., 1976, A6 60:99: a promising contrast agent for computed tomography of the liver and spleen, <u>Radiology</u>, 121:491.
Borowski, J., Roerdink, F. and Scherphof, G., 1977, Leakage of sucrose from phosphatidylcholine liposomes induced by interaction with serum albumin, <u>Biochim. Biophys. Acta</u>, 497:183.

Caride, V.J., 1981, Liposomes for diagnostic imaging, in: "Radio-
    pharmaceuticals: structure activity relationships", R.P.
    Spencer, ed., Grune & Stratton, New York.
Caride, V.J., Taylor, W., Cramer, J.A. and Gottschalk, A., 1976,
    Evaluation of liposomes-entrapped radioactive tracers as
    scanning agents. Organ distribution of liposome (99m Tc
    DTPA) in mice, J. Nucl. Med., 17:1067.
Caride, V.J, Sostman, H.D., Twickler, J., Zacharis, H., Orphanoudakis,
    S.C. and Jaffe, C.C., 1982, Brominated radiopaque liposomes:
    contrast agents for computed tomography of liver and spleen.
    A preliminary report, Invest. Radiol., 17:381.
Dean, P.B., Violante, M.R. and Mohoney, J.A., 1980, Hepatic CT con-
    trast enhancement: effect of dose, duration of infusion and
    time elapsed following infusion, Invest. Radiol., 15:158.
Espinola, L.G., Beaucaire, J., Gottschalk, A. and Caride, V.J.,
    1979, Radiolabeled liposomes as metabolic and scanning
    tracers in mice. (II) In III oxine compared with Tc 99m
    DTPA entrapped in multilamellar lipid vesicles, J. Nucl.
    Med., 20:434.
Fischer, H.W., 1977, Improvement in radiographic contrast media
    through the development of colloidal or particulate media.
    An analysis, J. Theor. Biol., 67:653.
Gregoriadis, G., 1978, Liposomes in therapeutic and preventive
    medicine. The development of the drug carrier complex,
    An. N.Y. Acad. Sci., 308:343.
Havron, A., Seltzer, S.E., Davis, M.A. and Shulkin, P., 1981,
    Radiopaque liposomes: a promising new contrast material for
    computed tomography of the spleen, Radiology, 140:507.
Juliano, R.L. and Stamp, D., 1975, The effect of particle size and
    charge on the clearance rates of liposomes and liposome-
    encapsulated drugs, Biochem. Biophys. Res. Comm., 63:651.
Lamarque, J.L., Bruel, J.M., Dondelinger, R., Vendrell, B.,
    Pelissier, O., Rouanet, J.P., Michel, J.L. and Boulet, P.,
    1979, The use of iodolipids in hepatosplenic computed
    tomography, J. Comput. Assist. Tomogr., 3:21.
McDougall, I.R., Dunnick, J.K., Goris, M.L. and Kriss, J.P., 1975,
    In-vivo distribution of vesicles loaded with radiopharma-
    ceutical: a study of different routes of administration,
    J. Nucl. Med., 16:488.
Pagano, R.E. and Weinstein, J.N., 1978, Interactions of liposomes
    with mammalian cells, Ann. Rev. Biophys. Bioeng., 7:435.
Proffitt, R.J., Williams, L.E., Presant, C.A., Tin, G.W., Uliana,
    J.A., Gamble, R.C. and Baldeschwieler, J.D., 1983, Liposomal
    blockade of the reticuloendothelial system: Improved tumor
    imaging with small unilamellar vesicles, Science, 220:502.
Richardson, V.J., Jeyasingh, K., Jewkes, R.F., Ryman, B.E. and
    Tattersall, M.H., 1978, Possible tumor localization of
    Tc 99m labeled liposomes: effects of lipid composition,
    charge and liposome size, J. Nucl. Med., 19:1049.
Seltzer, S.E., Adams, D.F., Davis, M.A., Hessel, S.J., Havron, A.,

Judy, P.F., Paskins-Hurlburt, A.J. and Hollenberg, N.K., 1981, Hepatic contrast agents for computed tomography: High atomic number particulate material, J. Comput. Assist. Tomogr., 5:370.

Senior, J. and Gregoriadis, G., 1982, Is half-life of circulating liposomes determined by changes in their permeability?, FEBS Lett., 145:109.

Vermess, M., Adamson, R.H., Doppman, J.L., Rabson, A.S. and Herdt, J.R., 1976, Intravenous hepatosplenography. Experimental evaluation of new contrast material, Radiology, 119:31.

Strand, S.E. and Persson, B.R.R., 1979, Quantitative lymphoscintigraphy I: Basic concepts for optimal uptake of radiocolloids in the parasternal lymph nodes of rabbits, J. Nucl. Med., 20:1038.

UNEXPECTED TISSUE DISTRIBUTION OF LIPOSOMES COATED WITH

AMYLOPECTIN DERIVATIVES AND SUCCESSFUL USE IN THE TREATMENT

OF EXPERIMENTAL LEGIONNAIRES' DISEASES

Junzo Sunamoto,* Mitsuaki Goto,*
Takaaki Iida,* Kohei Hara,** Atsushi
Saito,** and Akimitsu Tomonaga**

*Department of Industrial Chemistry
Faculty of Engineering and **The
Second Department of Internal Medicine
School of Medicine, Nagasaki University
Nagasaki 852, Japan

Bacterial and plant cell membranes are covered by cell walls of polysaccharide derivatives. The cell walls do not only serve to maintain the shape and stiffness of cells and to protect the cell membranes against external stimuli but also play an important role in various biological recognition processes, for instance antigen-antibody interaction, toxin recognition, and cell-cell adhesion.

Liposomes have gained wide acceptance in chemotherapy as potential carriers in introducing drugs into cells.[1-3] However, several problems to be overcome still remain. Among these, reduction of the leakage of water soluble drugs entrapped in the liposomes and direction of liposomes to specific target cells or tissues are prominent. Regarding the latter, we have succeeded in coating the outermost surface of liposomes with polysaccharide derivatives such as O-palmitoylamylopectin. The efficiency in coating liposomes has been ascertained by several ways.[4] The most interesting finding was that the polysaccharide-coated liposomes were highly distributed in the lungs when administered i.v. to guinea pigs.[5] Furthermore, effective phagocytosis of such liposomes by macrophages and monocytes

and lysis of liposomes in cells have been also confirmed. Because
of the results obtained, we have attempted to use our modified liposomes for the treatment of bacterial infectious diseases. For this
purpose, the effect of antibiotic sisomycin encapsulated in LUV
(large unilamellar liposomes) on Legionella pneumophila multiplication in human monocytes and the chemotherapeutic treatment of
experimental Legionnaires' disease in guinea pigs were evaluated.

PREPARATION OF THE POLYSACCHARIDE-COATED LIPOSOMES

We have prepared partly modified polysaccharides, O-palmitoylpullulan (OPP) and O-palmitoylamylopectin (OPA), as an artificial
cell wall for liposomes (Fig. 1).[4,5] The substitution degree of
the palmitoyl residue per 100 glucose units of the polysaccharides
has been determined by the [1]H-nmr method. O-palmitoylamylopectin
OPA-112(4.9), molecular weight 112,000, was substituted with 4.9
palmitoyl residues per 100 glucose units and O-palmitoylpullulan
OPP-50(3.4), molecular weight 50,000 was substituted with 3.4
palmitoyl residues per 100 glucose units, respectively. Egg phosphatidylcholine (egg PC) was isolated and purified according to
the method described by Singleton et al.[6]

Polysaccharide-coated small unilamellar liposomes (SUV) were
prepared by essentially the same procedure as that adopted for the
conventional liposomes without any polysaccharide.[7] LUVs (large
unilamellar vesicle) with or without artificial cell walls were
prepared by the method of Deamer and Bangham[8] and used for investigations on phagocytosis with macrophages, bacteria multiplication
in human monocytes incubated with antibiotics encapsulated in liposomes, and treatment of experimental Legionnaires' disease.

O-Palmitoyl Pullulan

O-Palmitoyl Amylopectin

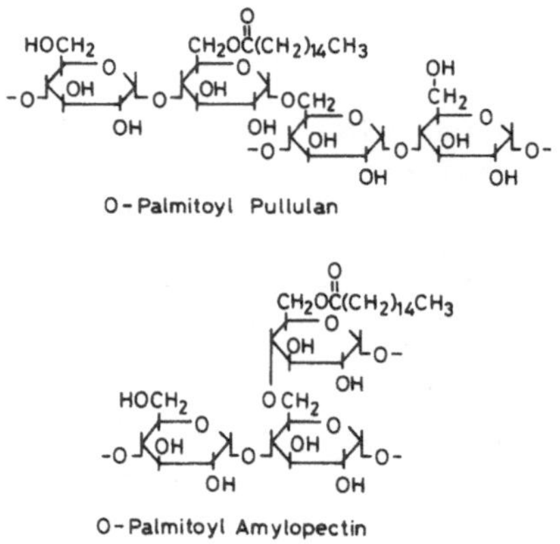

Fig. 1. Chemical structures of polysaccharide employed in this
         work as a component of artificial cell wall.

THE EFFICIENCY IN COATING LIPOSOMES WITH THE ARTIFICIAL CELL-WALL

POLYSACCHARIDE DERIVATIVES

The efficiency in coating liposomes with O-palmitoyl poly-saccharides was ascertained by several methods:[4],[5] (1) isolation of polysaccharide-coated liposomes by gel-filtration, (2) reduced permeability for 6-carboxyfluorescein (CF) encapsulated in the interior of liposomes, (3) increased resistance to phospholipase D, which was monitored by the fluorescence quenching of ANS upon the enzymatic lysis of liposomal membranes, (4) leakage of CF in the presence of serum, and (5) enzymatic digestion of polysaccha-rides adsorbed on the surface of liposomes. Results showed that the liposomes are effectively coated by O-palmitoyl polysaccha-rides and can be used as an improved drug carrier.

TISSUE DISTRIBUTION OF POLYSACCHARIDE-COATED LIPOSOMES CONTAINING

$[^{14}C]$-CoQ$_{10}$

Small unilamellar liposomes encapsulating $[^{14}C]$-CoQ$_{10}$ were pre-pared according to a method described previously.[7] The obtained liposome suspension of egg PC was concentrated to 2 ml by ultra-filtration. The concentrated liposome suspension containing 52.0 mg of egg PC was incubated with 5.2 mg of O-palmitoyl polysaccharides under stirring for 30 min at 20°C.

The metabolic fate of liposomes coated with polysaccharides was investigated by tracing the blood levels and tissue distribution of encapsulated $[^{14}C]$-CoQ$_{10}$. Liposomes were intravenously injected into the femoral vein of male guinea pigs (250-300 g body weight) at a dose of 0.6 mg lipid kg body weight. Blood samples were ob-tained by ear vein puncture at appropriate times. Animals were killed by decapitation 30 min and 24 h after injection. Just before decapitation, heart and brain were perfused with ice-cold saline, and then heart, brain, lung, liver, spleen, adrenal glands and kidneys were removed for scintillation counting. Blood (20 or 50 μl) was solubilized with 0.75 ml of Soluene-350/isopropanol (1/1 by vol.). Then, to the solubilized blood samples were added several drops of 30% H$_2$O$_2$. For tissues, a 100 mg homogenized sample of each tissue was incubated with 0.5 ml of Soluene-350 at 50°C for 2 h. Five ml of scintillation fluid consisting of Instagel and 0.5 N HCl (9/1 vol./vol.) was added to the solubilized blood or tissue samples and radioactivity was determined using the Aloka LSC 653 liquid scintillation counter.

To eliminate the size effect of liposomes on the tissue distrib-ution, all liposomes employed for the animal studies were controlled in terms of size, which was ascertained by electron microscopy.

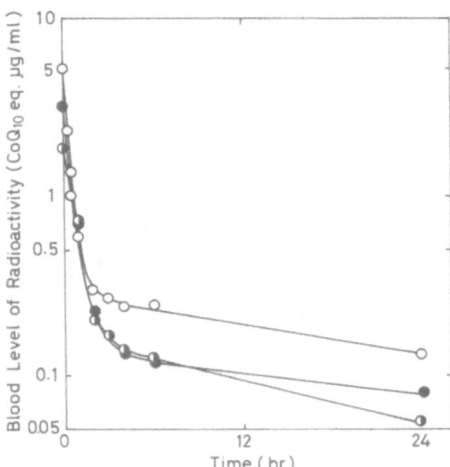

Fig. 2.   Radioactivity of $[^{14}C]$-CoQ$_{10}$ in blood as a function of
          time: -○-, conventional multilamellar liposomes (MLV);
          -●-, OPP-50(3.4)-coated MLV; -◐-, OPA-112(4.9)-coated
          MLV.

       Figure 2 shows radioactivity in blood after injection. Poly-
saccharide-coated liposomes were cleared more rapidly compared with
the conventional liposomes. In any event, the concentration of the
probe in blood decreases quite rapidly and is adsorbed by tissues
especially at an early stage within 2 h after injection. Table 1
shows the tissue distribution of radioactive CoQ$_{10}$ encapsulated in
polysaccharide-coated liposomes and conventional ones at 30 min and
24 h after administration. The most noteworthy finding in this work
is that the OPA-coated liposomes were highly distributed in the lung
and also liver compared with other kinds of liposomes. The reason
why only the OPA-coated liposome is highly distributed in the lung
is not clear at present. However, this specific distribution was
found to be correlated with the extent of aggregation of the polysaccharide-coated liposomes in the presence of concanavalin A: only
OPA-coated liposomes were significantly aggregated by con A, while
OPP-coated or O-palmitoyldextran-coated liposomes were not aggregated.
Fidler and his co-workers[9] have stated that liposomes bearing negatively charged surface are more easily distributed in the lung than
neutral ones. However, in this work anionic OPA and OPP bearing
phosphate or sulfate group did not show any priority in the lung
distribution compared with neutral polysaccharide derivatives.

Table 1. Tissue distribution of (^{14}C)-CoQ_{10} at 30 min and 24 h after intravenous injection

| Tissue | Control[a] | | OPA-112(4.9)-coated MLV | | OPP-50(3.4)-coated MLV | |
| | Radioactivity (% of the dose) | | | | | |
	30 min	24 h	30 min	24 h	30 min	24 h
Brain	0.04±0.01	0.03±0.002	0.03±0.01	0.02±0.002	0.03±0.002	0.01±0.003*
Heart	0.75±0.18	0.08±0.003	0.11±0.003*	0.04±0.008**	0.17±0.12	0.04±0.004**
Lung	5.50±0.56	2.40±0.53	27.2±4.9*	3.78±1.00	4.50±0.95	1.42±0.47
Kidney	0.33±0.03	0.18±0.05	0.47±0.09	0.16±0.02	0.32±0.07	0.14±0.05
Spleen	22.8±5.2	31.1±9.4	4.70±1.14*	9.27±2.48	13.2±4.4	11.5±3.3
Liver	38.2±2.8	56.3±5.1	54.1±4.4*	63.9±4.7	45.6±9.6	54.4±0.6
Adrenal glands	0.03±0.003	0.45±0.12	0.008±0.001**	0.13±0.01*	0.01±0.002*	0.24±0.05
Blood	9.09±1.54	1.56±0.19	2.39±0.09*	0.67±0.03*	7.74±2.80	0.93±0.17

Values are means ± S.E. of three guinea pigs.
[a] Conventional multilamellar liposomes (MLV) without polysaccharide.
* Significantly different from control liposomes, $P < 0.05$, Student's t-test.
** Significantly different from control liposomes, $P < 0.01$, Student's t-test.
Blood values are calculated by assuming that 7.3% of body weight is blood.

PHAGOCYTOSIS OF LIPOSOMES BY ALVEOLAR MACROPHAGES OF GUINEA PIGS

AND HUMAN MONOCYTES

    Internalization of liposomes into the phagocytes has been con-
firmed by three methods. First of all, the phagocytosis of MLV
coated with OPA-112(4.9) by alveolar macrophages from guinea pigs
was visualized by electron microscope. Macrophages were washed
out from the lungs of guinea pigs by injecting saline (10 ml x 10).
After washing of cells thrice with 10 ml of RPMI-1640 by centrifugation at 1800 rpm for 10 min (4°C) and subsequent decanting of
supernatants, the number of cells resuspended in RPMI medium containing 15% FCS (fetal calf serum) was counted and adjusted to
2 x 10 cells per 1 ml of the same medium. After incubating the
cells with MLV coated with OPA-112(4.9) at 37°C for 30 min or 1 h
on a gyratory shaker, sediments were examined by electron microscopy.
Figure 3 shows clearly the internalization of liposomes into alveolar
macrophages by phagocytosis upon incubation for 30 min (A) and the
subsequent disappearance of liposomes and vacuole in cells after
longer periods of incubation (B).

    By the same procedures, two LUV loaded with different kinds
of fluorescent probe, water soluble CF and water insoluble $Tb(acac)_3$
$(H_2O)_2$ (it emits at 531nm by exciting at 265nm), were incubated
respectively with alveolar macrophages. LUVs were prepared as
follows.[8] Egg PC (30 mg) was dissolved in 3 ml of ethyl ether and
mixed with 1 ml of saline. The resulting mixture was sonified in
a bath-type sonicator (Branson Ultrasonic Cleaner B-220) for 5 min
at 28 W and 0°C. Then, the homogeneous suspension obtained was
evaporated using a rotary evaporator under 350 mmHg until the

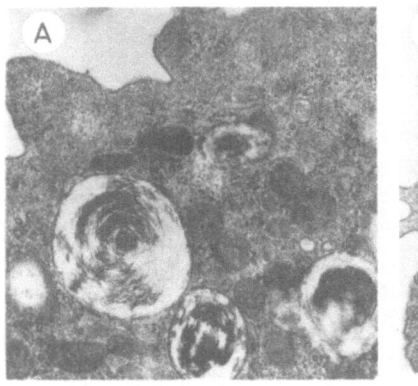 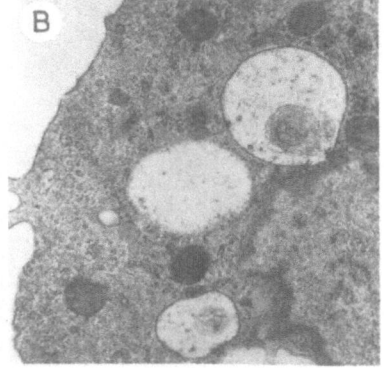

Fig. 3. Electron micrographs showing phagocytosis of egg PC MLV
         by alveolar macrophages of guinea pigs after incubation
         at 37°C for 30 min (A) and 1 h (B); x 85,000.

suspension became milky. A further 2 ml of saline was added and
the evaporation continued under 700 mmHg for about 1 h until ethyl
ether was completely removed. When the water insoluble Tb(acac)$_3$
(H$_2$O)$_2$ was to be encapsulated, it was dissolved in ether with the
lecithin, while water soluble compounds such as antibiotics or CF
were dissolved in saline and mixed with the ether solution at the
initial stage. Coating the LUV with polysaccharide was performed
by adding 1 ml of saline containing 3 mg of polysaccharide into the
LUV suspension obtained and keeping for 1 h at 20°C. Thereafter,
if necessary, water soluble unencapsulated materials were washed out
with saline by centrifugation repeated thrice at 4°C and 20,000 rpm
for 1 h. Figure 4 shows the pictures for the former probe (A) and
the latter probe (B). The most distinct difference for the two
types of liposomes is in the brightness of fluorescence in the cells:
for CF-loaded liposomes the whole cell fluoresces upon the lysis of
internalized liposomes and distribution of CF in the cell, while in
the case of the Tb-complex loaded liposomes no such distribution of
the fluorescent probe was observed. Instead, the probe was stained
as dots in the cells. Hence, present results suggest that when
water soluble drugs encapsulated in liposomes are used, they should
work effectively in the cells upon the phagocytosis and subsequent
lysis of liposomes. Even from direct observations in fluorescence
microscopy, it was apparent that OPA-coated liposomes were more
effectively internalized in phagocytes than conventional liposomes:
more cells were visible with fluorescent probes. This was confirmed
more clearly by the measurement of efficiency in the internalization
of LUV containing ^{14}C -dipalmitoylphosphatidylcholine (New England
Nuclear, U.S.A.). As shown in Table 2, in the case of both alveolar
macrophages and monocytes, OPA-112(4.9)-coated LUVs were more effectively internalized compared with the conventional liposomes without
OPA.

MULTIPLICATION OF LEGIONELLA PNEUMOPHILA IN HUMAN MONOCYTES

     Through the above investigations, it was confirmed that the
OPA-coated liposomes are useful as an improved drug carrier especially for the treatment of lung diseases such as pneumonia. The
best treatment of bacterial infectious diseases is considered to be
the use of antibiotics. In 1976, a new bacterial infectious disease,
Legionnaires' disease, was found at Philadelphia, U.S.A. There are
several problems in the treatment of Legionnaires' disease with
antibiotics: namely, the bacterium, Legionella pneumophila, grows
in macrophages and produces β-lactamase. There is, therefore, a
large discrepancy in the efficacy of antibiotics against L. pneumophila between in-vivo and in-vitro investigations. Thus, we
decided to utilize our OPA-coated LUV encapsulating an aminoglyco-
side antibiotic (sisomycin, Shering, U.S.A.) for the treatment of
experimental Legionnaires' disease in guinea pigs.

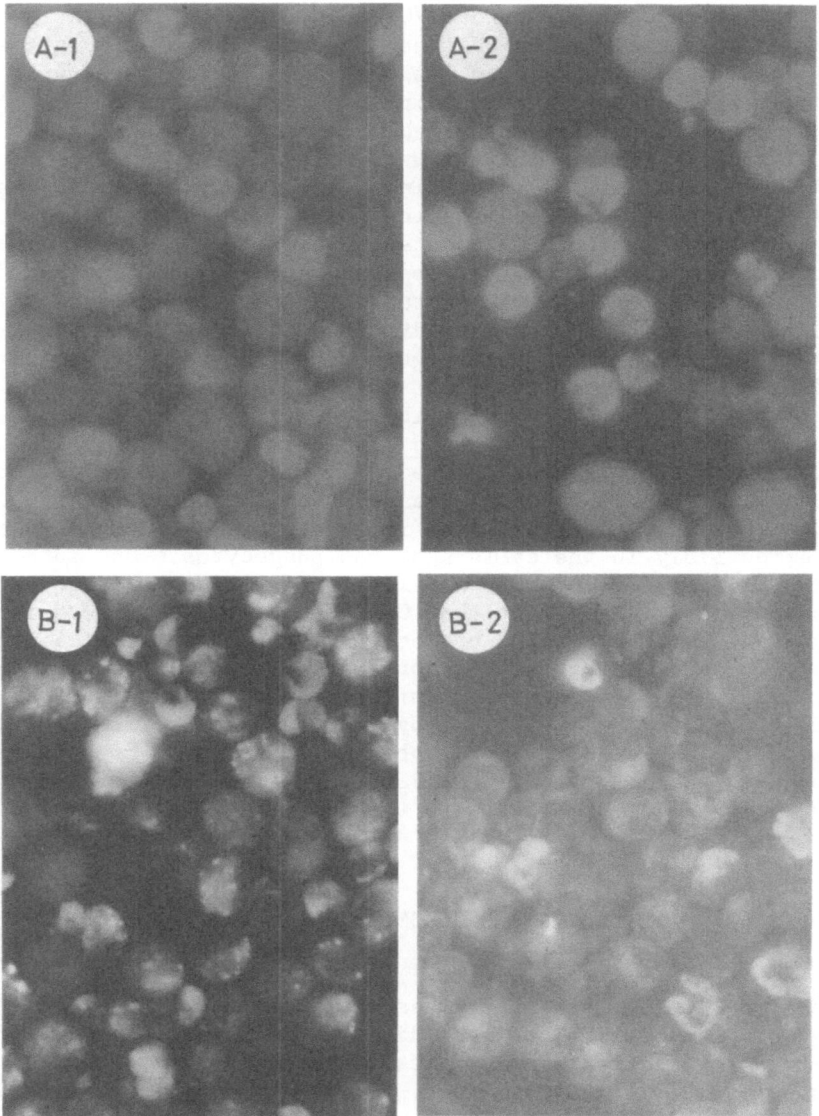

Fig. 4. Fluorescence microscopy pictures (x 700) of alveolar
macrophages from guinea pigs after incubation with the
OPA-coated (A-1) and conventional LUV (A-2) encapsulating the water soluble fluorescent probe, CF, and with
the OPA-coated (B-1) and conventional LUV (B-2) containing the water insoluble fluorescent probe, $Tb(acac)_3$
$(H_2O)_2$, at 37°C for 1 h.

Table 2. Phagocytosis of liposomes

Liposomes	Phagocytosis,[a] %[b]	
	Guinea pig alveolar macrophages	Human monocytes
OPA-112(4.9)-coated LUV[c]	19.1	80.8
Conventional LUV[c]	2.7	34.2

[a]Phagocytes (2 x 10^6/ml) were incubated with LUV at 37°C for 1 h in RPMI medium containing 15% FCS and the cells and supernatants were separated by centrifugation (see text).

[b]Calculated from cpm determined independently for both phagocytes and supernatants after separation by centrifugation.

[c]Liposomes were prepared by reversed phase method (see text).

Before starting the treatment, multiplication of L. pneumophila in human monocytes was investigated using sisomycin encapsulated in liposomes. Heparinized venous blood from healthy adults was mixed with an equal volume of 3% dextran in isotonic saline and red blood cells were separated by sedimentation at room temperature for 25 min. Leukocyte-rich supernatants obtained were centrifuged at 400 x g (room temperature) on Ficoll-Conray density gradients. Separated mononuclear cells were washed twice by and resuspended in RPMI-1640 medium containing 15% FCS. L. pneumophila, serogroup 1 (Philadelphia 1) (5 x 10^3 - 10^5 CFU (colony forming units)/ml), were incubated with monocytes (2 x 10^6 cells/ml) in a gyratory shaker for 1 h at 37°C and thereafter kept stationary in a 5% CO_2-95% air incubator. CFU of L. pneumophila were determined daily. Free and liposome-encapsulated antibiotics were added one day after infection unless otherwise stated. In some experiments, antibiotics were added at 1 h or 2 days after infection. Results are shown in Figs. 5 and 6. In Fig. 5, with the exception of the control (A) and experiment B (free antibiotics were administered), 36% of the antibiotics administered were encapsulated in OPA-112(4.9)-coated LUV, which means that 64% of the antibiotic was free and could kill bacteria present in the exterior of cells. The most noteworthy finding in this work is the CFU in monocytes as seen in the insert of Fig. 5. This shows that even if 50 γ/ml of sisomycin were administered, free sisomycin could not kill L. pneumophila in the cells, and no significant difference in CFU in monocytes was observed comparing with the control experiment. On the other hand, liposome-encapsulated sisomycin showed a drastic decrease in CFU in cells. In Fig. 6, except for the control (A) and experiment B, the antibiotic was encapsulated in OPA-112(4.9)-coated LUV and no free antibiotic was present in the system. Though the apparent germifuge activity was decreased compared with that of experiment 1

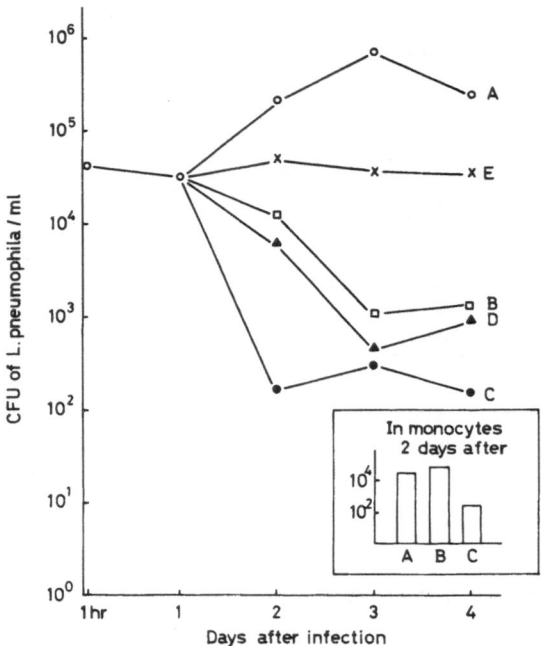

Fig. 5.   Effect of sisomycin encapsulated in OPA-coated LUV on
Legionella pneumophila multiplication in human monocytes
at 37°C (Experiment 1): A (-o-), control experiment without antibiotic; B (-□-), free antibiotic, 50𝑦/ml; C (-•-),
50 /ml, in the presence of liposomes; D (-▲-), 5𝑦/ml, in
the presence of liposomes; and E (-x-), 0.5𝑦/ml, in the
presence of liposomes. When the OPA-coated LUVs were
employed, 36% of the total dose were encapsulated in
liposomes and the rest was free (for details, see text).

(Fig. 5), the CFU value in monocytes was certainly decreased again
by antibiotics encapsulated in liposomes (see insert in Fig. 6).
These results show clearly that we were able to succeed in the
transfer of the water soluble antibiotic into the cells and kill
the bacteria growing in the cells.

TREATMENT OF EXPERIMENTAL LEGIONNAIRES' DISEASE IN GUINEA PIGS

     Following the method established by Pennington in 1978,[10]
intratracheal inoculation with tracheotomy was carried out in
guinea pigs (body weight 280-320 g). Treatment started at 24 h
after inoculation and 4 mg/kg body weight/day of antibiotic was
administered by intravenous and intramuscular injections twice a

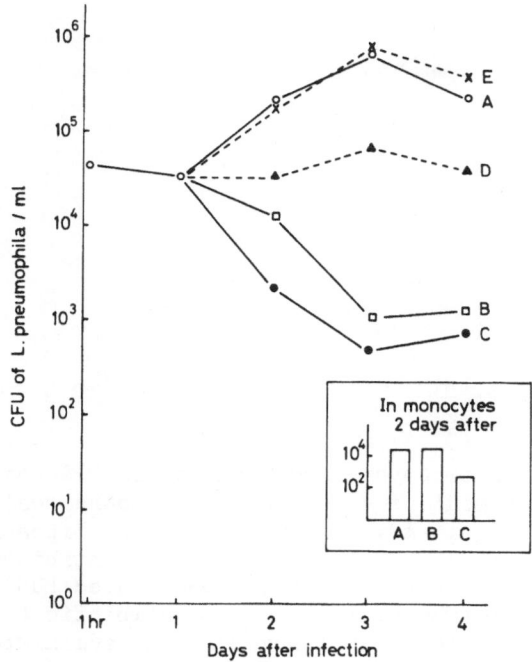

Fig. 6. Effect of sisomycin encapsulated in OPA-coated LUV on <u>Legionella pneumophila</u> multiplication in human monocytes at 37°C (Experiment 2): A (-o-), control; B (-□-), free antibiotic, 50γ/ml; C (-●-), 50γ/ml in liposomes; D (-▲-), 5 /ml in liposomes; and E (-x-), 0.5γ/ml in liposomes. In this experiment, all administered antibiotic was encapsulated in the OPA-coated LUV except in the cases of A and B (for details, see text).

day for 7 days. In this treatment, 36% of the antibiotic was encapsulated in OPA-coated LUV and the rest was free. Figure 7 shows that when the disease was treated with free antibiotic all guinea pigs died within 6 days, while 100% survival rate was attained when treated with a mixture of free and liposome-encapsulated antibiotic.

To our knowledge, this is the first success in the treatment of bacterial infectious diseases with antibiotics targeted to specific tissue using liposomes. We believe that the present success was attained by (1) targetable drug carrier of liposomes coated with the amylopectin derivatives as the recognition ligand, (2) increased efficiency in the internalization of such a drug carrier into the phagocytes, (3) transfer of the water soluble antibiotic into cells, and (4) decreased toxicity and increased germicidal activity of the antibiotic by its encapsulation in liposomes.

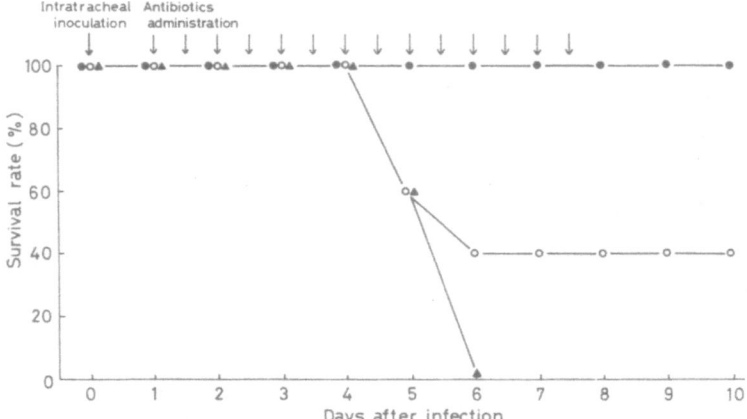

Fig. 7. Effect of sisomycin encapsulated in OPA-coated LUV in
the treatment of experimental Legionnaires' disease in
guinea pigs: -▲-, antibiotic without liposomes was
administered intramuscularly; -o-, antibiotic (of which
36% was encapsulated in the OPA-coated LUV) was administered intramuscularly; -●-, antibiotic (of which 36%
was encapsulated in liposomes) was administered intravenously. Five guinea pigs were treated in each experiment (for details, see text).

ACKNOWLEDGMENT

    We are grateful to Professor Lidia M. Vallarino of Virginia
Commonwealth University for the kind gift of terbium complex as a
fluorescent probe. Grants-in-Aid for Scientific Research (57219013
and 58211015) from the Ministry of Education, Science and Culture
of Japan are gratefully acknowledged.

REFERENCES

1. G. Gregoriadis, ed., "Drug Carriers in Biology and Medicine",
    Academic Press, London (1979).
2. D. Papahadjopoulos, ed., "Liposomes and Their Uses in Biology
    and Medicine", Ann. N.Y. Acad. Sci., vol. 308, New York (1978).
3. G. Gregoriadis, J. Senior and A. Trouet, eds., "Targeting of
    Drugs", Plenum, New York (1982).
4. J. Sunamoto, K. Iwamoto, M. Takada, T. Yuzuriha and K. Katayama,
    Improved drug delivery to target specific organ using lipo-
    somes as coated with polysaccharide, in: "Polymers in Medicine:

Biomedical and Pharmacological Applications", E. Chiellini, ed., Plenum Publishing Co., New York (1983).

5. J. Sunamoto, K. Iwamoto, M. Takada, T. Yuzuriha and K. Katayama, Polymer coated liposomes for drug delivery to target specific organs, in: "Recent Advances in Drug Delivery Systems", J.M. Anderson and S.W. Khim, eds., Plenum Publishing Co., New York (1984).

6. W.S. Singleton, M.S. Gray, M.L. Brown and J.L. White, Chromatographically homogeneous lecithin form egg phospholipids, J. Am. Oil Chem. Soc., 42:53 (1965).

7. J. Sunamoto, H. Kondo and A. Yoshimatsu, Liposomal membranes I. Chemical damage of liposomal membranes with functional detergent, Biochim. Biophys. Acta, 510:52 (1978).

8. D. Deamer and A.D. Bangham, Large volume liposomes by an ether vaporization method, Biochim. Biophys. Acta, 443:629 (1976).

9. I.J. Fidler, A. Ray, W.E. Fagler, R. Kirsh, P. Buyelski and G. Poste, Design of liposomes to improve delivery of macrophage-augmenting agents to alveolar macrophages, Cancer Res., 40:4460 (1980).

10. J.E. Pennington and M.G. Ehrie, Pathogenesis of Pseudomonas aeruginosa pneumonia during immunosuppression, J. Infect. Dis., 137:764 (1978).

ALTERATION OF LIPOSOMAL-SURFACE PROPERTIES

WITH SYNTHETIC GLYCOLIPIDS

M.M. Ponpipom and T.Y. Shen

Merck Sharp & Dohme Research Laboratories

Rahway, New Jersey, U.S.A.

INTRODUCTION

Liposomes have been proposed as carriers to introduce biologically active agents into cells, and their various applications have been extensively investigated.[1-3] The tissue distribution and clearance kinetics of drug-containing liposomes are known to be affected by phospholipid composition, size and surface charge.[4,5] Small liposomes, were shown to be better than larger liposomes for specific delivery of their contents to target cells in-vivo.[6] Although successes have been amply demonstrated for in-vitro targeting of liposomes including covalent attachment of proteins onto the surface of liposomes,[7,8] no in-vivo specificity has yet been reported for any liposomes thus modified. The interaction of liposomes with cells in-vivo occurs in quite a complex biological milieu and the binding of proteins from this milieu may also have important effects on liposome behaviour.[9] Our approach to improve liposomes as a drug-delivery system is to use small synthetic glycolipids as cell-surface ligands to alter the tissue distribution of the modified liposomes in various tissues and organs.

Carbohydrate determinants of glycolipids and glycoproteins present at the cell surface are known to participate in intercellular recognition processes (see Ponpipom et al., this volume). Such receptor recognition and uptake of carbohydrate derivatives have been extended to liposomes containing natural glycoproteins and glycolipids. In addition, small synthetic glycolipids are also effective for the introduction of carbohydrate determinants onto the surface of liposomes.

373

ADVANTAGES OF SYNTHETIC GLYCOLIPIDS

The general formula of synthetic glycolipids is represented by
S-R-L where S is the carbohydrate determinant, R is the spacer arm,
and L is the lipophile. The lipid-anchoring group may be a cholesterol, ceramide, diglyceride, phosphatidylethanolamine or any other
suitable lipid moiety. The synthesis of these glycolipids would
allow the optimization of:

a. The carbohydrate ligand S for desired cellular
   specificity.
b. The linkage R for receptor interaction and
   metabolic stability.
c. The lipophilic anchor L for membrane permeability
   and intercalation.
d. Physical characteristics of liposomes through the
   introduction of charged and/or other functional
   groups.

Compared with complex natural glycolipids and glycoproteins,
small synthetic glycolipids may also offer the following advantages:

a. A definitive indentification of the carbohydrate
   determinant(s) involved.
b. Ease of handling and more quantitative control of
   experimental systems.
c. Consistent preparation of large amounts of compounds
   in a pure state.
d. Convenience of studying molecular interactions of
   the ligand with receptors by spectroscopic methods.

In our studies, we chose cholesterol as a lipid-anchoring group
and the hexyl moiety as a spacer arm (Fig. 1). Thioglycosidic linkage was used because it is more stable towards acid and enzymatic
hydrolysis than O-glycosides.[10] It has been suggested[11] that carbohydrate groups which occur naturally in membrane oligosaccharides
consist of only D-glucose, D-galactose, D-mannose, 2-acetamido-2-
deoxy-D-glucose, 2-acetamido-2-deoxy-D-galactose, L-fucose, L-
arabinose, D-xylose, and N-acetyl-D-neuraminic acid (Fig. 2). Thus,
it was thought of interest initially to prepare glycolipids containing these saccharides for incorporation into liposomes and see
whether they would improve the selectivity, stability or any other
in-vivo characteristics of the modified liposomes. These glycolipids were prepared as outlined in Fig. 3 from Per-O-acetyl-1-
thioglycopyranoses and 6-(5-cholesten-3β-yloxy)hexyl iodide in
dichloromethane/triethylamine or in oxolane/1,5-diazabicyclo(5.4.0)
undec-5-ene. Both the blocked and deacetylated products were isolated in high yields.[12,13] These glycolipids (Fig. 2) can be
readily incorporated into liposomes. Some analogs were also found
to exert as immune adjuvant activity comparable to that of N-

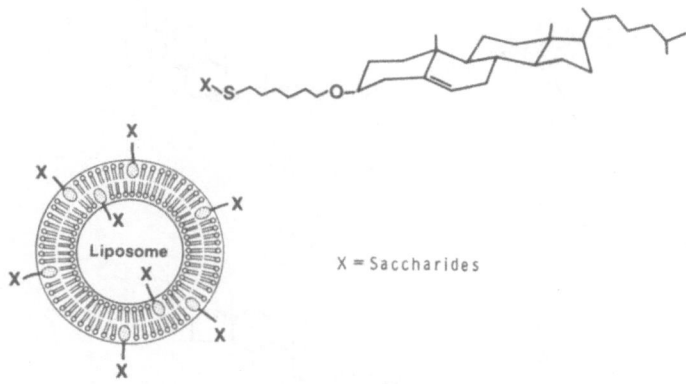

Fig. 1. Liposomes with synthetic determinants.

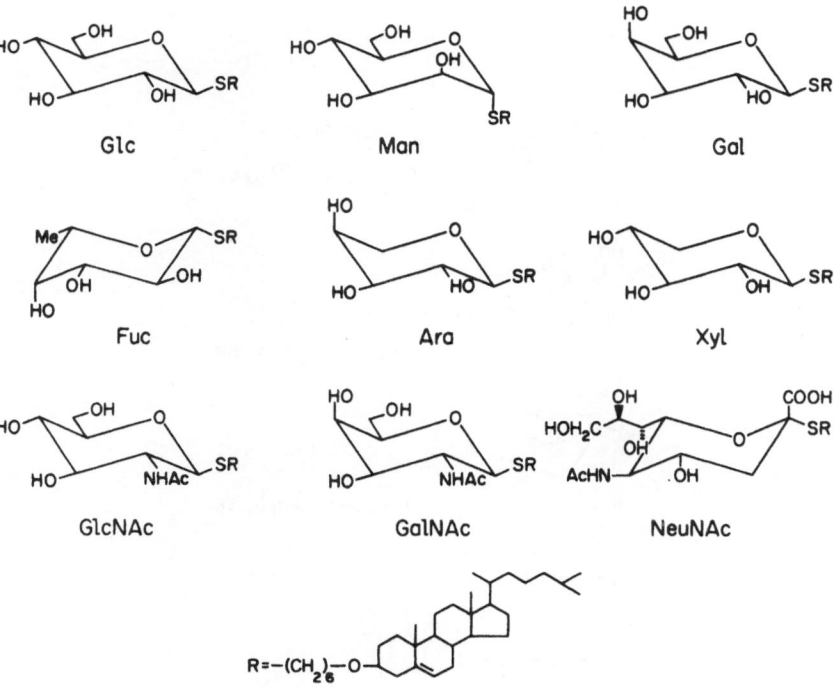

Fig. 2. Structures of synthetic glycolipids.

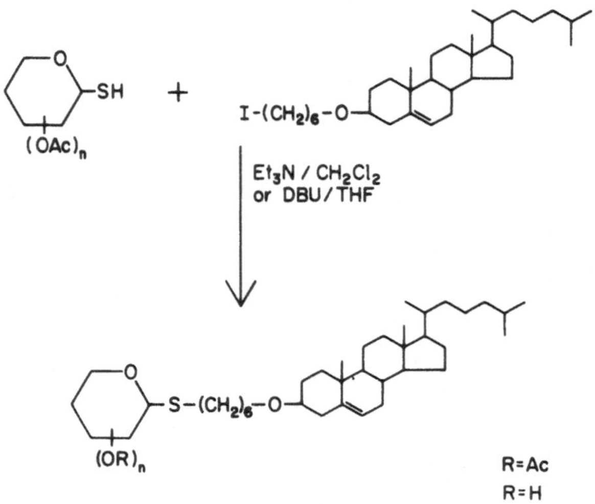

Fig. 3.   Synthesis of glycolipids.

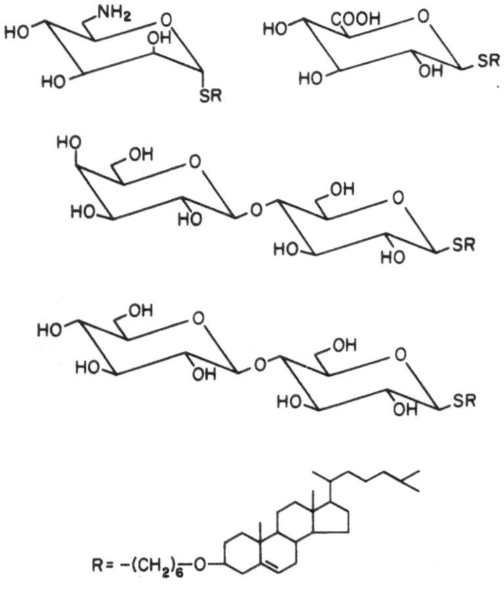

Fig. 4.   Disaccharides and charged glycolipids.

acetylmuramyl-L-alanyl-D-isoglutamine derivatives.[14]   6-Amino-6-
deoxy-D-mannose and D-glucuronic acid glycolipids were prepared to
assess the effects of surface charge (Fig. 4); glycolipids containing lactose and cellobiose were also prepared to examine the influence of disaccharides on liposomal surface properties as compared
to the influence exerted by monosaccharides such as D-galactose and
D-glucose.

INFLUENCE OF LIPID-ANCHORING GROUPS ON LIPOSOME STABILITY

     The stability in-vitro (in the presence of serum, plasma, or
whole blood) and in-vivo (blood circulation) is affected by the
amount of cholesterol incorporated in liposomes.[15]   Incorporation
of cholesterol in liposomes is known to pack the phospholipid molecules above their phase transition temperature and reduce permeability to solutes.[16]   It has been reasoned that the increased packing
of phospholipids by cholesterol can also prevent their removal by
plasma high density lipoproteins and preserve liposomal stability
in the presence of serum.[17,18]   It is thus noteworthy that liposomes
containing galactosylcholesterol are more stable than liposomes containing galactosylceramide (Fig. 5) both in in-vitro and in-vivo
systems (Fig. 6-8). The turbidity measurement of liposomes containing galactosyllipids is shown in Fig. 6. The increased turbidity
seen after sonication, for the GalCer liposomes, indicated that they
were either aggregating or undergoing fusion to form larger liposomes, at a faster rate than for the GalChol liposomes.[19]   This
observation was supported by the stability measurement of the

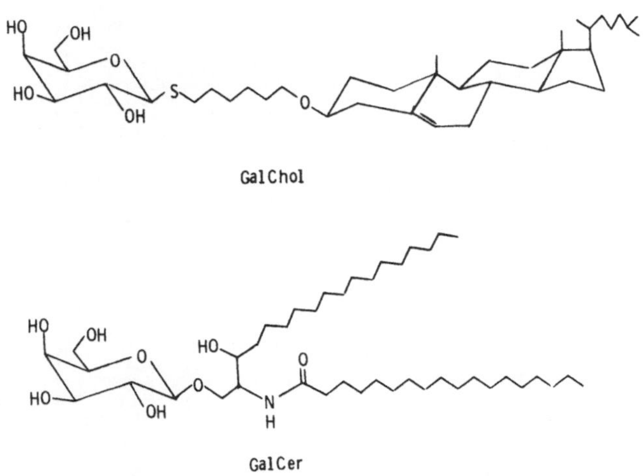

GalChol

GalCer

Fig. 5. D-Galactopyranosyllipids.

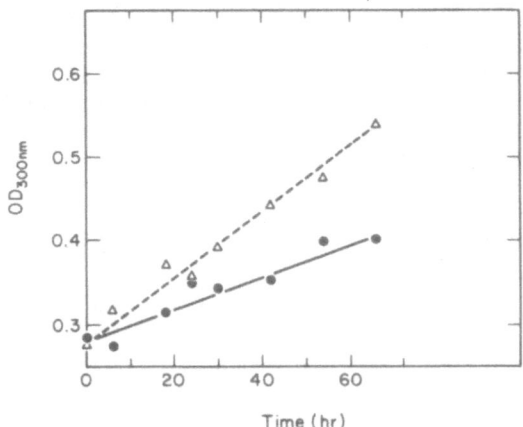

Fig. 6. Change in turbidity of D-galactose-modified
         liposomes stored at 37°C: ●, GalChol liposomes;
         △, GalCer liposomes.[19]

galactose-modified liposomes in serum using r-ray perturbed angular
correlation technique (Fig. 7). The different stabilities of GalChol
and GalCer liposomes in live mice injected intraperitoneally or
orally are shown in Figures 8A and 8B, respectively. Thirty hours
after oral administration, about 50% of the GalChol liposomes were
found to be intact, while less than 20% of the GalCer liposomes still

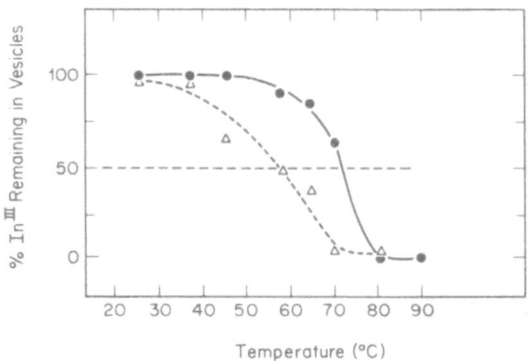

Fig. 7. Thermolability of D-galactose-modified lipo-
         somes in serum: ●, GalChol liposomes; △,
         GalCer liposomes.[19]

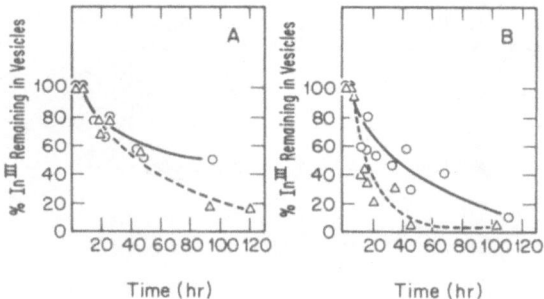

Fig. 8. Stability of GalChol and GalCer liposomes
         in live mice administered by intraperitoneal
         injection (A) or orally (B): O , GalChol
         liposomes; Δ , GalCer liposomes.[19]

retained their liposomal integrity (Fig. 8B). Similarly, when the
modified liposomes were administered intraperitoneally, the breakdown of GalCer liposomes was more rapid than that of GalChol liposomes[19] (Fig. 8A). These results indicate that glycosylcholesterols
are more suitable for in-vitro tissue distribution studies than
glycosylceramides.

UPTAKE OF LIPOSOMES BY MOUSE PERITONEAL MACROPHAGES

     Wu and co-workers[20] previously reported the high uptake of
liposomes modified with aminomannose by mouse peritoneal macrophages

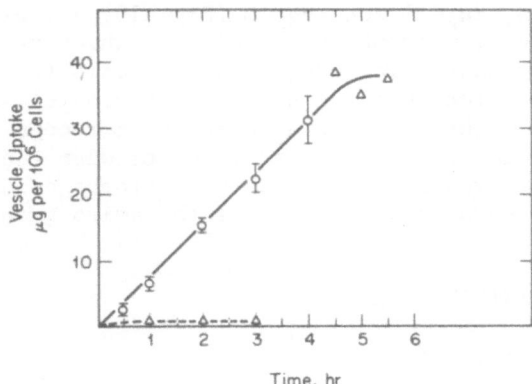

Fig. 9. Uptake of various liposomes by mouse peritoneal
         macrophage at 37°C as a function of incubation
         time: O and Δ , NH$_2$Man liposomes, Δ , steary-
         lamine, diacetylphosphate, Glc, Gal, Man, Fuc,
         and control liposomes.[20]

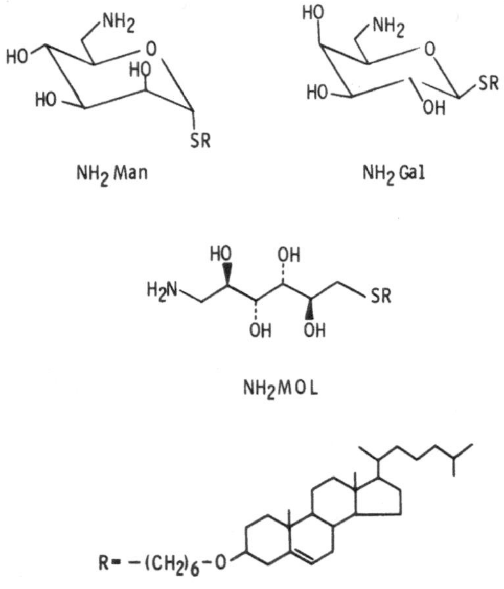

Fig. 10.   Structures of aminoglycolipids.

(Fig. 9). To investigate whether the high association of aminomannose liposomes with macrophages might simply be due to the
presence of large numbers of amino groups per unit area extending
above the liposomal surface, the uptake of other aminoglycolipid-
containing liposomes by macrophages was also studied. Liposomes
containing NH$_2$Man, NH$_2$Gal and NH$_2$MOL (Fig. 10) all exhibited a somewhat similar rate of macrophage association which was greater than
that of liposomes containing NH$_2$Chol and NH$_2$Carb (Fig. 11). These
observations indicated that the high association of NH$_2$Man liposomes
with macrophages might not be due to receptor recognition. It was
probably due to electrostatic interaction between the cationic liposomes and anionic groups present on the macrophage surface, and
might be a function of the distance of the amino group from the
liposomal surface.

IN-VIVO TISSUE DISTRIBUTION

(1) Intravenous Injection

To test directly whether small synthetic glycolipids would be
useful ligands for selective delivery of drugs entrapped in liposomes in-vivo, we investigated the tissue distribution of liposomes

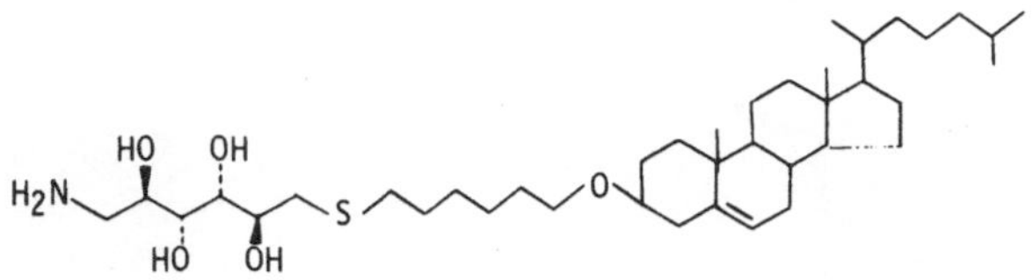

Fig. 11.   Structures of amino-containing lipids.

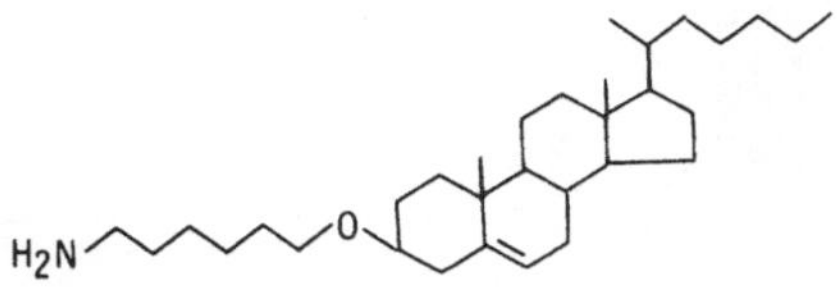

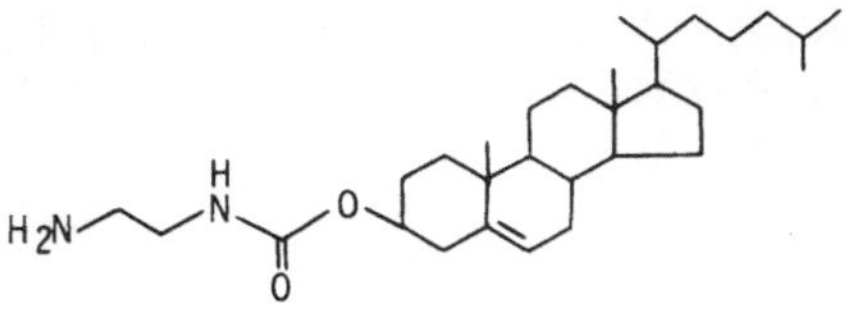

modified with the glycolipids shown in Fig. 2. Liposomes were prepared from DSPC and a glycolipid (14 mole %) with $^{51}CrO_4^=$ as an
internal marker. The modified liposomes were administered intravenously into rats and the radioactivity in each tissue was measured
with a Packard gamma-counter at time intervals. The tissue distribution of liposomes in rats was found to vary with different glycolipids (data not shown); however, no clear and useful direction of
liposomes to target organs and tissues was observed. In each case
the liposomes were removed from the bloodstream primarily by the
liver and spleen, as expected for intravenously injected liposomes.[21]
Failure to find tissue selectivity in-vivo (by the intravenous route)
could be due to a number of reasons: for example, insufficient exposure of the carbohydrate determinants above the liposomal surface,
non-specific masking of determinants by serum proteins, lack of liposomal access to the target cells because of the size of blood vessel
pores or other obstacles.[22]   It could also be due to non-recognition
of the tested glycolipids by the target cells.

The in-vivo tissue distribution of lactose-modified liposomes

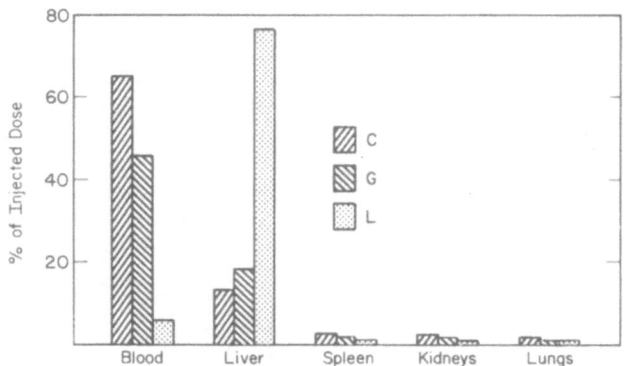

Fig. 12. Tissue distribution of recovered $^{51}CrO_4^=$ 2 h
after intravenous injection of liposomes in
rat: C, DSPC/PG/Chol, 45:5:50 liposomes; G,
DSPC/PG/Chol/Gal, 45:5:35:15; L, DSPC/PG/Chol/
Lac, 45:5:35:15. Each bar represents the mean
of six rats.

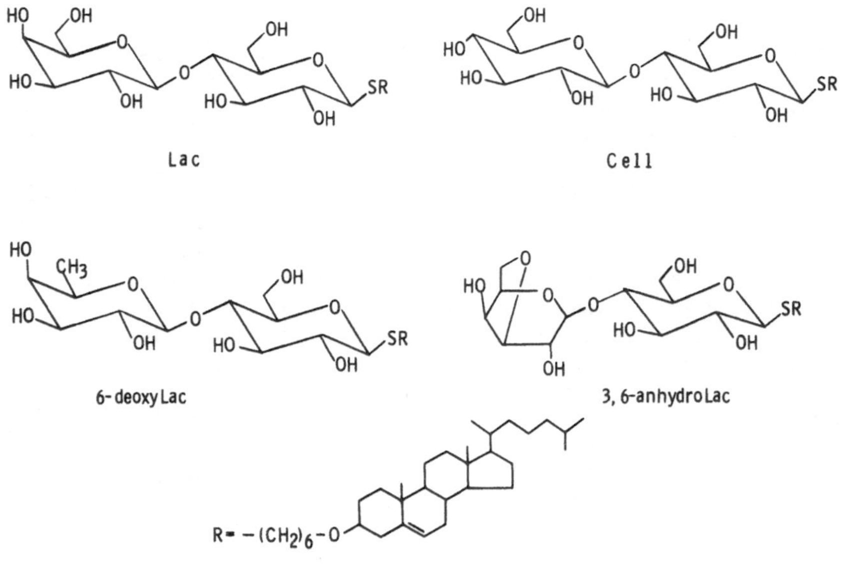

Fig. 13. Structures of disaccharide glycolipids.

Table 1. Tissue Distribution (%) of (^{51}Cr)-chromate 2 h after i.v. injection.

Tissue	Control	Fuc	Gal	Cell	Lac	6CH$_3$-Lac	3,6-Anhydro-Lac
Blood	56.32 ± 1.45	40.76 ± 0.45	47.90 ± 1.38	6.71 ± 1.98	5.75 ± 0.47	7.42 ± 1.71	11.09 ± 3.32
Liver	14.71 ± 1.60	15.59 ± 3.15	18.79 ± 1.68	75.68 ± 2.08	76.72 ± 2.90	81.36 ± 4.08	65.31 ± 5.25
Spleen	2.13 ± 0.21	2.96 ± 0.55	1.50 ± 0.15	1.78 ± 1.06	0.35 ± 0.04	1.34 ± 0.72	7.07 ± 3.48

Each set of values is an average of 2 experiments of 3 rats each and represents % ± SEM of injected radioactivity. Liposome compositions (molar ratios) were: Control liposomes, Chol/DSPC/PG (50:45:5); glycolipid-containing liposomes, glycolipid/Chol/DSPC/PG (15:35:45:5).

is more interesting. Liposomes containing lactose were rapidly
taken up by liver as compared to control liposomes or galactose-
modified liposomes (Fig. 12). For example, two hours after intravenous injection of liposomes containing lactose, 76-77% of the label
was found in the liver; whereas for control liposomes or galactose-
modified liposomes, only 15-19% of the label was found in the liver.
Similar findings were reported by Scherphof (this volume) and Szoka
and Mayhew[23] with lactosylceramide-containing liposomes. These
results may suggest better recognition of terminal galactose of the
disaccharide by hepatocytes, because it is more exposed above the
liposomal membrane for interactions with the cell-surface receptors.
To clarify the above observations, 6-deoxylactose and 3,6-anhydro-
lactose (Fig. 13) were also synthesized and incorporated into liposomes for in-vivo tissue distribution studies.[24] Interestingly,
these modified liposomes were also rapidly taken up by liver cells
(Table 1; 81% and 65% of the label were found in the liver 2 h after
intravenous injection). It is noteworthy that 3,6-anhydrolactose
also enhanced the uptake of liposomes by spleen. Since the terminal
saccharides, 6-deoxygalactose and 3,6-anhydrogalactose, are unnatural sugars, the enhanced uptake of liposomes modified with the disaccharides by the liver (Table 1) may not be due to receptor recognition. It appears that different glycolipids may be affecting the
results indirectly by influencing vesicle size, charge, or aggregation, rather than directly.

The tissue distribution of aminomannose liposomes injected
intravenously into mice is shown in Table 2. Liposomes modified with
NH_2Man were cleared more rapidly from the bloodstream than control
liposomes, and were taken up predominantly by the lungs at early
stages.[25,26] After a transient retention in the lungs, the liposomes passed to post-capillary venules and recirculated to other
organs. Most of the label eventually deposited in the liver. This
transient lung entrapment might result from recognition and binding
of NH_2Man liposomes by surface receptors of the lungs, or it might
simply be due to a physical entrapment of large aggregated liposomes
in the first capillary bed which they encountered after tail vein
injection. To further investigate this interesting observation,
NH_2Man liposomes were also injected into the abdominal aorta (Table
3). Liposomes introduced by this route gave less label in the
lungs but more in the lower extremities than those administered by
the abdominal vena cava or the tail route.[25] Thus, depending upon
the route of injection, NH_2Man liposomes can be retained by the
lungs or other capillaries. From these findings, it appears that
the uptake of NH_2Man liposomes by the lung cells may not be due to
receptor recognition.

(2) Footpad Injection

Liposomes modified with mannose and NH_2Man were also injected
into mice by the footpad route. The radioactivity remaining in the

Table 2. Percent of total injected radioactivity in mouse tissues after intravenous injection of liposomes containing $^{51}CrO_4^{=}$.

Time	Blood		Lungs		Liver	
	Control liposomes	Amino mannose liposomes	Control liposomes	Amino mannose liposomes	Control liposomes	Amino mannose liposomes
30 min	38	3	1	49	22	9
45 min	36	3	1	27	23	21
1 h	39	2	1	29	20	17
1.5 h	-	2	-	24	-	20
2 h	30	3	1	20	20	24
72 h	-	1	-	2	-	42

Table 3. Tissue distribution (%) of (^{51}Cr) chromate
10 min after intravenous injection of NH$_2$Man
liposomes.

Tissue	Route of Administration	
	Abdominal vena cava	Abdominal aorta
Lung	53.5%	26.5%
Lower body and viscera	22.5%	59.5%
Liver	16.5%	11.5%

Each set of values is an average of 2 rats. NH$_2$Man-containing
liposomes were prepared from DSPC/NH$_2$Man/Chol (2:0.5:0.5, molar
ratio).

footpad depended dramatically on the form of the injected formulation (Fig. 14). Free chromate and its mixture with preformed NH$_2$Man
liposomes were quickly carried away from the injection site. All
three liposome formulations significantly increased the retention of
the label at the injection site, with a clear saccharide effect:
NH$_2$Man>Man>Control liposomes.[25] The loss of the label from injection site was a biphasic process, most of the loss occuring in the
first 1.5-24 h and what remained at 24 h was removed at a much slower
rate.

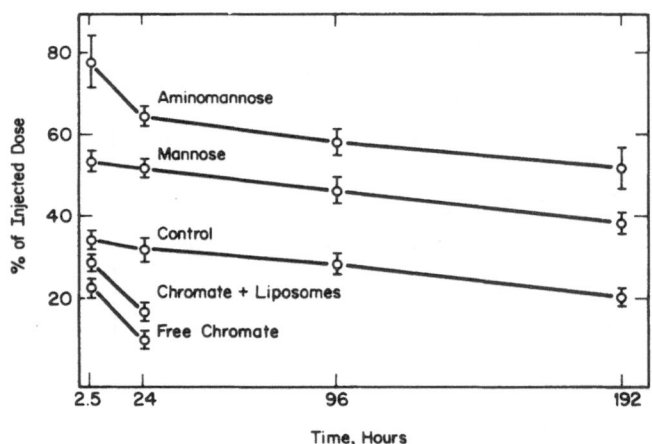

Fig. 14. Radioactivity remaining at the injection site
at various times after footpad administration
of ^{51}CrO$_4$= entrapped liposomes.[25]

Table 4. Radioactivity in popliteal lymph nodes after footpad injection of liposomes containing $^{51}CrO_4^=$.

Liposome Composition	% of Injected Dose			
	Time			
	2.5 h	1 d	4 d	8 d
Free chromate	0.3 ± 0.1	0.3 ± 0.1	-	-
Free chromate + Aminomannose liposomes	0.8	0.5	-	-
Control liposomes	0.5 ± 0.2	0.4 ± 0.1	0.2	0.2
Aminomannose liposomes	0.2 ± 0.1	0.6 ± 0.2	0.4 ± 0.1	0.8 ± 0.1
Mannose liposomes	1.8 ± 0.5	2.0 ±.0.3	1.9 ± 0.2	1.4

    Liposomes modified with mannose gave significant lymph node
labeling when they were administered by footpad injection (Table 4).
Although the lymphatic label was a small fraction of the total injected dose, it was actually quite concentrated as calculated on
weight basis. The popliteral node ($\simeq$5 mg per pair, net weight)
contained 2.0% of the injected label from mannose liposomes at 24 h
after footpad injection, i.e., a concentration of 0.4% per mg. This
concentration was much higher than the approximate 0.013% per mg in
the liver (2.0 g weight) of the same mice.[25]   The selectivity of
lymph node labeling vs. liver labeling was even more pronounced with
NH$_2$Man liposomes at later times because of the observed low label in
the liver. This lymphatic labeling can either be in transit or
bound, as by macrophage recognition of D-mannose residues.[26]

THE FATE OF MODIFIED LIPOSOMES IN-VIVO

    Alteration of liposomal surface properties with synthetic glyco-
lipids has marked influence on the in-vivo stability of liposomes.
The stability of liposomes administered subcutaneously has been shown
to be affected by surface charge (Fig. 15). Negatively charged lipo-

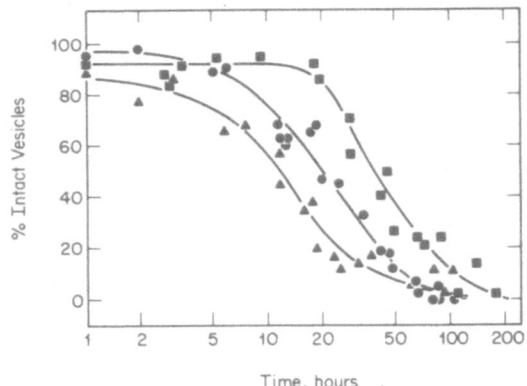

Fig. 15. Effect of charge on vesicle stability after
         subcutaneous administration of liposomes.
         Each subcutaneous injection of lipid vesicles
         typically contained 1.0 mg of total lipid,
         20 μCi of ^{111}In^{+3} bound to 1 mM NTA inside,
         and phosphate-buffered saline (0.9% NaCl, 5 mM
         sodium phosphate, pH 7.4) inside and outside
         of the vesicles. The total injection volume
         was 200 μl. The percentage of vesicles remain-
         ing intact was determined by PAC measurements
         of the skin. Symbols: ●, DSPC/Chol (2:1);
         ▲, DSPC/Chol/DCP (2:0.5:0.5); ■, DSPC/Chol/-
         Sa (2:0.5:0.5).[27]

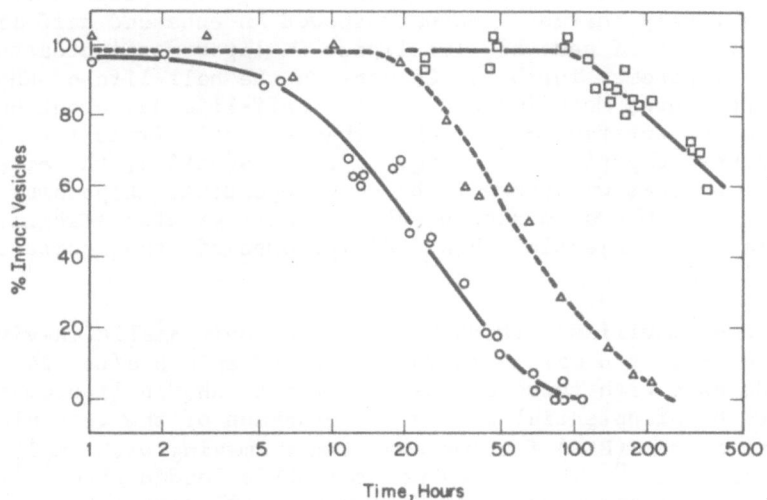

Fig. 16. Effect of NH$_2$Man on vesicle stability: O, DSPC/
Chol (2.1); □ , DSPC/Chol/NH$_2$Man (2:0.5:0.5), △ ,
DSPC/Chol/NH$_2$Man (2:0.9:0.1).[27]

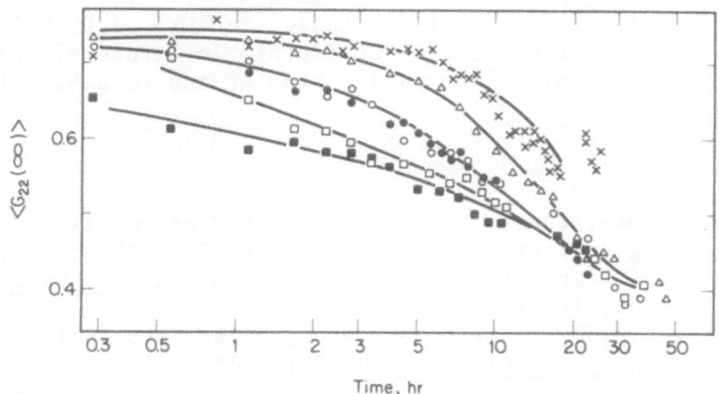

Fig. 17. Stability of vesicle system after intravenous
administration in live mice as measured by PAC.
Each set of points was the average from two
mice. Symbols: △ , DSPC/Chol; ■ , DSPC/Chol/DCP;
O, DSPC/Chol/Fuc; ● , DSPC/Chol/Gal; □ , DSPC/
Chol/SA; X, DSPC/Chol/NH$_2$Man.[28]

somes exhibited a shorter half-life than the control liposomes, whereas positively charged liposomes showed an enhanced half-life.[27] The incorporation of neutral glycolipids in liposomes increased the lifetime of liposomes further. Changes in the half-life of NH_2Man liposomes were even more impressive; the half-life was about 600 hours which was substantially longer than the half-lives for liposomes of other compositions[27] (Fig. 16). In addition, the effect of NH_2Man liposomes was shown to be dose dependent. Liposomes modified with NH_2Man also showed substantially greater stability after intravenous injection than did liposomes of other compositions[28] (Fig. 17).

Liposomes modified with NH_2Man were not only stable in-vivo, they also accumulated mainly in the liver and spleen after 24 hours.[29] Endowed with these unique properties, NH_2Man liposomes were reported to be of potential use in the blockade of the reticulo-endothelial system (RES) for improved tumor imaging with small unilamellar vesicles.[30] Thus, when neutral SUV's loaded with ^{111}In-NTA were injected into mice bearing EMT6 tumors, 50% more label was deposited in tumors of the animals with blocked RES than in controls. Tissue distribution studies showed that tumors from animals with blocked RES (by NH_2Man liposomes) had more than twice the radioactivity per g than any other tissue analyzed.[30] This finding may have significant therapeutic application to tumor diagnosis and therapy.

ACKNOWLEDGMENTS

    The authors thank Dr. J.C. Robbins and Mrs. M.S. Wu for some of the in-vivo tissue distribution studies and Drs. J.D. Baldeschwieler, P.S. Wu, and R.C. Gamble (California Institute of Technology, Pasadena) for our joint collaborative studies and valuable discussions.

REFERENCES

1.   G. Gregoriadis, Use of monoclonal antibodies and liposomes to improve drug delivery, Drugs, 24:261 (1982); The carrier potential of liposomes in biology and medicine, N. Engl. J. Med., 295:704, 765 (1976).
2.   S.B. Kay, Liposomes - Problems and promise as selective drug carriers, Cancer Treatm. Rev. 8:27 (1981).
3.   B.E. Ryman and D.A. Tyrrell, Liposomes - Bags of potential, Essays Biochem. 16:49 (1980).
4.   R.L. Juliano and D. Stamp, The effect of particle size and charge on the clearance rates of liposomes and liposome encapsulated drugs, Biochem. Biophys. Res. Commun., 63:651 (1975).
5.   Y.E. Rahman, E.A. Cerny, K.R. Patel, E.H. Lau, and B.J. Wright, Differential uptake of liposomes varying in size and lipid composition by parenchymal and Kupffer cells of mouse liver, Life Sci., 31:2061 (1982).

6. P. Machy and L.D. Leserman, Small liposomes are better than large liposomes for specific drug delivery in-vitro, Biochim. Biophys. Acta, 730:313 (1983).

7. L.D. Leserman, P. Machy, and J. Barbet, Cell-specific drug transfer from liposomes bearing monoclonal antibodies, Nature, 293:226 (1981).

8. T.D. Heath, J.A. Montgomery, J.R. Piper, and D. Papahadjopoulos, Antibody-targeted liposomes: Increase in specific toxicity of methotrexate-r-aspartate, Proc. Natl. Acad. Sci. USA, 80:1377 (1983).

9. R.L. Juliano and G. Lin, The interaction of plasma proteins with liposomes: Protein binding and effects on the clotting and complement systems, in "Liposomes and Immunobiology," B.H. Tom and H.R. Six, eds., Elsevier/North Holland, New York (1980).

10. M.D. Saunders and T.E. Timell, The acid hydrolysis of glycosides, Carbohydr. Res., 6:121 (1968).

11. S. Roseman, in "Cell Membranes, Biochemistry, Cell Biology and Pathology" G. Weissmann and R. Claiborne, eds., HP Publishing Co., New York (1975).

12. M.M. Ponpipom, R.L. Bugianesi, and T.Y. Shen, Cell surface carbohydrates for targeting studies, Can. J. Chem., 58:214 (1980).

13. J.C. Chabala and T.Y. Shen, The preparation of 3-cholesteryl 6-(glycosylthio)hexyl ethers and their incorporation into liposomes, Carbohyd. Rec., 67:55 (1978).

14. M.M. Ponpipom, R.L. Bugianesi, T.Y. Shen, and A. Friedman, Glycolipids as potential immunologic adjuvants, J. Med. Chem., 23:1184 (1980).

15. G. Gregoriadis, C. Kirby, and J. Senior, Targeting of drugs with liposomes: Studies on optimization in " Optimization of Drug Delivery", H. Bundgaard, A.B. Hansen, and H. Kofod, eds., Munksgaard, Copenhagen (1982).

16. R.A. Demel and B. De Kruyff, The function of sterols in membranes, Biochim. Biophys. Acta, 457:109 (1976).

17. L. Damen, J. Regts, and G. Scherphof, Transfer and exchange of phospholipid between small unilamellar liposomes and rat plasma high density lipoproteins: Dependence on cholesterol content and phospholipid composition, Biochim. Biophys. Acta, 665:538 (1981).

18. C. Kirby, J. Clarke, and G. Gregoriadis, Cholesterol content of small unilamellar liposomes controls phospholipid loss to high density lipoproteins in the presence of serum, FEBS Lett., 111:324 (1980).

19. P.S. Wu, H.M. Wu, G.W. Tin, J.R. Schuh, W.R. Croasmun, J.D. Baldeschwieler, T.Y. Shen, and M.M. Ponpipom, Stability of carbohydrate- modified vesicles in-vivo: Comparative effects of ceramide and cholesterol glycoconjugates, Proc. Natl. Acad. Sci. USA, 79:5490 (1982).

20. P.S. Wu, G.W. Tin, and J.D. Baldeschwieler, Phagocytosis of carbohydrate-modified phospolipid vesicles by macrophages,

Proc. Natl. Acad. Sci. USA, 78:2033 (1981).

21. G. Gregoriadis and A.C. Allison, eds.,"Liposomes in Biological Systems", Wiley, New York (1980).

22. G. Poste, Liposome targeting in-vivo: Problems and opportunities Biol. Cell, 47:19 (1983).

23. F.C. Szoka, Jr. and E. Mayhew, Alteration of liposomal disposition in-vivo by bilayer situated carbohydrates, Biochem. Biophys. Res. Commun., 110:140 (1983).

24. M.M. Ponpipom, M.S. Wu, J.C. Robbins, and T.Y. Shen, Synthesis of 6-(5-cholesten-3β-yloxy)hexyl 4-O-(6-deoxy-β-D-galactopyranosyl)-1-thio-β-Dglucopyranoside and derivatives thereof for in-vivo liposome studies, Carbohyd. Res., 118:47 (1983).

25. M.S. Wu, J.C. Robbins, R.L. Bugianesi, M.M. Ponpipom, and T.Y. Shen, Modified in-vivo behaviour of liposomes containing synthetic glycolipids, Biochim. Biophys. Acta, 674:19 (1981).

26. P.D. Stahl, J.S. Rodman, M.J. Miller, and P.H. Schlesinger, Evidence for receptor-mediated binding of glycoproteins, glycoconjugates, and lysosomal glycosidases by alveolar macrophages, Proc. Natl. Acad. Sci. USA, 75:1399 (1978).

27. M.R. Mauk, R.C. Gamble, and J.D. Baldeschwieler, Vesicle targeting: Timed release and specificity for leukocytes in mice by subcutaneous injection, Science, 207:309 (1980).

28. M.R. Mauk, R.C. Gamble, and J.D. Baldeschwieler, Targeting of lipid vesicles: Specificity of carbohydrate receptor analogue for leukocytes in mice, Proc.Natl. Acad. Sci. USA, 77:4430 (1980).

29 R.T. Proffitt, L.E. Williams, C.A. Presant, G.W. Tin, J.J. Uliano, R.C. Gamble, and J.D. Baldeschwieler, Tumor-imaging potential of liposomes loaded with ^{111}In-NTA: Biodistribution in mice, J. Nucl. Med., 24:45 (1983).

30. R.T. Proffitt, L.E. Williams, C.A. Presant, G.W. Tin, J.A. Uliano, R.C. Gamble, and J.D. Baldeschwieler, Liposomal blockade of the reticuloendothelial system: Improved tumor-imaging with small unilamellar vesicles, Science, 220:502 (1983).

ANTIBODY-BEARING LIPOSOMES AS PROBES OF

RECEPTOR-MEDIATED ENDOCYTOSIS

L.D. Leserman, D. Aragnol, J. Barbet,
P. Machy and A. Truneh

Centre d'Immunologie INSERM-CNRS de Marseille-Luminy
Case 906 - 13288 Marseille Cedex 9
France

INTRODUCTION

In 1981 this conference was called "Targeting of Drugs". That this year's conference is called "Receptor-Mediated Targeting of Drugs" indicates the pace of progress in the field, and emphasizes what are for us central themes, namely, that the nature of the receptor is critical in the efficiency of drug delivery, and that targeted drug delivery and related technologies constitute excellent probes for analysis of the dynamics of cell-surface proteins and receptors.

In this paper we review recent data from our laboratory showing that methotrexate delivered to cells in antibody bearing liposomes enters the cells by a process strongly suggestive of receptor-mediated endocytosis (Machy et al., 1982a, b; Machy and Leserman, 1983a). We additionally present methodology for the use of fluorescence quenching release (Weinstein et al., 1977, 1983; Leserman et al., 1979) to evaluate the kinetics of liposome uptake by cells (Truneh et al., 1983).

LIPOSOMES BOUND TO CELLS BY ANTIBODY ARE INTERNALIZED WITH AN EFFICIENCY WHICH DEPENDS ON THE TARGET DETERMINANT AND THE CELL TYPE

Liposomes containing methotrexate (MTX) and carboxyfluorescein (CF) were coupled to antibodies directed at determinants encoded by the murine major histocompatibility complex (MHC). When incubated with mitogen stimulated B and T lymphocytes from the appropriate strains, liposomes bound to cells (Table 1) in proportion to the

Table 1. Fluorescent liposomes (CF, pmoles) bound to
          selected determinants on B and T blast cells

	CBA		B6
Determinant	B	T	
H-2K^k	30	17	0.4
I-E^k	13	1.2	0.7
I-A^k	11	ND	0.6
LFA-1	2.6	2.3	2.9

Cells were incubated with liposomes (0.10 mM total
lipid) containing 213 pmol of CF. Results are the
mean of 4 experiments. SEM is less than 10% of the
value presented (Machy et al., 1982a).

Table 2. Numbers of binding sites for monoclonal anti-
          bodies on separated B and T CBA spleen cell
          blasts and on unseparated mitogen stimulated
          blasts from B6 spleens

B AND T BLAST EXPRESSION OF SELECTED DETERMINANTS (x 10^{-3})

	CBA		B6
Determinant	B	T	
H-2K^k	550	250	20
I-E^k	350	10	30
I-A^k	275	10	10
LFA-1	45	75	50

The numbers of binding sites were determined by Scatchard
plots only for anti-H-2K^k antibody on CBA B and T cell
blasts. The other numbers were determined by comparison
of the binding of radiolabeled protein A to unlabeled anti-
H-2K^k antibody and other mouse antibodies, and from an
evaluation of directly bound radiolabeled anti-LFA-1 antibody (Machy et al., 1982a).

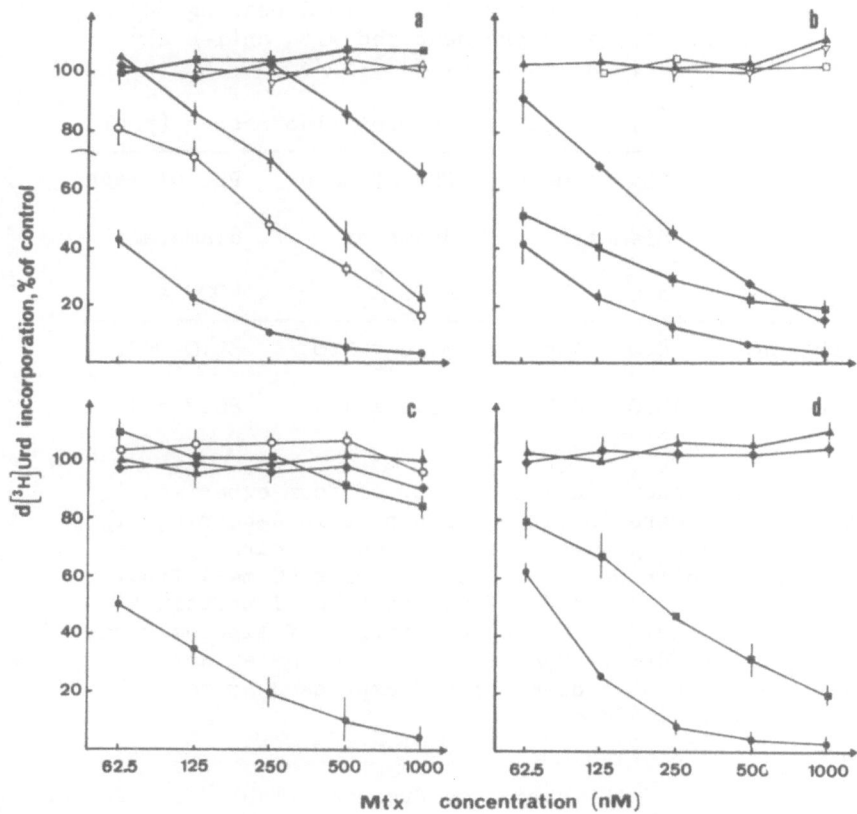

Fig. 1. Effect of MTX, free in solution or encapsulated in lipo-
         somes, on CBA T and B cell blasts and B6 Con A and LPS
         blasts. Blast cells (2 x 10^5 in 100 μl of medium) were
         incubated with 10 μl of medium or 10 μl of dilutions of
         free MTX or MTX encapsulated in liposomes. ●, Free MTX;
         ▲, MTX in anti-I-E^k antibody bearing liposomes; ■, MTX
         in anti-LFA-1 antibody bearing liposomes; O, MTX in anti-
         I-A^k antibody bearing liposomes; ◆, MTX in anti-H-2K^k
         antibody bearing liposomes; ▽, MTX in anti-human β2-
         microglobulin bearing liposomes; △, anti-I-E^k antibody
         bearing liposomes without MTX; □, anti-LFA-1 antibody
         bearing liposomes without MTX. Results are mean ± SD
         (for some points, SD was smaller than the symbol) (a) CBA
         B cell blasts. Control cells incorporated 180,000 ± 4,600
         cpm; control with Con A, 10,000 ± 1,000 cpm. (b) CBA T
         cell blasts. Control cells incorporated 120,000 ± 4,000
         cpm; control with LPS, 6,500 ± 300 cpm. (c) Unseparated
         B6 LPS blasts. Control cells incorporated 80,000 ± 2,000

Table 3. Specific binding of protein A-bearing SUV or
          REV to cells preincubated with anti-H-2K^k
          monoclonal antibody H100-30/33

	Cell-associated carboxyfluorescein (pmol)		
	SUV of mean	REV of mean	REV of mean
	diameter	diameter	diameter
	800 Å	2000 Å	4000 Å
RDM4 cells	8.0 ± 1.5	37.6 ± 2.0	35.0 ± 1.5
L cells	16.0 ± 2.0	32.0 ± 1.5	26.7 ± 1.2

Results are shown as mean ± S.E. of four experiments.
4×10^5 cells were incubated with protein A-bearing liposomes containing a total of 212 pmol of carboxyfluorescein. Cells were preincubated with 40 mg/1 final
of antibody and washed before addition of protein A-
bearing liposomes. The concentration of liposomes was
82.5 μM for SUV; 23.7 μM for 2000 Å diameter REV and
15.4 μM for 4000 Å diameter REV expressed as total lipid
(Machy et al., 1983a).

cells' expression of the target determinant (Table 2). Background
binding to cells bearing another allelic form of the target molecule
not recognized by the antibody was low, and those few liposomes which
bound had no effect on the uptake of radiolabeled (^{3}H) deoxyuridine,
which is an index of drug delivery to the cytoplasm (Leserman et al.,
1980a, b, 1981; Machy et al., 1982a, b; Machy and Leserman, 1983a).
Cells which specifically bound liposomes showed variable levels of
inhibition of (^{3}H) deoxyuridine uptake not necessarily related to
the number of liposomes which become cell-associated: B cells bear
twice as many target MHC molecules as T cells and bound twice as
many liposomes, but the drug effect was less than one fourth of that
seen for T cells. This was true despite the fact that both cell
types were equally sensitive to free MTX, and B cells were sensitive
to MTX in liposomes directed at other MHC-encoded cell-surface proteins (Fig. 1).

A second type of target molecule was even more striking in
showing this difference between the same determinants on different

cpm. (d) Unseparated, B6 Con A blasts. Control cells
incorporated 110,000 ± 4,500 cpm. Reprinted from Machy
et al., 1982a.

cell types. This is the LFA-1 molecule (Springer et al., 1982, Golstein et al., 1982) expressed by B and T cells at levels much lower than the MHC molecules. Despite the small amount of liposome-encapsulated drug which became bound to this determinant on T cells, the resulting drug effect was more important than for the larger quantity of liposomes directed at the MHC encoded determinant, and even more effective than the same amount of free methotrexate. However, approximately the same amount of liposomes bound to the target determinant on B lymphocytes had no effect on the cells (Fig. 1).

The fact that the drug effect is not proportional to the number of liposomes bound to cells rules out a non-specific process such as fusion or generalized phagocytosis. The absence of an identical effect from cell to cell for the same determinant (or, formally, molecules binding the same monoclonal antibody) shows that inform-ation for drug internalization does not reside entirely within the target molecule but also depends on the cell which expresses it. The explanation, in molecular terms, for this difference in drug effect as a function of the target determinant remains to be elucidated.

Similar functional differences have been reported for other cell surface molecules. Surface immunoglobulin of immature B cells is rapidly internalized when incubated in the presence of antibody to it; the internalization of similar molecules on mature B cells, expressed at the same density, is much slower (Raff et al., 1975). In addition, the receptor for the complement molecule C3b is ex-pressed in an identical form by both erythrocytes and nucleated cells. Obviously, only the latter cells are capable of internal-ization of particles opsonized with C3b (Fearon and Wong, 1983).

THE RELATION BETWEEN THE SIZE OF A LIPOSOME AND ITS EFFICIENCY IN DRUG DELIVERY

Since the volume of a sphere varies with the cube of its radius, liposomes of diameter 2x carry eight times more drug than liposomes of diameter x. This geometry provoked an interest in techniques for preparation of large liposomes, based on the assump-tion that their larger drug compartment would permit more efficient drug delivery, but this has not been supported by experimental evidence.

In our in-vitro model, in which it can be shown that, indeed, more drug becomes associated with cells when encapsulated in large than small liposomes (Table 3), the delivery of the drug is never-theless more effective when small liposomes are used (Fig. 2), despite the fact that liposomes are made of identical lipid com-position and are targeted to cells with the same antibody. We

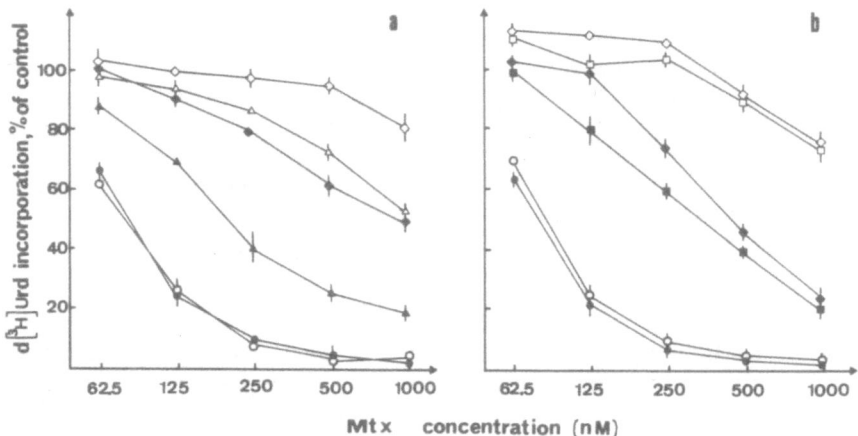

Fig. 2. MTX-mediated inhibition of d(^{3}H)Urd incorporation.
Cells (10^5) were preincubated with anti-H-2K^k monoclonal antibody before addition of SUV or REV coupled
to protein A. ○, free methotrexate; ■, methotrexate
in protein A-bearing SUV; ▲, methotrexate in protein A-bearing REV of 2000 Å in diameter; ●, methotrexate in protein A-bearing REV of 4000 Å in diameter,
□, protein A-bearing SUV without methotrexate; △,
methotrexate in protein A-bearing REV of 2000 Å in
diameter without preincubation of the cells with the
monoclonal antibody or with preincubation with an
irrelevant monoclonal antibody; ◇, methotrexate in
protein A-bearing SUV without preincubation of the
cells with anti-H-2K^k antibody or with preincubation
with an irrelevant antibody. The effect of methotrexate encapsulated in liposomes was totally inhibited by preincubation of the cells with free protein
A (50 mg/1) for 5 min before addition of liposomes
(data not shown). a. RDM4 tumor cell line: control
d(^{3}H)Urd incorporation, 152,000 ± 2000 cpm. b. L cells:
control d(^{3}H)Urd incorporation, 260,000 ± 10,000 cpm.
Each point plotted corresponds to the average of four
experiments. Reprinted by permission from Machy and
Leserman, 1983a.

cannot account for this difference by increased leakage of drug
from small liposomes, which are in fact more resistant to certain
stress, such as freezing (Machy and Leserman, 1983b). They are,
however, slightly less resistant to the effects of acid at 37°C
(Barbet et al., submitted for publication), but this small difference cannot account for their increased efficiency.

The differences of internalization of liposome contents seem, rather, to be related to the ability of the liposome to enter into the cell via an endocytic process permitting entry of 800 Å liposomes but excluding 2000 Å liposomes. This nicely corresponds to the known diameter of endocytic vesicles (1200-1500 Å) (Pastan and Willingham, 1981).

DRUG DELIVERY IS INHIBITED BY CHANGING THE pH OF ENDOCYTIC VESICLES

Antibody-bearing liposomes of the lipid composition used in these experiments do not leak the anionic molecules CF or MTX at neutral pH, but will rapidly do so at 37°C at pH and acid concentrations similar to that found in lysosomes and endocytic vesicles. The pH of lysosomes and endocytic vesicles can be increased by nearly 2 units in the presence of NH_4Cl. At a concentration of 10 mM, NH_4Cl has no effect on the number of liposomes which bind to cells, nor does it reduce the effect of free MTX. However, liposome-encapsulated MTX is significantly less active in the presence of NH_4Cl (Fig. 3), suggesting that the drug escapes from an acidic intracellular environment after endocytosis (Machy et al., 1982a). This has been directly verified for CF (Fig. 3).

TECHNOLOGY FOR THE DIRECT KINETIC EVALUATION OF ENDOCYTOSIS OF LIPOSOMES

The measurement of liposome uptake by the effect of encapsulated methotrexate offers numerous advantages over techniques such as electron microscopy; simplicity of the assay with objective results, possibility of measurement of many samples at a time, etc. It also offers some disadvantages, including the requirement that target cells are synthesizing DNA and are sensitive to the drug. In addition, the intracellular path followed by the drug before it binds to the enzyme dihydrofolate reductase is unknown. Because the assay is a bioassay, rapid kinetic studies are impossible, and, finally use of drugs (other than NH_4Cl) to probe the mechanism of liposome-encapsulated MTX uptake is precluded because of their potential effects on the action of MTX, rather than its delivery.

We have thus worked to develop an assay which would avoid the inconveniences detailed above, and which would in addition, permit objective evaluation on a single cell basis. This is a modification of techniques developed by Hagins and Yoshikami (1978) and by Weinstein and his collaborators (1977, 1983), which uses dequenching of the fluorophore carboxyfluorescein which follows as a consequence of its dilution from high concentration in liposomes into a measuring cuvette or the cytoplasm of a cell.

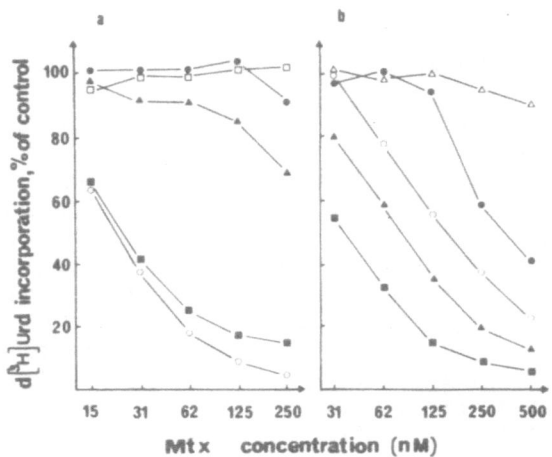

Fig. 3.   Action of NH₄Cl on the effect of free MTX or MTX
          encapsulated in liposomes bearing anti-H-2Kᵏ, anti-
          I-Eᵏ or anti-LFA-1 antibodies. Unseparated CBA LPS
          or Con A blasts (2 x 10⁵ in 100 μl of medium) were
          incubated with NH₄Cl (final concentration, 10 mM)
          for 15 min before incubation with free MTX or MTX
          encapsulated in liposomes. NH₄Cl was continuously
          present during the incubation. ●, free MTX; O,
          free MTX plus NH₄Cl; ▲, MTX in anti-I-Eᵏ antibody-
          bearing liposomes; △, MTX in anti-I-Eᵏ antibody-
          bearing liposomes plus NH₄Cl; ■, MTX in anti-LFA-1
          antibody-bearing liposomes; □, MTX in anti-LFA-1
          antibody-bearing liposomes plus NH₄Cl. Results are
          mean ± SD (for some points, SD was smaller than the
          symbol). (a) CBA LPS blasts. Control cells in-
          cubated with 10 mM NH₄Cl alone incorporated 140,000
          ± 1,700 cpm. (b) CBA Con A blasts. Control cells
          incubated with 10 mM NH₄Cl alone incorporated
          180,000 ± 5,300 cpm. Reprinted from Machy et al.,
          1982a.

Earlier attempts to use carboxyfluorescein as a quantitative marker of liposome-cell interactions resulted in ambiguous results (Szoka et al., 1979) which contributed to our decision to use MTX as a more reliable probe. However, we now appreciate that carboxyfluorescein leakage was a consequence of its contamination by hydrophobic impurities, which can be removed by simple chromatographic techniques (Ralston et al., 1981). In addition, CF leakage is markedly reduced by use of cholesterol containing liposomes (Scherphof et al., 1980).

When antibody-bearing liposomes are bound to the cell surface at 4°C, and non-bound liposomes are removed by washing, the liposomes remain stably bound for many hours if the incubation is maintained at 4°C. This can be confirmed both by fluorometry and by the fluorescence-activated cell sorter when the liposomes contain carboxyfluorescein. When the temperature of incubation is 37°C, the CF is at high concentration, and if the target determinant is subsequently endocytosed, the liposomes enter into the cell and the CF contained within them is released, presumably by exposure of the liposomes to the acid environment of endocytic vesicles or lysosomes. This dequenching process can be followed as a function of time, with kinetics depending on the rate of endocytosis of the target cell surface molecules, among other factors (Fig. 4).

Numerous details remain to be explored concerning this process; most importantly the intracellular path followed by the liposome and the released carboxyfluorescein. It is apparent that because CF fluorescence is pH dependent the resulting fluorescence signal is critically dependent on the pH of the compartment to which it is delivered (Weinstein et al., 1983). Similarly, by analogy with recent experiments in which liposomes were incubated with various acid species (Barbet et al., submitted for publication), the release will depend on the nature of the acid and its concentration in this compartment. It follows from this that a range of liposome-encapsulated fluorophores of different pH sensitivity can serve as probes of the diversity of intracellular vesicular compartments.

Conclusions and Perspectives

Liposomes are magnificent research tools when used in the appropriate context. Antibody targeted liposomes may never be used in the diagnosis or therapy of disease; but these reagents have amply served to demonstrate a previously unappreciated level of complexity in the function of cell surface receptors, and they will continue to serve as probes of cell physiology. Better understanding of this physiology will, in turn, lead to better methods for therapy, but that is a subject for a future conference.

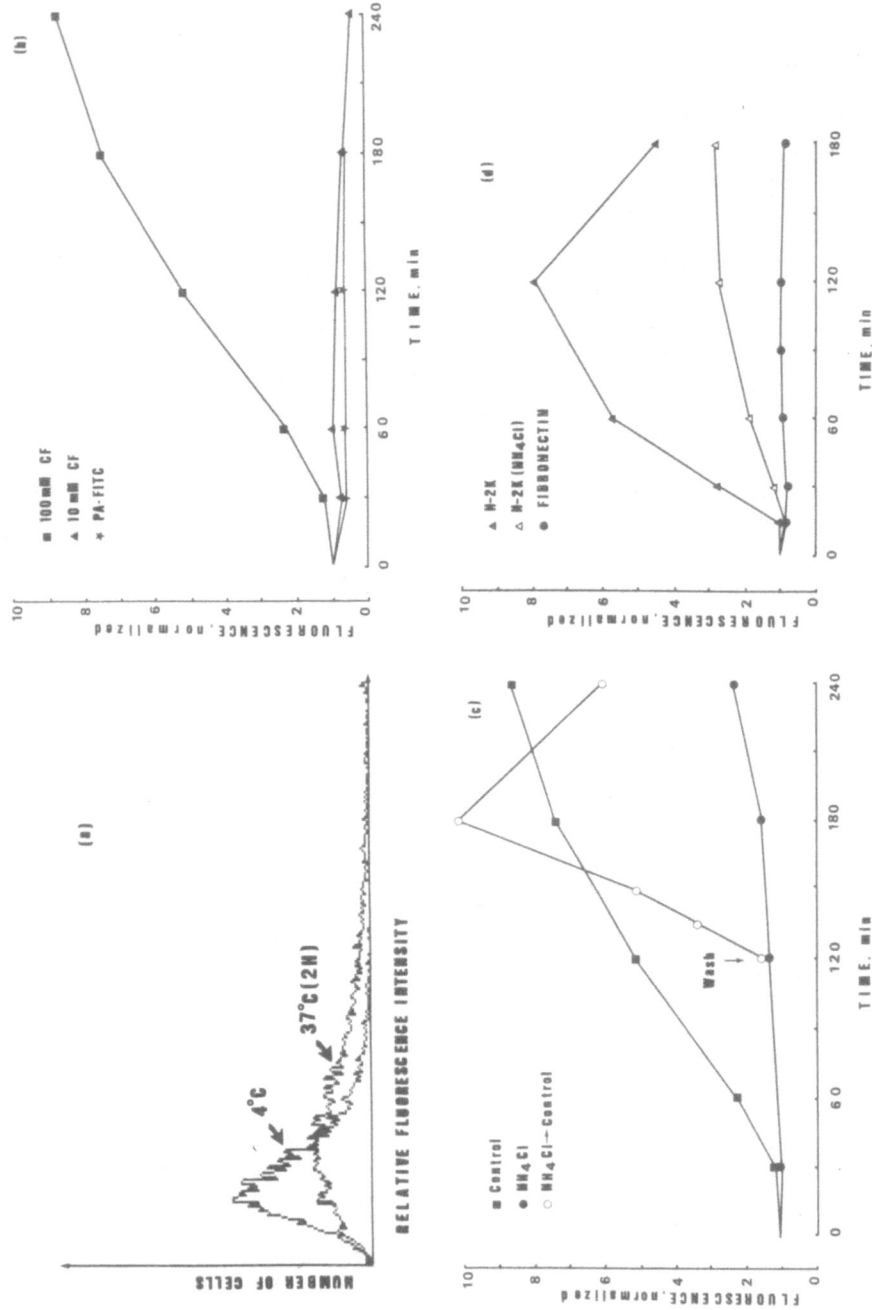

Fig. 4. Cytofluorometric measurements of the endocytosis of an MHC antigen H-2K and membrane surface fibronectin using liposome-encapsulated carboxyfluorescein. (a) Fluorescence histogram of L cells that had been stained with anti-H-2Kᵏ (100 μg/ml) and protein A–

liposomes (40 mM carboxyfluorescein) at 4°C or after incubation at 37°C for 2 h. The time course of change in fluorescence of T cells that had been similarly treated is shown in (b). Cells were stained either with liposomes containing (■) 100 mM or (▲) 10 mM carboxyfluorescein or (✱) with fluorescein isothiocyanate-labelled protein A. Calculations are based on setting the lower fluorescence window such as to exclude 90% of the stained cells at time zero as background. The shift towards higher fluorescence channels with incubation at 37°C is then analysed by determining the product of the number of cells that appear between these set windows and their median fluorescence channel. All values are then normalized by obtaining the ratio to the value at zero time. (c) illustrates the effect of NH$_4$Cl (■, control; ● + NH$_4$Cl) on T lymphoblasts that had been stained with anti-H-2K^k and protein A-liposomes containing 100 mM carboxyfluorescein. The washing step in one series of samples (○) involves incubating the cells in the presence of NH$_4$Cl for 2 h at 37°C, centrifuging (4°C), re-suspending in fresh medium and observing the changes at 37°C; in (d), L cells were stained with anti-H-2K^k and protein A-liposomes (40 mM carboxyfluorescein) and incubated (37°C) in the presence (▲) or absence (△) of NH$_4$Cl, or with anti-fibronectin serum (●). In all cases non-specific staining with protein A-liposomes (i.e. in the absence of antibody) represented less than 2% of the cell-associated fluorescence. Reprinted by permission from Truneh et al., 1983.

REFERENCES

Fearon, D.T. and Wong, W.W. 1983, Complement ligand-receptor
     interactions that mediate biological responses, Ann. Rev.
     Immunol., 1:243.
Golstein, P., Goridis, C., Schmitt-Verhulst, A.M., Hayot, B.,
     Pierres, A., Van Agthoven, A., Kaufmann, Y., Eshhar, Z.
     and Pierres, M. 1982, Lymphoid cell surface interaction
     structures detected using cytolysis-inhibiting monoclonal
     antibodies, Immunol. Rev., 68:5.
Hagins, W.A. and Yoshikami, S. 1978, Intracellular transmission of
     visual excitation in photoreceptors: Electrical effects of
     chelating agents introduced into rods by vesicle fusion, in:
     "Vertebrate photoreceptors", P. Fatt, H.B. Barlow, eds.,
     Academic Press, New York.
Leserman, L.D., Weinstein, J.N., Blumenthal, R., Sharrow, S. and
     Terry, W.D. 1979, Binding of antigen bearing fluorescent
     liposomes to the murine myeloma tumor MOPC 315, J. Immunol.,
     122:585.
Leserman, L.D., Weinstein, J.N., Blumenthal, R. and Terry, W.D.
     1980a, Receptor-mediated endocytosis of antibody opsonized
     liposomes by tumor cells, Proc. Natl. Acad. Sci. USA, 77:4089.
Leserman, L.D., Weinstein, J.N., Moore, J.J. and Terry, W.D. 1980b,
     Specific interaction of myeloma tumor cells with hapten-bearing
     liposomes containing methotrexate and carboxyfluorescein,
     Cancer Res., 40:4768.
Leserman, L.D., Machy, P. and Barbet, J. 1981, Cell specific drug
     transfer from liposomes bearing monoclonal antibodies, Nature,
     293:226.
Machy, P., Barbet, J. and Leserman, L.D. 1982a, Differential endo-
     cytosis of T and B lymphocyte surface molecules evaluated
     with antibody-bearing fluorescent liposomes containing metho-
     trexate, Proc. Natl. Acad. Sci. USA, 79:4148.
Machy, P., Pierres, M., Barbet, J. and Leserman, L.D. 1982b, Drug
     transfer into lymphoblasts mediated by liposomes bound to
     distinct sites on H-2 encoded I-A, I-E and K molecules,
     J. Immunol., 129:2098.
Machy, P. and Leserman, L.D. 1983a, Small liposomes are better than
     large liposomes for specific drug delivery in-vitro, Biochim.
     Biophys. Acta, 730:313.
Machy, P. and Leserman, L.D. 1983b, Freezing of liposomes, in:
     "Liposome Technology", Gregoriadis, G. ed., CRC Press, Boca
     Raton, in press.
Pastan, I.H. and Willingham, M.C. 1981, Journey to the center of
     the cell: the receptosome, Science, 214:504.
Raff, M.C., Owen, J.J.T., Cooper, M.D., Lawton, A.L. III, Megson, M.
     and Gathings, W.E. 1975, Differences in susceptibility of
     mature and immature mouse B lymphocytes to anti-immunoglobulin-
     induced immunoglobulin suppression in-vitro: possible impli-
     cations for B-cell tolerance to self, J. Exp. Med., 142:1052.

Ralston, E., Hjelmeland, L.M., Klausner, R.D., Weinstein, J.N. and
    Blumenthal, R. 1981, Carboxyfluorescein as a probe for lipo-
    some-cell interactions: Effect of impurities, and purification
    of the dye, Biochim. Biophys. Acta, 649:133.
Scherphof, G., Morselt, H., Regts, J. and Wilschut, J. 1980, The
    involvement of the lipid phase transition in the plasma-
    induced dissolution of multilamellar phosphatidylcholine
    vesicles, Biochim. Biophys. Acta, 620:90.
Springer, T.A., Davignon, D., Ho, M-K., Kurzinger, K., Martz, E.
    and Sanchez-Madrid, F. 1982, LFA-1 and Lyt-2, 3, molecules
    associated with T lymphocyte-mediated killing; and Mac-1, an
    LFA-1 homologue associated with complement receptor function,
    Immunol. Rev., 68:171.
Szoka, F.C., Jr., Jacobson, K. and Papahadjopoulos, D. 1979, The
    use of aqueous space markers to determine the mechanism of
    interaction between phospholipid vesicles and cells, Biochim.
    Biophys. Acta, 551:295.
Truneh, A., Mishal, Z., Barbet, J., Machy, P. and Leserman, L.D.
    1983, Endocytosis of liposomes bound to cell surface proteins
    measured by flow cytofluorometry, Biochem. J., 214:189.
Weinstein, J.N., Yoshikami, S., Henkart, P., Blumenthal, R. and
    Hagins, W.A. 1977, Liposome-cell interaction: Transfer and
    intracellular release of a trapped fluorescent marker, Science,
    195:489.
Weinstein, J.N., Ralston, E., Leserman, L.D., Klausner, R.D.,
    Dragsten, P., Henkart, P. and Blumenthal, R. 1983, Self-
    quenching of carboxyfluorescein fluorescence: uses in studying
    liposome stability and liposome-cell interaction, in: "Lipo-
    some Technology", Gregoriadis, G. ed., CRC Press, Boca Raton,
    in press.

ANTIBODY-TARGETED LIPOSOMES: DEVELOPMENT OF A CELL-SPECIFIC

DRUG DELIVERY SYSTEM

Timothy Heath, Keith Bragman, Katherine Matthay,
Ninfa G. Lopez and Demetrios Papahadjopoulos

Cancer Research Institute, M-1282
University of California Medical Center
San Francisco, CA 94143

SUMMARY

We have described the following important parameters for the
development of antibody-directed liposomes in drug delivery.

(i)   Liposome-dependent cytotoxic agents: Drugs such as methotrexate-
γ-aspartate which are unable to enter cells without a carrier
are essential since their use eliminates the non-specific
effects of drug which may leak from the liposomes.

(ii) Multivalency of liposome interactions: Liposomes conjugated to
antibody have a higher valency than the soluble antibody and
can bind to cells with up to 1,000-fold higher affinity constant.

(iii) Use of multiple ligands: Liposomes which interact with more
than one ligand on the cell surface show marked resistance to
inhibition of cell association by soluble ligands.

(iv) Optimal liposome size for drug delivery: The optimal liposome
size appears to vary from 0.05 to 0.1 μm depending on target
cell type.

These experiments have been performed with liposomes conjugated
to monoclonal anti H2K^k which bind to L929 cells and deliver drug 20
times more effectively to L929 cells than non-specific liposomes. We
have also used liposomes conjugated to antiglycophorin antibody.
These liposomes bind to K562 cells via an F_c receptor and glycophorin
A, and deliver drug 10 times more effectively than non-conjugated
liposomes.

INTRODUCTION

The production of antibody-directed cytotoxic agents is of
prime importance in cell biology and medicine. Use of such agents,
which can specifically kill cells bearing the appropriate antigen,
may give insight into the mechanisms of receptor-mediated internalization and ultimately may lead to production of tumor-specific
cytotoxic agents. We have developed efficient methods for conjugation of protein to liposomes (Heath et al., 1981; Martin et al.,
1981; Martin and Papahadjopoulos, 1980). We have proposed the use
of such liposomes with cytotoxic agents because targeted liposomes
bind to the appropriate cells with high efficiency (Heath et al.,
1980; Martin et al., 1981).

The efficient use of targeted liposomes in cancer therapy will
depend primarily on the difference between their toxicities for
target and non-target cells. Two factors must be optimized to
achieve this aim. First, the liposomes must deliver their contents
efficiently to the cytoplasmic compartment. Second, the non-specific
toxicity, which may arise through the leakage of the cytotoxic agent
from the liposome (Allen et al., 1981; Van Renswoude et al., 1979;
Scherphof et al., 1978), must be minimized. The toxicity of leaked
drug could be minimized by using a drug that cannot enter cells but
which would be toxic if delivered to the cytoplasm. Such a compound
would be analogous to viral nucleic acids, which become infectious
when encapsulated in liposomes.

DELIVERY OF METHOTREXATE-γ-ASPARTATE TO L929 CELLS

Methotrexate is a pteridine antifolate which inhibits cell
growth by inhibiting dihydrofolate reductase (Goldman et al., 1968).
Methotrexate has been used successfully in targeted liposomes by
Laserman et al. (1981) but it readily penetrates cells via the folate
transport system. Piper et al. (1982) have examined the effects of
structural alterations of methotexate on cytotoxicity, influx
kinetics, and ability to inhibit dihydrofolate reductase. Metho-
trexate-γ-aspartate proved to be virtually indistinguishable from
methotrexate in its ability to inhibit dihydrofolate reductase.
However, methotrexate-γ-aspartate is only 1/200th as toxic for L1210
cells compared to methotrexate becasue its influx k_m is at least 100
times greater. Methotrexate-γ-aspartate therefore fulfills our
criteria for a liposome-dependent drug because its ability to enter
cells is poor, but it should be equipotent to methotrexate if delivered intracytoplasmically. A detailed account of the experimental
methods is given elsewhere (Heath et al., 1983). Our first experiments to determine whether methotrexate-γ-aspartate could be a
liposome dependent drug involved liposomes targeted via monoclonal

Table 1. Interaction of antibody-targeted liposomes with L929
and Balb/c 3T6 fibroblasts

Cell	Serum[*]	Inhibitors[**]	% Bound		
			Uncoated	Non-specific	Targeted
L929	-	-	2.7	2.1	12.4
L929	+	-	1.2	0.6	10.6
L929	-	+	-	1.8	5.6
Balb/c 3T6	+	-	0.4	0.5	0.4

Liposomes were prepared from phosphatidylcholine/cholesterol/
gangliosides (45:45:10) and conjugated to anti-sheep erythrocyte
antibody (non-specific) or anti-H2K^k antibody (targeted) by the
method of Heath et al., 1981. The final antibody/lipid ratio was
80 g/mol, and the vesicles contained 2,000 cpm of (^{3}H) dipalmitoyl
phosphatidylcholine per nmol of lipid.

* Cells were incubated without or with 50%(vol/vol) newborn calf
serum.
** Cells were incubated without or with 5 mM sodium azide/50 mM
2-deoxy-D-glucose.

anti H2K^k (Oi et al., 1978). H2K^k is a protein of the mouse major
histocompatibility complex which is expressed at high frequency on
L929 fibroblasts. Table 1 summarizes the association of these targeted liposomes with L929 or Balb/c 3T6 fibroblasts. Non-specific
empty liposomes showed a low level of association that was not
affected by azide and 2-deoxy-D-glucose (Table 1). Liposomes coated
with monoclonal anti H2K^k showed a 6-fold higher interaction; this
interaction was decreased by 55% in the presence of azide and 2-
deoxy-D-glucose. The energy dependence of the targeted interaction
suggests that either targeted vesicles are internalized or optimal
interaction requires an energy-dependent redistribution of the H2K^k
antigen. The difference between targeted and non-specific or uncoated liposome interaction with cells was increased by the inclusion of serum. In the presence of 50% newborn calf serum, interaction of targeted vesicles with the cells was 20-fold greater than
interaction of liposomes coated with anti-sheep erythrocyte antibody.
This increase in the difference between targeted and non-specific
liposome interaction was largely due to a decrease in the interaction
on non-specific liposomes. The specificity of targeted liposomes is

confirmed by the low level of interaction of both vesicle preparations with Balb/c 3T6 fibroblasts. This cell line does not react
with either anti H2K^k or anti-sheep erythrocyte antibodies. The two
liposome preparations showed a low level of interaction comparable
to the interaction of the non-specific liposomes with L929 fibroblasts.

EFFECTS ON CELL GROWTH

Our early studies on growth inhibition of L929 were done with
targeted liposomes containing methotrexate (Papahadjopoulos et al.,
1982). In these studies, a comparison of the growth curves indicated
that the IC$_{50}$ for free methotrexate was 0.018μM whereas that of
methotrexate in uncoated liposomes but only 1/4 as effective as free
methotrexate. This result agrees qualitatively with the findings of
Leserman et al., (1981). We have examined the efficacy of targeted
and uncoated liposome preparations with various drug/lipid ratios.
The experiment discussed above (Drug/lipid ratio: 0.78 mmol/mol)
gave the greatest difference between targeted and non-targeted vesicles.

Table 2 summarizes our more recent experiments on the growth
inhibition of L929 fibroblasts by various preparations containing
methotrexate-γ-aspartate (Heath et al., 1983). The free drug has
an IC$_{50}$ of 0.68μM. Targeted liposomes were 10.3 times more affect-
ive than the free drug (IC$_{50}$ = 0.066μM). In contrast, non-specific
liposomes (bearing anti-thy 1.1 antibody) were 60% as effective as
the free drug (IC$_{50}$ = 1.2μM). Consequently, targeted liposomes were
18 times more effective than non-specific liposomes. Balb/c 3T6
fibroblasts showed comparable sensitivity to L929 for the free drug
(IC$_{50}$ = 0.58μM). Both vesicle preparations were less effective than
free drug and had similar IC$_{50}$ values (2.0 μM for non-specific and
2.5 μM for targeted). This demonstrates the superior efficacy of
targeted liposomes on L929 cells is due to the specificity of the
anti-H2K^k antibody.

Targeted empty liposomes showed no inhibitory effects in the
range of concentrations used. In addition, targeted empty liposomes
did not increase the inhibitory effects of the free drug. This confirms that the growth inhibitory effects of targeted liposomes are
due to the drug and demonstrate a requirement for encapsulation of
the drug. 58μM 5-formyltetrahydrofolate substantially reduced the
effects of encapsulated and free drug on cell growth. This confirms
that the inhibition of cell growth by targeted liposomes was due to
an inhibition of 5- formyltetrahydrofolate production, presumably
as a result of dihydrofolate reductase inhibition. Ammonium chloride
did not decrease the effects of the free drug but completely blocked
the growth inhibitory effects of both encapsulated preparations.
This suggests that the inhibitory effects of encapsulated metho-
trexate-γ-aspartate involve the endocytosis of the liposomes and may

Table 2. Growth inhibitory efficacy of methotrexate-γ-aspartate
in liposomes targeted with anti-H2K^k antibody

Preparation	Cell Line	Additions	IC$_{50}$ (μm)
free drug	L929	-	0.68
targeted	L929	-	0.066
non-specific	L929	-	1.2
free drug	Balb/c	-	0.58
targeted	Balb/c	-	2.5
non-specific	Balb/c	-	2.0
free drug	L929	folinic acid	30.0
targeted	L929	folinic acid	> 3.0
free drug	L929	NH$_4$Cl	0.70
targeted	L929	NH$_4$Cl	> 3.0
targeted	L929	non-specific empty liposomes	0.13
targeted	L929	targeted empty liposomes	0.66

be critically dependent on the low pH of the post-endocytic compartment.

Growth inhibition by methotrexate-γ-aspartate in targeted vesicles.
Fibroblasts were treated with methotrexate-γ-aspartate in various
preparations at the concentrations shown. Vesicles were prepared
from a 10:10:1 mixture of phosphatidylcholine/cholesterol/4 (p-
maleimidophenyl) butyrylphosphatidylethanolamine and conjugated with
thiolated antibody. Targeted liposomes contained 50 g of anti-H2K^k
per mol of phospholipid; non-specific liposomes contained 62 g of
anti-Thy 1.1 per mol of phospholipid; targeted empty liposomes contained 112 g of anti-H2K^k per mol phospholipid; non specific empty
liposomes contained 156 g of anti-sheep erythrocyte per mol of phospholipid.

Soluble anti-H2K^k antibody did not inhibit the effects of these targeted vesicles (not shown). This is consistent with the cell association data (Heath et al., 1983) which shows only a minimal inhibition of L929 cell binding of targeted vesicles by the soluble antibody. Non-specific empty liposomes inhibited targeted vesicle efficacy by 50%. However, inhibition of the effect to 10% was produced by reincubation with an excess of targeted empty liposomes. This further confirms the role of the anti-H2K^k antibody in vesicle delivery and also suggests that the liposomes bearing the antibody may be taken up more effectively to the cells than the soluble antibody itself.

THE ROLE OF MULTIVALENCY

The efficacy of liposome uptake may in part be due to the multivalent interaction of the liposomes with the cell surface, which presumably occurs because each liposome is conjugated to many antibodies. The effect of liposome multivalency has been demonstrated in this lab in the above experiments on drug delivery and in experiments which involve the agglutination of erythrocytes (Heath et al., 1980), and measurement of the affinity of targeted liposomes for the erythrocyte surface (Heath et al., 1984).

The multivalent interaction of antibody-targeted liposomes with the cell surface could be important in the in vivo use of targeted liposomes for several reasons. The proportion of targeted liposomes which bind to the target cells will be greater than the proportion of unconjugated antibody which would bind to cells under the same conditions, because the affinity of the liposomes for the cells if greater than that of the unconjugated antibody. For the same reason, the binding of targeted liposomes to cells may be partially insensitive to inhibition by the monovalent antigens which are frequently released into extracellular fluids by tumor cells.

It is possible to confer specificity on each liposome for more than one antigen by conjugation to a mixture of monoclonal antibodies. Liposomes capable of binding to several antigens on the cell surface may be valuable for several reasons. Such interactions may be even more insensitive to inhibition by soluble antigens than the interactions of liposomes directed to a single determinant. Moreover, targeting to several specificities on the tumor cell surface may allow liposomes to be targeted to tumors which do not possess a unique antigen, but rather display a unique combination of two or more antigens. The latter possibility may cicumvent the central problem in tumor immunology: the lack of tumor specific antigens.

With these possibilities in mind, we have further examined the interaction of targeted liposomes with human erythrocytes, and with K562 cells, a human cell line derived from a patient with chronic myelogenous leukemia in blast transformation (Lozzio and Lozzio,

1975). K562 cells have both erythroid and granulocytic characteristics and express both glycophorin A (Andersson et al., 1979) and
F_C receptors (Klein et al., 1976).

BINDING TO K562 CELLS: EVIDENCE FOR MULTIPLE RECEPTOR INTERACTION

We have examined the binding of liposomes conjugated with either
monoclonal anti-sheep erythrocyte or anti-human glycophorin A to K562
cells (Bragman et al., 1983). The greatest amount of vesicle-cell
association was seen in the low range of lipid addition. At 1 nmol
of lipid added, 58% and 52%, respectively, of the liposomes conjugated with anti-sheep erythrocyte and anti-glycophorin A antibody were
bound to cells. At 100 nmol of lipid added, 25% of both types of
antibody-conjugated liposomes were bound to the cells. Binding of
unconjugated liposomes to K562 cells was between 2 and 3% when the
cells were incubated with 1-100 nmol lipid. Therefore, high levels
of liposome binding to K562 cells is dependent on the presence of
antibody on the liposome surface. However, binding occurs regardless
of the specificity of the antibody, suggesting that the non-specific
antibody reacts via an F_C receptor.

We also examined the effect of soluble or aggregated human IgG
on the binding to K562 cells of liposomes conjugated with non-
specific (anti-sheep erythrocyte) antibody (Bragman et al., 1983).
10 μg aggregated or 30 μg soluble human IgG inhibits 90% of the
liposome binding, which suggests that the interaction of non-specific
antibody-coated liposomes with K562 cells is mediated by an F_C receptor. If liposomes conjugated to a non-specific antibody can interact
with an F_C receptor on K562 cells, we reasoned that liposomes conjugated to monoclonal anti-glycophorin A probably bind to both F_C
receptors and glycophorin in the K562 cell membrane. Such an interaction might be expected to show a marked resistance to inhibition
by soluble ligands which can only inhibit the binding to one of
these two membrane components. Ligand inhibition studies of K562
cell interactions for liposomes conjugated with monoclonal anti-
human glycophorin A reveal that 100μg soluble human IgG produce
only 10% inhibition of binding on targeted liposomes, while 100μg
anti-human glycophorin A inhibited liposomes binding by 50%. This
probably reflects the interaction of the soluble anti-glycophorin A
with both glycophorin A and the F_C receptor on the surface of K562
cells. The greatest inhibition of vesicle binding occurred when
both soluble anti-glycophorin A and 100μg of human soluble IgG were
added. Incubation with 10μg anti-glycophorin A and 100μg human
IgG inhibited 90% of liposome binding. Studies with lower concentrations of lipid (1-20 nmol per 0.2 ml) showed a similar pattern
for inhibition of liposome binding by soluble ligands. Prior incubation of K562 cells with the Fab fragment of anti-human glycophorin A did not inhibit the binding of anti-glycophorin A targeted
vesicles. However, in the presence of 100μg of human IgG, Fab
fragments inhibited liposome binding to K562 cells by up to 35%.

DRUG DELIVERY TO K562 CELLS

The growth inhibitory efficacy of antibody-targeted lipsomes containing methotrexate-γ-aspartate parallels the above results on liposome cell association and its inhibition by ligands (Bragman et al., in preparation). Methotrexate-γ- aspartate is up to 10 times more effective in targeted liposomes than it is in the free form, regardless of the antibody type. The efficacy of the drug in non-specific antibody-conjugated liposome is reduced 10-fold by the presence of aggregated human IgG. However, the efficacy of the drug encapsulated in liposomes conjugated to anti-glycophorin is not affected by either soluble anti-glycophorin or aggregated human IgG and is reduced only 2-fold by both ligands. Since the standard deviation for the IC_{50} in our experiments is only $\pm$ 20%, these results are signigicant and show that drug delivery parallels liposome-cell association in these experiments.

LIPOSOME SIZE

All the experiments described above were performed with liposomes prepared by reverse-phase evaporation which have a mean diameter of 0.4 μm (Szoka and Papahadjopoulos, 1978). These liposomes were used because they capture water soluble compounds in high efficiency and they proved effective for drug delivery to L929 and K562 cells. However, subsequent experiments with Rl.1 and AKR/J SL2 T-lymphomas (Matthay et al., in preparation) have shown that for these cells REV are not effective for drug delivery, while the smaller SUV (diameter 0.025μm) are much more effective. While drug in SUV conjugated to anti-H2K^k can be more than 50 times more effective than free drug, comparable preparations of REV show little growth inhibition and are more than 300 times less effective than SUV.

In recent experiments with negatively-charged liposomes (Heath et al., unpublished) which deliver drug non-specifically to many different cell types, we have also shown that liposome size is critical for optimal drug delivery, though for some cells (L929 fibroblasts, CV1-P cells) the optimal size is not 0.025μm but 0.1μm.

REFERENCES

Allen, T.M., McAllister, L., Mausolf, S. and Gyoffry, E., 1981, Liposome-cell interactions: A study of the interactions of liposomes containing entrapped anti-cancer drugs with the EMT-6, S49 and AE1 (transport defficient) cell lines, Biochem. Biophys. Acta, 643:346.

Andersson, L.C., Nilsson, K. and Gahmberg, C.G., 1979, K562 - A human erythroleukemic cell line, Int. J. Cancer, 23:143.

Bragman, K.S., Heath, T.D., and Papahadjopoulos, D., 1983, Simultaneous interaction of monoclonal antibody-targeted liposomes with two receptors on K562 cells, Biochim. Biophys. Acta, 730:187.

Goldman, I.D., Lichtenstein, N.S. and Oliveiro, V.Y., 1968, Carrier-
    mediated transport of the folic acid analogue methotrexate, in
    the L1210 leukemia cell, J. Biol. Chem., 243:5007.
Heath, T.D., Fraley, R.T. and Papahadjopoulos, D., 1980, Antibody
    targeting of liposomes: Cell specificity obtained by conjuga-
    tion of F(ab')$_2$ to vesicle surface, Science, 210:539.
Heath, T.D., Macher, B.A., and Papahadjopoulos, D., 1981, Covalent
    attachment of immunoglobulins to liposomes via glycosphingo-
    lipids, Biochim. Biophys. Acta, 640:66.
Heath, T.D., Montgomery, J.A., Piper, J.R. and Papahadjopoulos, D.,
    1983, Antibody-targeted liposomes: Increase in specific tox-
    icity of methotrexate-γ-aspartate, Proc. Natl. Acad. Sci. USA,
    80:1377.
Heath, T.D., Fraley, R.T., Bentz, J., Voss, E.W., Jr., Herron, J.
    and Papahadjopoulos, D., 1984, Antibody-directed liposomes:
    Determination of affinity constants for soluble and liposome-
    bound antifluorescein, Biochim. Biophys. Acta, (in press).
Klein, E., Ben-Bassat, H., Neuman, H., Ralph, P., Zeuthen, J.,
    Polliack, A. and Vanky, F., 1976, Properties of the K562 cell
    line, derived from a patient with chronic myeloid leukemia,
    Int. J. Cancer, 18:421.
Leserman, L.D., Machy, P. and Barbet, J., 1981, Cell-specific drug
    transfer from liposomes bearing monoclonal antibodies, Nature,
    (London) 293:226.
Lozzio, C.B. and Lozzio, B.B., 1975, Human chronic myelogenous
    leukemia cell-line with positive Philadelphia chromosome,
    Blood, 45:321.
Martin, F.J., Hubbell, W.L. and Papahadjopoulos, D., 1981, Immuno-
    specific targeting of liposomes to cells: a novel and efficient
    method for covalent attachment of Fab' fragments via disulfide
    bonds, Biochemistry, 20:4229.
Martin, F.J. and Papahadjopoulos, D., 1982, Irreversible coupling
    of immunoglobulin fragments to preformed vesicles, J. Biol.
    Chem., 257:286.
Oi, V.T., Jones, P.P., Goding, J.W. and Herzenberg, L.A., 1978,
    Properties of monoclonal antibodies to mouse Ig allotypes, H-2
    and Ia antigens, in: "Current Topics in Microbiology and
    Immunology: Lymphocyte Hybridomas," F. Melchers, M. Potter,
    and N. Warner, eds., Springer-Verlag, NY, 81:115.
Papahadjopoulos, D., Heath, T.D., Martin, F., Fraley, R. and
    Straubinger, R., 1982, Development of liposomes as an efficient
    carrier system: New methodologies for cell targeting and intra-
    cellular delivery of drugs and DNA, in: "Targeting of Drugs,"
    eds. G. Gregoriadis, J. Senior and A. Trouet, Plenum, NY, p.
    375.
Piper, J.R., Montgomery, J.A., Sirotnak, F.M. and Chello, P.L.,
    1982, Synthesis of α and γ-substituted amides, peptides and
    esters of methotrexate and their evaluation as inhibitors of
    folate metabolism, J. Med Chem., 25:182.
Scherphof, G., Roerdink, F., Waite, M. and Parks, J., 1978,

Disintegration of phosphatidylcholine liposomes in plasma as a result of interaction with high-density lipoproteins, <u>Biochim. Biophys. Acta</u>, 542:296.

Szoka, F.C. and Papahadjopoulos, D., 1978, Procedure for preparation of liposomes with large internal aqueous space and high capture by reverse-phase evaporation. <u>Proc. Natl. Acad. Sci. USA</u>, 75:145.

Van Renswoude, A.J.B.M., Westenberg, P. and Scherphof, G., 1979, <u>In vitro</u> interactions of Zajdela ascites hepatoma cells with lipid vesicles. <u>Biochem. Biophys. Acta</u>, 558:22.

TARGETING AND LYSOSOMAL HANDLING OF

POLYMETHACRYLAMIDE-OLIGOPEPTIDE CONJUGATES

J.B. Lloyd, R. Duncan, J. Kopeček[a] and P. Rejmanová[a]

Biochemistry Research Laboratory, Department of
Biological Sciences, University of Keele
Staffordshire, England

[a]Institute of Macromolecular Chemistry, Czechoslovak
Academy of Sciences, Prague, Czechoslovakia

INTRODUCTION

In the previous volume in this series, Trouet et al., (1982) explained the concept of the lysosomotropic drug-carrier conjugate and prescribed the necessary conditions for this approach to chemotherapy to succeed. We are engaged in the preparation and testing of such a conjugate, and here present a progress report on our work.

Our aim has been to design and evaluate a macromolecule to which a pharmacological agent could be attached by a linkage that is stable in plasma but susceptible to hydrolysis by the lysosomal enzymes. Targeting is to be achieved by incorporation into the macromolecule of features that will lead to receptor-mediated pinocytosis into the target cell (Fig. 1).

Other investigators pursuing this approach to chemotherapy have chosen proteins or other biopolymers as carrier. A novel feature of our work is the choice of a synthetic polymer, and it is first necessary to indicate the advantages we see in such polymers. A simple homopolymer has only one variable, molecular weight, but its properties can be systematically modified by conjugation with graded amounts of some chosen chemical substituent. In contrast, proteins are polymers that contain up to twenty different monomer units and the proteins readily available from natural sources represent a mere handful from the almost infinite number of theoretically possible soluble polypeptides. The complexity of protein structure makes it difficult to predict or indeed ascertain the effects of some chem-

417

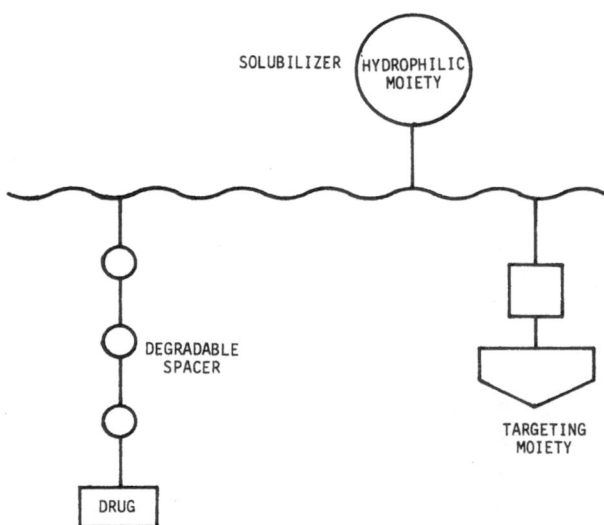

Fig. 1.   Design features required in a targetable
          lysosomotropic drug-carrier conjugate. The
          <u>targeting moiety</u> is designed to interact
          with cell-specific features on the plasma
          membrane of the target cell, thus leading
          to uptake by adsorptive pinocytosis and
          transfer into the lysosomes. The drug is
          attached covalently by a linkage stable in
          tissue fluids but degraded by enzymes
          resident in the lysosomes. The <u>degradable</u>
          'spacer' is designed so that the susceptible
          linkage is sufficiently distant from the
          polymer backbone to be accessible to enzymes.
          A <u>solubilizer</u> is necessary if the polymer
          backbone is too hydrophobic to be water-
          soluble, and comprises side-chains with
          hydrophilic character.

ical modification. Thus the design of a tailor-made carrier is much
easier if the starting material is a simple homopolymer rather than
a molecule as complex as a polypeptide.

        In two respects, however, synthetic polymers have disadvantages
when compared with proteins. The first disadvantage is their poly-
dispersity: in terms of molecular weight a population of molecules
of a synthetic polymer is much less well-defined than a population

of molecules of some purified protein. Secondly, as pointed out by
Trouet et al., (1982), a carrier should be biodegradable. Most synthetic polymers are not degradable by mammalian enzymes, but towards
the end of this paper we explain why this need not be an insuperable
obstacle.

A DEGRADABLE CONJUGATE

The lysosomal enzymes are a formidable array, capable together
of the degradation of a wide variety of biopolymers, including polypeptides, polynucleotides and complex polysaccharides. The individual enzymes, however, are amidases, esterases or glycosidases with
restricted substrate specificities, and it is far from self-evident
that a wholly unnatural synthetic polymer could provide the basis
of an acceptable substrate.

Poly(hydroxypropylmethacrylamide) (PHPMA) is a polymer to which
a range of side-chains can be attached by substituting various primary amines for some of the 1-amino-propan-2-ol residues (Fig. 2).
We have synthesized a series of compounds, each based on PHPMA of
$\bar{M}_w$ 50 000, in which approximately 2% of the monomer units are conjugated to an oligopeptide. The terminal amino acid is conjugated
by amide linkage to p-nitroaniline, a 'drug analogue' chosen for its
ease of detection by spectrophotometry. These compounds (Fig. 3)
each contain three types of amide linkage: that between methacrylic
acid and the most proximal amino acid, the peptide linkage between
adjacent amino acids, and that between the distal amino acid and
p-nitroaniline. It was our hope that some at least of these linkages
would prove susceptible to the lysosomal amidases, thus releasing the
drug analogue, either free or as a peptidyl or amino acyl derivative,

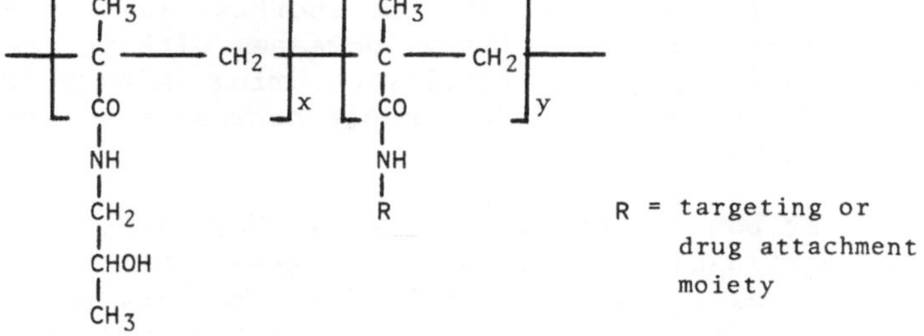

Fig. 2. Structure of poly N-(2-hydroxypropyl)
         metharcrylamide copolymers showing sites
         for attachment of drug and targeting
         moieties.

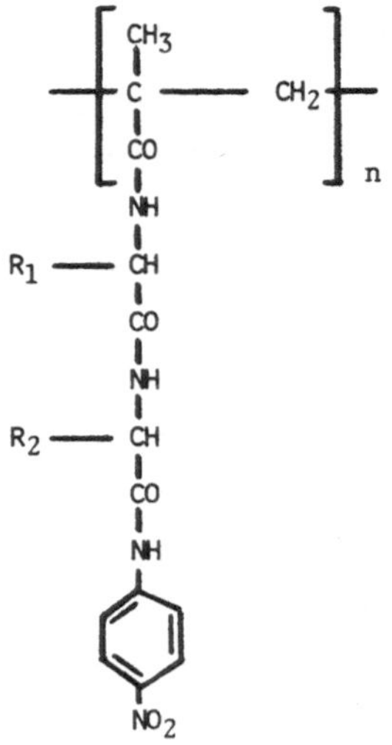

Fig. 3.   Oligopeptidyl p-nitroanilide side-chain of
          polymethacrylamide. A dipeptidyl side-chain
          is shown, but other side-chains synthesized
          contained up to six amino acids.

from the polymer. Our results, which have been reported in full
elsewhere (Duncan et al., 1980, 1982a, 1983a), demonstrate the
following:

1.   The enzymes of the rat liver lysosome can release p-nitroaniline from many of these conjugates. The rate of release varies
widely depending on the composition of the oligopeptide spacer.
The least susceptible are those with the shortest (dipeptide)
spacers and in general susceptibility increases with increasing
length. The inclusion of non-naturally occuring amino acids, e.g.
β-alanine or ϵ-aminocarproic acid, sharply decreases the rate of
release of p-nitroanaline.

2.   The first bond to be hydrolysed is that linking the terminal
amino acid to p-nitroaniline or, in a few cases, that between the two
most distal amino acids. We have no evidence for any hydrolysis of
any other linkages within the oligopeptide, although it is not im-
probable that more proximal peptide bonds are cleaved, particularly
in the longer side-chains. We have increasing evidence that the
amide linkage between methacrylamide and the first amino acid is not
degradable by lysosomal enzymes.

3. The amidase activity responsible for liberation of p-nitro-aniline is strongly dependent on the presence of a thiol, and is inhibited by leupeptin. These observations point to a cysteine-peptidase, a class of enzyme well represented in the lysosomes. The degradation of certain of the side-chains was less sensitive to leu-peptin inhibition or less thiol-dependent, indicating the involvement in such cases of another class of amidase.

DEGRADATION WITHIN LYSOSOMES FOLLOWING PINOCYTOSIS OF THE CONJUGATE

It has also been possible to demonstrate the pinocytosis of our PHPMA-oligopeptide-nitroaniline by living cells, and the subsequent release of side-chain degradation products. We used the 17.5-day rat yolk-sac in organ culture, an experimental system ideal for this purpose, and monitored degradation by a radioisotopic method, more sensitive than the spectrophotometric assay used in the experiments with isolated lysosomal enzymes. Four of the polymers from the previous series were selected, each with a side-chain that terminated in tyrosyl p-nitroanilide, and these were iodinated (^{125}I) on their tyrosine moieties using the chloramine T method. Side-chains of the four polymers represented both the susceptible and the resistant to isolated lysosomal enzymes. The results, which are reported by Duncan et al., (1981a), indicate the following:

1. The four polymers were all captured by yolk sacs at the same rate as ^{125}I-labelled polyvinylpyrrolidone, a reliable indicator of non-adsorptive (fluid-phase) pinocytosis. Thus PHPMA, even if sub-stituted to some 2% with oligopeptide side-chains, does not adsorb to the cell surface and enters the cell by fluid-phase pinocytosis.

2. Following pinocytosis of one of the polymers, radioactivity steadily accumulated in the yolk sacs, with no evidence of side-chains' degradation. The side-chain in question had the composition -Gly-βAla-Tyr-nitroanilide, and was one of those found resistant to degradation in the experiments with isolated lysosomal enzymes. In the case of the other three polymers, uptake by the yolk sac was accompanied by release into the culture medium of (^{125}-I) iodotyro-sine, clearly a side-chain degradation product. In the case of one such polymer, intralysosomal digestion was so efficient that accumu-lation of radiolabel by the tissue did not occur : outflow of (^{125}I) iodotyrosine from the lysosomes kept pace with inflow of ^{125}I-labelled polymer. It may be noted that release of (^{125}I)iodotyrosine implies the hydrolysis of both the terminal and the penultimate amide bonds of the side-chain.

3. Addition of leupeptin to the cultures of yolk sacs did not affect pinocytosis of the polymers, but inhibited intralysosomal digestion of side-chains to a greater or lesser extent. These data are consistent with an important but not exclusive role for the thiol-dependent amidases.

STABILITY IN BLOOD PLASMA

In contrast to their susceptibility to lysosomal enzymes, the polymer side-chains are not degraded to a significant extent when incubated in rat plasma or serum. This result no doubt reflects the presence within the circulation of powerful proteinase inhibitors.

TARGETING TO A CHOSEN CELL TYPE

As explained in the Introduction, selective uptake of a macromolecule requires an enhancement of the extent to which it is captured by adsorptive pinocytosis. Fortunately much has been learnt in the past decade about the features a macromolecule required in order to be an effective adsorptive substrate. These features may be classified as non-specific or as cell-specific. In the experiments reported PHPMA has been prepared with side-chain substituents designed to increase adsorptive pinocytosis. PHPMA is a particularly suitable substrate for such a study, since the unsubstituted molecule has no affinity for plasma membrane and, like polyvinylpyrrolidone, enters pinocytic vesicles entirely in the fluid phase (vide supra).

NON-SPECIFIC ADSORPTIVE UPTAKE

The plasma membranes of many mammalian cell types appear to have two functionally independent classes of binding sites recognizing respectively hydrophobic and cationic domains on macromolecules (Lloyd and Williams, 1984). It has been of interest to us to discover whether these properties of cells could be harnessed to increase the uptake of synthetic polymers. In recent experiments PHPMA has been synthesized with up to 20 mol % of tyrosinamide side-chains. Tyrosinamide was chosen for the hydrophobic moiety because it renders the polymer susceptible to ^{125}I-labelling; also a pilot experiment with another synthetic polymer substituted with tyramine (Duncan et al., 1982b) had given promising results. Our (unpublished) data show that uptake of PHPMA was predominantly by fluid-phase pinocytosis, with tyrosinamide contents up to 10 mol %. Beyond 10 % tyrosinamide, uptake was strongly by adsorptive pinocytosis. We have not yet attempted a PHPMA with cationic substituents, but Pratten et al., (1982) reported adsorptive pinocytic uptake by macrophages when polyvinylpyrrolidone was modified to incorporate some cationic character.

CELL-SPECIFIC ADSORPTIVE UPTAKE

One of the most striking discoveries of the seventies was the existence on various cell types of highly specific binding sites for glycoproteins. Many cell types bind and internalize glycoproteins bearing a 6-phosphomannose residue, cells of the mononuclear phagocyte family recognize an exposed mannose or N-acetylglucosamine, whereas hepatocytes have a receptor specific for oligosaccharide chains terminating in galactose.

We have recently demonstrated that PHPMA can be specifically targeted to the liver by incorporation of glycylglycyl-D-galactos-amine side-chains (Duncan et al., 1983b). Uptake by the liver was enhanced at least six-fold. In the same experiment we also showed the potential for drug delivery : by incorporating into the same polymer some side-chains comprising glycylglycyl-L-tyrosinamide, and labelling the aromatic ring with ^{125}I, first hepatic uptake and then release of (^{125}I)iodotyrosinamide (here acting as a drug analogue) was demonstrated.

MOLECULAR SIZE AS A DISCRIMINATOR

It is worthy of mention that molecular size has some targeting potential. Some cell types appear to be unable to pinocytize macro-molecules above a certain size, whereas others preferentially capture the larger molecules. Using polyvinylpyrrolidone of different mole-cular size-ranges, preferential uptake of the larger molecules by macrophages and of the smaller by rat yolk sac has been demonstrated (Duncan et al., 1981b).

POLYMERS WITH DEGRADABLE CROSS-LINKS

A polymer such as PHPMA, if used as a drug carrier, would accu-mulate steadily within lysosomes after releasing its charge of drug. When the cell dies, either as a result of intended toxicity from a cytotoxic agent or in the natural course of events, the carrier would be released into the extracellular compartment to be pinocytized again by another cell. Thus, chronic administration of such a con-jugate would result in a steady accumulation of polymer within the tissues. This feature is probably of little importance if the disease being treated is a life-threatening one, but nevertheless it is better avoided.

We aim to break this cycle of release and re-uptake by using PHPMA of low molecular weight cross-linked by lysosomally degradable oligopeptides. The size of the individual chains would be below the glomerular filtration threshold, but the cross-linked polymers would be too large to be filtered. Such cross-linked polymers, on uptake into lysosomes, would not only release their drug but themselves be processed. The short polymer chains would be retained within the lysosomes until the death of the cell. Once released into the extracellular space they would rapidly be cleared into the glomerular filtrate and so eliminated from the body.

Our experiments with such polymers are at an early stage. We have however prepared short ($\bar{M}_w$= 30 000) chains of PHPMA cross-linked by di-oligopeptidyl-hexamethylenediamine, and demonstrated that following pinocytosis by rat yolk sacs, some intralysosomal degradation of the cross-links takes place (Cartlidge et al., 1982).

CONCLUSIONS AND PROSPECT

The experiments reported above confirm the feasibility of the approach to drug targeting outlined in the Introduction. We have demonstrated that it is possible to attach a drug analogue covalently to a synthetic polymer by a spacer moiety that is degraded in the lysosomes to release the drug analogue. We have shown that such macromolecular conjugates are stable in plasma and are not excreted by the kidney. Also that they are pinocytozed by cells that have access to the circulation and that features may be incorporated into the macromolecule that lead to enhanced uptake by chosen cells, through receptor-mediated pinocytosis. Finally that the use of degradable cross-links between relatively short polymer chains can overcome the problem of tissue accumulation of undegraded polymer.

There is clearly much to be done before a clinically useful product becomes available. The susceptibility to enzymic hydrolysis of the linkage binding the drug is likely to be profoundly affected by the chemical nature of the drug. There are no predictable extrapolations from drug analogue to drug, nor from one drug to the next. Secondly, and more seriously, we still have no means of targeting to many cell types of the body, nor to neoplastic cells while avoiding uptake by their normal counterparts. Nevertheless there is encouragement in the undoubted fact that our knowledge of the distinctive features of cell surfaces is still in its infancy. It seems certain that additional determinants for cell-specific receptor-mediated pinocytosis will yet be discovered, and that study of these determinants will lead to an increasing simplification of the moiety seen to be required. There is also much yet to be discovered about the control of endocytosis, and it may well be that pinocytosis will be found to be a finely modulated phenomenon susceptible to stimulation and inhibition. Thus the problems that remain do not appear insoluble.

We believe this approach to chemotherapy has a bright future, particularly for the treatment of cancer. There are undoubtedly major problems, such as tumor heterogeneity, but even a degree of improvement in the cell specificity of anti-tumor agents would be a worthwhile advance.

ACKNOWLEDGEMENT

We thank the Cancer Research Campaign for their generous support of the work reported above. The collaboration between Keele and Prague is supported by the Royal Society under their cultural agreement with the Czechoslovak Academy of Sciences and by the British Council under their Academic Links with Eastern Europe Scheme.

REFERENCES

Cartlidge, S.A., Duncan, R., Lloyd, J.B., Rejmanová, P., and
    Kopeček, J., 1982, Pinocytic capture and intracellular
    degradation of poly N-(2-hydroxypropyl)-methacrylamide
    chains connected by oligopeptide sequences, Proc. Internat.
    Conf. Biomedical Polymers, Durham,   289.
Duncan, R., Lloyd, J.B., and Kopeček, J., 1980, Degradation of
    side chains of N-(2-hydroxypropyl)methacrylamide copolymers
    by lysosomal enzymes, Biochem. Biophys. Res. Commun., 94:284.
Duncan, R., Rejmanová, P., Kopeček, J., and Lloyd, J.B., 1981a,
    Pinocytic uptake and intracellular degradation of N-(2-
    hydroxypropyl)methacrylamide copolymers. A potential drug
    delivery system, Biochem. Biophys. Acta, 678:143.
Duncan, R., Pratten, M.K., Cable, H.C., Ringsdorf, H., and Lloyd,
    J.B., 1981b, Effect of molecular size of ^{125}I-labelled poly
    (vinylpyrrolidone) on its pinocytosis by rat visceral yolk
    sacs and rat peritoneal macrophages, Biochem. J., 196:49
Duncan, R., Cable, H.C., Lloyd, J.B., Rejmanová, P., and Kopeček,
    J., 1982a, Degradation of side-chains of N-(2-hydroxypropyl)
    methacrylamide copolymers by lysosomal thiol-proteinases,
    Bioscience Reps., 2:1041.
Duncan, R., Starling, D., Rypáček, F., Drobnik, J., and Lloyd, J.B.,
    1982b, Pinocytosis of poly $(\alpha,\beta$ -(N-2-hydroxyyethyl)-DL-
    aspartamide and a tyramine derivative by rat visceral yolk
    sacs cultured in vitro. Ability of phenolic residues to
    enhance the rate of pinocytic capture of a macromolecule,
    Biochem. Biophys. Acta, 717:248.
Duncan, R., Cable, H.C., Lloyd, J.B., Rejmanová, P., and Kopeček,
    J., 1983a, Polymers containing enzymatically degradable bonds.
    Design of oligopeptide side-chains in poly(N-(2-hydroxypropyl)
    methacrylamide)copolymers to promote efficient degradation by
    lysosomal enzymes, Makromol. Chem., in press.
Duncan, R., Kopeček, J., Rejmanová, P., and Lloyd, J.B., 1983b,
    Targeting of N-(2-hydroxypropyl)methacrylamide copolymers to
    liver by incorporation of galactose residues, Biochem. Biophys.
    Acta, 755:518.
Lloyd, J.B., and Williams, K.E., 1984, Non-specific adsorptive
    pinocytosis, Biochem. Soc. Trans. in press.
Pratten, M.K., Cable, H.C., Ringsdorf, H., and Lloyd, J.B., 1982,
    Adsorptive pinocytosis of polycationic copolymers of
    vinylpyrrolidone with vinylamine by rat yolk sac and rat
    peritoneal macrophage, Biochem. Biophys. Acta, 719:424.
Trouet, A., Baurain, R., Deprez-de-Campaneere, D., Masquelier, M.,
    and Pirson, P., 1982, Targeting of antitumour and anti-
    protozoal drugs by covalent linkage to protein carriers, in:
    "Targeting of Drugs", G. Gregoriadis, J. Senior and A. Trouet,
    eds., Plenum Press, New York and London.

DRUG TARGETING IN CANCER THERAPY

George Poste

Smith Kline and French Laboratories
Philadelphia, Pennsylvania and Department
of Pathology and Laboratory Medicine
University of Pennsylvania Medical School
Philadelphia, Pennsylvania, USA

Progress in cancer chemotherapy over the last twenty years has produced important gains in the treatment of testicular germ cell tumors, choriocarcinoma, Burkitt's lymphoma, Hodgkins disease, and several childhood cancers.[1] Progress in treating the three most common solid malignancies of man arising in breast, lung and colon is less impressive, however, and the majority of patients with these tumors die of metastatic disease, despite assault by ever larger combinations of drugs. Two factors are responsible for limiting the effectiveness of current therapeutic approaches. First, the presence within the same tumor of subpopulations of cells that differ widely in their responses to cytotoxic drugs and other agents used in cancer treatment means that therapy will be successful only if this diversity can be circumvented.[2,3] Second, the lack of selectivity of most anticancer agents for tumor cells causes significant toxicity to normal tissues with resulting problems in clinical management, patient compliance and reduced quality of life.

In seeking new approaches for the therapy of neoplastic disease three different, non-exclusive, strategies are being pursued. The first involves continuation of painstaking clinical studies that seek to improve the effectiveness of existing drugs by exploring their use in new combinations and dosing regimens. Such studies are constrained, however, by shortcomings inherent in the pharmacology of current anti-neoplastic agents. The second strategy involves the search for new drugs with entirely novel pharmacologic actions. Although the need for such drugs is obvious, their identification is by no means guaranteed. The cellular and subcellular

changes identified to date in tumor cells compared with their normal
counterparts appear to be quantitative rather than qualitative and
the search for targets for drug action unique to tumor cells may be
unrealistic. The recent fascinating observations on oncogenes have
generated the inevitable hyperbole regarding a "breakthrough" that
will lead to dramatic gains in cancer treatment.[4] Even if future
research reveals that such genes play a role in either the initiation, progression and/or metastasis of human tumors, it will be many
years before drugs acting on these elements will be available for
clinical use. This delay provides the rationale for the final strategy of improving the effectiveness of existing drugs using pharmaceutical techniques to alter drug disposition, pharmacokinetics and/
or dose-response relationships to achieve a higher therapeutic index.
Drug delivery systems have been in the vanguard of this effort and
a considerable part of research on this important topic has been
devoted to the problem of "targeted" or site-specific drug delivery.

The ability to target drugs to specific cells within the body
has been one of the most sought after goals in therapeutics. Ever
since Paul Ehrlich[5] foresaw the use of "bodies which possessed a
particular affinity for a certain organ ... as carriers by which to
bring therapeutic active groups to the organ in question", considerable ingenuity has been devoted to the study of this challenging
problem. A wide variety of cellular, macromolecular and particulate
carriers have been investigated as potential drug delivery systems
with the stated objective of improving cancer chemotherapy. These
include: erythrocytes, leukocytes, antibodies, nucleic acids, heat-
or chemically-denatured plasma proteins and a diverse array of particulate carriers of differing sizes and biodegradability prepared
from various polymeric materials, dextran, gelatin, agarose, cellulose, albumin or phospholipids.[6-11] Many have failed to fulfill
their initial promise, but interest in drug delivery systems remains
high in both academic and industrial laboratories.

This article surveys the current status of efforts to target
drugs to tumor cells. The emphasis will be on general principles
of drug delivery to tumors and their metastases. No attempt will
be made to review information on specific classes of drugs or tumor
types. The merits and drawbacks of different delivery systems are
discussed in the context of how far their proposed application is
consistent with current concepts in tumor cell biology, cancer
metastasis and the pathophysiology of the tumor-bearing host.

SITE-SPECIFIC DRUG DELIVERY

Site-specific drug delivery requires completion of several
sequential, independent steps: (a) localization of drug and its
carrier within the desired target organ; (b) recognition and interaction of the carrier with specific target cell(s) and (c) delivery
of therapeutic concentrations of drug to target cell(s) with little

or no uptake by non-target cells. For widespread clinical use a carrier should be capable of incorporating a broad spectrum of drugs without loss or alteration in activity. A carrier must protect drug from degradation or inactivation during transit to the target and release of drug during transit should be minimal or nonexistent. Once at the desired tissue site(s), transfer of drug to target cells must occur in a controlled and predetermined manner without impairing drug action or inducing toxic reactions not associated with the conventional dosage form. To minimize host toxicity, particularly during chronic therapy, drug carriers must be biodegradable, biochemically inert, non-immunogenic (unless specifically desired) and have no adverse effects on tumor behaviour or host response to therapeutic agents. Finally, preparation of the drug-carrier complex should be straightforward, reproducible, cost-effective and yield homogeneous stable dosage forms that can be administered with minimum inconvenience to the patient. To date no carrier has fulfilled all of these criteria.

Site-specific drug delivery can be classified according to the level of selectivity achieved in the delivery process: delivery to individual organs or tissues (organ targeting); targeting to a specific cell type(s) within a tissue (cellular targeting); and delivery to different intracellular compartments in target cells by engineering the internalization of drug or drug-carrier complex via specific transport pathways (subcellular targeting). A distinction can also be made between active and passive targeting systems.[11] "Passive targeting" involves therapeutic exploitation of the natural (passive) distribution pattern of a drug carrier in-vivo. For example, the role of the reticuloendothelial system (RES) in clearing foreign particulate materials from the blood permits drugs encapsulated in particulate drug carriers such as liposomes to be passively targeted to RES macrophages. In contrast, "active targeting" attempts to alter the natural disposition pattern of a drug or a carrier and achieve targeting to specific cells, tissues or organs. Current efforts to achieve this goal typically involve coupling drugs to antibodies or other cell-specific ligands that interact with molecules specific to the surface of the desired target cell. A third strategy, referred to as "physical targeting", employs drug-carrier complexes that release drug only when exposed to specific microenvironments such as changes in pH or temperature.[12] The use of external magnetic fields to direct drug carriers containing magnetic particles to different organs also falls within this classification.[13]

The present chapter will consider only the issue of targeting in-vivo. Development of methods for targeting drug carriers to homogeneous cultured cell populations in-vitro is considered to be of less value since in most cases the convenience of working with well characterized cells in culture enables the carrier to be added directly to the desired target cell. This criticism does not apply,

of course, to efforts to achieve carrier-mediated targeting to
specific cells or subsets of cells in mixed, heterogeneous cell
populations in-vitro. Such experiments may yield valuable information for subsequent clinical applications in ex-vivo treatment protocols in which tissues such as bone marrow are removed from cancer
patients, treated with targeted drug-carrier complexes in-vitro to
eliminate tumor cells and then returned to the patient.[14]

In-vitro experiments will also be useful in developing methods
that might eventually be used in-vivo. Drug carriers bearing cell-
specific ligands may interact with cells in an entirely different
fashion to carriers without ligands and it is important to understand how such differences may affect drug uptake and activity.[11,15]
Analysis of such questions is logically undertaken using cultured
cells in-vitro.

ANTIBODIES AND OTHER MACROMOLECULAR LIGANDS AS CARRIER SYSTEMS
FOR SITE-SPECIFIC DRUG DELIVERY

The appeal of antibodies for active targeting of drugs to
specific cell types in-vivo has long been recognized. Until recently, the feasibility of this approach could not be explored fully
because of the variable properties of polyclonal antisera obtained
from immunized animals. With the advent of in-vitro hybridoma technology this situation has changed dramatically. Highly purified
monoclonal antibodies of defined class and antigen specificity can
now be produced in virtually unlimited quantities. Antibodies to
tumor cell surface antigens have attracted considerable interest
because of their possible value in cancer diagnosis and in targeting
drugs, toxins and other therapeutic agents to tumor cells.[11,16-23]

Monoclonal antibody-drug conjugates offer the advantage that the
clinical pharmacology of the active agent is known.[20] However, since
most drugs act stoichiometrically the efficacy of antibody-drug com-
lexes is determined by the number of antibody molecules that bind
and release active drug. Since major cell surface antigens are not
usually present at densities greater than 10^5 molecules/cell, and
many antigens are present at much lower densities, the probability
of achieving therapeutic drug levels using antibodies directed
against such targets is low.[11,22] This problem has favored an alternative strategy in which antitumor antibodies are covalently bound to
potent toxin molecules of plant and bacterial origin (immunotoxins)[24]
The best known are from the bacterium Corynebacterium diphtheriae
(diphtheria toxin) and seeds of the plants Abrus precatorius (abrin)
and Ricinis communis (ricin).[24] Although the precise number of toxin
molecules needed to kill cells is still debated, it is accepted that
on a molar basis their cytotoxicity is significantly greater than
most antitumor drugs and thus can be used for targeted chemotherapy
directed against tumor antigens present at very low densities. These
toxins consist of two subunits, designated A (active) and B (binding).

Opinion differs as to whether the entire molecule (holotoxin) or
only the A subunit is more suitable for targeted chemotherapy.[24,25]
There is some evidence that the binding capacity of the B subunit
reduces the specificity of antibody-holotoxin conjugates for tumor
cells. Also, premature release of holotoxin during transit could
create serious toxicity problems. In contrast, release of the isolated A subunit would be of no consequence because of its inability to
bind and be internalized by cells in the absence of the B subunit.[24]
Other data suggest, however, that the B subunit plays an important
role in the penetration of the A subunit into the cytoplasm after
internalization within endosomes.[25] This has prompted proposals
that isolated A and B subunits could be administered together so
that simultaneous uptake of both subunits by a cell would facilitate
cytoplasmic penetration of the A subunit and that cells taking up
either subunit alone would not be at risk.[26] In the author's opinion the probability of achieving uptake of both subunits into a
sufficient number of tumor cells to achieve therapeutic effects is
extremely low because of dilution effects that will occur when the
subunits are injected.

The availability of high affinity antibodies that recognize
tumor-associated antigens and have no cross-reactivity with normal
cells is an obvious prerequisite for the successful use of antibodies
in the delivery of therapeutic agents. With the exception of idio-
types on B cell tumors such antigens have not yet been found.[19] A
large number of monoclonal antibodies that react with tumor-associated antigens on histologically diverse human and animal tumors have
been reported in the last two years, but all exhibit varying cross-
reactivities with epitopes on normal tissues.[11,16-23] Clinically,
the crucial question is whether the cross-reacting antigen is expressed on normal cells that are essential for life or whose damage
would produce unacceptable toxicity. These considerations assume
particular importance when using highly potent toxic molecules such
as toxins.

The requirements for chemical coupling of antibodies to drugs
and toxins are also conflicting in that stable bonding is needed to
prevent premature release of cytotoxic molecules during transit to
the target but once bound to the target cell the linkage must be
sufficiently labile to permit release of active molecules. Preferably, the latter step should occur only after internalization of the
carrier-drug/toxin complex to reduce the risk of the toxic moiety
interacting with normal bystander cells. The most common method for
coupling antibodies to therapeutic agents is via disulphide bridges
created by the SPDP reaction. Since this is a natural linkage it
is reasoned that this will allow breakdown and release of the active
agent at the target site.[27] More information is needed, however,
about the stability of this linkage in-vivo to determine what fraction of the total conjugates survive to reach the target site.

A diverse array of other naturally occurring macromolecules such as hormones, lectins and growth factors that interact with specific cell types in-vivo have been proposed as potential carriers for site-specific drug delivery.[8-11] However, in the current climate of enthusiasm for monoclonal antibodies interest in these ligands has apparently waned.

PARTICULATE CARRIERS AND SITE-SPECIFIC DRUG DELIVERY

A heterogeneous collection of natural and artificial particles has been examined as potential drug carriers (Fig. 1). Of these, liposomes and microspheres prepared from albumin or ethylcellulose have attracted most interest.[8,9,11,21,28,29]

Liposomes are closed vesicles prepared from phospholipids comprising either one (unilamellar) or several lipid bilayers (multilamellar) surrounding an internal aqueous space or spaces. Their potential as drug carriers stems from the ability to encapsulate water-soluble drugs within their aqueous interior or to incorporate hydrophobic drugs within the lipid bilayer(s).[8,28] Antibody molecules can be covalently linked to the outer surface of liposomes to act as cell recognition ligands that bind to surface molecules on target cell(s).[8,28] Methods for the preparation and characterization of liposomes are reviewed in detail elsewhere.[8,28] The majority of published work on liposome-mediated drug delivery has been done with two classes of liposomes: small sonicated unilamellar liposomes (SUV); and large multilamellar liposomes (MLV). These were developed in the 1960's as model membranes for analyzing the properties of lipid bilayers[29] and have been adopted for drug delivery, largely without modification. Current preparation methods suffer from low efficiency of drug capture, high rates of leakage of water-soluble drugs on exposure to body fluids, poor storage stability and inadequate control of particle size.[8,28,30] Significant methodological improvements in liposome preparation, drug encapsulation efficiency, formulation homogeneity and stability must occur before the full potential of liposomes as a delivery system can be realized.[11]

The physiologic processes that affect drug delivery by liposomes in-vivo apply to particulate carriers in general.[28,30,31] It is thus surprising that liposomes have attracted so much attention when many of the systems listed in Fig. 1 not only exhibit higher drug carrying capacity but can be produced as homogeneous, stable preparations.[9-11] In the author's opinion this emphasis will change and biodegradable nanoparticles, and microspheres will assume increasing importance.[11] In addition to their superior formulation and stability characteristics much more is known about the physicochemical basis of drug capture and release from microparticles prepared from albumin and biodegradabe polymers.[9,11,32] However, data on these issues are derived almost exclusively from in-vitro studies and in common with liposomes, correlative information on the kinetics

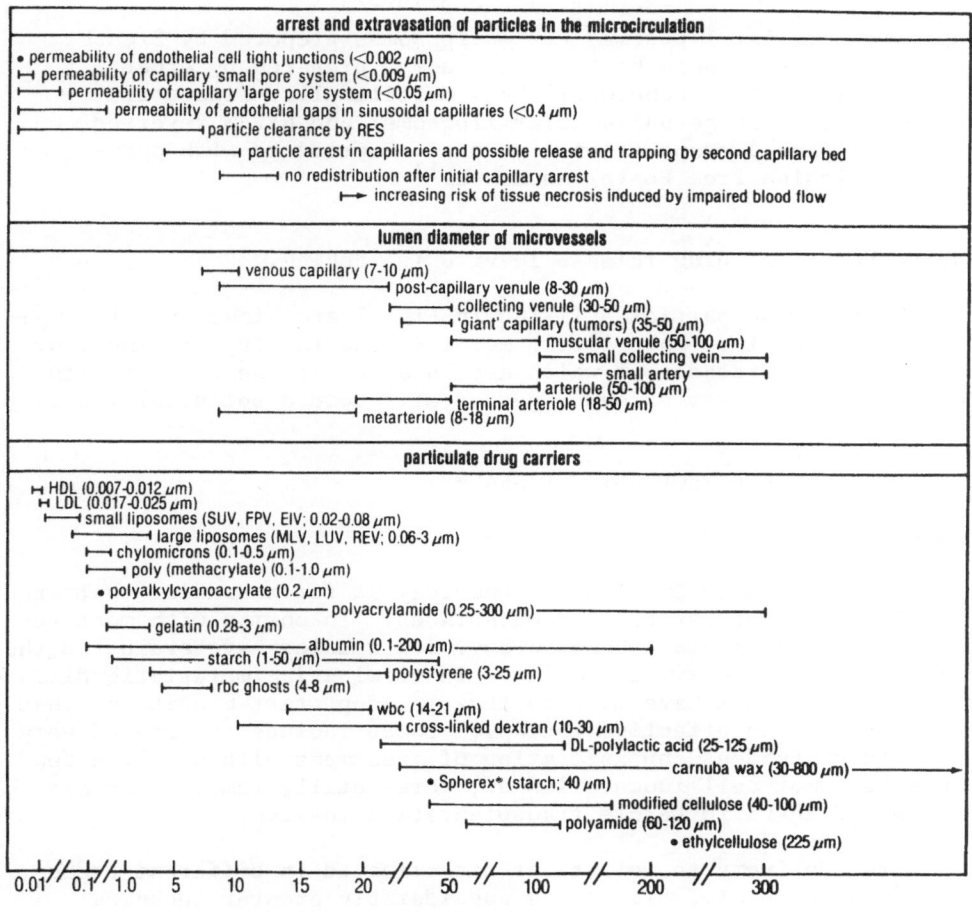

arrest and extravasation of particles in the microcirculation

- permeability of endothelial cell tight junctions (<0.002 μm)
- permeability of capillary 'small pore' system (<0.009 μm)
- permeability of capillary 'large pore' system (<0.05 μm)
- permeability of endothelial gaps in sinusoidal capillaries (<0.4 μm)
- particle clearance by RES
- particle arrest in capillaries and possible release and trapping by second capillary bed
- no redistribution after initial capillary arrest
- increasing risk of tissue necrosis induced by impaired blood flow

lumen diameter of microvessels

- venous capillary (7-10 μm)
- post-capillary venule (8-30 μm)
- collecting venule (30-50 μm)
- 'giant' capillary (tumors) (35-50 μm)
- muscular venule (50-100 μm)
- small collecting vein
- small artery
- arteriole (50-100 μm)
- terminal arteriole (18-50 μm)
- metarteriole (8-18 μm)

particulate drug carriers

- HDL (0.007-0.012 μm)
- LDL (0.017-0.025 μm)
- small liposomes (SUV, FPV, EIV; 0.02-0.08 μm)
- large liposomes (MLV, LUV, REV; 0.06-3 μm)
- chylomicrons (0.1-0.5 μm)
- poly (methacrylate) (0.1-1.0 μm)
- polyalkylcyanoacrylate (0.2 μm)
- polyacrylamide (0.25-300 μm)
- gelatin (0.28-3 μm)
- albumin (0.1-200 μm)
- starch (1-50 μm)
- polystyrene (3-25 μm)
- rbc ghosts (4-8 μm)
- wbc (14-21 μm)
- cross-linked dextran (10-30 μm)
- DL-polylactic acid (25-125 μm)
- carnuba wax (30-800 μm)
- Spherex® (starch; 40 μm)
- modified cellulose (40-100 μm)
- polyamide (60-120 μm)
- ethylcellulose (225 μm)

0.01 0.1 1.0 5 10 15 20 50 100 200 300

Particle diameter (μm)

Fig. 1. Sizes of particulate drug carriers in relation to their arrest within and extravasation from the vascular bed after intravenous injection. Top panel: the effect of particle size on particle arrest in the microcirculation, clearance by the reticuloendothelial system (RES) and the limiting sizes for particle extravasation. Middle panel: the internal lumenal diameter of different microvessels in the venous and arterial circulations which define at which level in the vascular bed arrest of particles of different sizes will occur. Bottom panel: the sizes of natural (lipoproteins; chylomicrons; cellular carriers) and synthetic particles evaluated as potential drug carriers in-vivo. Abbreviations: HDL = high density lipoprotein; LDL = low density lipoprotein; SUV = small (sonicated) uni-

(continued)

and mechanism of drug release in-vivo is lacking.

Most of the particles listed in Fig. 1 are biodegradable. Non-
degradable carriers have little merit in the therapy of cancer or
other chronic diseases in which accumulation of the carrier with
repeated dosing over extended period will pose a potential toxicity
hazard.

DRUG DELIVERY AND NEOPLASTIC DISEASE

Choice of Tumor Models

The major need in clinical oncology is for therapies that are
effective against established metastases. In common with most research in experimental chemotherapy, few studies have evaluated the
efficacy of drug-carrier complexes in eliminating metastatic disease.
Most investigators have been content to adopt test conditions that
are conducive to effective therapy. These include the use of very
small tumor burdens and initiation of treatment within only a few
hours of tumor cell inoculation or, worse still, tumor cells are
pretreated in-vitro before transplantation in-vivo.

Drug delivery to metastases disseminated in different organs,
including the brain, presents a considerably greater technical
challenge than therapy of localized tumors.[3,11] Few of the animal
tumor(s) used in evaluating drug delivery systems address the problem of metastasis. Intraperitoneal therapy of a minimal ascites
tumor burden localized in the peritoneal cavity is of little relevance since the drug or the drug carrier gains immediate access to
tumor cells without interference from anatomic barriers. Similarly,
well vascularized, transplanted tumors growing s.c. or i.m., though
more difficult to eradicate than i.p. tumors, fail to provide a
sufficiently demanding model for evaluating the effectiveness of
agents in treating metastases. All animal tumors are flawed in some
respect as models of human neoplasms.[3,33] This does not mean, however, that efforts should not be made to employ models that share
as many features as possible with human malignancies. This requires
that drug delivery systems be evaluated for activity against established metastases in an adjuvant therapy setting comparable to that
undertaken in the clinic following removal/reduction of the primary
tumor. The rationale for proposing that experimental therapies be
tested for activity against established metastases is not based on

a belief that the therapeutic responses of cells in metastases may differ from tumor cells in the primary lesion. There is no a priori reason to assume that the chemosensitivities of metastatic tumor cell subpopulations cannot be determined by assaying the response of a heterogeneous tumor cell population containing both metastatic and non-metastatic cell subpopulation implanted s.c. or i.m.[3,11] The rationale for using metastatic tumor models is to confirm that the disposition and pharmacokinetics of a drug and its delivery system are consistent with achieving therapeutic drug concentrations in organs typically affected by metastases in the nepolasm of interest. [11] Since metastases in the same host vary significantly in their size, growth rates, cell growth fraction and vascular supply[3,33-35] testing in metastatic models also yields useful information about the efficacy of a drug delivery system in circumventing these factors.[11]

Localization of Drug Carriers in Tumors and their Metastases

Intravenous (i.v.) drug administration presently offers the only practical route of administration for most agents used in treating widely disseminated metastatic disease. The localization and retention of drug carriers administered via this route in tumors in-vivo are determined by: the anatomic location(s) of the tumor(s); their blood supply; the behaviour of the carrier in the systemic circulation; and the structure and permeability of the tumor microvasculature. For solid tumors and their metastases located in extravascular tissues, targeted drug delivery to tumor cells requires that the carrier must leave the circulation. For carriers equipped with cell recognition ligands such as antitumor antibodies, retention will also be influenced by the distribution of antigen-positive tumor cells and the properties of the target antigen.[11,16-23]

Whole body scintigraphy reveals marked variation in the efficiency with which antitumor antibodies localize in animal and human tumors.[18,36,37] Impressive reports of antitumor monoclonal antibodies that localize in tumors in-vivo at five fold or greater concentrations than a normal tissue are now commonplace.[17,18,23] Reports are equally common, however, in which monoclonal antibodies with similar high affinities for tumor cells in-vitro fail to label tumors in-vivo.[17,18,23] Despite the disappointing precedent of studies on the labeling of tumors in-vivo with heteroantisera, research using monoclonal antibodies is still in its infancy and detailed analysis of the effect of antibody class, the properties of the target antigen, coupling of cytotoxic agents to antibodies and investigation of the optimum dose, route and schedule of antibody administration may yield important therapeutic gains. It is important, however, that such studies assay localization within metastases rather than large tumor masses implanted s.c. or i.m.

Effective drug delivery by antitumor antibodies or other target-

ing ligands requires that they penetrate uniformly throughout the
tumor mass. Autoradiography of histologic sections of tumors ex-
cised after i.v. administration of radiolabeled antibodies or anti-
body-conjugates reveals considerable variation in labeling patterns
in different tumors.[11,18,36,37] Limited localization of antibodies
in either the peripheral or central portions of the tumor is the
most common pattern seen to date but in certain tumors uniform antibody distribution throughout the lesion has been achieved.[11,18,36,37]

    Vascular hemodynamics and microvessel permeability (Fig. 2A,B,C)
are of obvious importance in determining the uptake of circulating
drug-carrier complexes by tumors. Total blood flow, rates of perfusion and vessel permeability may vary significantly in different
regions of the same tumor.[36] The vascular supply in tumors transplanted s.c. or i.m. may differ substantially from metastases that
arise from them. For example, implantation of a large bolus of tumor
cells into host tissue disrupts the local microvasculature and
surrounding tissues and enhances vascular permeability at the injection site which persists until vascular repair is achieved within
2-3 days. Penetration of i.v. injected materials is facilitated at
such sites during this period and administration of drug or a drug-
carrier complex within 24-48 hours of tumor implantation can result
in unusually high concentrations of drugs within the tumor.[11] Also,
the blood supply of transplanted tumors is established by emigration
of new capillary sprouts from surrounding host vessels. These
sprouts are highly permeable and allow virtually unlimited passage
of materials, including erythrocytes, into the extravascular tissue.[38]
Penetration of liposomes, antibodies or other circulating materials
into transplanted tumors may thus be artificially high[11] and provide
little or no insight into the behaviour of these materials in the
vascular bed of human tumors or spontaneous animal tumors in which
the vascular supply will evolve in an entirely different fashion in
concert with progressive enlargement of the tumor cell population.
In transplanted tumors new vessels grow in from the periphery of the
implant. In newly formed spontaneous tumors, which contain far
fewer cells, the topographic relationship of tumor cells to blood
capillaries will be completely different. Similarly, in formation
of hematogenous metastases, a single tumor cell, or at most a few
cells, extravasate and grow as pericapillary colonies[2] before the
angiogenic response needed to support additional cell mass occurs.[38]
These events suggest that localization of blood borne materials in
newly transplanted tumors (Fig. 2B) will more likely occur in the
periphery but in spontaneous tumors (Fig. 2A) and hematogenous
metastases (Fig. 2C) it will be central or uniform throughout the
lesion.

    In contrast to antibodies and other macromolecular carriers
whose initial distribution is uniform throughout the capillary circulation, the disposition of particulate carriers injected i.v. is
non-uniform and is determined by particle size, vascular anatomy

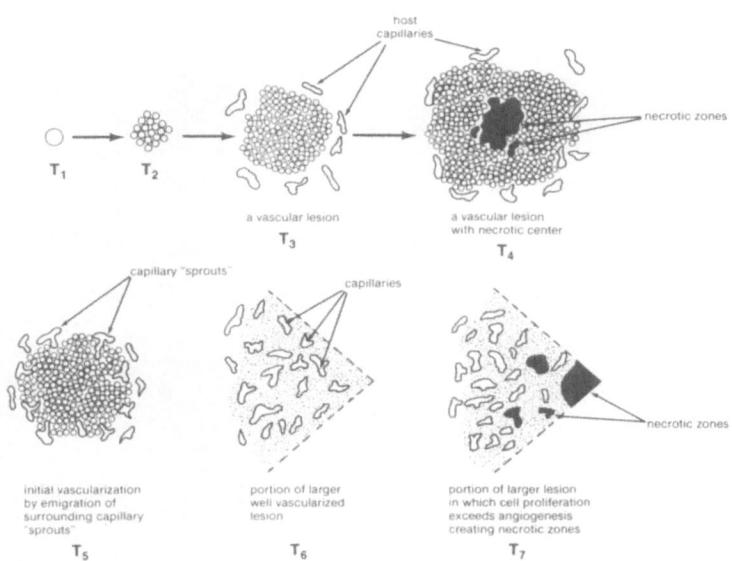

Fig. 2A. Vascularization of spontaneous tumors. Exposure of a
susceptible target cell to a carcinogenic insult(s)
results in its transformation to a tumor cell (T_1).
Studies using karyotypic, immunologic and biochemical
(isozyme) markers indicate that the majority of human
solid neoplasms are of single cell origin. Subsequent
cell proliferation then increases the tumor mass (T_2
and T_3). Once the tumor cell mass has reached a size
in which cellular nutrition cannot be satisfied by
diffusion of nutrients from the blood circulating in
the adjacent microcirculation, cell proliferation will
either cease or cell growth will be accompanied by
death and necrosis (■ in T_4) of cells in the interior
of the nodule. Significant expansion in tumor cell mass
to form a clinically detectable lesion depends on the
migration of new vessels into the tumor by budding of
new capillary sprouts (angiogenesis) from the nearby
microcirculation(T_5). Assuming that the angiogenic response proceeds in parallel with cell proliferation new
vessels will be constantly recruited so that vessels
will be distributed uniformly throughout the expanding
tumor mass (T_6 : this depicts only a segment of the
total lesion). If tumor cell proliferation outstrips
angiogenesis cell death will occur in areas of limited
nutrient supply, generating focal or central necrotic
zones (T_7). The relative sizes of the tumor foci at
different times and the sizes of the blood vessels
relative to tumor mass are not to scale and are meant
to merely depict the general distribution of vessels
within the lesion.

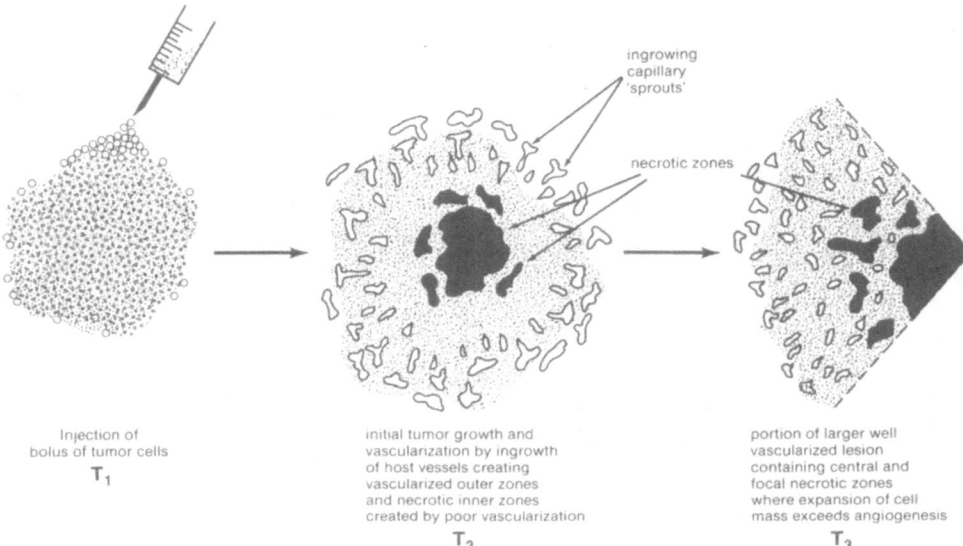

Fig. 2B. Vascularization in transplanted tumors. A bolus of
        tumor cells is injected into a host tissue (T_1).
        Typically, several thousand or tens of thousands of
        tumor cells will be injected. The angiogenic response
        needed to vascularize this relatively large cell mass
        again proceeds by emigration of new capillaries into
        the tumor from the surrounding host tissue (T_2). How-
        ever, depending on the size of the implanted bolus cells
        in the deeper internal regions may not be vascular-
        ized effectively and undergo necrosis (■ in T_2 and T_3).
        The relative sizes of the tumor foci at different
        times and the sizes of the blood vessels relative to
        tumor mass are not to scale and are meant to merely
        depict the general distribution of vessels within the
        lesion.

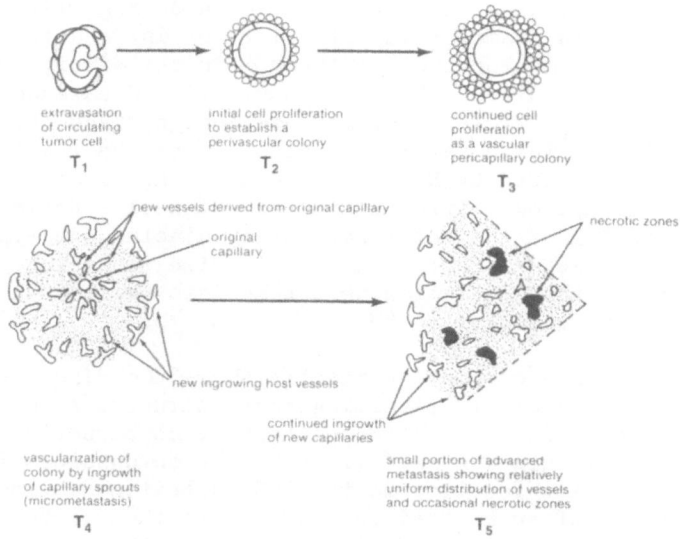

extravasation
of circulating
tumor cell
T_1

initial cell proliferation
to establish a
perivascular colony
T_2

continued cell
proliferation
as a vascular
pericapillary colony
T_3

new vessels derived from original capillary

original
capillary

necrotic zones

new ingrowing host vessels

continued ingrowth
of new capillaries

vascularization of
colony by ingrowth
of capillary sprouts
(micrometastasis)
T_4

small portion of advanced
metastasis showing relatively
uniform distribution of vessels
and occasional necrotic zones
T_5

Fig. 2C. Vascularization of hematogenous metastases. Formation
of hematogenous metastases involves the arrest and
extravasation of a single circulating tumor cell or,
at most, a clump of few tumor cells (T_1). Following
extravasation, these cells survive and proliferate as
avascular pericapillary colonies (T_2). Diffusion of
nutrients from the blood in the capillary can support
limited cell proliferation and tumor cells can survive
when located up to 150μm from the capillary wall. Subsequent growth and vascularization of these micrometastases by ingrowth of capillary sprouts from host capillaries (T_4) is then similar to the process shown in
Figure 1A for primary tumors. Assuming that comparable
levels of tumor cell mass evoke similar angiogenic
responses, expansive growth of metastases will result
in distribution of new vessels throughout the tumor
mass (T_5) except in those regions in which cell growth
exceeds neovascularization where necrosis will occur
(■ in T_6). The structure of new blood vessels formed
in tumors may vary substantially in different regions
of the same tumor and may also vary between a primary
tumor and its metastases and between different metastases. As discussed in the text, the type of structural defect(s) and their frequency will have major
effects on vessel permeability and uptake of circulating
materials by tumor cells located in the extravascular
compartment. The relative sizes of the tumor foci at
different times and the sizes of the blood vessels
relative to tumor mass are not to scale and are meant
to merely depict the general distribution of vessels
within the lesion.

and interaction with the RES.[11,21,34] (see also Fig. 1). Particles
of 0.1 to 5.0μm in diameter injected i.v. or intra-arterially are
cleared rapidly by macrophages of the RES in the liver, spleen and
bone marrow and circulating blood monocytes.[39] Clearance kinetics
are influenced by particle size, aggregation, surface charge and
composition.[11,39] These parameters also affect the fraction of
total particles cleared by RES components in the liver versus the
spleen and bone marrow. Irrespective of which site dominates,
between 80 and 95% of the total particles injected are typically
taken up by the RES.[39] Particles in this size range injected i.p.
also localize in the RES following uptake into peritoneal lymphatics
and return to the circulation.[40]

     Rapid clearance of small particles (0.1-5μm diameter) by the
RES is a major obstacle to targeting these structures to other sites.
Efforts to alter this disposition pattern by constructing particles
with prolonged circulating half-lives have been unsuccessful.[39]
Such particles still localize in the RES, albeit more slowly.[39] The
slower clearance of such particles may enable them to be used as
rate controlled drug release devices in cancer therapy, particularly
for labile drugs that are rapidly inactivated or degraded by blood
components.[41] This requires that particles can be constructed with
reproducible permeability characteristics to ensure that the rate of
drug release is consistent with sustaining therapeutic levels of drug
in the blood. Ironically, the search for highly stable liposomes
and other particles with very low permeability characteristics and
long circulating half-lives may be of limited value if the rate of
drug release is so slow that therapeutic drug levels are never
achieved.[42]

     Blockade of the ability of the RES to clear particles by pre-
dosing with particles without drug before injecting of particles
containing drug has been advocated as a way of diverting particulate
drug carriers away from the RES.[43] Although this strategy enhances
liposome localization of small (< 800Å),[43]   but not large liposomes[39] in some,[43] but not all,[39] tumors, this technique has no
effect on liposome localization in metastases.[21] Even under conditions of RES blockade a higher concentration of liposome-associated
drug may still accumulate in macrophages of the RES than in unblockaded animals treated with free drug.[43] This may be of little consequence for many classes of drugs but for highly cytotoxic agents
of the kind used in cancer therapy this can cause destruction of
the RES.[39] Also, repeated RES blockade can impair RES function.[39]

     The key assumption in proposals to use liposomes or other part-
iculate carriers to target drugs to solid tumors and their metastases located in extravascular tissues is that these structures can

extravasate to reach tumor cells in the extravascular compartment.

In view of the crucial importance of this issue in establishing the feasibility of perhaps the most frequently cited advantage of liposomes as a drug carrier, it is remarkable that so little attention has been given to this basic question. This deficiency probably refelcts the formidable technical difficulties involved in answering this question.

Even though a more critical perspective about the feasibility of liposome targeting in-vivo is emerging, the legacy of earlier speculation about liposome targeting still survives in portions of the current literature. Even in 1983, uncritical generalizations and proposals for liposome targeting continue to be published which have little rationale when examined in terms of current knowledge of anatomy, physiology and pathology. The myriad problems that must be overcome if liposome targeting to cells in extravascular compartments is to every become a practical reality have received scant attention from proponents of the concept. These merit critical review.

Liposome Capillary Interactions in Normal Tissues

The anatomy of the microcirculation in different tissues and organs can reasonably be expected to be of crucial importance in determining whether liposomes can escape into the surrounding extravascular tissue.[30,39] In common with the extravasation of circulating blood cells, extravasation of liposomes can reasonably be expected to be restricted to capillaries and small diameter post-capillary venules which possess minimal adventitial elements.

Blood capillaries are classified according to the architecture of the lining endothelium and the underlying subendothelial basement membrane (basal lamina) into three different groups (Fig. 3): continuous capillaries; fenestrated capillaries; and discontinuous or so-called sinusoidal capillaries.

In continuous capillaries adjacent endothelial cells adhere via tight junctions to form a "continuous" lining and they also typically possess an uninterrupted subendothelial basal lamina or basement membrane. In fenestrated capillaries, the endothelium is interrupted by fenestrae which vary from 300Å to 800Å in diameter. However, with the exception of fenestrated endothelial cells in the renal glomeruli, the fenestrae do not represent simple openings and the fenestrae are spanned by a thin membranous diaphragm (40-60Å thick). [39,44] As in continuous capillaries, the subendothelial basal lamina in fenestrated capillaries is continuous. The third class of

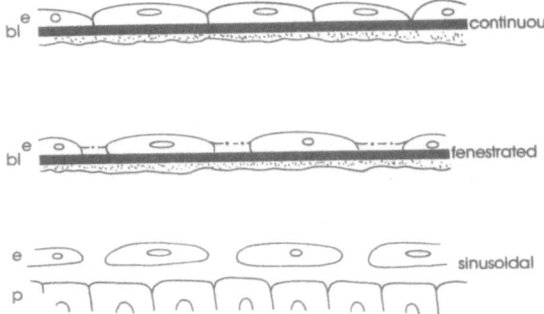

Fig. 3.   Schematic illustration of the structure of different
           classes of blood capillaries. A. Continuous capillary.
           The endothelium is continuous with tight junctions
           between adjacent endothelial cells. The subendothelial
           basement membrane is also continuous. B. Fenestrated
           capillary. The endothelial exhibit a series of fen-
           estrae which are sealed by a membranous diaphragm. The
           subendothelial basement membrane is continuous. C.
           Discontinuous (sinusoidal) capillary. The overlying
           endothelium contains numerous gaps of varying size
           enabling materials in the circulation to gain access
           to the underlying parenchymal cells. The subendoth-
           elial basement is either absent (liver) or present
           as a fragmented interrupted structure (spleen; bone
           marrow).   Reproduced with permission from Post et
           al.[21]

capillaries, the discontinuous (sinusoidal) capillaries, are thin-
walled vessels found only in the liver, spleen and bone marrow. The
endothelium in these vessels has large gaps which may be as large
as several thousand Å in diameter. In most species the sinusoidal
capillaries of the liver lack a basement membrane but an interrupted
basement membrane is present in these vessels in the spleen and the
bone marrow.[44]

    Viewed simply from a mechanical standpoint, continuous and
fenestrated capillaries present a major barrier to the escape of
liposomes from the microcirculation. In contrast, liposomes might
be expected to penetrate the relatively large gaps in the endothelium in sinusoidal capillaries and thus come into immediate contact
with the underlying organ parenchymal cells.[39] Extravasation of
liposomes in organs lined by sinusoidal capillaries would therefore
appear to be limited only the diameter of the gaps in the endoth-

elium. Typically, these gaps in hepatic sinusoids are less than
0.1μm in diameter. Small sonicated SUV liposomes should thus be
able to pass through such openings whereas larger liposomes would
be retained within the sinusoid. This is consistent with reports
showing that uptake of SUV by hepatocytes is greater than with MLV
or LUV, though for all classes of liposomes uptake into the RES
predominates.[39] The endothelium of hepatic sinusoids contains openings larger than 0.1μm in diameter and which are large enough to
allow penetration of MLV, REV or LUV liposomes.[39] However, the lower
frequency of these openings, and their irregular distribution within
sinusoids, dictate that extravasation of large liposomes is less
efficient than for SUV liposomes.

Extravasation of materials from the bloodstream in continuous
and fenestrated capillaries is more complicated and typically occurs
by one of two pathways.[39,44-46] The so-called "small pore" pathway
is limited to materials of <90Å diameter and thus does not represent
an obvious pathway for liposome extravasation since the smallest SUV
are 250-300Å in diameter.[39] In contrast, the "large pore" system
is permeable to materials up to 700Å in diameter. In fenestrated
capillaries this system is believed to be represented by the fenestrated openings within endothelial cells but, as emphasized above,
the presence of a membranous diaphragm in the fenestrations dictates
that permeability is not defined simply by the diameter of the fen-
estration. The permeability of diaphragmed fenestrae appears to
vary signigicantly in different tissues.[45] Fenestrated capillaries
in tissues such as submaxillary gland, renal medulla, pancreas and
intestine are permeable to molecules such as ferritin (Einstein-
Stokes radius (ESR) 61Å). However, the proportion of intestinal
capillaries that are permeable to dextrans of differing sizes (ESR
62.5-100Å) and glycogens (ESR 110-150Å) can vary from between 20-
70%.[45] Some fenestrated capillaries, however, such as those in the
choriocapillaries, are not permeable to tracers with ESR of >32Å.[47]

In continuous capillaries, the "large pore" permeability pathway is provided by the endothelial vesicle system.[45,46] Until recently, the predominant view was that vesicles formed at the lumenal
surface of endothelial cells and then passed across to the cell to
fuse with the ablumenal plasma membrane and discharge their contents.
Since endothelial vesicles do not exceed 700Å diameter they would
not accommodate large MLV or REV liposomes. It is not certain, however, that SUV liposomes of less than 700Å diameter could be transported across capillaries in this system. There is no evidence that
the lumenal plasma membrane of endothelial cells is able to "flow"
around a liposome (or other particle) bound to the endothelial cell
surface in the way that actively phagocytic cells such as macrophages and PMNs engulf particles by surrounding them with pseudopo-
dial extensions. Endothelial cells in the liver, spleen and in
newly formed capillaries exhibit significant phagocytic activities
but most endothelial cells do not.[44] Consequently, uptake and

transport of materials via the endothelial vesicle system may be limited to materials whose diameter does not exceed the width of the "neck" (60-100Å diameter) at the opening of a vesicle at the lumenal surface. If this is correct then even the smallest SUV liposomes would be too large to be transported by this pathway.

Additional doubts about the size of materials that can enter endothelial vesicles have been raised as a result of recent studies which suggest that endothelial vesicles may not be a dynamic system as proposed originally in which new vesicles are formed constantly at the lumenal surface, internalized and translocated to the abluminal surface to undergo exocytotic fusion.[46] Careful three-dimensional morphometric studies using serial sections suggest that vesicles are static structures.[46] Additional evidence suggesting that lumenal vesicles are static and open to the vessel lumen via a constricted channel comes from studies showing that tracers are taken up by vesicles under conditions where energy dependent membrane internalization processes are inhibited as, for example, in fixed tissues or in tissues exposed to anoxia, meabolic poisons or low temperature.[46]

Although these new findings do not exclude the contribution of endothelial vesicles as a pathway for macromolecular transport, the possibility of particles such as liposomes being transported in this system seems remote since access to such vesicles will be limited by the size of the channel (60-100Å diameter) by which the vesicles open to the vessel lumen.

From this brief review of capillary structure, it is evident that for continuous and fenestrated capillaries, it might be predicted based on theoretical considerations alone, that there are major anatomic barriers to the escape of circulating liposomes into the surrounding extravascular tissues. Experimental evidence to support this interpretation has been obtained recently in our laboratory.[30,39]

The capacity of liposomes to cross continuous capillaries has been studied by analyzing extravasation of small sonicated SUV liposomes (mean diameter 600Å) and MLV liposomes (0.4μm diameter) containing encapsulated ^{125}I-BSA in the microcirculation of the mouse lung.[30,31,39] Extravasation of MLV liposomes injected i.v. was assayed by their presence within pulmonary macrophages recovered from pulmonary alveoli by bronchial lavage. However, studies in which the blood monocyte populations of mice were depleted experimentally demonstrated that liposomes per se did not exit from pulmonary capillaries but were engulfed within the bloodstream by blood monocytes which subsequently migrated into the alveoli.[31] Similar passive translocation of MLV within macrophages could presumably provide a mechanism for liposome extravasation in other tissues in which the anatomy of the microcirculation frustrates extravasation of free liposomes.

Further evidence for the inability of liposomes to cross continuous capillaries in the lung has been obtained using classical methods of monitoring capillary permeability. Extravasation of materials from circulating blood in the lung can be measured by recovery and analysis of lung lymph.[30,39] By measuring the concentration of a marker molecule in the blood and in pulmonary lymph, capillary permeability can be measured and dynamic changes in capillary permeability monitored to distinguish whether alterations in permeability are due to enhanced hydrostatic pressure (filtration) or leakage. Using a heart-lung perfusion preparation studies in my laboratory examined the ability of SUV and MLV liposomes containing ^{125}I-BSA to cross pulmonary capillaries and be recovered in the lymph.[30,39] In this system extravasation of liposome-associated ^{125}I-BSA was not detected, even under conditions in which significant extravasation of unencapsulated (free) ^{125}I-BSA occurred as a result of elevated hydrostatic pressure induced by increasing left atrial pressure via an indwelling balloon catheter or administration of vasoactive drugs that increase capillary leakage.[30,39]

Analogous organ perfusion techniques have also been used to evaluate the permeability of continuous capillaries in skeletal muscle and fenestrated capillaries in dog colon and cat submandibular gland to SUV or MLV liposomes containing ^{125}I-BSA. Liposome extravasation was not detected in any of these tissues.[30,39]

These observations suggest that liposomes are probably unable to exit from the microcirculation in organs lined with either continuous capillaries (skeletal, smooth and cardiac muscles; connective tissue; central nervous system; exocrine pancreas; gonads; lung) or fenestrated capillaries (most exocrine and endocrine glands; gastrointestinal tract; renal glomeruli and peritubular capillaries; choroid plexus). Although permeability measurements have not been made on capillaries in all of these tissues, the data from studies on capillaries in lung, skeletal muscle, gut and salivary gland at least confirm experimentally what might reasonably be predicted on mechanical considerations, namely, that the endothelial lining and/or the continuous basal lamina found in these classes of capillaries present major anatomic barriers to liposome extravasation.[30,39]

Hwang et al.,[49] have claimed that SUV (200Å diameter; sphingomyelin/cholesterol 2:1) can pass across capillaries in the stomach, intestine and skin. This claim was advanced on the basis of evidence showing gradual accumulation of liposome-derived 111 In in the intestine, stomach and skin over a period of several days. However, this study did not establish whether the radioactivity in these tissues was associated with intact liposomes and the question of whether intact liposomes had entered the extravascular compartment cannot therefore be answered. Furthermore, the increase in liposome-derived radioactivity in the gut is equally as likely to represent biliary excretion of encapsulated material from liposomes taken up and broken down within the liver. In agreement with previous findings[30],[39] these investigators found no evidence of liposome extravasation in the lung or in skeletal muscle.

The most zealous protagonists of liposome targeting will prob-ably continue to argue that until similar capillary permeability measurements are made on all tissues using liposomes of the smallest size and of endlessly variable composition the feasibility of target-ing liposomes to extravascualr tissues cannot be totally discounted. This is undoubtedly true but the evidence available to date argues against this possibility. Until experimental data are obtained show-ing that liposomes can successfully cross continuous or fenestrated capillaries in a particular tissue, proposals for targeting liposomes to parenchymal cells in the extravascular compartment of tissues lined by these two classes of capillaries should be viewed with con-siderable skepticism. Extravasation of liposomes from continuous or fenestrated capillaries may be occurring on a limited scale below the detection limits of the assay methods used to date. Even if this is the case, the more important issue is whether such low levels of extravasation could be exploited therapeutically.

The preceding discussion refers only to situations in which the endothelial lining retains its structural integrity. In inflammation and ischemia this is not guaranteed and such conditions may alter endothelial cell architecture and, in turn, affect vessel permeabil-ity. Arrest of very large liposomes or large aggregates of liposomes within a capillary could conceivably impair blood flow downstream. The resulting hypoxia could evoke reversible retraction of endoth-elial cells downstream[34] and allow particles of comparable size to liposomes to extravasate.

That liposomes can occasionally pass through the junctional spaces between organized endothelial cells and reach the underlying basement membrane is demonstrated by the work of Lubec et al.[50] These investigators presented electronmicroscopic evidence for pen-etration of positively-charged MLV (PC/SA/Chol) between adjacent glomerular endothelial cells in the rat kidney. However, the fre-quency of this phenomenon was not quantified and occurrence of the same phenomenon in capillaries of other organs was not examined. Once again, however, the issue is not so much whether penetration of the endothelium may occur in particular tissues and/or certain con-ditions, but whether it occurs with sufficient frequency to be useful therapeutically.

Even if penetration of endothelial cells by liposomes were tak-ing place on a significant scale, the question then arises as to whether liposomes can penetrate the subendothelial basement membrane. Lubec et al.,[50] observed liposomes between, and beneath, endothelial cells, but no examples were seen of liposomes that had penetrated the basement membrane or the surrounding adventitial elements to reach the extravascular compartment and liposomes remained on the lumenal side of the basement membrane.

Ultrastructural studies using electron-dense tracers of differing molecular weights and known Einstein-Stokes radii have shown convincingly that the subendothelial basement membrane in continuous and fenestrated capillaries provides a highly efficient barrier to the penetration of materials whose ESR are considerably lower than the diameter of the smallest SUV liposomes.[45,51] Clearly, blood-borne materials, including cells and motile parasitic larvae, cross subendothelial basement membranes. Current concepts of basement membranes view them as fully hydrated gels with thixotropic properties that undergo gel-sol transformations in response to localized deforming pressures to create 'sol' channels that allow materials to pass.[52] As fully hydrated gels, basement membranes would be freely permeable to water, small solutes and low molecular weight drugs which can pass through the lattice in its fully hydrated 'gel' state. In contrast, passage of larger molecules and cells requires that they induce localized deformation of the lattice and transformation to the 'sol' phase. In the case of cells this would be induced by the cell's own motile activity. However, for macromolecules and inert particles such as liposomes that lack active, energy-dependent motility systems, passage would be possible only if pulsatile intracapillary pressure was adequate to initiate local deformation of the basement membrane.[52]

The energy of deformation needed to induce a gel-sol transformation in a local area of sufficient size to permit entry of even the smallest liposomes would be greater than that needed to create similar 'sol' zones for smaller plasma proteins. Consequently, it seems likely that the intracapillary pressure conditions necessary to induce successful transit of liposomes across the basement membrane would simultaneously render the vessel highly permeable to a variety of blood macromolecules. The experiments described earlier using the heart-lung preparation indicates that leakage of this kind does not occur since the lymph: plasma protein ratio remains stable after i.v. infusion of liposomes.[30,39] This suggests that normal vascular hemodynamics are not sufficient to serve as a "driving force" to induce thixotropic deformation of endothelial basement membrane on the scale needed to permit extravasation of liposomes.

Liposome-Capillary Interaction at Sites of Inflammation and Ischemia

Capillary permeability increases significantly at sites of inflammation.[33] However, recent studies in my laboratory using the granuloma pouch assay to monitor the permeability of vessels in an inflammatory lesion have revealed that liposomes do not extravasate even under conditions in which extensive macromolecular leakage occurs.[30,39]

There is evidence, however, that capillaries at sites of tissue ischemia become permeable to liposomes. Occlusion of the coronary arteries in dogs or guinea pigs for various intervals has been shown to cause preferential accumulation of i.v. injected liposomes in the resulting areas of ischemic myocardium.[54,55] Similar accumulation of circulating liposomes at sites of ischemia has been reported in the rat intestine following experimental ligation of the mesenteric artery.[56] Electronmicroscopic studies of ischemic myocardium reveal liposomes in both intravascular and extravascular locations.[55] Damage to capillaries was evident as shown by swelling, distortion and retraction of endothelial cells and structural abnormalities in the underlying basement membrane. These changes, albeit of a non-specific nature, would be expected to enhance capillary permeability to liposomes and many other circulating materials. This is consistent with previous studies showing that a wide range of low and high molecular weight solutes, drugs and particulate materials such as colloidal carbon will accumulate in the ischemic myocardium.[56]

Liposome-Capillary Interactions in Tumors

Even though extravasation of circulating liposomes from the microcirculation may prove to be a major limitation in targeting liposomes in tissues with a normal vascular supply, proponents of liposome-mediated drug delivery in cancer therapy have speculated that targeted drug delivery to tumors may still be possible because of the enhanced vascular permeability found in tumors. This argument is not confined to advocates of liposome targeting and is voiced with equal conviction by proponents of other targeted drug delivery systems such as immunotoxins and drug-antibody conjugates.

The factors affecting the localization and retention of circulating materials in tumors are complex, variable, unpredictable and poorly understood.[11,36,37] Most of the available information on this topic has come from studies of the localization of radiopharmaceuticals and anti-tumor antibodies employed in tumor imaging and from studies with labeled cells and particulates used in estimating the volume of the vascular bed in tumors.[37] The following are among the more important factors in determining the efficiency with which materials introduced into the circulation localize in tumors: total blood flow; tissue perfusion rates and pressure; the properties of the circulating material(s); microvascular architecture and permeability; and the properties of the tumor cells.

The only safe conclusion that can be drawn from the extensive experimental data accumulated over the last twenty years is that any generalization about the role of these factors is likely to be overly simplistic and that predictions about vascular hemodynamics in any given tumor are highly imprecise. Total blood flow, rates of perfusion and the efficiency with which circulating materials localize in tumors can vary considerably in tumors of similar histologic

origin and clinical staging in different patients, in different
regions of the same tumor and in different metastases in the same
host.[36],[37] Also, as emphasized earlier, the structure and function
of the vascular supply in tumors transplanted s.c. or i.m. may be
completely different from spontaneous tumors and metastases. Furthermore, the vascular system in these different types of lesions is not
static and will change during progressive tumor growth and after
therapeutic assault.[3]

    In common with other tissues, tumors have an arterial supply,
a capillary network to distribute blood and a venous drainage system.
Except where they are invaded by neoplastic cells, or compressed by
expanding masses of tumor cells, the arteries of tumors show minimal
or no structural changes from their normal counterparts. Structural
changes in venous structures of tumors are more common. Invasion of
veins is an almost universal finding in malignant tumors and foci of
necrotic endothelial cells are not uncommon in tumor venules.[36],[57]

    It is in the structure of capillaries that tumors show the most
dramatic departure from the vascular anatomy of normal tissues.[21],[58]
The microvessels found in tumors fall into four general classes:
continuous; discontinuous; sinusoidal blood channels; and giant capillaries (Fig. 4). Continuous capillaries of tumors are the most
like normal microvessels. They are lined by an uninterrupted sheet
of endothelial cells joined by typical tight junctions and resting
on a well-formed basement membrane which is sometimes reduplicated.
Continuous capillaries in tumors are usually supported by incomplete
layers of pericytes and bundles of collagen fibers. An increased
frequency of fenestrated endothelial cells has also been reported in
certain tumor microvessels, but such vessels usually possess a continuous subendothelial basement membrane which will present a sub-

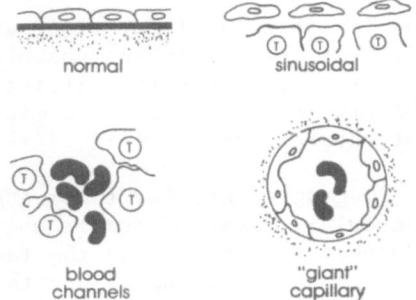

Fig. 4.   Schematic illustration of the structure of diff-
           erent types of capillaries and other micro-
           vessels commonly found in neoplastic tissue.
           Reproduced with permission from Poste et al.[21]

stantial mechanical barrier to extravasation of liposomes and other particulate materials.

Discontinuous capillaries have an incomplete endothelial lining with the endothelial cells lying on a continuous or fragmented basement membrane reminiscent of capillaries in areas of inflammation. Alternatively, the lumen can be lined almost exclusively by tumor cells, with only scattered endothelial cells. Newly forming capillaries may also be grouped in this category because they commonly have large gaps between adjacent endothelial cells. Blood can also circulate within tumors through vascular channels lined exclusively by tumor cells rather than endothelial cells. The circulation pattern in this type of vessel is such that blood simply percolates around solid cords of tumor cells. So called giant capillaries are also found at the growing edge of tumors and their lumen may exceed 50 μ m in diameter. Although such vessels may easily be confused with venules on histologic analysis, ultrastructural studies have shown that they are composed only of a single layer of endothelial cells with little or no supporting connective tissue and thus warrant classification as capillaries.

On mechanical grounds the discontinuous capillaries and blood channels found in tumors may represent sites at which circulating materials can penetrate into the extravascular space. What is not known, and cannot be predicted with any measure of accuracy for any given tumor, is what fraction of the total tumor vasculature is composed of these highly permeable vessels. Even within the same tumor, structurally normal capillaries can be observed in close proximity to abnormal vessels.[58] Although structural defects in a proportion of vessels might be useful in allowing liposomes containing antitumor agents to gain access to tumor cells, the presence of many normal vessels in other portions of the same tumor will probably ensure that a significant number of tumor cells will remain inaccessible to liposomes injected i.v.

In discussing the feasibility of targeting any drug carrier to tumors it is important that changes in the vascular permeability within tumors not be overstated since the alterations are relative and unlimited permeability is found only in a minor group of microvessels with marked structural changes.[11,36,37,58]

Success in using imaging agents for tumor angiography and measurements of the vascular volume of tumors using radiolabeled erythrocytes and microspheres both rely on the fact that these materials will not escape from tumor vessels into the extravascular compartment. This is well illustrated by the behaviour of sodium diatrizoate (M_r 600), an x-ray contrast medium commonly used in angiography, which circulates unbound to plasma proteins yet does not rapidly extravasate into tumor tissue.[59]

The kinetics of partition of molecules between the intra - and extravascular compartments are also relevant to the present discussion. Once again, great variability is the rule, as illustrated by the following examples. Goldacre and Sylven[60] found that the dye Lissamine Green extravasated within as little as 30 seconds in both normal and tumor tissues following i.v. injection. However, areas were commonly seen in tumors that contained large numbers of viable tumor cells but failed to stain at any time after dye infusion. Gullino and Grantham[61] found that high molecular weight (500,000) dextran took between 1 and 5 hours to appear in the interstitial fluid of a series of rat tumors. Underwood and Carr[62] observed that extensive leakage of Evans Blue into the extravascular compartment of a transplanted rat sarcoma occurred within 1 hour after i.v. injection but extravasation of particles of colloidal carbon (30 nm diameter) or saccharated iron (3-10 nm diameter) did not occur. Extravasation of the latter, but not the former, could be elicited by intratumoral injection of histamine.[62]

The most definitive and comprehensive analysis of the permeability of tumor blood vessels completed to date is provided by the work of Peterson and his colleagues.[36] Their elegant studies on the partition between blood and tumor tissue of radiolabeled molecules of differing molecular weights indicate that although tumors in general may display a higher permeability to plasma proteins than normal tissues, considerable barriers to free permeation exist in a significant fraction of the tumor vasculature.[36]

Even tumor microvessels that lack an endothelial cell lining do not necessarily represent sites of unrestricted permeability. Martinez-Palomo[63] examined the permeability of junctions between adjacent tumor cells in the blood channels within several different rat and mouse tumors. Using colloidal lanthanum as an ultrastructural probe of junctional permeability he found that while many intercellular junctions allowed greater penetration of lanthanum than endothelial junctions in normal vessels, others were identified which completely excluded lanthanum. Although termed collodial lanthanum, the probe used by Martinez-Palomo has been reported to be largely ionic in nature.[64] That some junctional areas are able to exclude an ionic probe also argues against the likelihood of free passage of particulates. Penetration of circulating particulate materials into the extravascular compartment of tumors can occur, however, as shown by Long et al[65] for fluorocarbon emulsions and by Bugelski et al[66] for aggregates formed between ionic lanthanum and serum proteins.

Detailed information on the permeability of tumor vessels to liposomes is lacking. Several groups have reported significant localization of i.v. administered liposomes in tumors implanted s.c. or i.m.[43,67-70] but a systematic analysis of the effect of liposome size, composition dose and tumor characteristics (i.e., spontaneous versus transplanted; primary versus metastases; tissue of origin) on

the localization of liposome localization and rentention in solid
tumors has still to be undertaken.

The available data indicate that the permeability of the microcirculation in tumors is often greater than that found in normal
tissues but this is a highly variable and unpredictable element of
tumor physiology. Enhanced vascular permeability in tumors is still
relative, however, and free access of circulating materials to the
extravascular interstitium is not found, even for serum proteins.
Inorganic colloids and proteins which have effective diameters of
the order of 30 nm, e.g., colloidal carbon and IgM, have only limited
access to the extracellular space of tumors and localize rapidly
within the RES.[71] Since the smallest SUV liposomes have typical
diameters of 25-35 nm it seems likely that only a small fraction of
such liposomes would have the opportunity to extravasate in tumors
before being cleared by the RES. Larger liposomes are cleared even
more rapidly by the RES and will thus have an even lower probability
of extravasation. Blockade of the RES by predosing with liposomes
or other particulate materials can increae the opportunities for
the localization of SUV liposomes in tumors.[43] However, this strategy has a number of drawbacks. First, effective therapy of tumors
will almost certainly involve a multiple dosing protocol irrespective
of whether liposome-encapsulated drug or some other treatment is
employed. If RES blockade must be imposed before each liposome
treatment this introduces additional risk since repeated blockade
can impair both RES function and bone marrow hematopoiesis (see
below). Second, even under conditions of RES blockade a higher concentration of liposome-associated drug may still accumulate in macrophages of the RES than in unblockaded animals treated with free drug.
This altered drug disposition may be of little consequence for many
classes of drugs but for the highly cytotoxic agents used in cancer
therapy this can cause destruction of the RES.[30,39]

In debating the fading prospects for successful targeting of
circulating liposomes to tumor cells at extravascular sites, advocates of the concept argue that ways might still be found to render
vessels permeable to lipsomes. Such methods could probably be devised, but for what purpose, and at what risk? It is difficult to
envisage how any selectivity could be imposed on this process and
how it could be localized to the microcirculation in interest.
Vessels "permeabilized" for this purpose may instead become highly
leaky for a wide array of macromolecules that do not ordinarily
penetrate into the extravascular space (and should not). In our
opinion such proposals are naive and pose the risk of significant
toxicity. Even if this questionable strategy were pursued and
methods developed to allow safe "permeabilization" of limited
regions of the microvasculature in specific tissues and organs,
this remarkable achievement would virtually eliminate the need for
using liposomes as a drug carrier since free drug would also be able
to extravasate in larger amounts.

TUMOR CELL HETEROGENEITY AND CANCER THERAPY

The phenotypic heterogeneity of tumor cell subpopulations co-existing within the same tumor is probably the single most important factor contributing to the current lack of success in treating many solid malignancies. Malignant tumors are not uniform entities populated by cells with identical properties but instead contain multiple subpopulations of cells with diverse phenotypes, including metastatic properties and responses to anti-neoplastic agents.[2,3,34,35,72] Carrier-mediated drug delivery presently offers no tangible advantage over other approaches in addressing this vexing problem. Even if a carrier containing a single drug or several antitumor drugs is more effective than the free drug(s) in treating certain tumors, the probability is still high that drug-resistant tumor cell subpopulations will be present in the tumor and survive to produce recurrent disease.

Consequently, irrespective of which carrier system is selected, the only successful therapeutic strategy will be one which can circumvent the diverse chemosensitivities of different tumor cell subpopulations present in any given tumor. Current strategies to achieve this goal fall into two non-exclusive categories.

The first involves the use of multiple antineoplastic agents in an effort to address the presence of multiple subpopulations of cells with widely differing sensitivities to drugs and other therapeutic modalities coexisting in the same lesion. Logically, the trend is toward the use of increasingly large combinations of drugs.

The second strategy attempts to limit the emergence of drug-resistant subpopulations by reducing the rate at which new subpopulations of cells with variant phenotypes are generated. Among the concepts being explored in this context are accelerated staging of multiagent therapy and the use of short cycles of alternating multiagent therapy instead of administering the same agents in non-alternating, sequential fashion.[3,34,35,72,73]

Neither of these strategies preclude the use of carrier-associated drugs. However, given the trend towards multiagent therapy, to be of value a drug delivery vehicle must be compatible with multi-agent and/or multi-modality treatment protocols.

Tumor cell heterogeneity has profound implications for the use of antibodies as targeting ligands, either as direct conjugates to drugs and toxins or as cell recognition ligands coupled to particulate carriers. Qualitative and quantitative differences in antigen expression have been identified between tumor cells in the same primary tumor, between cells in the primary tumor and metastases and between cells in individual metastases arising from the same primary tumor.[11,16-23] Effective targeting will thus likely require

a "cocktail" of different antibodies and different "cocktails" may
be needed to treat tumors in different patients. If the latter
applies, prospects for routine clinical use and commercial development of this approach are bleak since the antigenic profile of a
patient's metastases cannot be defined. In addition, the target
antigen(s) selected must be expressed by the clonogenic stem cell
fraction in a form accessible to antibodies and binding to the target
antigen must permit subsequent drug delivery.[11,22]

Availability of a large panel of anti-tumor antibodies is no
guarantee of success. Antigenic modulation and/or immunoselection
of antigenic-negative tumor cells remain real risks.[11,22,74] The
factors regulating antigen modulation after exposure to antibody are
poorly understood. Some antigens modulate slowly or not at all,
while other modulate rapidly.[11,16,18,22,74] Also, surface antigens
that are unaffected by exposure to a single antibody may modulate
when exposed to multiple antibodies directed against unrelated antigens.[75] The mechanism of modulation and its kinetics will also be
important. In the more common form of modulation, antigen-antibody
complexes are internalized by endocytosis.[16,18]

The complexity of signal transduction mechanisms in the internalization of antigen-antibody complexes is illustrated by studies
on the internalization of drug-containing liposomes bearing monoclonal antibodies directed against different surface antigens on
mouse B and T lymphocytes.[76] Liposome binding and internalization
by the same cell type differed significantly depending on the target
antigen used. Also, liposomes targeted to the same antigen expressed
at similar densities on B and T cells showed different internalization patterns. Finally, monoclonal antibodies directed against different epitopes on the same target antigen induced different internalization responses.[77] Although these experiments were done with antibodies covalently bound to liposomes there is no a priori reason to
conclude that similar mechanisms will not affect the uptake of antibody-drug conjugates and immunotoxins. It thus becomes necessary to
establish that targeting ligands not only display suitable specificities but can also evoke internalization of the cell receptor-
ligand-drug (carrier) complex. Once internalized, effective therapy
requires that the drug/toxin can dissociate from the complex and
exit from the endosome.

In targeting to surface determinants that are modulated by
shedding of ligand-receptor complexes from the cell surface, effective therapy will require that the drug be released from the complex,[11]
preferably before shedding occurs, and that the concentration of
drug in the biophase, and its residence time, are sufficient for
killing. Selectivity is compromised, however, since drugs released
extracellularly will also affect normal cells in the vicinity.

Free antigen blockade poses another potential drawback to anti-

body-mediated drug targeting in-vivo.[11,16-18] Preliminary evidence
suggests that antibody molecules covalently bound to liposomes are
less susceptible than free antibody to inhibition by blocking antigens,[71] but studies using a range of antibodies are needed to establish if this is a general feature of antibodies bound to particles.
Even if blocking antigens are not present at the onset of therapy,
tumor cell destruction induced by initial treatment may quickly
elevate the level of free antigen in the circulation and reduce the
effectiveness of subsequent treatment cycles.[11,16-18]

EMERGING OPPORTUNITIES FOR TARGETED DRUG DELIVERY IN CANCER THERAPY

Even though the anatomy of the microcirculation poses a major
obstacle to the targeting of drug carriers, and particulate carriers
in particular, to cells outside of the vasculature in most tissues,
significant opportunities nonetheless exist for targeting drugs to
cell types within the vasculature. These fall into four categories:
1) passive targeting to the RES; 2) active targeting to circulating
tumor cells and tumor cells in the bone marrow; 3) active targeting
to specific subsets of circulating blood cells; and 4) active targeting to vascular endothelial cells.

Passive Targeting of Drug Carriers to Mononuclear Phagocytes

The discouraging results obtained in both experimental and clinical efforts to develop specific active immunotherapeutic regimens
for the treatment of neoplastic diseases has stimulated a renaissance
in therapeutic approaches for augmenting immunologically nonspecific
host responses to tumors. Specifically, the growing evidence that
cells of the mononuclear phagocyte series (MPS) play a significant
role in host defence against both microbial and neoplastic diseases
has lead to a re-evaluation of the potential therapeutic value of
augmenting host defences by the selective activation of macrophages
in-vivo. Once activated by a variety of synthetic or naturally
occurring agents, macrophages appear capable of selectively recognizing tumor cells irrespective of their degree of phenotypic diversity, and in addition, macrophage mediated tumoricidal activity
appears devoid of the problem of generation of resistance which is
often observed in therapy with cytotoxic drugs.[78,79] For these
reasons, a significant effort is now being undertaken in many laboratories to identify novel agents (biological response modifiers,
BRM) that can selectively enhance macrophage-mediated tumoricidal
activity in-vivo.[81,82]

The localization of liposomes and other particular drug carriers
within fixed macrophages of the RES and circulating blood monocytes
after i.v. injection offers an efficient method for targeting BRM
agents, albeit passively, to these cells.

Studies in my laboratory, in collaboration with Dr.I.J. Fidler

and his colleagues at the Frederick Cancer Research Facility of the
National Cancer Institute in Maryland, have shown that systemic
administration of liposomes containing immunomodulators that activate
macrophages to render them cytotoxic for tumor cells significantly
enhances host resistance to cancer metastases and augments macrophage-mediated destruction of tumor cells in-vitro and established
lung metastases in-vivo.[78-81] A full description of the use of liposomes containing immunomodulators in the treatment of experimental
animal tumors is provided in recent reviews.[30,39,78-81]

The optimal conditions for therapy with liposome-encapsulated
immunomodulators and the efficacy of this modality in treating metastatic tumor burdens of increasing size have still to be defined.
It is considered unlikely that liposome-encapsulated immunomodulators
could serve as a single modality in treating advanced metastatic
disease. In common with many other anti-tumor therapies optimal
application will probably involve its use in combination with other
anti-tumor agents. For example, if the ratio of macrophages to tumor
cells required for optimal macrophage-mediated tumoricidal activity
in-vivo is similar to that operating in-vitro, then, even allowing
for maximum recruitment of monocytes from the blood, there are insufficient macrophages in the lung to permit effective destruction
of pulmonary metastatic tumor burdens exceeding 10^8 tumor cells/
lung.[78,80] Since tumor burdens of this size are easily attained,
it is clear that therapeutic stimulation of macrophage-mediated anti-
tumor activity will be unsuccessful in treating large metastatic
foci, no matter how effective this modality at activating macrophages. Therapeutic regimens designed to stimulate the anti-tumor
properties of macrophages will thus almost certainly have to be used
in combination with other treatments such as chemotherapy which
would be used to reduce the "bulk" tumor burden to a sufficient level
that activated macrophages could kill the surviving tumor cells.

Evaluation of the efficacy of therapeutic agents in augmenting
macrophage-mediated reactions to tumors has been hindered by the lack
of methods for quantifying the macrophage content of metastases.
Recently, my laboratory has described a new double-label histochemical method that not only enables macrophages to be identified reliably in histologic sections of metastases but also allow macrophages
that have entered lesions during therapy to be distinguised from
those present at the onset of therapy.[66] In evaluating macrophage
responses to tumors it is also necessary to quantify changes in
macrophage content that can occur at different stages in tumor
growth. Morphologic analysis of several hundred lung metastases
produced by the B16 melanoma using serial sections of each metastasis has revealed marked heterogeneity in the macrophage content
of individual metastases present in the same animal.[83] This study

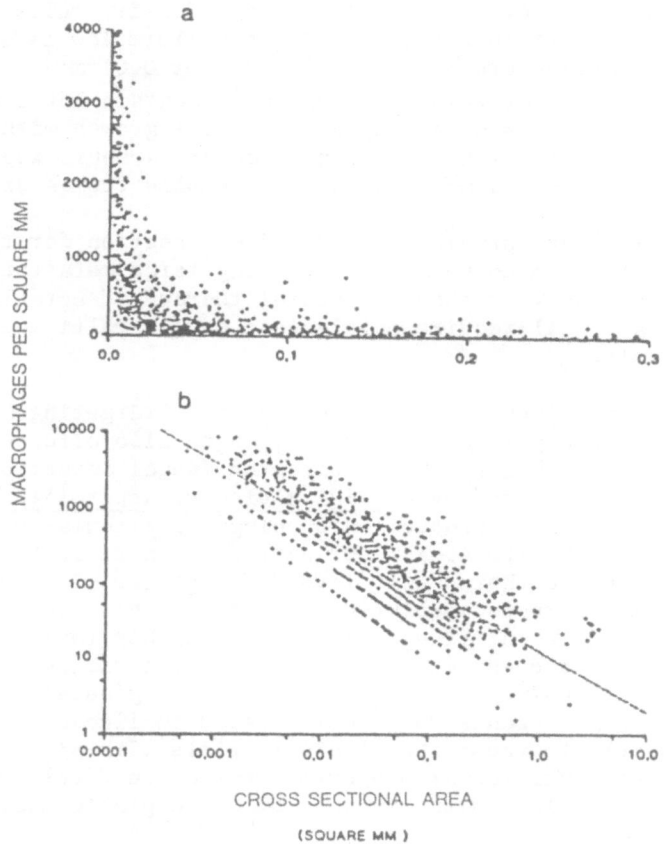

CROSS SECTIONAL AREA

(SQUARE MM)

Fig. 5. Scatter diagram of macrophage density versus cross sectional area of lung micrometastases. C57BL/6J mice were
injected with 1 x 10^5 B16-F cells (tail vein) and subsequently injected with colloidal iron-dextran, i.v.
(tail vein) 8,14, or 22 days after injection of tumor
cells, to label tumor associated macrophages. Mice were
sacrificed 24 hours after injection of colloidal iron
and their lungs fixed in formalin. Paraffin sections
were first treated with potassium ferrocyanide to convert colloid iron to Prussian blue and then stained for
pseudoperoxidase activity with diaminobenzidine.[66]
Staining cells were counted and the cross-sectional area
of individual metastases determined and macrophages per
unit area calculated using a Zeiss Videoplan. Each
point represents an individual metastases. Data from
335 separate metastasis sections are shown. Metastases
with cross sectional areas up to $3mm^2$ have been examined and show uniformly low macrophage density. Repro-
duced with permission from Poste et al.[21]

revealed that the intratumoral macrophage density falls rapidly as metastases increase in size, reaching a uniform low level in metastases with median cross-sectional areas of 0.01 mm^2 or greater (Fig. 5). Lesions of this size typically contain between 400 and 1000 tumor cells. Assuming exponential cell growth within metastases, tumor cells with a doubling time of 24 hours would take only one week to generate a metastasis of this size from a single cell.[83]

These findings provide a possible explanation for the failure of non-specific immunotherapy in treating large metastatic burdens and reinforce the view that successful therapy of established metastatic disease will require multi-agent and/or multi-modality treatment protocols.

Recent experiments suggest that passive targeting of liposomes containing antifungal agents to macrophages also offers a powerful approach for the therapy of disseminated fungal infections.[84-86] Systemic mycoses caused by Candida albicans, Aspergillus sp. and Mucor sp. are a major problem in oncology, particularly in patients with leukemias and lymphomas. Of particular note is the marked improvement in the therapeutic index of amphotericin B when administered in liposome-encapsulated form.[84,85] This agent, in common with many systemic anti-fungal agents, is highly toxic. The ability to reduce toxic side-effects without loss of efficacy is an important therapeutic gain. If this proves to be a general feature of other anti-fungal agents when administered in liposomes, it may become possible to undertake clinical trials with other agents that display potent anti-fungal properties but whose development was previously curtailed on the grounds of unacceptable host toxicity.

Liposomes containing biological response modifier agents are also highly effective in stimulating macrophage-mediated host resistance to bacterial infections.[21] Systemic administration of MLV liposomes containing muramyl dipeptide produces a significant increase in the ability of experimental animals to resist acute infections by a variety of bacteria (Listeria sp.; Salmonella sp.; Klebsiella sp.; Pseudomonas aeruginosa).[21] Evidence has also been obtained for synergy in the action of liposome-encapsulated MDP and antibiotics in the therapy of experimental infections produced by these classes of microorganisms (unpublished observations).

Finally, passive localization of systemically administered particulate drug carriers in cells of the mononuclear phagocyte series also offers a strategy for delivering cytotoxic drugs to destroy malignancies arising in these cells such as histiocytic medullary reticulosis, monocytic leukemia, hairy cell leukemia and certain forms of Hodgkins disease. This strategy would require, however, that neoplastic cells retain a high level of phagocytic activity and that toxic destruction of normal phagocytes remains within clinically acceptable limits.

Active Targeting of Drugs to Lymphoid Cell Subsets

The ability to target drugs to specific subsets of circulating blood cells and their neoplastic counterparts would hold substantial clinical promise. Once again, the key issue is how drug uptake by the RES will be avoided. The availability of antibodies or other ligands that react with non-RES cells is a lesser problem compared to the difficulty of avoiding localization in the RES.

Remarkable progress has been made in the last few years in identifying monoclonal antibodies that react with specific subsets of B and T lymphocytes[22] and continued isolation of new antibodies and their use to develop even more sophisticated immunotaxonomy schemes for these cells can be confidently expected. Systemic administration of monoclonal antibodies directed against surface antigens on leukemia cells has been shown to temporarily reduce the level of circulating leukemia cells in both man and animals.[16] However, insufficient information is available to permit critical evaluation of the effect of target cell heterogeneity, blocking antigens and host immune responses to the injected antibodies in determining clinical efficacy. Although antibody-drug conjugates and immunotoxins are effective in limiting the growth of murine leukemias, clinical studies have not been reported. In the author's opinion such studies should not proceed until adequate assurance can be provided that non-specific uptake of toxic moieties by the RES does not pose a serious hazard.

Site-Specific Drug Delivery Via Targeting to Determinants on Endothelial Cells

Ligand-directed targeting of drug carriers to determinants expressed on the surface of vascular endothelial cells may merit investigation. The antigens and drug receptors expressed by endothelial cells may vary quantitatively and qualitatively in different regions of the vascular tree.[30,39] These observations raise the possibility that these molecules could serve as targets in ligand-directed localization of liposomes to specific regions of the vasculature. As in the case of circulating blood cells, new information on the expression of differentiation antigens and other cell surface molecules on endothelial cells will emerge in the next few years. This knowledge could create fascinating opportunities for zone-specific or perhaps tissue-specific targeting of drugs within the vascular system.

Ex-Vivo Drug Therapy

Bone marrow toxicity is a major problem in cancer chemotherapy. Heterologous bone marrow transplants have been of limited value in overcoming this problem. Interest is now directed to autologous bone marrow grafting in which bone marrow is removed from a patient before therapy and re-infused to reconstitute bone marrow function in patients exposed to supralethal chemotherapy and/or irradiation

to destroy tumor cells at other sites. This approach demands that
tumor cells can be reliably eliminated from the bone marrow before
return to the patient. Targeted drug delivery using antibodies
directed against tumor cells is anticipated to play an important role
in the success of this technique.[22,88,89] Selective elimination of
tumor cells from bone marrow in-vitro poses fewer problems than
efforts to achieve the same goal in-vivo. Marrow can be exposed to
the drug-carrier complex in-vitro under low temperature conditions
that limit antigenic modulation in the target tumor cells. Blocking
antigens can be removed and multiple treatment cycles used if needed.
Also, if antibodies are used as the targeting ligand, heterologous
complement can be added to kill antibody-coated cells. Finally, the
specificity requirements for the targeting ligands may be less
demanding than for in-vivo applications. The ligand(s) selected
must be capable of reacting with all tumor cells while leaving hematopoietic stem cells intact. Destruction of normal marrow cells
that are irrelevant to successful engraftment is presently viewed
as acceptable.[11,16,22] However, further experience may modify this
view since certain stromal cells are needed to create the appropriate
microenvironment for stem cell function.[22]

Lymph Node Metastases

Early diagnosis and therapy of lymph node metastases is considered important in limiting further tumor spread to form distant metastases in major organs. Unfortunately, localization of most i.v.
injected imaging agents and antitumor drugs in regional lymph nodes
is poor.[11] Lymphatic uptake of such agents after s.c. or i.m. injection results in transport to the regional nodes but the transport
process kinetics of the lymphatic clearance are erratic and/or slow
and the residence times of most agents within nodes are limited. It
is thus difficult to achieve effective concentrations of imaging or
therapeutic agents within nodes. Drug delivery systems show promise
in overcoming these problems. Liposomes[90,91] or lipid micro-
mulsions[92] containing cytotoxic drugs injected s.c. or i.m. are
retained in regional lymph nodes significantly longer than conventional drug formulations, with accompanying gain in the therapy of
nodal metastases.[90,92] Similarly, radiolabeled monoclonal antibodies
to tumor cells injected s.c. show longer nodal retention times and
produce superior immunoscintigraphic imaging of tumor deposits than
higher doses of antibodies injected i.v.[93] Theoretically, this
approach could be used to target imaging or therapeutic agents to
any cell type in lymph nodes. Filtration of circulating lymph
through the reticulum cell network of the node offers drug carriers
direct access to a variety of cell types. Macrophages may still be
a major "sink" for injected materials, however, because of their
well developed phagocytic capacities, their presence in large numbers
and their location in the node.[11,39] However, growth of metastatic
tumor cells can alter or stop lymph circulation in the node and the
utility of these approaches will probably be limited to the diagnosis

and therapy of small metastases that do not impair lymph flow.

PHYSICO-CHEMICAL TARGETING OF DRUG CARRIERS IN CANCER CHEMOTHERAPY

Yatvin et al[94] and Magin and Weinstein[95] have explored the intriguing concept of constructing liposomes from phospholipids or
phospholipid mixtures that have a transition temperature (T_c)
slightly above normal physiological temperature. Such liposomes
should only become permeable and release their contents in tissues
warmed to the T_c by local hyperthermia. By localized heating of a
tumor it might be possible to confine drug release to the tumor.
This strategy assumes, of course, that liposomes retain their temperature-sensitive properties after interaction with various blood
components and that they remain in the circulation for a sufficient
time to allow maximum drug-release by hyperthermia before being
cleared by the RES. Initial experiments using 'physical' targeting
of this kind to enhance release of methotrexate from temperature
sensitive liposomes in L1210 tumors implanted s.c. were successful
in achieving tumor-associated drug ratios in heated versus unheated
tumors of between 4:1 and 14:1. However, in reviewing an extensive
series of experiments using this strategy, Yatvin and Lelkes[94] indicate that this approach may be less-effective with other drugs and
different tumor models.

A further major drawback is that this method is not applicable
to the therapy of metastatic disease. Whole body hyperthermia cannot
be used since uniform elevation of the body'core' temperature will
induce generalized drug release throughout the body. Enthusiasm for
the use of hyperthermia is also dampened by reports showing that
local and whole body hyperthermia significantly enhance metastasis
of experimental animal tumors.[96,97]

In an effort to apply the concept of 'physical' targeting of
liposomes to the treatment of established metastases, Yatvin and his
colleagues have also constructed "pH-sensitive" liposomes that
release their contents only when exposed to tissue environments of
low pH.[94] The rationale for this approach comes from reports showing
that the pH of interstitial fluids in a variety of human and animal
tumors is lower (pH 6.0-6.5) than in normal tissues (pH>6.5)..
Yatvin has succeeded in developing liposomes that release encapsulated material 5 to 6 times faster at pH 6.0 than at pH 7.4 but their
effectiveness in achieving site-specific drug delivery to metastases
has still to be determined.

It is unclear as to how far the assumption on which this strategy is based reflects a general feature of neoplastic tissue. For
example, Jahde et al[98] were unable to detect any significant difference in the pH of interstitial fluids sampled from normal and neoplastic tissues. In this study, tissue fluid was sampled using pH
microelectrodes with a diameter of <10μm. In contrast, earlier

measurements reporting lower pH values in neoplastic tissues employed sampling electrodes with tip diameters of <0.1 mm. The tissue compression and trauma accompanying insertion of these larger electrodes may produce significant cell death and release of lysosomal enzymes with resulting artifactual lowering of tissue pH values.

Physical targeting of particulate carriers using external magnets to direct liposomes containing ferromagnetic particles to specific organs has also been investigated.[99,100] Small particulate carriers such as liposomes and microspheres are suitable for drug delivery to capillary beds in specific organs. However, the much larger microcapsules (20 μm) used by Kato[100] arrest in larger vessels and are unable to return to the venous circulation. The smaller size of liposomes may be a disadvantage in controlling liposomes that are circulating within major supply vessels. For example, a magnetic field of 8000 Oe was required to achieve localization of 50% of an injected dose of small microspheres within the rat tail artery with a flow rate of only 0.6 ml/min.[99] In contrast, a similar 50% level of magnetic control efficiency could be achieved for large microcapsules at magnetic fields of only 250 Oe/cm in the dog abdominal aorta in which the blood flow rates as high as 240 ml/min.[100] These data suggest that magnetic control and targeting of liposomes containing ferromagnetic particles may be feasible only if combined with arterial catheterization techniques to achieve initial delivery to the organ of interest.

In its present state of development drug release from magnet-directed magnetized particles occurs within the circulation and selective destruction of tumor cells is thus not possible. In the case of the large microspheres (40-225 μm diameter) used by Kato[100] significant tissue necrosis occurs due to embolization. Tissue necrosis can probably be controlled to some extent by using particles with rapid dissolution rates. However, even with such technical refinements, this approach to cancer treatment is non-selective, and is technically cumbersome and not devoid of risk becasue of the need for organ catheterization.

The safety of a drug carrier must be evaluated from two standpoints: a) the toxicity of the carrier itself; and b) the risk of novel drug-induced toxicities arising from alterations in the disposition, pharmacokinetics and metabolism of carrier-associated drug dompared with conventional drug formulations. These criteria must be evaluated under test conditions that mimic as far as possible the dose, frequency and route of administration envisaged for clinical use.

Repeated infusion of antibodies, particularly xenogeneic antibodies, carries the risk of eliciting anti-idiotypic antibodies that will block interaction of the original antibody with target cells and will also increae the risk of adverse hypersensitivity reactions in subsequent antibody treatments.[11,16,22,101] Observations in renal transplant and cancer patients injected with large doses of rodent monoclonal antibodies indicate that formation of anti-idiotypic antibodies occurs rapidly.[16,22] Fortunately, the incidence of severe hypersensitivity reactions to date has been low. The future availability of human monoclonal antibodies may further reduce such risks. However, reports of lethal anaphylaxis reactions in rhesus monkeys injected with relatively low doses of monoclonal antibodies emphasizes the need for caution.

Single doses of biodegradable particles such as liposomes and albumin microspheres are well tolerated by a variety of animal species.[30,39] However, repeated dosing with particulate carriers can impair RES clearance functions.[30,39] The onset, extent and duration of RES failure is affected by particle size, dose, number and frequency of doses and, for liposomes, phospholipid composition is also relevant.[30,39] Impairment in RES function results from sequential saturation and exhaustion of particle clearance capacities in the liver, spleen and bone marrow.[30,39] Histologic evidence of bone marrow hypoplasia and alterations in hematopoiesis have been observed in extended i.v. dosing with liposomes (unpublished observations).

The disposition, pharmacokinetics and metabolic fate of drugs administered in association with a carrier may differ substantially from conventional formulations of the same agents and the risk of novel toxicities must therefore be considered. Perhaps the most obvious example concerns the use of liposomes and other particulate carriers to deliver anticancer drugs. By delivering high concentrations of cytotoxic drugs to mononuclear phagocytes in the blood and the RES this approach may induce toxic ablation of a vital element of host defence. Inhibitors of DNA synthesis might be expected to have little toxic effect on non-dividing macrophage populations but drugs that impair RNA and protein synthesis may be toxic to these cells. Recent studies have shown that this fear is justified. Systemic administration of several anti-tumor drugs encapsulated within liposomes was found to enhance the metastatic spread of mouse tumors.[21,30,39] This effect was not induced by liposomes injected s.c. or i.m. and was reversed by injection of syngeneic macrophages 12 hours after each treatment cycle, indicating that the iatrogenic enhancement of metastases was probably caused by toxic destruction of mononuclear phagocytes. The case for using liposomes or other particulate carriers for drug delivery in cancer treatment may thus be seriously flawed if the drug in question can destroy host macrophages.

The ability of certain cytotoxic anti-tumor drugs to destroy the RES when administered in association with liposomes was perhaps predictable in the light of existing knowledge identifying the RES as the major site of liposome localization in-vivo. This phenomenon

is not unique to liposomes and impairment of RES function has been
reported in mice injected i.v. with erythrocyte ghosts containing
encapsulated bleomycin.[102]

Drug-antibody conjugates and immunotoxins are not exempt from
this complication. Scintigraphic analyses of the disposition of
radiolabeled anti-tumor antibodies in both animals and man indicate
that even when significant labeling of the tumor is achieved, high
levels of non-specific antibody uptake by the liver and spleen are
also occurring.[18,37] Chemical modification and/or denaturation of
antibodies during coupling to drugs or toxins, and physical changes
such as antibody aggregation, also predispose to clearance by the
RES.[37]

In the author's opinion, the potential risk posed by drug
carriers that show appreciable localization in the RES is of sufficient magnitude to preclude their clinical use as carriers for cytotoxic antineoplastic drugs until extensive toxicology studies are
completed to show that ablation of the RES is not induced by the
specific drug(s) to be used in clinical studies. Multiple dosing
with particulate carriers for long periods also presents the risk
of additional RES toxicity induced by the carrier itself.

Antibody uptake by the RES, whether as free antibodies, antibody-drug conjugates or antibodies coupled to particulate carriers,
can be reduced by using Fab antibody fragments that lack the F
region recognized by receptors on RES cells. Fab fragments have
several other desirable properties as carriers compared with intact
IgG molecules: equilibrium distribution in extracellular fluids is
more rapid; the volume of distribution is greater; elimination time
is more rapid; immunogenicity is lower; and immune complexes that
may be formed are smaller than those that cause nephrotoxicity and
complement cannot be fixed because of the absence of the F_c region.

COMMERCIAL DEVELOPMENT OF NOVEL DRUG DELIVERY SYSTEMS

Routine application of targeted drug delivery in cancer treatment will occur only with successful commercialization. As with any
pharmaceutical product, the decision to undertake the lengthy,
costly and high risk development process demands clear definition
of the medical need and the technical feasibility of the proposed
approach.[11] Drug carriers present certain additional unique features. By patenting a drug carrier, patent protection will extend
to the drug-carrier complex. Assuming that the complex has demon-
strable therapeutic advantage over other formulations of the drug,
carrier systems provide a powerful strategy for countering generic
competition to drugs with limited or no patent protection. Also,

if the carrier is suitable for use with several drugs, the impact
of the cost of carrier development on the price of individual products is diluted.

The need and therapeutic benefits of targeted drug delivery in
cancer treatment need not be debated. However, the technical and
economic feasibility of commercial development of the delivery syst-
tems reviewed in this article is far from certain.

A carrier must be amenable to commercial scale production using
convenient, cost effective methods that do not require protracted
process development, substantial capital investment in equipment or
plant, or pose novel or unacceptable chemical or biological hazards
to production workers and which allows manufacture under conditions
acceptable to regulatory authorities. The final product must be
uniform, have a shelf-life of at least 12 months under diverse
storage conditions, offer the physician and patient maximum convenience and also meet the increasing demand for cost-containment being
set by Government-sponsored and private health care systems.[11]

The response of the national agencies that regulate the production and marketing of pharmaceuticals to drug carriers is largely
untested but will inevitably vary in different countries. There
are no clear precedents. The process adopted in the USA for approval
of prodrugs and controlled-release drug formulations involving drug-
polymer complexes suggests that carrier-associated drugs will probably be treated as new drug substances and required to fulfill the
complex range of tests needed for approval of any new chemical entity
(NCE) administered to patients. However, ambiguity is introduced by
a recent proposal from the Food and Drug Administration (FDA) which
states "coupled antibodies, i.e., products that consist of an antibody component coupled with a drug or radionuclide component in which
both components provide a pharmacological effect but the biological
component determines the site of action" will be reviewed by the
Office of Biologics rather than the Office of New Drug Evaluation
which presently reviews NCEs.[104] Commercially, this is not a trivial
issue. Unless the Office of Biologics establishes new guidelines
for product evaluation, the advantages of submission to this office
are substantial since the complexity, time and cost of developing a
"biological" product are far less than for a NCE.

Confusion also surrounds the type of tests that would be needed
for products employing monoclonal antibodies as targeting ligands.
Considerable delay has occurred in the FDA's stated intention to pub-
lish formal guidelines for testing monoclonal antibodies for clinical
use. A preliminary FDA position paper[105] describing "points to consider" in the manufacture of antibodies for clinical use lists the
following criteria: hybridoma origin; identification of all heavy
and light chains synthesized and secreted by the parent myeloma and
the hybridoma; immunoglobulin class and subclass; electrophoretic
and IEF profiles; contamination problems (microorganisms, nucleic
acids, processing materials and solvents); immunogenicity/anaphy-

lactogenicity (antibodies, aggregates, complexes, denatured proteins); cross-reactivity with blood cells or components and cells of vital organs; biodistribution; pharmacokinetics; catabolism and excretion rates. No mention was made of possible requirements for novel pre-clinical toxicity tests. Reference is made only to the use of anti-bodies alone or as radiolabeled molecules and additional assays that might be required for testing drug-antibody conjugates, immunotoxins or antibodies coupled to particulate carriers were not addressed.

The principal issues to be considered in commercial development of a systemically administered drug carrier for clinical use are summarized in Table 1. Liposomes in particular present substantial development problems. Notwithstanding the lack of methods for effi-cient drug capture and retention within liposomes, size heterogeneity and poor storage properties, the cost and technical demands of pre-paring highly purified phospholipids on a very large scale are daun-ting. For example, if regulatory authorities mandate that long-term (2 year) toxicity testing be conducted with liposome-drug complexes, synthesis of several hundred kilograms of purified phospholipid would be required for every drug tested.

Many of the properties of liposomes are shared by multiphase emulsions and biodegradable microparticles.[106] Commercial factors currently favor these alternative carriers since they can be produced as uniform, pharmaceutically acceptable formulations at far less cost than liposomes using methods and manufacturing equipment already widely available within the pharmaceutical industry.

The next few years will determine whether any of the systems reviewed here will be developed commercially or will join the ranks of other drug carriers that have failed to fulfill their theoretical promise. The technical problems that must be solved for any of the systems reviewed here to achieve commercial success are substantial. However, the unprecedented scope and pace of advances in tumor biol-ogy, immunology, biochemistry and pharmaceuticals have greatly en-hanced the prospect of success for at least some examples of site-specific drug delivery. Realistic opportunities now exist for passive targeting of particulate carriers to the RES and for active ligand-directed targeting of drugs to specific cell types within the bloodstream, the bone marrow and lymph nodes. It seems likely, how-ever, that for the foreseeable future the role of anatomic and phy-siologic factors in determining the behaviour and fate of drug carriers will dominate, and largely frustrate, efforts to achieve the ultimate objective of targeted drug delivery to cells in extra-vascular locations. Paul Ehrlich's vision of targeting systemically administered drugs to diseased cells in any location is still far from becoming reality.

Table 1. Criteria for the characterization and development activities required for commercialization of a novel systemically administered drug carrier.

1. Pharmaceutical formulation and characterization of the carrier

   development of pilot scale production methods
   (evaluate novel worker safety hazards)
   drug entrapment efficiency
   particle size and tolerance limits (particulate carriers)
   surface charge
   sterility
   pyrogenicity
   stability (± cell recognition ligand)
      chemical stability
      in diluent
      lyophilized and after reconstitution
      size changes, aggregation or sedimentation
      characterize degradation products
      product lifetime (shelf life)
   cell recognition ligand (if used)
      homogeneity
      stability in biological fluids
      density/particle
      elution during storage
   specifications and process reproducibility

2. Preclinical studies

   metabolism
      dose-response relationships
      absorption site(s)/rate
      stability in body fluids
      bioavailability
      organ/tissue/cell disposition
      active species (parent drug or metabolites)
      identification of metabolites (compare with conventional drug formulation)
      excretion

   toxicology
      major organ function tests
      acute and subacute toxicity studies
      novel carrier-induced toxicities (e.g. RES paralysis due to saturation by particles)
      novel toxicities due to altered tissue disposition of carrier-associated drug (e.g. RES toxicity)
      immunogenicity/anaphylactogenicity of targeting ligand

3. Clinical studies

   Phase I
      define dose-response relationships
      identify novel or unexpected adverse reactions
      preliminary metabolic data
   Phase II/III
      complete metabolic profile of carrier and drug (absorption, distribution, metabolism and excretion)
      therapeutic efficacy
      evaluate adverse reactions

4. Extended development studies (non-clinical) conducted in parallel with phase II/III clinical studies

   reproductive toxicology
   chronic toxicity
   selection of final formulation and define commercial specifications
   develop manufacturing scale production methods
   confirm suitability of batches produced on manufacturing scale

Modified from Poste and Kirsh.[11]

REFERENCES

1. V.T. DeVita, Jr., S. Hellman and S.A. Rosenberg (eds) "Cancer: Principles and Practice of Oncology", Lipincott, Philadelphia (1982).
2. G. Poste and I.J. Fidler, The pathogenesis of cancer metastasis, Nature 283:139 (1980).
3. G. Poste, Experimental systems for analysis of the malignant phenotype, Cancer Met. Rev. 1:141 (1982).
4. P. Newmark, Priority by press release. Nature 304:108 (1983)
5. P. Ehrlich,"Collected Studies on Immunity", Vol 2. Wiley, New York (reprinted 1906).
6. Y.W. Chien (ed.) "Novel Drug Delivery Systems", M. Dekker, New York (1982).
7. R.E. Counsell and R.C. Pohland, Lipoproteins as potential site-specific delivery systems for diagnostic and therapeutic agents, J. Med. Chem. 25:1115 (1982).
8. G. Gregoriadis, J. Senior and A. Trouet (eds.) "Targeting of Drugs", Plenum Press, New York (1982).
9. S.D. Bruck (ed.) "Controlled Drug Delivery", CRC Press, Boca Raton (1982).
10. E.P. Goldberg (ed.) "Targeted Drugs", Wiley-Interscience, New York (1983).
11. G. Poste and R. Kirsh, Site-specific (targeted) drug delivery in cancer therapy, Biotechnology 1:869 (1983).
12. M.B. Yatvin, H. Muhlensiepen, W. Porschen, J.N. Weinstein and L.E. Feinendegen, Selective delivery of liposome associated cis-dichloro-diamine platinum II by heat and its influence on tumor drug uptake and growth, Cancer Res. 41:1602 (1981).
13. K.J. Widder, A.E. Senyei, and B. Sears, Experimental methods in cancer therapeutics, J. Pharm. Sci. 71:379 (1982).
14. Anon., GVH disease after marrow transplantation, Lancet 1:491 (1984).
15. L.D. Leserman, P. Machy, C. Devaux and J. Barbet, Antibody-bearing liposomes: targeting in-vivo, Biol. Cell 47:111 (1983)
16. R. Levy and R.A. Miller, Tumor therapy with monoclonal antibodies, Fed. Proc. 42:2650 (1983).
17. K. Sikora and M. Smedley, Clinical potential of monoclonal antibodies, Cancer Surveys 1:521 (1983).
18. K.A. Foon, M.I. Bernhard and R.K. Oldham, Monoclonal antibody therapy: assessment by animal tumor models, J. Biol. Resp. Modifiers 1:277 (1982).
19. I.S. Trowbridge and D.L. Domingo, Prospects for the clinical use of cytotoxic monoclonal antibody conjugates in the treatment of cancer, Cancer Surveys 1:543 (1982).
20. R. Arnon and M. Sela, Targeted chemotherapy:drugs conjugated to anti-tumor antibodies, Cancer Surveys 1:429 (1982).
21. G. Poste, R. Kirsh and P. Bugelski, Liposomes as a drug delivery system in cancer therapy in "Novel Approaches to Cancer Therapy", P. Sunkara, ed., Acadamic Press, New York (1984).

22. M.F. Greaves, "Target" structures and "target" cells for cancer therapy with monoclonal antibodies: finding the candidates, Cancer Surveys 1:451 (1982).

23. R.W. Baldwin and M.V. Pimm, Antitumor monoclonal antibodies for radioimmunodetection of tumors and drug targeting, Cancer Met. Rev. 2:89 (1983).

24. S. Olsnes and A. Pihl, Cytotoxic proteins with intracellular site action: mechanism of action and anti-cancer properties, Cancer Surveys 1:467 (1982).

25. R.E. Oldham, Current status of monoclonal antibodies in cancer therapy, Clin. Immunol. News 4:131 (1983).

26. G. Kolata, The magic in magic bullets, Science 222:310 (1983).

27. E.L. Dobson and H.B. Jones, The behaviour of intravenously injected particulate material. Its rate of disappearance from the blood stream as a measure of liver blood flow, Acta Med. Scand. 273:1 (1982).

28. C. Nicolau and G. Poste (eds.) "Liposomes in-vivo", Biol. Cell. Vol 47, No. 1. (1983).

29. A.D. Bangham, (ed.) "Liposome Letters", Acadamic Press, London (1983).

30. G. Poste, R. Kirsh and T. Koestler, The challenge of liposome targeting in-vivo, in "Liposome Technology" G. Gregoriadis, ed., CRC Press, Boca Raton (1984).

31. G. Poste, C. Bucana, A. Raz, P. Bugelski, R. Kirsh and I.J. Fidler, Analysis of the fate of systemically administered liposomes and implications for their use in drug delivery, Cancer Res. 42:1412 (1982).

32. G. Tomlinson, Microsphere drug delivery systems for drug targeting and controlled release, Int. J. Pharm.Technol. Prod. Manuf. in press.

33. G. Poste and G.L. Nicolson, Experimental systems for analysis of the surface properties of metastatic tumor cells, in: "Biomembranes", Volume II, A. Nowotny, ed., Plenum, New York (1983).

34. G.L. Nicolson and G. Poste, Tumor cell diversity and host responses in cancer metastasis. Host immune responses and therapy of metastases, Curr. Concepts Cancer VII:3 (1983).

35. G.L. Nicolson and G. Poste, Tumor cell diversity and host responses in cancer metastasis. 1. Properties of metastatic cells, Curr. Problem Cancer 7:1 (1982).

36. H.I. Peterson (ed.) "Tumor Blood Circulation: Angiogenesis, Vascular Morphology and Blood Flow of Experimental and Human Tumors", CRC, Boca Raton (1979).

37. L.J. Anghileri (ed.) "General Processes of Radiotracer Localization" CRC, Boca Raton (1982).

38. J. Folkman and C. Haudenschild, Induction of capillary growth in-vitro, in "Cellular Interactions", J.T. Dingle and J.L. Gordon, eds., Elsevier, Amsterdam (1981).

39. G. Poste, Liposome targeting in-vivo: Problems and opportunities. Biol. Cell. 47:19 (1983).

40. R.J. Parker, S.M. Sieber and J.N. Weinstein, Effect of liposome encapsulation of a fluorescent dye on its uptake by the lymphatics of the rat, Pharmacology 23:128 (1981).

41. E. Mayhew, Y.M. Rustum and F. Szoka, Therapeutic efficacy of cytosine arabinoside trapped in liposomes, in "Targeting of Drugs", G. Gregoriadis, J. Senior and A. Trouet, eds., Plenum, New York (1982).

42. C.A. Hunt, Liposomes disposition in-vivo. V. Liposome stability in plasma and implications for drug carrier function, Biochim. Biophys. Acta 719:450 (1982).

43. R.T. Profitt, L.E. Williams, C.A. Presant, G.W. Tin, J.A. Uliana, R.C. Gamble and J.D. Baldeschwieler, Liposomal blockade of the reticuloendothelial system: Improved tumor imaging with small unilamellar vesicles, Science 220:502 (1983).

44. L. Weiss and R.O. Greep, "Histology", McGraw-Hill, San Francisco (1977).

45. G.G. Schneeberger, Proteins and vesicular transport in capillary endothelium, Fed. Proc. 42:2419 (1983).

46. M. Bundgaard, Vesicular transport in capillary endothelium. Does it occur?, Fed. Proc.42:2425 (1983).

47. R.M. Pino and E. Essner, Permeability of rat choriocapillaries to hemeproteins. Restriction of tracers by a fenestrated endothelium, J. Histochem. and Cytobiochem. 29: 281 (1981).

48. G. Poste, C. Bucana and I.J. Fidler, Stimulation of host response against metastatic tumors by liposome-encapsulated immunomodulators in: "Targeting of Drugs", G. Gregoriadis, J. Senior and A. Trouet, eds., Plenum, New York (1982).

49. K.J. Hwang, K.F.S. Luk and P.L. Beaumier, Volume of distribution and transcapillary passage of small unilamellar vesicles, Life Sci. 31:949 (1982).

50. G. Lubec, K. Kuhn, U. Latzka and E. Reale, Glomerular permeability for proteins of high molecular weight entrapped in liposomes, Renal Physiol. 4:131 (1981).

51. A.P. Fishman, (ed.), "Endothelium" N.Y. Acad. Sci., New York (1982).

52. L.O. Simpson, Biological thixotropy of basement membranes: the key to the understanding of capillary permeability, in: "The Committee in Postgraduate Medical Education", D. Garlick, ed., The University of New South Wales (1981).

53. G. Gabbiani and G. Majno, Pathophysiology of small vessel permeability, in: "Microcirculation", G. Kaley and B.M. Altura, eds., Vol III (1980).

54. V.J. Caride and B.L. Zaret, Liposome accumulation in regions of experimental myocardial infarction, Science 198:735 (1977).

55. T.M. Mueller, M.L. Marcus, H.E. Mayer, J.K. Williams and K. Hermsmeyer, Liposome concentration in canine ischemic myocardium and depolarized myocardial cells, Cir. Res. 49:405 (1981).

56. T.N. Palmer, V.J. Caride, L.A. Fernandez, and J. Twickler, Liposome accumulation in ischaemic intestine following experi-

mental mesenteric occlusion, Bioscience Reports 1:337 (1981).

57. R.A. Willis, "Pathology of Tumors", Butterworths, London (1973).
58. B.A. Warren, The vascular morphology of tumors, in: "Tumor
     Blood Circulation: Angiogenesis, Vascular Morphology and Blood
     Flow of Experimental and Human Tumors", H.I. Peterson, eds.,
     CRC Press Inc., Boca Raton (1979).
59. H. Abrams, "Angiography", Little Brown and Co., Boston (1961).
60. R.J. Goldacre, and B. Sylven, On the access of blood-borne
     dyes to various tumor regions, Br. J. Cancer 16:306 (1981).
61. P. Gullino and F.H. Grantham, The vascular space of growing
     tumors, Cancer Res. 24:1727 (1964).
62. J.C.E. Underwood and I. Carr, The ultrastructure and perm-
     eability characteristics of the blood vessels of a trans-
     plantable rat sarcoma, J. Pathol. 107:157 (1972).
63. A. Martinez-Palomo, Ultrastructural modifications of intra-
     cellular junctions in some epithelial tumors, Lab. Invest.
     22:605 (1970).
64. P.F. Schatski and A. Newsome, Neutralized lanthanum solution:
     A largely noncolloidal ultrastructural tracer, Stain Tech.
     50:171 (1975).
65. D.M. Long, F.K. Multer, A.G. Greenburg, G.W. Peskin, E.C.
     Lasser, W.G. Wickham and C.M. Sharts, Tumor imaging with
     x-rays using macrophage uptake of radiopaque fluorocarbon
     emulsions, Surg. 84:104 (1978).
66. P. Bugelski, R. Kirsh and G. Poste, A new histochemical
     method for measuring intratumoral macrophages and macrophage
     recruitment into experimental metastases, Cancer Res., in press.
67. G. Gregoriadis and E.D. Neerunjun, Homing of liposomes to
     target cells, Biochem Biophys. Res. Commun. 65:537 (1975).
68. G. Gregoriadis, E.D. Neerunjun and R. Hunt, Fate of liposome-
     associated agents injected into normal and tumor-bearing
     rodents; Attempts to improve localization in tumor tissues,
     Life Sci. 21:357 (1977).
69. M.B. Yatvin, T.C. Cree and J.I. Gipp, Hyperthermia-mediated
     targeting of liposome-associated anti-neoplastic drugs, in:
     "Targeting of Drugs", G. Gregoriadis, J. Senior and A Trouet,
     eds., Plenum Press, New York (1982).
70. K.R. Patel, M.P. Li and J.D. Baldeschwieler, Suppression of
     liver uptake of liposomes by dextran sulfate 500, Proc. Natl.
     Acad. Sci USA, 80:6518 (1983).
71. E.L. Dobson and H.B. Jones, The behaviour of intravenously
     injected particulate material. Its rate of disappearance
     from the bloodstream as a measure of liver blood flow,
     Acta Med. Scand. 273:1 (1982).
72. G. Poste and R. Greig, On the genesis and regulation of
     cellular heterogeneity in malignant tumors, Invasion and
     Metastasis 2:137 (1982).
73. J.H. Goldie, New thoughts on resistance to chemotherapy,
     Hosp. Practice 18:165 (1983).
74. L. Olsson, Phenotypic diversity in leukemia cell populations,
     Cancer Met. Rev. 2:153 (1983).

75. K.S. Webb, J.L. Ware S.F. Parks, W.H. Briner and D.F. Paulson, Monoclonal antibodies of different epitopes on a prostate tumor-associated antigen: Implications for immunotherapy, Cancer Immunol. Immunother. 14:155 (1983).

76. P. Machy, J. Barbet and L.D. Leserman, Differential endocytosis of T and B lymphocyte surface molecules evaluated with antibody-bearing fluorescent liposomes containing methotrexate, Proc. Natl. Acad. Sci. USA, 79:4148 (1982).

77. K.S. Bragman, T.D. Heath and D. Papahadjopoulos, Simultaneous interaction of monoclonal antibody-targeted liposomes with two receptors on K562 cells. Biochim Biophys. Acta 730:187 (1983).

78. G. Poste and I.J. Fidler, Stimulation of macrophage-mediated destruction of lung metastases by administration of immunomodulators encapsulated in liposomes, in "Liposomes, drugs and immunocompetent cell functions", C. Nicolau and A Paraf, eds., Academic Press, New York (1981).

79. I.J. Fidler and G. Poste, Macrophage-mediated destruction of malignant tumor cells and new strategies for the therapy of metastatic disease. Springer Semin. Immunopathol. 5:161 (1982).

80. G. Poste and R. Kirsh, Liposome-encapsulated macrophage activation agents and active non-specific immunotherapy of neoplastic disease, in: "Cell Function and Differentiation", Part A, G. Akoyounoglou, A.E. Evangelopoulos, J. Georgatsos, G. Palaiologos, A. Trakatellis, and C.P. Tsiganos, eds., A.R. Liss, New York (1982).

81. G. Poste and I.J. Fidler, Therapeutic amplification of macrophage-mediated destruction of tumor cells: an approach to cancer therapy that addresses the problem of tumor cell heterogeneity, in: "Design of Models for Testing Cancer Therapeutic Agents", I.J. Fidler and R.J. White, eds., Van Nostrand, New York (1982).

82. R.K. Oldham, Biologicals: new horizons in pharmaceutical development, J. Biol. Resp. Modifiers 2:199 (1983).

83. P. Bugelski, R. Kirsh, J. Sowinski and G. Poste, Changes in the macrophage content of lung metastases at different stages in tumor growth, Cancer Res., in press.

84. J.R. Graybill, P.C. Craven, R.L. Taylor, D.M. Williams and W.E. Magee, Treatment of murine cryptococcosis with liposome-associated amphotericin B, J. Infect. Dis. 145:748 (1982).

85. G.Lopez-Berestein, R. Mehta, R.L. Hopfer, K. Mills, L. Kasi, K. Mehta, V. Fainstein, M. Luna, E.M. Hersh and R. Juliano, Treatment and prophylaxis of disseminated infection due to Candida albicans in mice with liposome-encapsulated amphotericin B, J. Infec. Dis. 5:939 (1983).

86. R.L. Taylor, D.M. Williams, P.C. Craven, J.R. Graybill, D.J. Drutz and W.E. Magee, Amphotericin B in liposomes: a novel therapy for histoplasmosis, Am. Rev. Respir. Dis. 125:610 (1982).

87. M.W. Fountain, C. Dees, and R.D. Shultz, Enhanced intra-cellular killing of Staphylococcus aureus by canine mono-cytes treated with liposomes containing amikacin, gentamicin, kanamycin, and tobramycin, Current Microbiol. 6:373 (1981).

88. D.W. Mason, P.E. Thorpe and W.C.J. Ross, Elimination of leukaemic cells from rodent bone marrow in-vitro with anti-body-ricin conjugates: implications for autologous marrow transplantation in man, Cancer Surveys 1:389 (1982).

89. M. Muirhead, P.J. Martin, B. Torok-Storb, J.W. Uhr and S. Vitetta, Use of an antibody-ricin A-chain conjugate to delete neoplastic B cells from human bone marrow, Blood 62:327 (1983).

90. J. Khato, E.R. Priester, and S.M. Sieber, Enhanced lymph node uptake of melphalan following liposomal entrapment and effects on lymph node metastasis in rats, Cancer Treatment Rep. 66:517 (1982).

91. B. Ryman and G.M. Barratt, Liposomes - further considerations of their possible role as carriers of therapeutic agents, in: "Targeting of Drugs", G. Gregoriadis, J. Senior and A. Trouet, eds., Plenum Press, New York.(1982).

92. M. Hashida, S. Muranishi, S. Sezaki, N. Tanigawa, K. Satomura and Y. Hikasa, Increased lymphatic delivery of bleomycin by microsphere in oil emulsion and its effect on lymph node metastasis, Int. J. Pharm. 2:245 (1979).

93. J.N. Weinstein, R.J. Parker, A.M. Keenan, S.K. Dower, H.C. Morse, and S.M. Sieber, Monoclonal antibodies in the lymphatics: toward the diagnosis and therapy of tumor metastases, Science 218:1334 (1982).

94. M.B. Yatvin and P.I. Lelkes, Clinical prospects for lipo-somes, Med. Phys. 9:149 (1982).

95. R.L. Magin, and J. Weinstein, Delivery of drugs in temperature-sensitive liposomes, in: "Targeting of Drugs" G. Gregoriadis, J. Senior and A. Trouet, eds., Plenum Press, New York (1982).

96. A. Walker, H.M. McCallum, T.E. Wheldon, A.H. Nias and A.S. Abdelaal, Promotion of metastasis of C3H mouse mammary carcinoma by local hyperthermia, Br. J. Cancer 38:561 (1978).

97. M. Urano, L. Rice, R. Epstein, H.D. Suit and A.M. Chu, Effect of whole-body hyperthermia on cell survival, met-astasis frequency, and host immunity in moderately and weakly immunogenic murine tumors, Cancer Res. 43:1039 (1983).

98. E. Jahde, M.F. Rajewsky and H. Baumgartl, pH distributions in transplanted neural tumors and normal tissues of BDIX rats as measured with pH micoelectrodes, Cancer Res. 42:1498 (1982).

99. K.J. Widder, A.E. Senyei and B. Sears, Experimental methods in cancer therapeutics, J. Pharm. Sci. 71:379 (1982).

100. T. Kato, Encapsulated drugs in targeted cancer therapy in: "Controlled Drug Delivery", S.D. Bruck, ed., CRC Press Boca Raton (1982).

101. J.L. Marx, Suppressing autoimmunity in mice, Science 221:843 (1983).

102. W.E. Lynch, G.P. Sartiano and A. Ghaffar, Erythrocytes as carriers of chemotherapeutic agents for targeting to the reticuloendothelial system, Am. J. Hematol. 9:249 (1980).

103. E. Haber, Antibodies as models for rational drug design, Biochem. Parmacol. 32:1967 (1983).

104. Federal Register Food and Drug Administration. Proposed new drug, antibiotic and biologic drug product regulations, 48:26720 (1983).

105. Office of Biologics, Food and Drug Administration, Points to consider in the manufacture of in-vitro monoclonal antibody products subject to licensure (1983).

106. H. Hauser, Methods of preparation of lipid vesicles assessment of their suitability for drug encapsulation, Trends Pharm. Sci. 3:274 (1982).

Participants of the NATO ASI "Receptor-Mediated Targeting of Drugs" held in Cape Sounion, Greece during 20 June–1 July, 1983. The Organizing Committee included Gregory Gregoriadis (ASI Director and Chairman, Andre Trouet (ASI Co-director), A, Evangelopoulos, George Poste and Phillip Thorpe. Judith Senior served as ASI Coordinator.

CONTRIBUTORS

Abarca, J., International Institute of Cellular and Molecular
    Pathology, 75 Avenue Hippocrate, B1200, Brussels, Belgium.

Aboud-Pirak, E., International Institute of Cellular and Molecular
    Pathology, 75 Avenue Hippocrate, B1200, Brussels, Belgium.

Allen, L.V., University of Oklahoma Health Sciences Center,
    940 Northeast 13th Street, Oklahoma City, Oklahoma, 73190,
    U.S.A.

Alving, C.R., Department of Membrane Biochemistry, Walter Reed
    Army Institute of Research, Washington, DC. 20307, U.S.A.

Aragnol, D., Centre d'Immunologie INSERM-CNRS de Marseille-Luminy,
    Case 906, 13288 Marseille Cedex 9, France.

Barbet, J., Centre d'Immunologie INSERM-CNRS de Marseille-Luminy,
    Case 906, 13288 Marseille Cedex 9, France.

Basala, M., Department of Pathology, Harvard Medical School and
    the Dana-Farber Cancer Institute, Boston, Mass. 02115, U.S.A.

de Bault, L.E., University of Oklahoma Health Sciences Center,
    940 Northeast 13th Street, Oklahoma City, Oklahoma, 73190,
    U.S.A.

Baurain, R., International Institute of Cellular and Molecular
    Pathology, 75 Avenue Hippocrate, B1200, Brussels, Belgium.

Blackstock, R., University of Oklahoma Health Sciences Center,
    940 Northeast 13th Street, Oklahoma City, Oklahoma, 73190,
    U.S.A.

Blytham, H.E., Centre de Recherche Clin-Midy, Section Immunologie,
    Rue du Prof. Joseph Blayac, 34082, Montpellier, France

Bourrie, B., Centre de Recherche Clin-Midy, Section Immunologie,
    Rue du Prof. Joseph Blayac, 34082, Montpellier, France

477

Bragman, K., Department of Pharmacology, Cancer Research Institute
    M-1282, University of California, San Francisco, CA 94143,
    U.S.A.

Bugianesi, R.L., Merck Sharp and Dohme Research Laboratories,
    Rahway, N.J. 07065, U.S.A.

Caride, V.J., Department of Nuclear Medicine, Hospital of St.
    Raphael and Department of Diagnostic Imaging, Yale University
    Medical School, New Haven, CT., U.S.A.

Carriere, D., Centre de Recherche Clin-Midy, Section Immunologie,
    Rue du Prof. Joseph Blayac, 34082, Montpellier, France.

Casellas, P., Centre de Recherche Clin-Midy, Section Immunologie,
    Rue du Prof. Joseph Blayac, 34082, Montpellier, France.

Cawley, D., Department of Microbiology and Immunology, Box 8093,
    Washington University Medical School, St. Louis, Mo., 63110,
    U.S.A.

Ceulemans, F., International Institute of Cellular and Molecular
    Pathology, 75 Avenue Hippocrate, B1200, Brussels, Belgium.

Comp, P.C., University of Oklahoma Health Sciences Center,
    940 Northeast 13th Street, Oklahoma City, Oklahoma, 73190,
    U.S.A.

Damen, J., The Netherlands Cancer Institute, Division of Cell
    Biology, Plesmanlaan 121, 1066 CX Amsterdam, The Netherlands.

Deprez-de Campeneere, D., International Institute of Cellular
    and Molecular Pathology, 75 Avenue Hippocrate, B1200,
    Brussels, Belgium.

Dijkstra, J., Department of Pharmacology, School of Pharmacy,
    University of California, San Francisco, CA 94143, U.S.A.

Doebber, T.W., Merck Sharp and Dohme Research Laboratories,
    Rahway, N.J., 07065, U.S.A.

Duncan, R., Biochemistry Research Laboratory, Department of
    Biological Sciences, University of Keele, Staffordshire, U.K.

Dussossoy, D., Centre de Recherche Clin-Midy, Section Immunologie,
    Rue du Prof. Joseph Blayac, 34082, Montpellier, France.

Duzgunes, N., Department of Pharmacology, Cancer Research Institute,
    M-1282, University of California, San Francisco, CA 94143,
    U.S.A.

Eiklid, K., Norsk Hydro's Institute for Cancer Research and the
  Norwegian Cancer Society, Montebello, Oslo, Norway.

Epenetos, A.A., Royal Postgraduate Medical School, Hammersmith
  Hospital, Du Cane Road, London, W12 OHS, U.K.

Esmon, C.T., University of Oklahoma Health Sciences Center,
  940 Northeast 13th Street, Oklahoma City, Oklahoma, 73190,
  U.S.A.

Friend, D.S., Department of Pathology, Cancer Research Institute,
  M-1282, University of California, San Francisco, CA 94143,
  U.S.A.

Goto, M., Department of Industrial Chemistry, Faculty of Engineering,
  Nagasaki University, Nagasaki 852, Japan.

Gregoriadis, G., Division of Clinical Sciences, Clinical Research
  Centre, Harrow, Middlesex, HA1 3UJ, U.K.

Griffin, T.W., Division of Oncology, Department of Medicine,
  University of Massachusetts Medical School, Worcester, Mass.
  01605, U.S.A.

Gros, O., Centre de Recherche Clin-Midy, Section Immunologie,
  Rue du Prof. Joseph Blayac, 34082, Montpellier, France.

Hara, K., The Second Department of Internal Medicine, School of
  Medicine, Nagasaki University, Nagasaki 852, Japan.

Hanson, W.L., Department of Parasitology, College of Veterinary
  Medicine, University of Georgia, Athens, GA., 30602, U.S.A.

Harriman, A., University of Oklahoma Health Sciences Center,
  940 Northeast 13th Street, Oklahoma City, Oklahoma, 73190,
  U.S.A.

Haynes, L.R., Division of Oncology, Department of Medicine,
  University of Massachusetts Medical School, Worcester, Mass.
  01605, U.S.A.

Heath, T., Department of Pharmacology, Cancer Research Institute
  M-1282, University of California, San Francisco, CA. 94143,
  U.S.A.

Herschman, H.R., Department of Biological Chemistry and Laboratory
  of Biomedical and Environmental Sciences, UCLA Center for the
  Health Sciences, Los Angeles, CA. 90024, U.S.A.

Holloway, F.A., University of Oklahoma Health Sciences Center, 940 Northeast 13th Street, Oklahoma City, Oklahoma, 73190, U.S.A.

Hong, K., Department of Pharmacology, Cancer Research Institute, M-1282, University of California, San Francisco, CA. 94143, U.S.A.

Humphrey, G B., University of Oklahoma Health Sciences Center, 940 Northeast 13th Street, Oklahoma City, Oklahoma, 73190, U.S.A.

Iida, T., Department of Industrial Chemistry, Faculty of Engineering, Nagasaki University, Nagasaki 852, Japan.

Jansen, F.K., Centre de Recherche Clin-Midy, Section Immunologie, Rue du Prof. Joseph Blayac, 34082, Montpellier, France.

Kirby, C., Division of Clinical Sciences, Clinical Research Centre, Harrow, Middlesex, HA1 3UJ, U.K.

Kopeček, J., Institute of Macromolecular Chemistry, Czechoslovak Academy of Sciences, Prague, Czechoslovakia.

Krous, H.F., University of Oklahoma Health Sciences Center, 940 Northeast 13th Street, Oklahoma City, Oklahoma, 73190, U.S.A.

Laurent, J.C., Centre de Recherche Clin-Midy, Section Immunologie, Rue du Prof. Joseph Blayac, 34082, Montpellier, France.

Leserman, L.D., Centre d'Immunologie INSERM-CNRS de Marseille-Luminy, Case 906, 13288 Marseille Cedex 9, France.

Lesur, B., International Institute of Cellular and Molecular Pathology, 75 Avenue Hippocrate, B1200, Brussels, Belgium.

Levin, L.V., Division of Oncology, Department of Medicine, University of Massachusetts Medical School, Worcester, Mass. 01605, U.S.A.

Liance, M.C., Centre de Recherche Clin-Midy, Section Immunologie, Rue du Prof. Joseph Blayac, 34082, Montpellier, France.

Lloyd, J.B., Biochemistry Research Laboratory, Department of Biological Sciences, University of Keele, Staffordshire, U.K.

Lopez, N.G., Department of Pharmacology, Cancer Research Institute M-1282, University of California, San Francisco, CA. 94143, U.S.A.

Machy, P., Centre d'Immunologie INSERM-CNRS de Marseille-Luminy, Case 906, 13288 Marseille Cedex 9, France.

Masquelier, M., International Institute of Cellular and Molecular Pathology, 75 Avenue Hippocrate B1200, Brussels, Belgium.

Matthay, K., Department of Pharmacology, Cancer Research Institute M-1282, University of California, San Francisco, CA. 94143, U.S.A.

McIntosh, D., Institute of Cancer Research, Chester Beatty Laboratories, Fulham Road, London SW3 6JB, U.K.

Mojarad, M., University of Oklahoma Health Sciences Center, 940 Northeast 13th Street, Oklahoma City, Oklahoma, 73190, U.S.A.

Munnell, J.F., Department of Anatomy and Radiology, College of Veterinary Medicine, University of Georgia, Athens, GA. 30602, U.S.A.

Neville Jr., D.M., National Institute of Mental Health, Laboratory of Neurochemistry, Section on Biophysical Chemistry, 9000 Rockville Pike, Bethesda, Maryland 20205, U.S.A.

Olsnes, S., Norsk Hydro's Institute for Cancer Research and the Norwegian Cancer Society, Montebello, Oslo, Norway.

Otte-Slachmuylder, C., International Institute of Cellular and Molecular Pathology, 75 Avenue Hippocrate B1200, Brussels, Belgium.

Papahadjopoulos, D., Department of Pharmacology, Cancer Research Institute, M-1282, University of California, San Francisco, CA. 94143, U.S.A.

Pihl, A., Norsk Hydro's Institute for Cancer Research and the Norwegian Cancer Society, Montebello, Oslo, Norway.

Poncelet, P., Centre de Recherche Clin-Midy, Section Immunologie, Rue du Prof. Joseph Blayac, 34082 Montpellier, France.

Ponpipom, M.M., Merck Sharp and Dohme Research Laboratories, Rahway, N.J., 07065, U.S.A.

Poste, G., Smith Kline and French Laboratories, 1500 Spring Garden Street, P.O. Box 7929, Philadelphia, PA. 19101, U.S.A.

Raso, V., Department of Pathology, Harvard Medical School and the Dana-Farber Cancer Institute, Boston, Mass. 02115, U.S.A.

Rejmanova, P., Institute of Macromolecular Chemistry, Czechoslovak
    Academy of Sciences, Prague, Czechoslovakia.

Richer, G., Centre de Recherche Clin-Midy, Section Immunologie,
    Rue du Prof. Joseph Blayac, 34082 Montpellier, France.

Robbins, J.C., Merck Sharp and Dohme Research Laboratories,
    Rahway, N.J. 07065, U.S.A.

Roerdink F., Laboratory of Physiological Chemistry, University of
    Groningen, Bloemsingel 10, 9712 KZ Groningen, The Netherlands.

Rolin-van Swieten, D., International Institute of Cellular and
    Molecular Pathology, 75 Avenue Hippocrate B1200, Brussels,
    Belgium.

Saito, A., The Second Department of Internal Medicine, School of
    Medicine, Nagasaki University, Nagasaki 852, Japan.

Sandvig, K., Norsk Hydro's Institute for Cancer Research and the
    Norwegian Cancer Society, Montebello, Oslo, Norway.

Scherphof, G., Laboratory of Physiological Chemistry, University
    of Groningen, Bloemsingel 10, 9712 KZ Groningen, The
    Netherlands.

Schneider, Y.-J., International Institute of Cellular and Molecular
    Pathology, 75 Avenue Hippocrate B1200, Brussels, Belgium.

Senior, J., Division of Clinical Sciences, Clinical Research Centre,
    Harrow, Middlesex, HA1 3UJ, U.K.

Shen, T.Y., Merck Sharp and Dohme Research Laboratories, Rahway,
    N.J. 07065, U.S.A.

Simpson, D.L., Department of Biochemistry, Meharry Medical College,
    1005 D.B. Todd Boulevard, Nashville, Tennessee 37208, U.S.A.

Spanjer, H., Laboratory of Physiological Chemistry, University of
    Groningen, Bloemsingel 10, 9712 KZ Groningen, The Netherlands.

Stone, H.L., University of Oklahoma Health Sciences Center, 940
    Northeast 13th Street, Oklahoma City, Oklahoma, 73190, U.S.A.

Straubinger, R.M., Department of Pharmacology, Cancer Research
    Institute, M-1282, University of California, San Francisco,
    CA. 94143, U.S.A.

Sunamoto, J., Department of Industrial Chemistry, Faculty of
    Engineering, Nagasaki University, Nagasaki 852, Japan.

Sundan, A., Norsk Hydro's Institute for Cancer Research and the
    Norwegian Cancer Society, Montebello, Oslo, Norway.

Taylor-Papadimitriou, J., Imperial Cancer Research Fund, Lincoln's
    Inn Fields, London WC2A 3PX, U.K.

Thorpe, P., Imperial Cancer Research Fund, Lincoln's Inn Fields,
    London WC2A 3PX, U.K.

Tomonaga, A., The Second Department of Internal Medicine, School
    of Medicine, Nagasaki University, Nagasaki 852, Japan.

Trouet, A., International Institute of Cellular and Molecular
    Pathology, 75 Avenue Hippocrate B1200, Brussels, Belgium.

Truneh, A., Centre d'Immunologie INSERM-CNRS de Marseille-Luminy,
    Case 906, 13288 Marseille Cedex 9, France.

Vidal, H., Centre de Recherche Clin-Midy, Section Immunologie,
    Rue du Prof. Joseph Blayac, 34082 Montpellier, France.

Vitetta, E.S., Department of Microbiology, University of Texas
    Southwestern Medical School, Dallas, Texas 75235, U.S.A.

Uhr, J.W., Department of Microbiology, University of Texas
    Southwestern Medical School, Dallas, Texas 75235, U.S.A.

Weldon, J.S., Department of Anatomy and Radiology, College of
    Veterinary Medicine, University of Georgia, Athens, GA. 30602,
    U.S.A.

Wierimaa, R., University of Oklahoma Health Sciences Center,
    940 Northeast 13th Street, Oklahoma City, Oklahoma 73190,
    U.S.A.

Wolff, B., Division of Clinical Sciences, Clinical Research Centre,
    Harrow, Middlesex, HA1 3UJ, U.K.

Youle, R.J., National Institute of Mental Health, Laboratory of
    Neurochemistry, Section on Biophysical Chemistry, 9000
    Rockville Pike, Bethesda, Maryland 20205, U.S.A.

Abrin
  antibody conjugates, 105
  endocytic uptake, 90
  receptors for, 90
  ribosomes, interaction with,
    90
Albumin
  linkage to daunorubicin, 2
Amines
  toxins, protection against, 96
Aminomannose
  in liposomes, 379
    tissue distribution, 384
Ammonium chloride
  as lysosomotropic agent
    effect on liposome uptake
      in vitro, 302
    effect in vivo, 170
    effect on liposome-cell
      interaction, 298
    endocytosis, effect on, 399
  in endocytosis
    effects on antibody-toxin
      conjugate, 155
    liposome endocytosis, effect
      on, 410
    toxins, protection against, 96
Amylopectin
  coating of liposomes, 359
Antibodies
  differentiation, study of, 216
  epithelial cells, against, 201
  imaging with, 435
  malignancy, study of, 216
  targeting of liposomes with,
    253, 407
Antibodies, monoclonal
  against α-foetoprotein, 4

applications of, 221
as drug carriers of toxins, 119
cell identification with, 221
epithelial membrane antigens,
  against, 212
hepatitis B surface antigen,
  against, 4
human
  production of, 208
human epithelial cells, against
  tumour specificity of, 212
mouse, 209
naturally occurring, 202
production of, 203
rat, 209
selection of, 207
Antibody conjugates
  abrin, with, 105
  ricin, with, 105
  ricin A, with, 109, 114
  ricin B, with, 109
  toxins, with, 106, 113
    pinocytosis of, 113
Antibody drug conjugates
  in cancer chemotherapy, 430
  transcapillary passage, 448
Antibody, radiolabelled
  tumour growth, effect on, 236
    in immunodeficient rats, 240
Antibody ricin A conjugate
  polyclonal antibodies, and,
    192
  preparation of, 188, 189
  radioimmunoassay of, 125
  ricin B, synergism with, 106
Antibody ricin conjugates, 154
  kinetics of
    lactose, effect of, 143

Antibody toxin conjugates, 119
  as antitumour agents, 140
  cells, binding to
    effect of lactose, 133
  clinical potential
    in vivo, 170
    ex vivo, 170
  cytotoxicity
    factors affecting, 162
    in vitro, 154
    kinetics, 148
    to tumour cells, 148
  endocytosis
    effect of lysosomotropic
      amines, 155
  galactose, interaction with
    abrogation of, 110
  in cancer treatment, 147, 179,
    180
  in vitro stability, 165
  in vivo clearance, 165
  interaction with cells
    efficiency of killing, 158
    kinetics of, 147
  kinetics of clearance, 128
  long term kinetics, 139
  non-specific activity
    effect of lactose, 147
    removal of, 110
  preparation, 121
  uptake by cells
    kinetics, 124
Antigens, organotypic
  in targeting, 3
Antigens, tumour associated
  in targeting, 3
Asialofetuin
  derivatization
    with SPDP, 30, 31
  linkage to primaquine, 2
  ricin derivative
    toxicity, 33-36
  ricin, linkage with, 42
  targeting of liposomes with,
    253
  toxic conjugates, 27 42
    hepatocytes, effect on, 42
Asialoorosomucoid
  targeting of liposomes with,
    255

B cells
  differentiation, 202
  Ig-secreting clones, 202
  targets for immunotoxins, 179
Bone marrow
  cancer chemotherapy
    ex vivo treatment, 459
    toxicity, 459
  ex vivo treatment of, 162, 170,
    430
  uptake of liposomes by, 252

Calcein
  in liposomes, 298
Cancer treatment
  drug targeting in, 427
  immunotoxins in, 187, 195
  liposomes in
    macrophage activation, 455
  lymph nodes, of, 460
  tumour cell heterogeneity, in
    453
  tumour models, 434
Carboxyfluorescein
  in liposomes, 298
Carcinoma, human breast
  antibodies against, 15
  as target, 15
  milk fat globules, 15, 16
Carcinoembryonic antigen
  antibodies against, 187
    action of, 196
  monoclonal antibodies against
    ricin A chain conjugate, 191
  nature of, 209
  polyclonal antibodies against
    ricin A chain conjugate, 192
Carriers
  glycopeptides, 2
  neoglycoproteins, 2
  reticuloendothelial system,
    uptake by
  control of, 440
Chloroquine
  as lysosomotropic agent
  effect on liposome-cell
    interactions, 298, 302
  toxins, protection against, 96
Coated pits
  endocytosis of toxins in, 92

Cross-linking reagents, 74

Daunorubicin
  biliary excretion, 11
  linkage to IgG, 18
  linkage to galactosylated
    albumin, 7
  tissue distribution, 11
  toxicity, 13
Daunorubicin-galactosylated
    albumin complex
  biliary excretion, 11
  chemotherapy with, 13, 14
  Hep G2 cells, uptake by, 9
  hepatocytes, uptake by, 9
  tissue distribution, 11
  toxicity, 12
Diphtheria toxin
  cell entry, mode of, 94
  fetuin, linkage with, 38
  properties, 29
  resistance to, 29
  structure, 87
  toxicity of, 29, 36
Drug carriers
  tumour metastases, treatment
    of, 435
Drug conjugates
  transferrin-ricin A chain, 73
    effect on protein synthesis,
    76
Drug delivery systems
  commercial development of, 464
  microcirculation and, 441
Drug targeting
  in cancer therapy, 427
  macrophage activation by, 455
  to lymphoid subsets, 458

Endocytosis
  antibody-toxin conjugates, of
    155
Epidermal growth factor
  as a mitogen, 28
  derivatization, 30, 31
  diphtheria toxin, linkage with
    toxicity, 36-41
  endocytosis, 32
  linkage to ricin, 31, 32, 42
    toxicity, 44

properties, 28
purification, 28
ricin derivative
  toxicity, 33-36
target cells for, 32
toxic conjugates, 27, 42
  hepatocytes, effect on, 42
Epithelial cell membranes
  carbohydrate antigens in, 215
Epithelial proliferating antigen
  monoclonal antibodies, against,
    235

Fluorescein isothiocyanate
    dextran
  as liposomal marker, 299

Galactosylceramide
  in liposomes, 377
Galactosylcholesterol
  in liposomes, 377
Galactosylated albumin
  daunorubicin, linkage with, 8
  preparation of, 5, 6
  tritiated, 6
  uptake by hepatocytes, 6
Galactose receptor
  asialoglycoproteins, uptake of
    intracellular fate, 28
  detection of, 4
  in Hep G2 cells, 5
  in hepatocytes, 28
  in hepatoma cells, 4
  in pulmonary metastases, 5
  purification, 28
Glycolipids
  as drug carriers, 53
  liposome-associated, 374
Glycopeptides
  as drug carriers, 3, 53, 55
  uptake by macrophages, 61
Glycophorin
  in liposome targeting, 407
Gold
  as liposomal marker, 299

Hepatocarcinoma
  in drug targeting, 3
Hepatocytes
  ricin, interaction with, 42

Hepatocytes (continued)
  ricin, sensitivity to, 42
Hybrid antibody producing cells
  production of, 223-228
Hybrid delivery systems
  antibodies in, 129
  ricin subunits in, 129
Hybridomas
  cloning, 206
  growth, 206
  production
    myeloma cell lines, 205

Imaging agents
  use in liposomes, 348
Immunotoxins (see also antibody-
      toxin conjugates)
  A-chain, 183
  bone marrow, treatment of
    specificity, 180
  in cancer therapy
    ex vivo, 180
    in vivo, 182
  in vivo kinetics, 193

Lactosylceramide
  in liposome targeting, 287
Lecithins
  coagulation, effect on, 343
Legionella pneumophila
  multiplication of, 367
  use of liposomes in, 367
Legionnaires' disease
  liposomes in, 359
Leishmaniasis
  experimental model, 320
  drug therapy of, 318
  forms, 317
  liposomal therapy of, 317
  visceral
    liposomal drug therapy of,
      318
Leupeptin
  as enzyme inhibitor, 421
Liposomes
  amylopectin-coated
    clearance of, 362
    efficiency of coating, 361
    tissue distribution of,
      361, 362

antibody-bearing
  in endocytosis, 393
  multivalency, 412
antibody-targeted
  cells, interaction with, 407
as carriers of macromolecules,
  297
as imaging agents, 348
behavioural effects, 337
blood components, interaction
  with, 267
cells, delivery of molecules
  into, 309
cells, interaction with
  ammonia, effect of, 298
  chloroquine, effect of, 298
  effect of lysosomotropic
    agents, 302
  fate of markers, 303
  inhibition, 414
  mechanism, 297
cells, uptake by
  ammonia, effect of, 283
  chloroquine, effect of, 283
  evaluation of kinetics, 400
clearance from circulation
  effect of cholesterol, 250
  effect of phospholipid
    composition, 251
  lipid composition and, 286
coagulation, effect on, 343
dehydration-rehydration
    vesicles
  preparation of, 252
diagnostic imaging, in, 343
  gastrointestinal tract, 352
distribution in vivo
  after subcutaneous
    administration, 351
drug delivery
  effect of vesicle size, 407
  efficiency, 298
drug leakage
  effect of blood, 244
  effect of pH, 299
  fluorescent markers, 299
  HDL, role of, 244, 271
drug retention
  cholesterol, role of, 244,
    271

Liposomes (continued)
  drug retention (continued)
    galactosylcholesterol,
      effect of, 377
    polysaccharides in, 359
  drug targeting
    with polysaccharides, 359,
      369
  endocytosis
    ammonium chloride, effect of
      399
    in vitro, 305
    in vivo, 306
    vesicle size, effect of, 397
  fate of entrapped molecules,
    297
  fate in vivo, 267-271
    glycolipids, effect of, 388
  gold-containing
    preparation of, 304
  hepatic uptake
    lysosomotropic agents in,
      280
  in lympn node imaging, 351
  intracellular fate of markers
    in vitro, 365
  in macrophage activation, 455
  intracellular fate of markers
    in vitro, 365
  Kupffer cells, interaction
    with, 278
  Legionnaires' disease, in, 359
  Leishmaniasis, in, 317, 318
    by alveolar macrophages, 364
    by human monocytes, 364
  plasma, interaction with, 271
  plasma lipoproteins,
    interaction with, 269, 271
    effect of vesicle size, 274,
      277, 278
  polysaccharide-coated
    preparation of, 360
  radiopaque, 353
  site specific drug delivery,
    432
  size determination, 334
  stability in blood
    galactosylcholesterol, effect
      of, 377

  surface properties
    glycolipids, role of, 373
  targeting, 252, 408
    antibodies, with, 253, 257,
      393
    asialofetuins, with, 253
    glycolipids, with, 373
    glycoproteins, with, 253, 256
    lactosylceramide, with, 287
    to cells
      in vivo, 254, 261
      in vitro, 258
      effect of serum, 409
  thrombogenic potential
    phosphatidylserine, role of,
      333
  tissue distribution
    glycolipids, effect of, 380
  transcapillary passage, 444
    in inflammation, 447
    in ischemia, 447
    in tumours, 448
  uptake by bone marrow, 252
Liposome-entrapped
  antibiotics, 367, 369
  bromine
    in diagnostic imaging, 354
  carboxyfluorescein, 393
  CoQ10, 361
  fluoresceinated dextran
    intracellular delivery of,
      309
  In^{111}-8-hydroxyquinoline
    in diagnostic imaging, 351
  lactose, 384
  methotrexate, 393
  methotrexate-γ-aspartate, 407
  radiolabels, 348
    in diagnostic imaging, 349
  radiopaque agents
    in diagnostic imaging, 353
  sisomycin
    in Legionnaires' disease,
      360
  Tb(acac)$_3$(H$_2$O)$_2$, 364
  Tc99m
    in diagnostic imaging, 349
Lysosomotropic agents
  effect on liposome-cell
    interactions, 298

Macromolecules
  linkage to anthracyclines, 2
  linkage to vinca alcaloids, 2
Malaria
  treatment
    with asialofetuin
      conjugates, 3
    with primaquine, 3
Mannose-lysine conjugates
  as drug carriers, 55
  biological activity, 59
  Gaucher's disease, in, 63
Methotrexate
  in liposomes, 393, 407
Methotrexate-γ-aspartate
  cell uptake of
    role of liposomes, 408
Microcirculation
  in drug delivery, 441
Microspheres
  albumin, 432
  ethylcellulose, 432
  site-specific drug delivery,
    432
Milk fat globule membranes
  monoclonal antibodies against,
    17
    preparation of, 17, 18
Modeccin
  endocytic uptake, 90
  receptor for, 90
  ribosomes, interaction with,
    90
Monoclonal antibodies
  radiolabelled
    in cancer treatment, 235

Neoglycoproteins
  as drug carriers, 3
Nanoparticles
  site-specific drug delivery
    in, 432

Pinocytosis
  carriers of, polymeric, 417
Poly(hydroxypropylmethacrylamide)
  drug targeting, in, 419
Polymeric carriers
  biodegradable, 423
  enzymic hydrolysis, 421

  in drug delivery, 417, 432
  targeting, 423
Polymethacrylamide-oligopeptide
    conjugates
  targeting with, 417
Polyvinylpyrrolidone
  pinocytosis, 422

Radiopaque agents
  in liposomes
    imaging of kidneys, 353
    imaging of liver, 354
    imaging of spleen, 354
Raffinose
  protein, linked to, 62
  radioactive tracer, 62
Receptors
  for galactose, 2
  for N-acetyl-β-glucosamine, 2
  for mannose, 2
  for saccharide, drug delivery,
    53
  for transferrin, 73, 135
  on lymphoma cells, 79
Reticuloendothelial system
  blockade
    diagnostic imaging after,
      351, 440
  particle clearance by
    control, 440
Ricin
  A chain, 73, 89, 154, 187
    mechanism of action, 29
    toxicity, 29
  antibody conjugates, 105
  B chain, 73, 89, 154
  composition, 28
  endocytic uptake, 90
  linkage to EGF, 31, 32
  purification, 29
  receptors for, 90
  ribosomes, interaction with, 90
  structure, 87
Ricin A chain
  antibody conjugate, 120
    cell entry, 114
  in antibody-mediated cell
    killing, 179
  preparation, 75
  radioimmunoassay, 125

Ricin B chain
  in toxin binding, 179

Sisomycin
  liposome-entrapped, 360

Targeting
  conditions for, 243
  of liposomes, 243
    to cells in vitro, 258
    with antibodies, 253. 257
    with glycoproteins, 253, 256
Toxins
  A chain, 105
    conjugated to antibody, 147
  B chain, 105
  cells, binding to, 90
  cells, entry into
    amines, effect of, 96
  Diphtheria, 89
  endocytosis
    coated pits, 92

intracellular action, 98
Pseudomonas aeruginosa,
    exotoxin A, 89
Shigella dysenteriae, 90
Transferrin
  cell specific carrier, 130
  hybrid with anti-ricin A chain,
    135
  on non-haemoglobin
    synthesising cells, 83
  receptor for, 73, 135
  ricin, linkage with, 73
  ricin A, linkage with
    protein synthesis, effect
    on, 76
Tumour models
  cancer treatment, in, 235

Viscomin
  endocytic uptake, 90
  receptors for, 90
  ribosomes, interaction with, 90